Theoretical Elasticity

A. E. Green
University of Newcastle Upon Tyne

W. Zerna
Ruhr-Universität, Bochum

Second Edition

Dover Publications, Inc.
Mineola, New York

This Dover edition, first published in 1992 and reissued in 2012, is an unabridged and unaltered republication of the second edition (1968) of the work first published at The Clarendon Press, Oxford, England, in 1954. It is reprinted by special arrangement with Oxford University Press, 200 Madison Avenue, New York, N.Y. 10016.

Library of Congress Cataloging-in-Publication Data

Green, A. E. (Albert Edward)
Theoretical elasticity / A.E. Green, W. Zerna. — 2nd ed.
p. cm.
Originally published: London : Oxford University Press, 1968.
Includes bibliographical references and indexes.
ISBN-13: 978-0-486-67076-8
ISBN-10: 0-486-67076-7
1. Elasticity. I. Zerna, Wolfgang. II. Title.
QA931.G7 1992
531′.382—dc20 92-7456
CIP

Manufactured in the United States by Courier Corporation
67076703
www.doverpublications.com

Preface to the Second Edition

THE main plan of the book is unaltered. Changes have been made in the presentation of the general theory of non-linear elasticity in Chapter 2 and additional sections have been added on thermodynamics. Small changes have been made in Chapters 4 and 5. There are considerable alterations in the chapters on linear shell theory. Chapter 10 contains exact results for shells. Chapters 11 and 12 are concerned with membrane theory, and a bending theory of shells which is satisfactory provided that stresses arising from inextensional deformations are unimportant. The equations of these chapters are given in general curvilinear coordinates and the special forms the equations assume for cylindrical shells and for shells of revolution are studied in Chapters 14 and 15. The general theory of shallow shells, based on the bending theory of Chapter 12, is contained in Chapter 13. Finally, a new Chapter 16 is added, which deals with methods of deriving membrane theory, inextensional theory, and bending theory, by asymptotic expansions of the three-dimensional linear elastic equations.

We thank the Press for their care and attention in getting the book into print and Miss J. Edger for help in preparing the manuscript.

A. E. G.
W. Z.

Preface to the First Edition

THIS book is mainly concerned with three aspects of elasticity theory which have attracted considerable attention in recent years, and is not intended to be an exhaustive treatise. Many important topics, such as the torsion and flexure of beams, energy methods, and the theory of elastic stability, are omitted because they have already been extensively discussed in other books. The three main topics considered here are finite elastic deformations, complex variable methods for two-dimensional problems for both isotropic and aeolotropic bodies, and shell theory, the latter topics being confined to classical infinitesimal elasticity. In addition, some mention is made of three-dimensional problems for isotropic and transversely isotropic bodies. Throughout the book emphasis is placed on the use of general tensor notations in which general theories can be expressed in an elegant and compact form, and which are of considerable help in the solution of special problems, particularly for finite deformations. Vector notations are also used with tensors whenever appropriate.

A number of books on tensor analysis are available, but since workers in the field of elasticity are still often unfamiliar with these notations we have included a summary of tensors in Chapter 1. By restricting the discussion to a three-dimensional Euclidean space it is possible to present a comparatively simple account of the relevant theory. The main properties of two-dimensional surfaces are then deduced by regarding such surfaces as being embedded in three-dimensional Euclidean space. Tensor analysis can then be extended to general Riemannian spaces without difficulty, but this extension is not required here. The advantages of the presentation of tensor analysis from a restricted point of view seem to be sufficient to justify the sacrifice of some generality.

Chapter 2 contains an account of the general theory of elasticity for finite deformations, using the notations of Chapter 1. Special attention is given to the formulation of stress–strain relations for an isotropic body. Chapter 3 contains solutions of a number of special problems, mostly for incompressible isotropic bodies, the majority being obtained in a general form which is independent of the choice of strain-energy function. A number of these solutions have been found to have practical application for rubber-like materials, but discussion of such applications is not included in the book.

In Chapter 4 a theory of small deformations, which are superposed on finite deformations, is given, again making no assumptions about the form of the strain-energy function. This theory is available for both compressible and incompressible bodies and a number of special problems are solved. The advantages of tensor notations are again evident in this chapter.

Chapter 5 contains the classical infinitesimal theory of elasticity which is deduced as a special case of the general theory developed in Chapter 2. Once again, tensor notations are used, so that specialization to particular coordinate systems is then a straightforward matter. In this chapter stress–strain relations are deduced for aeolotropic as well as isotropic bodies. The chapter closes with the solution of some three-dimensional problems for both isotropic and transversely isotropic bodies.

Chapters 6 and 7 deal with plane strain and plate theories for both isotropy and aeolotropy and are first developed in tensor notation. By specialization of this general form of the two-dimensional theories it is possible to introduce complex variable notations in a consistent and natural manner, so that complex combinations of stresses appear by using tensor transformations from rectangular-cartesian to complex coordinates. Two-dimensional problems are discussed in Chapter 8 for isotropic bodies, and in Chapter 9 for aeolotropic bodies for plane strain, and for plates deformed by forces in their planes, using some powerful and elegant tools of complex function theory. Problems of transverse flexure of plane plates are not discussed in detail since many of these problems are analytically similar to those occurring in plane strain. In Chapter 7, however, an extension of the classical theory of flexure of isotropic plates, due to Reissner, is considered, and a special problem is solved in Chapter 8, using this theory.

Some of the general methods of solution of two-dimensional problems given in Chapter 8 are due to Muskhelishvili and other Russian writers, and this chapter (and § 1.14 to § 1.21) owes much to a book by Muskhelishvili, *Singular Integral Equations* (Moscow, 1946), which was translated by J. R. M. Radok and W. G. Woolnough (Australia, 1949). At the time of writing, this book, translated by J. R. M. Radok, has received wider publication by P. Noordhoff (Groningen, Holland, 1953). In addition, another important book by Muskhelishvili, *Some Basic*

Problems of the Mathematical Theory of Elasticity, translated by J. R. M. Radok, has just been published by P. Noordhoff. Reference is made to the Russian version of this book in footnotes to Chapter 6 but the book has not been available to the present writers.

In footnotes on page 183 of Chapter 6 references are given to pioneer work by Lechnitzky in two-dimensional problems for aeolotropic bodies, and references to work by other Russian writers are contained in the paper by Sokolnikoff which is quoted on page 183. The work on aeolotropic materials by Russian writers has not been available to the present authors so that it has not been possible to refer to it adequately in this book. For this reason reference is often made to papers which have appeared in British journals, but it is recognized that Russian authors may frequently claim priority.

The last chapters of the book, 10–16, are devoted to the theory of shells; here emphasis is placed on the formulation of a general theory and only a few special problems are discussed. Once more the value of tensor notations is evident. The theory of shells given in Chapter 10 differs in some respects from existing theories. Attention is restricted to a first approximation, which, it is believed, is satisfactory for many problems which arise in practice.

It is a pleasure to thank Professor I. N. Sneddon for help in proof-reading, Miss M. J. Haining for typing all the manuscript, and the Press for their care and attention in getting the book into print.

Thanks are due to the Royal Society for permission to reproduce Figs. 9.1–9.4 from their Proceedings.

A. E. G.
W. Z.

January 1954

Contents

1 Mathematical Preliminaries

THIS chapter contains a summary of some important definitions, relations, and formulae of vector and tensor calculus, functions of a complex variable, and Fourier integrals, which are essential for our treatment of the theory of elasticity. We have not included proofs of all theorems and formulae but it is hoped that sufficient details are given so that the reader may understand the remaining chapters without being forced to make constant reference to other mathematical books.†

1.1. Indicial notation. Summation convention. Kronecker delta

Consider symbols which are characterized by one or several indices which may be either subscripts or superscripts,‡ such as A_i, B^i, A_{ij}, B^i_j, etc. Sometimes it is necessary to decide the order of the indices when subscripts and superscripts occur together, and then, for example, we write $A^i_{.j}$ where the dot before j shows that j is the second index and i the first.

Unless otherwise stated Latin indices represent the numbers 1, 2, 3. Thus, A^i represents any one of the three elements A^1, A^2, A^3, and A_{ij} represents any one of the nine elements A_{11}, A_{12}, A_{13}, A_{21}, A_{22}, A_{23}, A_{31}, A_{32}, A_{33}.

† More detailed treatments of vector and tensor calculus are given, e.g. by A. Duschek and A. Hochrainer, *Grundzüge der Tensorrechnung in Analytischer Darstellung*, Vol. i (1948), Vol. ii (1950), Vienna; L. A. Eisenhart, *An Introduction to Differential Geometry* (Princeton, 1947); A. J. McConnel, *Applications of the Absolute Differential Calculus* (London and Glasgow, 1947); F. D. Murnaghan, *Introduction to Applied Mathematics* (New York, 1948); J. L. Synge and A. Schild, *Tensor Calculus* (Toronto, 1949); I. S. Sokolnikoff, *Tensor Analysis* (New York, 1951); C. E. Weatherburn, *An Introduction to Riemannian Geometry and the Tensor Calculus* (Cambridge, 1938). The method of presentation of tensor calculus given in this chapter differs, however, in some respects from that of other writers. §§ 1.14 to 1.21 on complex variable theory were written with the help of a book by N. I. Muskhelishvili, *Singular Integral Equations* (Moscow, 1946), translated by J. R. M. Radok and W. G. Woolnough (Australia, 1949). See also J. R. M. Radok, *Singular Integral Equations* (Groningen, Holland, 1953).

Reference may be made to E. C. Titchmarsh, *Theory of Fourier Integrals* (Oxford, 1937), for the results of § 1.22.

‡ It is understood that indices as superscripts are not taken as powers.

Systems of elements which, like A^i, depend on one index only, are called systems of the first order, and the separate terms A^1, A^2, A^3 are called the components of the system. Systems of the first order have one or other of the two forms

$$A^i,\ B_i.$$

Systems of the second order depend on two indices and can be of the three types

$$A_{ij},\ A^i_{\cdot j} \text{ or } A_j^{\cdot i},\ A^{ij},$$

and there are nine components in each system. Similarly, we have systems of the third and higher orders.

A single element ϕ, which has no indices, is called a system of order zero.

Expressions which consist of a sum are formed by the following summation convention, unless stated otherwise, Any term in which the same index (subscript or superscript) appears twice stands for the sum of all such terms obtained by giving this index its complete range of values. The following examples illustrate this convention:

$$\left.\begin{aligned} A^iB_i &= \sum_{i=1}^{3} A^iB_i = A^1B_1+A^2B_2+A^3B_3 \\ A^i_{\cdot i} &= \sum_{i=1}^{3} A^i_{\cdot i} = A^1_{\cdot 1}+A^2_{\cdot 2}+A^3_{\cdot 3} \end{aligned}\right\}, \tag{1.1.1}$$

$$\left.\begin{aligned} A_{ij}x^ix^j = \sum_{i=1}^{3}\sum_{j=1}^{3} A_{ij}x^ix^j &= A_{11}x^1x^1+A_{12}x^1x^2+A_{13}x^1x^3+ \\ &\quad +A_{21}x^2x^1+A_{22}x^2x^2+A_{23}x^2x^3+ \\ &\quad +A_{31}x^3x^1+A_{32}x^3x^2+A_{33}x^3x^3 \end{aligned}\right\}. \tag{1.1.2}$$

Since the repeated index is summed it follows that we may substitute for the particular letter used any other letter without altering the value of the expansion. Thus

$$A^iB_i = A^jB_j.$$

No summation is carried out if the same index is repeated more than twice as, for example, in $A^{ij}B_{ii}$. On the other hand, when more than one summation is necessary, different summation indices are taken, as in example (1.1.2).

It may be mentioned that it does not, of course, follow from an equation of the type

$$A^iB_i = A^iC_i$$

that $B_i = C_i$, since both sides of the equation represent the sums of three different terms.

A special meaning is given to the symbols δ^i_j, which are called Kronecker deltas. The Kronecker deltas have the following values:

$$\left.\begin{aligned} \delta^i_j &= 0 \quad (i \neq j) \\ \delta^i_j &= 1 \quad (i = j,\, j \text{ not summed}) \end{aligned}\right\}. \tag{1.1.3}$$

We therefore have

$$\delta^1_2 = \delta^1_3 = \delta^2_1 = \delta^2_3 = \delta^3_1 = \delta^3_2 = 0,$$
$$\delta^1_1 = \delta^2_2 = \delta^3_3 = 1.$$

The Kronecker delta is sometimes called the substitution operator since, for example,

$$\delta^i_j a^j = a^i, \qquad \delta^i_k a_{ij} = a_{kj}. \tag{1.1.4}$$

1.2. Transformations of coordinates

We denote by θ^i three independent variables whose differentials are $d\theta^i$. We also introduce the convention that partial derivatives of a function with respect to the independent variables are denoted by a comma. For example,

$$A_{,i} = \frac{\partial A}{\partial \theta^i},$$

$$A_{jk,i} = \frac{\partial A_{jk}}{\partial \theta^i},$$

$$A^{jk}{}_{,i} = \frac{\partial A^{jk}}{\partial \theta^i}.$$

Let us now suppose that the variables θ^i are transformed into a set of new variables $\bar{\theta}^i$ by any arbitrary single-valued functions of the form

$$\bar{\theta}^i = \bar{\theta}^i(\theta^1, \theta^2, \theta^3). \tag{1.2.1}$$

We assume that the arbitrary functions possess derivatives up to any order required and also that the transformation (1.2.1) is reversible. We therefore have the inverse transformation

$$\theta^i = \theta^i(\bar{\theta}^1, \bar{\theta}^2, \bar{\theta}^3), \tag{1.2.2}$$

and we assume that the functions θ^i are also single-valued.

The transformation of differentials $d\bar{\theta}^i$ and $d\theta^i$ follows immediately from (1.2.1) and (1.2.2), so that

$$\left.\begin{aligned} d\bar{\theta}^i &= \bar{c}^i_j\, d\theta^j \\ d\theta^i &= c^i_j\, d\bar{\theta}^j \end{aligned}\right\}, \tag{1.2.3}$$

where

$$\bar{c}^i_j = \frac{\partial \bar{\theta}^i}{\partial \theta^j}, \qquad c^i_j = \frac{\partial \theta^i}{\partial \bar{\theta}^j}. \tag{1.2.4}$$

The functions in (1.2.4) are related by the equations

$$\bar{c}^i_k c^k_j = c^i_k \bar{c}^k_j = \delta^i_j, \tag{1.2.5}$$

from which the values of $\bar{c}^i_j$ can be calculated when c^i_j are known, and vice versa, provided that the functional determinant

$$c = |\bar{c}^i_j| \neq 0.$$

This last condition holds because our transformations are assumed to be reversible.

From (1.2.3) we see that the transformation of the differentials is a linear one, while the transformation of the variables θ^i in (1.2.1) is, of course, not linear in general.

1.3. Invariants. Tensors

Consider a system T of functions (of any order) whose components are defined in the general set of variables θ^i and are functions of $\theta^1, \theta^2, \theta^3$. If the variables θ^i can be changed to $\bar{\theta}^i$ by equations (1.2.1) we can define new components of T in the general variables $\bar{\theta}^i$ which are functions of $\bar{\theta}^1, \bar{\theta}^2, \bar{\theta}^3$, and if the components of T in the two sets of variables are related by certain rules, which we now examine, the system of functions T is called a tensor.

I. A system of order zero may be defined to have a single component ϕ in the variables θ^i, and a single component $\bar{\phi}$ in the variables $\bar{\theta}^i$. If

$$\bar{\phi}(\bar{\theta}^1, \bar{\theta}^2, \bar{\theta}^3) \equiv \phi(\theta^1, \theta^2, \theta^3), \tag{1.3.1}$$

that is, numerically equal at corresponding values of $\bar{\theta}^i$, θ^i, then the functions ϕ and $\bar{\phi}$ of the θ^i's and $\bar{\theta}^i$'s respectively are the components in their respective variables of a *tensor of order zero.* This system is also called a *scalar invariant* or *scalar.* For brevity we shall say that ϕ (or $\bar{\phi}$) is a scalar.

II. A system of order one may be defined to have three components A^i in the variables θ^i and three components $\bar{A}^i$ in the variables $\bar{\theta}^i$. If

$$\bar{A}^i = \bar{c}^i_j A^j, \tag{1.3.2}$$

then the functions $A^i(\theta^1, \theta^2, \theta^3)$ and $\bar{A}^i(\bar{\theta}^1, \bar{\theta}^2, \bar{\theta}^3)$ are the components in their respective variables of a *contravariant tensor of order one.* For brevity we shall say that A^i (or $\bar{A}^i$) is a contravariant tensor of order one.

III. A system of order one may be defined to have three components A_i in the variables θ^i and three components $\bar{A}_i$ in the variables $\bar{\theta}^i$. If

$$\bar{A}_i = c^j_i A_j, \tag{1.3.3}$$

then the functions $A_i(\theta^1, \theta^2, \theta^3)$ and $\bar{A}_i(\bar{\theta}^1, \bar{\theta}^2, \bar{\theta}^3)$ are the components in their respective variables of a *covariant tensor of order one*. For brevity we shall say that A_i (or $\bar{A}_i$) is a covariant tensor of order one.

IV. A system of order two may be defined to have nine components A^{ij} in the variables θ^i and nine components $\bar{A}^{ij}$ in the variables $\bar{\theta}^i$. If

$$\bar{A}^{ij} = \bar{c}^i_m \bar{c}^j_n A^{mn}, \tag{1.3.4}$$

then the functions $A^{ij}(\theta^1, \theta^2, \theta^3)$ and $\bar{A}^{ij}(\bar{\theta}^1, \bar{\theta}^2, \bar{\theta}^3)$ are the components in their respective variables of a *contravariant tensor of order two*. For brevity we shall say that A^{ij} (or $\bar{A}^{ij}$) is a contravariant tensor of order two.

V. A system of order two may be defined to have nine components A_{ij} in the variables θ^i and nine components $\bar{A}_{ij}$ in the variables $\bar{\theta}^i$. If

$$\bar{A}_{ij} = c^m_i c^n_j A_{mn}, \tag{1.3.5}$$

then the functions $A_{ij}(\theta^1, \theta^2, \theta^3)$ and $\bar{A}_{ij}(\bar{\theta}^1, \bar{\theta}^2, \bar{\theta}^3)$ are the components in their respective variables of a *covariant tensor of order two*. For brevity we call A_{ij} (or $\bar{A}_{ij}$) a covariant tensor of order two.

VI. A system of order two may be defined to have nine components $A^i_{.j}$ (or $A_j^{.i}$) in the variables θ^i and nine components $\bar{A}^i_{.j}$ (or $\bar{A}_j^{.i}$) in the variables $\bar{\theta}^i$. If

$$\left.\begin{aligned} \bar{A}^i_{.j} &= \bar{c}^i_m c^n_j A^m_{.n} \\ \bar{A}_j^{.i} &= \bar{c}^i_m c^n_j A_n^{.m} \end{aligned}\right\}, \tag{1.3.6}$$

then the functions $A^i_{.j}(\theta^1, \theta^2, \theta^3)$ {or $A_j^{.i}(\theta^1, \theta^2, \theta^3)$} and $\bar{A}^i_{.j}(\bar{\theta}^1, \bar{\theta}^2, \bar{\theta}^3)$ {or $\bar{A}_j^{.i}(\bar{\theta}^1, \bar{\theta}^2, \bar{\theta}^3)$} are the components in their respective variables of a *mixed tensor of order two*. For brevity we call $A^i_{.j}$ (or $\bar{A}^i_{.j}$), and $A_j^{.i}$ (or $\bar{A}_j^{.i}$), mixed tensors of order two.

The order of the tensor is denoted by the number of indices and the type of tensor (contravariant, covariant, or mixed) by the position of the indices, so that we frequently refer to tensors (e.g. A^i, A_{ij}, $A_j^{.i}$) and omit explicit reference to the order or type.

In a similar way tensors of higher orders may be formed. For example, mixed components $A^i_{.jk}(\theta^1, \theta^2, \theta^3)$ and $\bar{A}^i_{.jk}(\bar{\theta}^1, \bar{\theta}^2, \bar{\theta}^3)$, in their respective variables, of a tensor of order three, are related by the law

$$\bar{A}^i_{.jk} = \bar{c}^i_r c^s_j c^t_k A^r_{.st}. \tag{1.3.7}$$

Again, we often omit the word component and refer to a mixed tensor $A^i_{.jk}$ (or $\bar{A}^i_{.jk}$) of order three, or simply, a tensor $A^i_{.jk}$, and similarly for tensors of any order or type.

From (1.2.3) we see that the differentials $d\theta^i$ transform according to the law for contravariant tensors, so that the position of the upper index is justified. The variables θ^i themselves are in general neither contravariant nor covariant and the position of their index must be recognized as an exception. In future the index in non-tensors will be placed either above or below according to convenience. For example, we shall use either θ^i or θ_i.

We notice that each component of a tensor in the new variables is a linear combination of the components in the old variables. Consequently, if all components of a tensor are zero in one system of variables they are also zero in all other systems of variables which can be obtained by transformations of the type (1.2.1) and (1.2.2).

It also follows immediately from the definitions (1.3.1) to (1.3.7) and the relations (1.2.4) and (1.2.5) that all functions of the form

$$A^iB_i, \qquad A^{ij}B_{ij},$$

etc., are invariants, if, for example, using an obvious notation, we define

$$\overline{A^iB_i} = \bar{A}^i\bar{B}_i, \qquad \overline{A^{ij}B_{ij}} = \bar{A}^{ij}\bar{B}_{ij}.$$

We observe that components of tensors satisfy the group property under general transformations of variables. In other words, if the transformation relations exist when the variables are changed from θ_i to $\bar{\theta}_i$, and when they are changed from θ_i to θ'_i, then they also exist when the variables are changed from θ'_i to $\bar{\theta}_i$. This property follows easily from the forms of the transformations.

We may notice at this point the *arbitrary* character of tensors. We may take as the components of a tensor in a given set of variables any set of functions of the requisite number. We then define components in any *general* system of variables by the equations expressing the law of transformation for that particular tensor. Because of the group property we then know that the components of our tensor expressed in a general system of variables will always transform according to our tensor rules.

1.4. Addition, multiplication, and contraction of tensors

The operations of addition and subtraction of tensors apply only to tensors of the same order and type and lead to tensors of the same order and type. For example, if A^i and B^i are contravariant tensors of order one then C^i defined by the equation

$$C^i = A^i+B^i$$

is also a contravariant tensor of order one. Or again, the difference of $A^i_{.jk}$ and $B^i_{.jk}$ leads to a mixed tensor $C^i_{.jk}$ of the third order where

$$C^i_{.jk} = A^i_{.jk}-B^i_{.jk}.$$

The equations $$A^{ij} = B^{ij}$$

are said to form a *tensor* equation, that is, if they are true in one system of variables they are true in all other systems. This follows at once from the definition of a tensor, since components in one set of variables define one unique tensor.

The *multiplication* of tensors leads to tensors of higher orders. For example, if we multiply A_i and B_{jk} we get the tensor

$$C_{ijk} = A_i B_{jk}$$

of order three.

Another operation applied to tensors is the operation of *contraction*. Let us consider the mixed tensor $A^{ij}_{\cdot\cdot k}$ of order three. If we make the indices k and i the same so that the tensor becomes $A^{ij}_{\cdot\cdot i}$, and if we remember that the repeated index is to be summed, we see that the new system has its order reduced by two. It can be shown from (1.3.2) to (1.3.7) that the new system forms a contravariant tensor of order one.

We see that any combination of the operations of addition, subtraction, multiplication, and contraction on tensors produces new tensors, and these operations are called tensor operations. We often recognize the tensorial character of a system of functions by observing that they are formed by a combination of these operations on known tensors.

Finally, we mention another important rule concerning tensors. If, for example, we have a system of nine functions $A_{(ij)}$ such that

$$A_{(ij)} a^i = B_j$$

for *every* contravariant tensor a^i, where B_j is known to be a covariant tensor, then

$$A_{(ij)} = A_{ij}$$

is a tensor. This result can be generalized to apply to tensors of any order and type.

1.5. Symmetric and skew-symmetric tensors

If we are given the tensors A^{ij}, A_{ij} it may happen that each component is unaltered in value when the indices are interchanged, so that

$$A^{ij} = A^{ji}, \qquad A_{ij} = A_{ji}. \tag{1.5.1}$$

The tensors are then said to be *symmetric*. More generally, a tensor of any order is said to be symmetric in two subscripts or superscripts if it is unaltered when the two indices are interchanged, and the tensor is completely symmetric if the interchange of any subscripts or superscripts leaves it unaltered. The tensor A_{ijk} of the third order which is completely symmetric will, for example, satisfy the relations

$$A_{ijk} = A_{ikj} = A_{jik} = A_{jki} = A_{kij} = A_{kji}. \tag{1.5.2}$$

On the other hand, a tensor is skew-symmetric if the interchange of two indices alters only the sign of the components so that, for example,

$$A^{ij} = -A^{ji}, \qquad A_{ij} = -A_{ji}. \tag{1.5.3}$$

In this case we deduce immediately that

$$A^{ij} = 0, \qquad A_{ij} = 0 \quad (i = j, j \text{ not summed}). \tag{1.5.4}$$

Similarly, a tensor of any order may be skew-symmetric with respect to two indices, or may be completely skew-symmetric in all indices.

It can be shown that if a tensor is completely symmetric or skew-symmetric in one system of variables it is so in every system.

Every second order contravariant or covariant tensor may be expressed as the sum of a symmetric and a skew-symmetric tensor. For example, the tensor A_{ij} may be written

$$A_{ij} = \tfrac{1}{2}(A_{ij}+A_{ji})+\tfrac{1}{2}(A_{ij}-A_{ji}), \tag{1.5.5}$$

where the tensor $\frac{1}{2}(A_{ij}+A_{ji})$ is symmetric and the tensor $\frac{1}{2}(A_{ij}-A_{ji})$ is skew-symmetric.

1.6. Curvilinear coordinates

The coordinates x^i or x_i referred to a right-handed orthogonal cartesian system of axes define a three-dimensional Euclidean space (see Fig. 1.4). We introduce general coordinates θ_i by the transformation

$$\theta_i = \theta_i(x_1, x_2, x_3), \tag{1.6.1}$$

where θ_i are arbitrary single-valued functions of the cartesian coordinates x_i which we suppose possess derivatives up to any required order and which are independent of each other. We assume that the transformation is reversible so that we can solve for x_i, giving

$$x_i = x_i(\theta_1, \theta_2, \theta_3), \tag{1.6.2}$$

where the functions in (1.6.2) are single-valued. Since the transformation is reversible the determinant of the transformation

$$\left|\frac{\partial x^i}{\partial \theta^j}\right|$$

is not zero.†

With the above assumptions we see that to each set of values of x_i corresponds a unique set of values of θ_i, and vice versa. Hence the variables θ_i determine points in the defined three-dimensional Euclidean space. We may therefore represent our space by the variables θ_i instead

† For some transformations, however, exceptions to this may be allowed at certain points.

of by the cartesian system x_i, but the space remains, of course, Euclidean.

We now consider the geometrical significance of the coordinates θ_i. The relation $\theta_i(x_1, x_2, x_3) =$ constant is the equation of a surface and as the value of the constant varies we get a family of surfaces. In other words, corresponding to $i = 1, 2, 3$ we have three families of surfaces and the point of intersection of one member from each of these families determines a point P of the space. The conditions imposed on the functions θ_i ensure that the three surfaces obtained by taking a member of each family intersect in one and only one point, thus defining the position of the point uniquely.

We call the surfaces $\theta_i =$ constant *coordinate surfaces* and refer to them briefly as the θ_i-surfaces. The intersections of these surfaces also give us three curves through every point P, two of them lying on each coordinate surface. These curves are called *coordinate curves*, and θ_i are called *curvilinear coordinates* (Fig. 1.5).

1.7. Special tensors

We recall from § 1.3 that general tensors may be obtained by defining their components in a given set of coordinates and then transforming to a *general* system of coordinates. Here we define components in a cartesian system x_i and transform to the general curvilinear coordinates θ_i by equations of the types (1.3.2) to (1.3.7), using the transformations (1.6.1) and (1.6.2).

Consider first the differentials dx_i (or dx^i) which form the components of either a covariant or contravariant tensor of order one. In general coordinates these become

$$d\theta^j = \frac{\partial\theta^j}{\partial x^i}dx^i, \qquad dx^i = \frac{\partial x^i}{\partial\theta^j}d\theta^j, \tag{1.7.1}$$

$$d\theta_j = \frac{\partial x^i}{\partial\theta^j}dx_i, \qquad dx_i = \frac{\partial\theta^j}{\partial x^i}d\theta_j, \tag{1.7.2}$$

so that $d\theta^j$ and $d\theta_j$ are contravariant and covariant tensors respectively. From (1.7.1) we see that we can identify $d\theta^j$ with the usual differential of the variables θ_j, so that the use of the upper index for the differentials is justified.†

We next introduce tensors of the second order which are derived from the Kronecker deltas.‡ In cartesian coordinates Kronecker deltas

† See also § 1.3. ‡ The Kronecker deltas δ_j^i have already been defined in (1.1.3).

are defined by the equations

$$\left.\begin{aligned}\delta_{ij} = \delta^{ij} = \delta^i_j = \delta^j_i = 0 \quad (i \neq j)\\ \delta_{ij} = \delta^{ij} = \delta^i_j = \delta^j_i = 1 \quad (i = j,\, j \text{ not summed})\end{aligned}\right\}. \tag{1.7.3}$$

We take these Kronecker deltas to define, in cartesian coordinates x_i, the components of covariant, contravariant, and mixed tensors of the second order. Then, using our tensor transformation rules, we obtain at once the components of these tensors in general coordinates θ_i. Thus

$$g_{ij} = \frac{\partial x^m}{\partial \theta^i}\frac{\partial x^n}{\partial \theta^j}\delta_{mn} = \frac{\partial x^m}{\partial \theta^i}\frac{\partial x^m}{\partial \theta^j}, \tag{1.7.4}$$

$$g^{ij} = \frac{\partial \theta^i}{\partial x^m}\frac{\partial \theta^j}{\partial x^n}\delta^{mn} = \frac{\partial \theta^i}{\partial x^m}\frac{\partial \theta^j}{\partial x^m}, \tag{1.7.5}$$

$$g^i_j = \frac{\partial \theta^i}{\partial x^m}\frac{\partial x^n}{\partial \theta^j}\delta^m_n = \frac{\partial \theta^i}{\partial x^m}\frac{\partial x^m}{\partial \theta^j} = \delta^i_j. \tag{1.7.6}$$

The tensors whose components are defined in (1.7.4) to (1.7.6) are called the covariant, contravariant, and mixed *metric* tensors, the significance of which will be seen in § 1.9. We note that the mixed tensor g^i_j is identical with δ^i_j and is therefore constant in all coordinate systems. It also follows from (1.7.4) to (1.7.6) that the three metric tensors are symmetric and that

$$g_{im}g^{mj} = \delta^j_i. \tag{1.7.7}$$

Also the determinants of the coefficients g_{ij} and g^{ij} are both positive and equal to

$$\left.\begin{aligned} g = |g_{ij}| = \left|\frac{\partial x^i}{\partial \theta^j}\right|^2\\ \frac{1}{g} = |g^{ij}| = \left|\frac{\partial \theta^i}{\partial x^j}\right|^2\end{aligned}\right\}. \tag{1.7.8}$$

Equation (1.7.7) may be solved to give

$$g^{ij} = \frac{D^{ij}}{g}, \tag{1.7.9}$$

where† D^{ij} is the cofactor of g_{ij} in the determinant g.

Finally in this section we consider special skew-symmetric tensors called the ϵ-systems, whose components in general coordinates are denoted by ϵ_{ijk} and ϵ^{ijk}. In cartesian coordinates x_i we denote these

† D^{ij} is not a tensor in our sense but is sometimes known as a relative or weighted tensor.

components by e_{ijk} and e^{ijk} respectively, and these are defined by the equations

$$e_{ijk} = e^{ijk} = 0 \quad \text{when any two of the indices are equal;}$$
$$= +1 \quad \text{when } i, j, k \text{ is an even permutation of the numbers } 1, 2, 3;$$
$$= -1 \quad \text{when } i, j, k \text{ is an odd permutation of the numbers } 1, 2, 3.$$

For example

$$e^{112} = e_{112} = e^{133} = e_{133} = e^{222} = e_{222} = 0,$$
$$e^{123} = e_{123} = e^{231} = e_{231} = e^{312} = e_{312} = 1,$$
$$e^{213} = e_{213} = e^{132} = e_{132} = e^{321} = e_{321} = -1.$$

In general coordinates, therefore,

$$\epsilon_{rst} = \frac{\partial x^i}{\partial \theta^r}\frac{\partial x^j}{\partial \theta^s}\frac{\partial x^k}{\partial \theta^t} e_{ijk}, \tag{1.7.10}$$

$$\epsilon^{rst} = \frac{\partial \theta^r}{\partial x^i}\frac{\partial \theta^s}{\partial x^j}\frac{\partial \theta^t}{\partial x^k} e^{ijk}, \tag{1.7.11}$$

or, alternatively,

$$e_{ijk} = \frac{\partial \theta^r}{\partial x^i}\frac{\partial \theta^s}{\partial x^j}\frac{\partial \theta^t}{\partial x^k} \epsilon_{rst}, \tag{1.7.12}$$

$$e^{ijk} = \frac{\partial x^i}{\partial \theta^r}\frac{\partial x^j}{\partial \theta^s}\frac{\partial x^k}{\partial \theta^t} \epsilon^{rst}. \tag{1.7.13}$$

The right-hand sides of (1.7.10) and (1.7.11) are, however, well known† and can be expressed in terms of the determinants

$$\left|\frac{\partial x^i}{\partial \theta^j}\right| \quad \text{and} \quad \left|\frac{\partial \theta^i}{\partial x^j}\right|$$

respectively. Hence, using (1.7.8), we find that

$$\left.\begin{aligned} \epsilon_{rst} &= e_{rst}\sqrt{g} \\ \epsilon^{rst} &= \frac{e^{rst}}{\sqrt{g}} \end{aligned}\right\}. \tag{1.7.14}$$

Also, remembering that e_{rst} and e^{rst} are identical, it is a simple matter to prove that

$$\epsilon^{rst} = g^{ri}g^{sj}g^{tk}\epsilon_{ijk}, \qquad \epsilon_{rst} = g_{ri}\,g_{sj}\,g_{tk}\,\epsilon^{ijk}. \tag{1.7.15}$$

† If $|A^i_j|$ is the determinant with elements A^i_j then

$$e^{ijk}A^r_i\,A^s_j\,A^t_k = |A^i_j|e^{rst},$$
$$e_{ijk}\,A^i_r\,A^j_s\,A^k_t = |A^i_j|e_{rst}.$$

We close this section by adding some useful formulae

$$\left.\begin{aligned}\epsilon^{rst}\epsilon^{ijk}g_{ri}g_{sj}g_{tk} &= 6\\ \epsilon_{rst}\epsilon_{ijk}g^{ri}g^{sj}g^{tk} &= 6\end{aligned}\right\}, \tag{1.7.16}$$

$$\left.\begin{aligned}\epsilon^{rst}\epsilon^{ijk}g_{sj}g_{tk} &= 2g^{ri}\\ \epsilon_{rst}\epsilon_{ijk}g^{sj}g^{tk} &= 2g_{ri}\end{aligned}\right\}, \tag{1.7.17}$$

which may be verified by observing that they are tensor equations which are clearly satisfied in cartesian coordinates. We denote the product of ϵ^{rst} and ϵ_{ijk} by

$$\delta^{rst}_{ijk} = \epsilon^{rst}\epsilon_{ijk}. \tag{1.7.18}$$

In particular we find from (1.7.15) that

$$\delta^{rst}_{ist} = \epsilon^{rst}\epsilon_{ist} = g_{im}g_{sn}g_{tp}\,\epsilon^{mnp}\epsilon^{rst} = 2g_{im}g^{mr} = 2\delta^r_i, \tag{1.7.19}$$

on using (1.7.17) and (1.7.7).

1.8. Vectors

In this section we restrict our attention to right-handed orthogonal cartesian coordinate axes and we consider transformations from one such system of axes Ox_i to another $\bar{O}\bar{x}_j$ which (apart from additional constants) can be written

$$\bar{x}_i = \frac{\partial\bar{x}^i}{\partial x^j}x_j, \qquad x_i = \frac{\partial x^i}{\partial\bar{x}^j}\bar{x}_j. \tag{1.8.1}$$

For such transformations

$$\frac{\partial x^i}{\partial\bar{x}^j} = \frac{\partial\bar{x}^j}{\partial x^i}, \tag{1.8.2}$$

since each of these expressions represents the cosine of the angle between Ox_i and $\bar{O}\bar{x}_j$. It follows from (1.3.2) and (1.3.3) that the transformations defining covariant and contravariant tensors of order one are identical in rectangular cartesian coordinate systems, and there is no longer any distinction between covariant and contravariant tensors of order one. The position of the indices is therefore immaterial and we may write

$$A^i = A_i.$$

In the same way covariant, contravariant, and mixed tensors of any order are no longer distinct. Since

$$\frac{\partial\bar{x}^i}{\partial x^k}\frac{\partial x^k}{\partial\bar{x}^j} = \delta^i_j \tag{1.8.3}$$

it follows from (1.8.2) that

$$\left|\frac{\partial\bar{x}^i}{\partial x^j}\right| = \pm 1 \tag{1.8.4}$$

and the sign can be shown to be + by considering the identical transformation $\bar{x}_i = x_i$.

We now define a *three-dimensional vector* or *space vector* to be a system of functions A_i (or A^i) of order one, one system corresponding to each rectangular cartesian frame of reference, and the components for any two rectangular reference frames Ox_i, $\bar{O}\bar{x}_i$, being related by the equation

$$\bar{A}_i = \frac{\partial x^j}{\partial \bar{x}^i} A_j. \tag{1.8.5}$$

This vector may be represented by a single symbol $\mathbf{A}$ where

$$\mathbf{A} = (A_1, A_2, A_3) = (A_i),$$

and A_i are the components of $\mathbf{A}$ with reference to the rectangular cartesian frame Ox_i. The three components (A_1, A_2, A_3) for *any* given reference frame Ox_i may be arbitrarily assigned and then the vector is determined, the components for *any* other reference frame $\bar{O}\bar{x}_i$ being given by (1.8.5).

The zero vector $\mathbf{0} = (0, 0, 0)$, obtained by putting $A_i = 0$, is the same in all reference frames.

If P_1, P_2 are two points whose coordinates referred to the frame Ox_i are $x_i^{(1)}$ and $x_i^{(2)}$ respectively, and $A_i = x_i^{(2)} - x_i^{(1)}$, then A_i are the projections of the line segment $\overrightarrow{P_1 P_2}$ on the axes of the reference frame. Because of the transformation (1.8.5), $\bar{A}_i$ are the projections of the same line segment on the axes of any other rectangular reference frame $\bar{O}\bar{x}_i$. There are many line segments $\overrightarrow{P_1 P_2}$ which have the same projections on the axes Ox_i and the collection of all these equally long and equally directed line segments gives a geometrical representation of the vector $\mathbf{A}$, and we write

$$\mathbf{A} = \overrightarrow{P_1 P_2}. \tag{1.8.6}$$

A particular representation of the vector occurs when P_1 coincides with the origin O and P_2 coincides with the point P $(A_i = x_i)$. In this case the vector of which the line segment OP is a representation, is called the *position vector* of the point P (with respect to the origin O of the Ox_i reference frame). Thus

$$\mathbf{A} = (A_1, A_2, A_3) = \overrightarrow{P_1 P_2} = \overrightarrow{OP}. \tag{1.8.7}$$

The magnitude of the vector $\mathbf{A}$ is defined to be

$$A = |\mathbf{A}| = \sqrt{(A_i A_i)} = P_1 P_2 = OP, \tag{1.8.8}$$

and is the common length of any line segment which is a representation of the vector. We observe that the magnitude is independent of the reference frame since

$$\bar{A}_i \bar{A}_i = \frac{\partial x^m}{\partial \bar{x}^i} \frac{\partial x^n}{\partial \bar{x}^i} A_m A_n$$

$$= \delta^m_n A_m A_n$$

$$= A_n A_n.$$

A vector of unit length is called a *unit vector*.

Let $\mathbf{A} = (A_i)$, $\mathbf{B} = (B_i)$ be any space vectors (referred to the frame Ox_i). Then

I. $$\mathbf{A} = \mathbf{B}$$

if $A_i = B_i$.

II. The vector sum of **A** and **B** is the vector

$$\mathbf{A}+\mathbf{B} = (A_i+B_i). \tag{1.8.9}$$

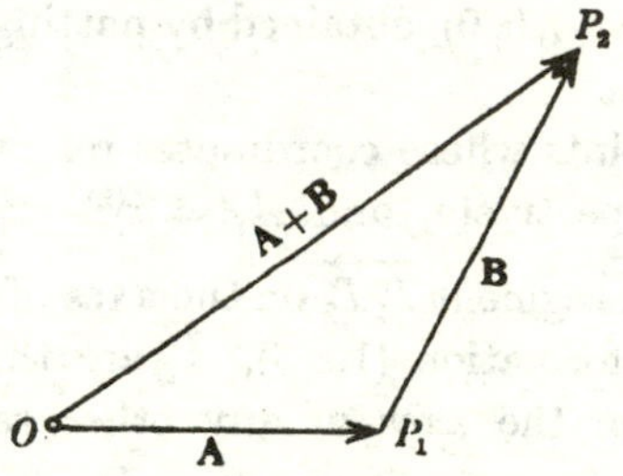

FIG. 1.1. Vector addition.

If the line segment $\overrightarrow{OP_1}$ is a representation of **A** and the line segment $\overrightarrow{P_1P_2}$ is a representation of **B** then the line segment $\overrightarrow{OP_2}$ is a representation of $\mathbf{A}+\mathbf{B}$ (see Fig. 1.1).

If $\mathbf{A} = (A_i)$, $\mathbf{B} = (B_i)$,..., $\mathbf{N} = (N_i)$ are n vectors their sum is

$$\mathbf{A}+\mathbf{B}+\ldots+\mathbf{N} = (A_i+B_i+\ldots+N_i). \tag{1.8.10}$$

To obtain a representative line segment of the sum of any number of vectors we lay representative segments of individual vectors end to end; the line segment from the initial point of the first segment to the terminal point of the last segment is a representation of the sum of the vectors. Addition of vectors is commutative and associative.

III. The product of a vector $\mathbf{A} = (A_i)$ by a scalar k is the vector

$$k\mathbf{A} = (kA_i). \tag{1.8.11}$$

This is represented by any line segment parallel to the line segment representations of **A** and of length $|k||\mathbf{A}|$, the direction being the same or opposite to that of **A** according as k is positive or negative.

IV. The *scalar product* of **A** and **B** is denoted by **A.B** and is defined as

$$\mathbf{A}.\mathbf{B} = A_i B_i. \tag{1.8.12}$$

Since
$$(A_i+B_i)^2 = A_i A_i + B_i B_i + 2A_i B_i$$

is independent of the coordinate axes it follows that the definition of scalar product is independent of the choice of axes. On choosing a reference frame whose positive x_1-axis has the direction of **A** we find that

$$\mathbf{A}.\mathbf{B} = AB\cos\theta, \tag{1.8.13}$$

where θ is the angle between **A** and **B**. In particular

$$\mathbf{A}.\mathbf{A} = A_i A_i = |\mathbf{A}|^2 = A^2. \tag{1.8.14}$$

V. The *vector product* of **A** and **B**, denoted by $\mathbf{A}\times\mathbf{B}$, is the vector

$$\begin{aligned}\mathbf{A}\times\mathbf{B} &= (A_2 B_3 - A_3 B_2,\ A_3 B_1 - A_1 B_3,\ A_1 B_2 - A_2 B_1)\\ &= (e_{ijk} A_i B_j).\end{aligned} \tag{1.8.15}$$

This can be shown to be a vector with the help of the transformations (1.8.1), and equation (1.8.4), since

$$\begin{aligned}\bar{e}_{rst}\bar{A}_r\bar{B}_s &= e_{rst}\left|\frac{\partial x^i}{\partial \bar{x}^j}\right|\bar{A}_r\bar{B}_s\\ &= e_{ijk}\frac{\partial x^i}{\partial \bar{x}^r}\frac{\partial x^j}{\partial \bar{x}^s}\frac{\partial x^k}{\partial \bar{x}^t}\bar{A}_r\bar{B}_s\\ &= \frac{\partial x^k}{\partial \bar{x}^t}(e_{ijk}A_i B_j).\end{aligned}$$

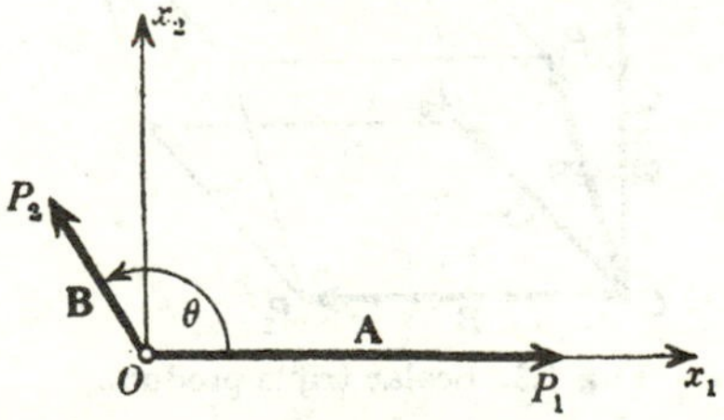

FIG. 1.2.

Suppose the vectors **A**, **B** are represented by the line segments $\overrightarrow{OP_1}$, $\overrightarrow{OP_2}$ (Fig. 1.2) and let θ be the angle between $\overrightarrow{OP_1}$, $\overrightarrow{OP_2}$ where $0 \leqslant \theta \leqslant \pi$. If we choose the reference frame so that Ox_1 has the direction of **A** while Ox_2 lies in the plane determined by OP_1, OP_2, its direction being such that the x_2-component of **B** is positive, we then have

$$\mathbf{A} = (A, 0, 0), \qquad \mathbf{B} = (B\cos\theta, B\sin\theta, 0).$$

If $\theta = 0$ or π the line segments $\overrightarrow{OP_1}$, $\overrightarrow{OP_2}$ do not determine a plane and the x_2-axis may have any direction perpendicular to **A**. Hence

$$\mathbf{A}\times\mathbf{B} = (0, 0, AB\sin\theta).$$

Thus the vector $\mathbf{A}\times\mathbf{B}$ is of magnitude $AB\sin\theta$ and is represented by the area of the parallelogram with adjacent sides $\overrightarrow{OP_1}$, $\overrightarrow{OP_2}$. If it is not the zero vector it is directed along the positive x_3-axis. We may therefore write

$$\mathbf{A}\times\mathbf{B} = \mathbf{n}AB\sin\theta \quad (0 \leqslant \theta \leqslant \pi), \tag{1.8.16}$$

where **n** is a unit vector perpendicular to the plane of **A** and **B**. Also **A**, **B**, **n** form a right-handed system.

VI. Let $\mathbf{A} = (A_i)$, $\mathbf{B} = (B_i)$, and $\mathbf{C} = (C_i)$ be three vectors. The *scalar triple product* of these vectors, which is written as $[\mathbf{A}\,\mathbf{B}\,\mathbf{C}]$, is defined to be

$$\mathbf{A}.(\mathbf{B}\times\mathbf{C}) = \begin{vmatrix} A_1 & A_2 & A_3 \\ B_1 & B_2 & B_3 \\ C_1 & C_2 & C_3 \end{vmatrix} = e_{ijk}A_iB_jC_k, \tag{1.8.17}$$

and we observe that this may also be written $(\mathbf{A}\times\mathbf{B}).\mathbf{C}$. Also

$$[\mathbf{A}\,\mathbf{B}\,\mathbf{C}] = -[\mathbf{A}\,\mathbf{C}\,\mathbf{B}] = [\mathbf{B}\,\mathbf{C}\,\mathbf{A}] = -[\mathbf{C}\,\mathbf{B}\,\mathbf{A}]$$
$$= [\mathbf{C}\,\mathbf{A}\,\mathbf{B}] = -[\mathbf{B}\,\mathbf{A}\,\mathbf{C}]. \tag{1.8.18}$$

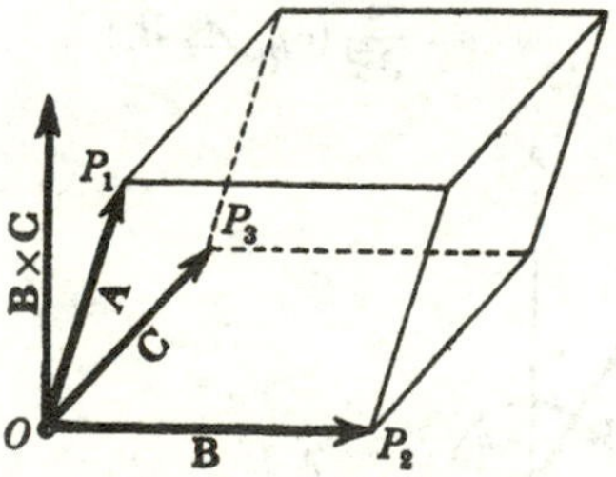

FIG. 1.3. Scalar triple product.

Since the magnitude of $\mathbf{B}\times\mathbf{C}$ is represented by the area of the parallelogram of which $\overrightarrow{OP_2}$, $\overrightarrow{OP_3}$ are adjacent sides (Fig. 1.3), $[\mathbf{A}\,\mathbf{B}\,\mathbf{C}]$ is the volume of the box of which $(\overrightarrow{OP_1}, \overrightarrow{OP_2}, \overrightarrow{OP_3})$ are adjacent sides. This volume is positive when the angle between **A** and $\mathbf{B}\times\mathbf{C}$ is acute, and negative when this angle is obtuse.

VII. A set of three vectors **A**, **B**, **C** are linearly independent if and only if

$$[\mathbf{A}\,\mathbf{B}\,\mathbf{C}] \neq 0,$$

that is, the only linear combination of **A**, **B**, **C** of the form

$$\lambda_1 \mathbf{A}+\lambda_2 \mathbf{B}+\lambda_3 \mathbf{C}$$

which is the zero vector, is when all the constants λ_i are zero.

Every set of four (or more) space vectors are linearly dependent. That is, a linear combination

$$\lambda_1 \mathbf{A}+\lambda_2 \mathbf{B}+\lambda_3 \mathbf{C}+\lambda_4 \mathbf{D}$$

of the space vectors exists which is the zero vector, where not all the λ's are zero. Thus if **A**, **B**, **C** are linearly independent,

$$\mathbf{D} = \lambda_1 \mathbf{A}+\lambda_2 \mathbf{B}+\lambda_3 \mathbf{C}, \tag{1.8.19}$$

where
$$\lambda_1 = \frac{[\mathbf{D}\,\mathbf{B}\,\mathbf{C}]}{[\mathbf{A}\,\mathbf{B}\,\mathbf{C}]}, \quad \lambda_2 = \frac{[\mathbf{D}\,\mathbf{C}\,\mathbf{A}]}{[\mathbf{A}\,\mathbf{B}\,\mathbf{C}]}, \quad \lambda_3 = \frac{[\mathbf{D}\,\mathbf{A}\,\mathbf{B}]}{[\mathbf{A}\,\mathbf{B}\,\mathbf{C}]}. \tag{1.8.20}$$

VIII. Every linearly independent set of space vectors forms a *basis* for space vectors. The simplest basis is formed by the three unit vectors

$$\mathbf{i}^1 = \mathbf{i}_1 = (1,0,0); \quad \mathbf{i}^2 = \mathbf{i}_2 = (0,1,0); \quad \mathbf{i}^3 = \mathbf{i}_3 = (0,0,1), \tag{1.8.21}$$

which are such that $[\mathbf{i}_1\,\mathbf{i}_2\,\mathbf{i}_3] = 1$ and

$$\mathbf{i}_r.\mathbf{i}_s = \mathbf{i}_r.\mathbf{i}^s = \mathbf{i}^r.\mathbf{i}_s = \mathbf{i}^r.\mathbf{i}^s = \delta^r_s, \tag{1.8.22}$$

$$\mathbf{i}_r \times \mathbf{i}_s = \mathbf{i}^r \times \mathbf{i}^s = e_{rst}\,\mathbf{i}^t = e^{rst}\,\mathbf{i}_t. \tag{1.8.23}$$

Hence any vector **A** can be written in the form

$$\mathbf{A} = (\mathbf{A}.\mathbf{i}_r)\mathbf{i}_r = A_r\,\mathbf{i}_r, \tag{1.8.24}$$

where
$$A_r = \mathbf{A}.\mathbf{i}_r. \tag{1.8.25}$$

In particular, if
$$\mathbf{R} = x_r\,\mathbf{i}_r \tag{1.8.26}$$

is the position vector of the point x_r, then

$$x_r = \mathbf{R}.\mathbf{i}_r. \tag{1.8.27}$$

We also observe that, using (1.8.22) to (1.8.24),

$$\mathbf{A}.\mathbf{B} = A_r\,\mathbf{i}_r.B_s\,\mathbf{i}_s = A_r B_s \delta^r_s = A_r B_r, \tag{1.8.28}$$

$$\mathbf{A}\times\mathbf{B} = A_r\,\mathbf{i}_r \times B_s\,\mathbf{i}_s = e_{rst} A_r B_s\,\mathbf{i}_t, \tag{1.8.29}$$

in agreement with (1.8.12) and (1.8.15).

IX. If a vector **A** is a function of some scalar parameter t, the derivative of **A** with respect to t is

$$\frac{\partial \mathbf{A}}{\partial t} = \left(\frac{\partial A_i}{\partial t}\right), \tag{1.8.30}$$

or

$$\frac{\partial \mathbf{A}}{\partial t} = \frac{\partial A_r}{\partial t}\mathbf{i}_r. \tag{1.8.31}$$

In particular, from (1.8.26),

$$\frac{\partial \mathbf{R}}{\partial x^r} = \mathbf{i}_r. \tag{1.8.32}$$

1.9. Line element. Base vectors. Metric tensors

Let **R** be the position vector of a point P whose coordinates are x_i and let $d\mathbf{R}$ denote the infinitesimal vector $\overrightarrow{PQ}$, where Q has coordinates

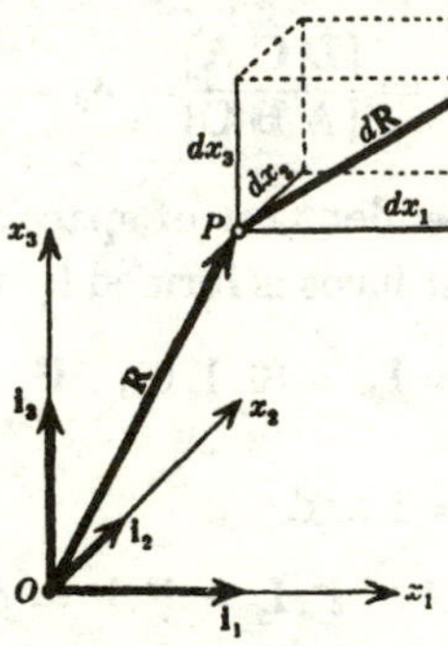

FIG. 1.4. Position vector and its differential.

x_i+dx_i (see Fig. 1.4). Then since **R** is a function of the coordinates x_i and is assumed to be differentiable,

$$d\mathbf{R} = \frac{\partial \mathbf{R}}{\partial x^r}\,dx_r = dx_r\,\mathbf{i}_r \tag{1.9.1}$$

on using (1.8.32). If $ds = |d\mathbf{R}|$ is the length of the vector $\overrightarrow{PQ}$

$$ds^2 = d\mathbf{R}.d\mathbf{R} = dx_r\,dx_r. \tag{1.9.2}$$

Using (1.6.2) we can express the position vector as a function of the general coordinates θ_i, thus

$$\mathbf{R} = \mathbf{R}(\theta_1, \theta_2, \theta_3) \tag{1.9.3}$$

and, from (1.7.1), (1.7.2), and (1.9.1),

$$d\mathbf{R} = \mathbf{g}_r\,d\theta^r = \mathbf{g}^r\,d\theta_r, \tag{1.9.4}$$

where

$$\mathbf{g}_r = \frac{\partial x^s}{\partial \theta^r}\mathbf{i}_s, \qquad \mathbf{g}^r = \frac{\partial \theta^r}{\partial x^s}\mathbf{i}^s. \tag{1.9.5}$$

We therefore see that the vectors $\mathbf{g}_r$, $\mathbf{g}^r$ are obtained from the constant unit vectors $\mathbf{i}_r$ (or $\mathbf{i}^r$) by transformations which are respectively of the same form as the transformations for covariant and contravariant tensors of order one. In any further transformation of coordinates we can always associate two new triplets of vectors with the new coordinates which are derived from $\mathbf{g}_r$ and $\mathbf{g}^r$ by covariant and contravariant types of transformations respectively. We observe that the differential $d\mathbf{R}$ therefore remains invariant, as it should do. The inverse transformations to (1.9.5) are

$$\mathbf{i}_r = \mathbf{i}^r = \frac{\partial x^r}{\partial \theta^s}\mathbf{g}^s = \frac{\partial \theta^s}{\partial x^r}\mathbf{g}_s. \tag{1.9.6}$$

The vectors $\mathbf{g}_r$ are called *covariant base vectors* and $\mathbf{g}^r$ are *contravariant base vectors*. They are connected with tensors g_{rs}, g^{rs}, g^r_s by the relations

$$\left.\begin{aligned} \mathbf{g}_r.\mathbf{g}_s = g_{rs}, \quad \mathbf{g}^r.\mathbf{g}^s = g^{rs}, \quad \mathbf{g}^r.\mathbf{g}_s = g^r_s = \delta^r_s \\ \mathbf{g}^r = g^{rs}\mathbf{g}_s, \quad \mathbf{g}_r = g_{rs}\mathbf{g}^s \end{aligned}\right\}, \tag{1.9.7}$$

which may be immediately verified with the help of (1.7.4) to (1.7.6), (1.8.22), and (1.9.5). Also, the magnitudes of the covariant and contravariant base vectors are

$$\left.\begin{aligned} |\mathbf{g}_r| = \sqrt{(\mathbf{g}_r.\mathbf{g}_r)} = \sqrt{g_{rr}} \\ |\mathbf{g}^r| = \sqrt{(\mathbf{g}^r.\mathbf{g}^r)} = \sqrt{g^{rr}} \end{aligned}\right\}, \tag{1.9.8}$$

where the index is not summed.

Using (1.9.4) the line element ds, which is given by (1.9.2), may now be written in the form

$$\begin{aligned} ds^2 = d\mathbf{R}.d\mathbf{R} = g_{rs}\,d\theta^r d\theta^s = g^{rs}d\theta_r d\theta_s \\ = \delta^r_s d\theta_r d\theta^s = d\theta_r d\theta^r. \end{aligned} \tag{1.9.9}$$

The reason for the term *metric tensor* is now apparent.

The line elements along the coordinate curves may be represented by vectors

$$d\mathbf{s}_i = \mathbf{g}_i\,d\theta^i \quad (i \text{ not summed}), \tag{1.9.10}$$

with magnitudes

$$ds_i = \sqrt{(g_{ii})}\,d\theta^i. \tag{1.9.11}$$

Referring to (1.9.4) we see that the base vectors $\mathbf{g}_i$ may also be written as the three partial derivatives

$$\mathbf{g}_i = \mathbf{R}_{,i} = \frac{\partial \mathbf{R}}{\partial \theta^i}, \tag{1.9.12}$$

and the vectors characterize the change of the position vector $\mathbf{R}$ as we move along the coordinate curves from a given point P. These vectors are directed tangentially along the coordinate curves and may be represented geometrically by directed lines† as shown in Fig. 1.5. We may verify from (1.9.12) that the covariant base vectors transform according to the covariant type of transformation.

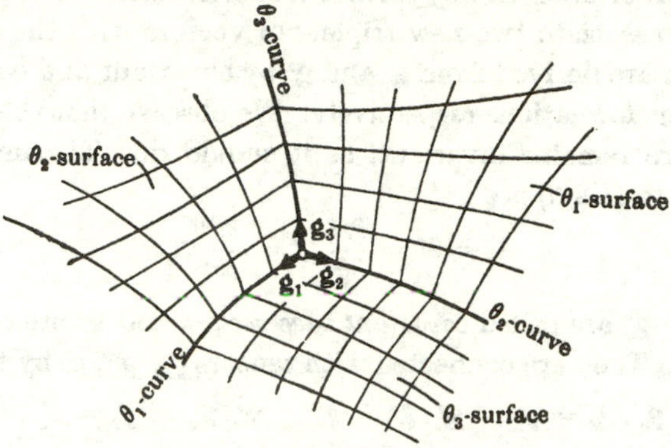

FIG. 1.5. Coordinate surfaces.

From (1.9.7) we see that the contravariant base vectors $\mathbf{g}^1$, $\mathbf{g}^2$, $\mathbf{g}^3$ are respectively perpendicular to the planes of $\mathbf{g}_2\,\mathbf{g}_3$, $\mathbf{g}_3\,\mathbf{g}_1$, $\mathbf{g}_1\,\mathbf{g}_2$.

We record some other useful results. The vector product of $\mathbf{g}_r$ and $\mathbf{g}_s$ is obtained from (1.9.5) and (1.8.23) and is

$$\mathbf{g}_r \times \mathbf{g}_s = \epsilon_{rst}\,\mathbf{g}^t. \tag{1.9.13}$$

Similarly

$$\mathbf{g}^r \times \mathbf{g}^s = \epsilon^{rst}\mathbf{g}_t. \tag{1.9.14}$$

The combination of scalar and vector products of base vectors leads to the following scalar triple products:

$$\left.\begin{aligned}[\mathbf{g}_r\,\mathbf{g}_s\,\mathbf{g}_t] = [\mathbf{g}_s\,\mathbf{g}_t\,\mathbf{g}_r] = [\mathbf{g}_t\,\mathbf{g}_r\,\mathbf{g}_s] = \epsilon_{rst}\\ [\mathbf{g}^r\,\mathbf{g}^s\,\mathbf{g}^t] = [\mathbf{g}^s\,\mathbf{g}^t\,\mathbf{g}^r] = [\mathbf{g}^t\,\mathbf{g}^r\,\mathbf{g}^s] = \epsilon^{rst}\end{aligned}\right\}. \tag{1.9.15}$$

In particular

$$[\mathbf{g}_1\,\mathbf{g}_2\,\mathbf{g}_3] = \sqrt{g}, \qquad [\mathbf{g}^1\,\mathbf{g}^2\,\mathbf{g}^3] = \frac{1}{\sqrt{g}}. \tag{1.9.16}$$

The element of area dS_1 on the θ_1-surface is

$$dS_1 = |d\mathbf{s}_2 \times d\mathbf{s}_3| = |\mathbf{g}_2 \times \mathbf{g}_3|\; d\theta^2 d\theta^3,$$

and using (1.9.13) and (1.9.8) this becomes

$$dS_1 = \sqrt{(gg^{11})}\, d\theta^2 d\theta^3. \tag{1.9.17}$$

In general, the element of area dS_i on the θ_i-surface is

$$dS_i = \sqrt{(gg^{ii})}\, d\theta^j d\theta^k \quad (i \text{ not summed, } i \neq j \neq k). \tag{1.9.18}$$

† The vectors, of course, depend on the point P so that they are functions of θ_1, θ_2, and θ_3.

The volume element

$$d\tau = d\mathbf{s}_1 . d\mathbf{s}_2 \times d\mathbf{s}_3 = [\mathbf{g}_1\, \mathbf{g}_2\, \mathbf{g}_3]\, d\theta^1 d\theta^2 d\theta^3 = \sqrt{(g)}\, d\theta^1 d\theta^2 d\theta^3. \tag{1.9.19}$$

Finally, if the coordinate curves are orthogonal, then

$$\left.\begin{array}{c} g^{rs} = g_{rs} = 0 \quad (r \neq s) \\ g^{11} = \dfrac{1}{g_{11}}, \quad g^{22} = \dfrac{1}{g_{22}}, \quad g^{33} = \dfrac{1}{g_{33}}, \quad g = g_{11}\, g_{22}\, g_{33} \end{array}\right\}. \tag{1.9.20}$$

Consider the expression $v^r \mathbf{g}_r$, where v^r is a contravariant tensor of order one transforming according to the rule (1.3.2), and $\mathbf{g}_r$ are the covariant base vectors at a point P of our space which transform according to the covariant form of transformation (1.3.3). This expression is therefore an *invariant* and since it is a linear combination of the base vectors $\mathbf{g}_r$ it is a vector. We denote this vector by

$$\mathbf{v} = v^r \mathbf{g}_r. \tag{1.9.21}$$

Also, using (1.9.7), we have

$$\mathbf{v} = v^r \mathbf{g}_r = v_s\, \mathbf{g}^s, \tag{1.9.22}$$

where

$$v_r = g_{rs} v^s, \tag{1.9.23}$$

and v_r is therefore a covariant tensor of order one. From (1.9.22) and (1.9.7), or from (1.9.23) and (1.7.7), we have

$$v^r = g^{rs} v_s. \tag{1.9.24}$$

In view of (1.9.22) the three elements v^r are sometimes called the *contravariant components* of the vector $\mathbf{v}$, and the three elements v_r are called the *covariant components* of $\mathbf{v}$. In our Euclidean space, therefore, we may regard the contravariant tensor v^r and the associated covariant tensor v_r, which is related to v^r by (1.9.23) or (1.9.24), as two different representations or components of the space vector $\mathbf{v}$. With the help of (1.9.23) and (1.9.24) we are able to calculate the covariant components of a vector when the contravariant components are known, and vice versa. This means that we have established a process for the lowering or raising of indices. The word 'component' is frequently omitted and we refer to v_r and v^r as covariant and contravariant vectors respectively, but we emphasize that from this point of view they are not different vectors but two aspects of the single space vector $\mathbf{v}$.

If

$$\mathbf{w} = w^r \mathbf{g}_r = w_s\, \mathbf{g}^s \tag{1.9.25}$$

is a second vector the scalar product of $\mathbf{v}$ and $\mathbf{w}$ is

$$\mathbf{v} . \mathbf{w} = v^r w^s \mathbf{g}_r . \mathbf{g}_s = v_r\, w_s\, \mathbf{g}^r . \mathbf{g}^s, \tag{1.9.26}$$

and using (1.9.7) we see that

$$\mathbf{v}.\mathbf{w} = g_{rs}v^r w^s = g^{rs}v_r w_s = v_r w^r = v^s w_s. \tag{1.9.27}$$

The magnitude of $\mathbf{v}$ is

$$|\mathbf{v}| = \sqrt{(\mathbf{v}.\mathbf{v})} = \sqrt{(g_{rs}v^r v^s)} = \sqrt{(g^{rs}v_r v_s)} = \sqrt{(v_r v^r)}. \tag{1.9.28}$$

The vector product of two vectors $\mathbf{v}$ and $\mathbf{w}$ is

$$\mathbf{v}\times\mathbf{w} = v^r w^s \mathbf{g}_r \times \mathbf{g}_s = v_r w_s \mathbf{g}^r \times \mathbf{g}^s \tag{1.9.29}$$

and, using (1.9.13) and (1.9.14), this becomes

$$\mathbf{v}\times\mathbf{w} = \epsilon_{rst} v^r w^s \mathbf{g}^t = \epsilon^{rst} v_r w_s \mathbf{g}_t. \tag{1.9.30}$$

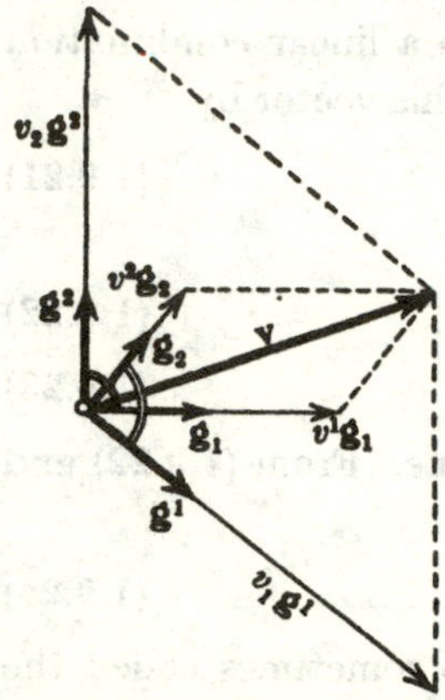

FIG. 1.6. Contravariant and covariant components of a vector (in two dimensions).

When the vector $\mathbf{v}$ is represented by a directed line from a point P at which the values of the vector are calculated, the contravariant components v^r are the components of $\mathbf{v}$ in the direction of the covariant base vectors, while the covariant components v_r are the components of $\mathbf{v}$ in the direction of the contravariant base vectors. This is illustrated in Fig. 1.6, where we have restricted ourselves to two dimensions for the sake of clarity.

If θ is the angle between the vectors $\mathbf{v}$ and $\mathbf{w}$ then from (1.8.13), (1.9.27), and (1.9.28), we find that

$$\cos\theta = \frac{g_{rs}v^r w^s}{\sqrt{(g_{rs}g_{mn}v^r v^s w^m w^n)}}, \tag{1.9.31}$$

with similar forms in terms of the other components of the vectors. Hence, if $\mathbf{v}$ and $\mathbf{w}$ are perpendicular, $\cos\theta = 0$ and

$$g_{rs}v^r w^s = g^{rs}v_r w_s = v^r w_r = v_s w^s = 0. \tag{1.9.32}$$

1.10. Associated tensors

The lowering or raising of indices as applied to the components of vectors in (1.9.23) and (1.9.24) can be extended to tensors of any order. Consider the covariant tensor A_{rs} of the second order. If we multiply it by g^{mn} and contract with respect to the first index we get the mixed tensor

$$A^{n}_{.s} = g^{rn}A_{rs}. \tag{1.10.1}$$

The dot is placed before the s in order to indicate that s is the second index and that it is the first index which has been raised. The process of raising the second index is written similarly

$$A_r^{\cdot m} = g^{ms} A_{rs}. \tag{1.10.2}$$

Lastly we can raise both indices by the formula

$$A^{rs} = g^{rm} g^{sn} A_{mn}. \tag{1.10.3}$$

If A_{rs} is symmetric it is a simple matter to see that $A^r_{\cdot s}$ and $A_s^{\cdot r}$ are equal and can both be written as A^r_s, the dot being unnecessary. Also it follows from (1.10.3) that when A_{rs} is symmetric A^{rs} is also symmetric.

Corresponding formulae for the lowering of indices are

$$\left.\begin{aligned} A^r_{\cdot s} &= g_{sm} A^{rm} \\ A_r^{\cdot s} &= g_{mr} A^{ms} \end{aligned}\right\}, \tag{1.10.4}$$

and other useful results are

$$\left.\begin{aligned} A^{rs} &= g^{rm} A_m^{\cdot s} = g^{sm} A^r_{\cdot m} \\ A_{rs} &= g_{sm} A_r^{\cdot m} = g_{rm} A^m_{\cdot s} \end{aligned}\right\}. \tag{1.10.5}$$

These processes may obviously be extended to tensors of any order. We illustrate the extension by one further example. If A^{rst} is a tensor of the third order then the second index is lowered by the formula

$$A^{r\cdot t}_{\cdot s \cdot} = g_{sm} A^{rmt}. \tag{1.10.6}$$

Sufficient dots are inserted to show that the index has been lowered vertically.

All tensors obtained from each other in this way are called *associated* tensors. We close this section with two results which are of interest in this connexion. The equations

$$A_{rst} = B_{rs} C_t \tag{1.10.7}$$

are equivalent to the equations

$$A^r_{\cdot st} = B^r_{\cdot s} C_t \tag{1.10.8}$$

in which the index r has been raised on both sides of the equation. A corresponding result holds for lowering an index. Again,

$$A^r_{\cdot s} B_r \equiv A_{rs} B^r, \tag{1.10.9}$$

where one summation index is raised and another is lowered. Similar formulae hold in a corresponding way for tensors of any order.

1.11. Canonical form of symmetric tensors

Consider the system of equations

$$(A_{ij} - k g_{ij}) n^i = 0, \tag{1.11.1}$$

where g_{ij} is the metric tensor and A_{ij} a symmetric tensor. The values of n^i given by (1.11.1) are non-zero, if and only if

$$|A_{ij}-kg_{ij}| = 0. \tag{1.11.2}$$

Under a general transformation of coordinates this equation is unaltered in form so that the roots of the equation, i.e. the three values of k, are invariants. This justifies the use of the contravariant index for n^i in (1.11.1).

We now wish to prove that the roots of (1.11.2) are all real. Suppose that $k = \alpha+i\beta$ and that the corresponding values of n^i are $\lambda^i+i\mu^i$. If we then equate real and imaginary parts in equation (1.11.1) we obtain

$$(A_{ij}-\alpha g_{ij})\lambda^i+\beta g_{ij}\mu^i = 0, \tag{1.11.3}$$

$$(A_{ij}-\alpha g_{ij})\mu^i-\beta g_{ij}\lambda^i = 0. \tag{1.11.4}$$

If we multiply (1.11.3) by μ^j and (1.11.4) by λ^j, and sum j for 1 to 3, we obtain by subtraction

$$\beta(g_{ij}\lambda^i\lambda^j+g_{ij}\mu^i\mu^j) = 0$$

if we remember that A_{ij} is symmetric. But λ^i, μ^i are not all zero and every expression of the form $g_{ij}\nu^i\nu^j$ is positive, hence $g_{ij}\lambda^i\lambda^j$ and $g_{ij}\mu^i\mu^j$ cannot both be zero, so that $\beta = 0$. Hence the roots of (1.11.2) are always real and the corresponding values of n^i from (1.11.1) are also real.

We consider further the case in which the roots of (1.11.2) are distinct. If k_r, k_s are two unequal roots and $n^i_{(r)}$, $n^i_{(s)}$ are the corresponding values of n^i, then

$$(A_{ij}-k_r g_{ij})n^i_{(r)} = 0 \quad (r \text{ not summed}),$$

$$(A_{ij}-k_s g_{ij})n^i_{(s)} = 0 \quad (s \text{ not summed}),$$

and therefore, since A_{ij} is symmetric,

$$(k_r-k_s)g_{ij}n^i_{(r)}n^j_{(s)} = 0 \quad (r, s \text{ not summed}).$$

But $k_r \neq k_s$, hence

$$g_{ij}n^i_{(r)}n^j_{(s)} = 0,$$

so that $n^i_{(r)}$, $n^i_{(s)}$ are contravariant components of orthogonal vectors. The three equations (1.11.1) determine the contravariant components of three vectors $\mathbf{n}_t$, corresponding to the three values of k, and these vectors form a triply orthogonal system.

We now take a system of coordinates such that at one point P in space their covariant base vectors are in the directions of the three vectors $\mathbf{n}_t$. At P equations (1.11.1) then have three solutions which are proportional to $(1,0,0)$, $(0,1,0)$, $(0,0,1)$. Also $\bar{g}_{mn} = 0$ if $m \neq n$. It follows from (1.11.1) that

$$\bar{A}_{mn} = 0 \quad (m \neq n),$$

the bars denoting the values of g_{mn}, A_{mn} in these special coordinates. In addition, the three roots of (1.11.2) are

$$k_r = \frac{\bar{A}_{rr}}{\bar{g}_{rr}} \quad (r \text{ not summed}),$$

so that the non-zero components of the symmetric tensor in our special coordinate system are $\bar{g}_{11}k_1$, $\bar{g}_{22}k_2$, $\bar{g}_{33}k_3$.

We have seen, therefore, that a set of axes may always be chosen so that, at a given point P, the symmetric tensor A_{ij} of the second order reduces to components $\bar{A}_{11}$, $\bar{A}_{22}$, $\bar{A}_{33}$ which are related to the roots of (1.11.2). This is the canonical form of the symmetric tensor. The corresponding directions of the vectors $\mathbf{n}_i$ are called the principal axes of the symmetric tensor at P.

When the roots of (1.11.2) are all distinct the three principal directions are unique. Otherwise it may be shown that there exists an infinity of principal directions.

1.12. Covariant differentiation

In previous sections we have been dealing with scalars, vectors, and tensors at the same point in space but we now study their properties when comparing them at different points. If to every point of space there correspond scalars, vectors, and tensors then we say that these form respectively scalar, vector, and tensor fields. This leads us naturally to the study of the differentiation of scalars, vectors, and tensors.

Consider first differentiation of the base vectors $\mathbf{g}_i$, which may be written

$$\mathbf{g}_{i,j} = \frac{\partial^2 \mathbf{R}}{\partial\theta^j \partial\theta^i} = \mathbf{R}_{,ij} = \mathbf{R}_{,ji}. \tag{1.12.1}$$

It follows that†

$$\mathbf{g}_{i,j} = \mathbf{g}_{j,i}.$$

These derivatives may be put in a more convenient form. From (1.9.5), remembering that $\mathbf{i}_r$ form a set of constant base vectors, we obtain

$$\mathbf{g}_{i,j} = \frac{\partial^2 x^r}{\partial\theta^i \partial\theta^j}\mathbf{i}_r, \tag{1.12.2}$$

and using (1.9.6) this becomes

$$\mathbf{g}_{i,j} = \Gamma_{ijs}\mathbf{g}^s = \Gamma^r_{ij}\mathbf{g}_r, \tag{1.12.3}$$

where

$$\left.\begin{aligned} \Gamma_{ijs} &= \frac{\partial^2 x^r}{\partial\theta^i \partial\theta^j}\frac{\partial x^r}{\partial\theta^s} \\ \Gamma^r_{ij} &= g^{rs}\Gamma_{ijs} \end{aligned}\right\}. \tag{1.12.4}$$

† This also follows from (1.12.3).

It follows from (1.7.4) and (1.12.4) that

$$\Gamma_{ijs} = \tfrac{1}{2}(g_{is,j}+g_{js,i}-g_{ij,s}). \tag{1.12.5}$$

The symbols Γ_{ijs} and Γ^r_{ij} are called the Christoffel symbols of the first and second kind respectively.

In a similar way from (1.9.6) we find for the derivative of the contravariant base vectors

$$\mathbf{g}^i{}_{,j} = -\Gamma^i_{jr}\,\mathbf{g}^r. \tag{1.12.6}$$

We notice the following results:

$$\left.\begin{aligned} \Gamma_{ijr} &= \Gamma_{jir} = \mathbf{g}_r\cdot\mathbf{g}_{i,j} = \mathbf{g}_r\cdot\mathbf{g}_{j,i} \\ \Gamma^r_{ij} &= \Gamma^r_{ji} = \mathbf{g}^r\cdot\mathbf{g}_{i,j} = \mathbf{g}^r\cdot\mathbf{g}_{j,i} = -\mathbf{g}_i\cdot\mathbf{g}^r{}_{,j} \end{aligned}\right\}. \tag{1.12.7}$$

Also

$$\begin{aligned} \Gamma^i_{ir} &= \tfrac{1}{2}g^{is}(g_{is,r}+g_{rs,i}-g_{ir,s}) \\ &= \tfrac{1}{2}g^{is}g_{is,r} \\ &= \frac{1}{2g}\frac{\partial g}{\partial g_{is}}\frac{\partial g_{is}}{\partial\theta^r} = \frac{1}{\sqrt{g}}\frac{\partial\sqrt{g}}{\partial\theta^r}, \end{aligned} \tag{1.12.8}$$

and

$$\Gamma_{irj}+\Gamma_{jri} = g_{ij,r}. \tag{1.12.9}$$

In (1.12.8) we have used the result

$$\frac{\partial g}{\partial g_{is}} = D^{is} = gg^{is}.$$

The Christoffel symbols are not, in general, the components of a tensor. The transformation law for Γ^r_{ij} is found to be

$$\bar{\Gamma}^r_{ij}\frac{\partial\theta^m}{\partial\bar{\theta}^r} = \Gamma^m_{rn}\frac{\partial\theta^r}{\partial\bar{\theta}^i}\frac{\partial\theta^n}{\partial\bar{\theta}^j}+\frac{\partial^2\theta^m}{\partial\bar{\theta}^i\partial\bar{\theta}^j}.$$

When the coordinate curves are orthogonal we have the following formulae for the Christoffel symbols:

$$\left.\begin{aligned} \Gamma_{ijr} &= 0 \quad (i\neq j\neq r\neq i) \\ \Gamma_{iir} &= -\frac{1}{2}\frac{\partial g_{ii}}{\partial\theta^r} \quad (r\neq i) \\ \Gamma_{iri} &= \Gamma_{rii} = \frac{1}{2}\frac{\partial g_{ii}}{\partial\theta^r} \\ \Gamma^r_{ij} &= 0 \quad (i\neq j\neq r\neq i) \\ \Gamma^r_{ii} &= -\frac{1}{2g_{rr}}\frac{\partial g_{ii}}{\partial\theta^r} \quad (r\neq i) \\ \Gamma^i_{ij} &= \Gamma^i_{ji} = \frac{1}{2g_{ii}}\frac{\partial g_{ii}}{\partial\theta^j} = \frac{1}{2}\frac{\partial\ln g_{ii}}{\partial\theta^j} \end{aligned}\right\}. \tag{1.12.10}$$

Repeated indices in the above formulae are not to be summed.

Consider now the derivative of a scalar ϕ, which we write as

$$\phi_{,i} = \frac{\partial\phi}{\partial\theta^i}. \tag{1.12.11}$$

If $\bar{\phi}$ is the value of ϕ when the coordinates are changed to $\bar{\theta}_i$ then

$$\frac{\partial\bar{\phi}}{\partial\bar{\theta}^i} = \frac{\partial\bar{\phi}}{\partial\theta^r}\frac{\partial\theta^r}{\partial\bar{\theta}^i} = c_i^r\frac{\partial\phi}{\partial\theta^r} = c_i^r\phi_{,r}. \tag{1.12.12}$$

In other words, the partial derivatives of a scalar transform according to the covariant type of transformation and in this case form a covariant tensor.

If $\mathbf{v}$ is a vector then

$$\frac{\partial\mathbf{v}}{\partial\bar{\theta}^i} = \frac{\partial\mathbf{v}}{\partial\theta^r}\frac{\partial\theta^r}{\partial\bar{\theta}^i} = c_i^r\frac{\partial\mathbf{v}}{\partial\theta^r} = c_i^r\mathbf{v}_{,r},$$

so that $\mathbf{v}_{,i} = \partial\mathbf{v}/\partial\theta^i$ transforms according to the covariant type of transformation. Also, from (1.9.22),

$$\left.\begin{aligned}\mathbf{v}_{,i} &= v^r{}_{,i}\,\mathbf{g}_r + v^r\mathbf{g}_{r,i}\\ &= v_{r,i}\,\mathbf{g}^r + v_r\,\mathbf{g}^r{}_{,i}\end{aligned}\right\}. \tag{1.12.13}$$

Introducing (1.12.3) and (1.12.6) we find that

$$\mathbf{v}_{,i} = v^r|_i\,\mathbf{g}_r = v_r|_i\,\mathbf{g}^r, \tag{1.12.14}$$

where

$$\left.\begin{aligned}v^r|_i &= v^r{}_{,i} + \Gamma^r_{si}v^s\\ v_r|_i &= v_{r,i} - \Gamma^s_{ri}v_s\end{aligned}\right\}. \tag{1.12.15}$$

The expressions $v^r|_i$, $v_r|_i$ are called the *covariant derivatives* of the components v^r, v_r respectively of the vector $\mathbf{v}$. Covariant differentiation is denoted by a vertical line. Since $\mathbf{v}_{,i}$ transforms according to the covariant type of transformation it follows immediately from (1.12.14) that the covariant derivatives of the components of a vector form a tensor of order two.

In a similar way we can form the covariant derivatives of a tensor of order two by considering the derivative of the invariant

$$T = A_{ij}B^iC^j,$$

where B^i, C^j are components of arbitrary vectors. These covariant derivatives are also tensors and are given by

$$\left.\begin{aligned}A_{ij}|_r &= A_{ij,r} - \Gamma^m_{ir}A_{mj} - \Gamma^m_{jr}A_{im}\\ A^i_{\cdot j}|_r &= A^i_{\cdot j,r} + \Gamma^i_{rm}A^m_{\cdot j} - \Gamma^m_{jr}A^i_{\cdot m}\\ A^{ij}|_r &= A^{ij}{}_{,r} + \Gamma^i_{rm}A^{mj} + \Gamma^j_{rm}A^{im}\end{aligned}\right\}. \tag{1.12.16}$$

This process of forming covariant derivatives is quite general and can be applied to obtain covariant derivatives of tensors of any type and order. For example, the covariant derivatives of the mixed tensor $A^{r}_{\cdot st}$ of order three are given by

$$A^{r}_{\cdot st}|_i = A^{r}_{\cdot st,i}+\Gamma^{r}_{mi}A^{m}_{\cdot st}-\Gamma^{m}_{si}A^{r}_{\cdot mt}-\Gamma^{m}_{ti}A^{r}_{\cdot sm}. \tag{1.12.17}$$

Since covariant derivatives are also tensors, all their indices can be raised or lowered by the rules developed in § 1.10. For example,

$$\left.\begin{aligned} A_{ij}|^r &= g^{rm}A_{ij}|_m \\ A^{ij}|^r &= g^{rm}A^{ij}|_m \end{aligned}\right\}, \quad \text{etc.} \tag{1.12.18}$$

It can be shown that covariant differentiation of sums or products follows the usual rules for ordinary differentiation. Thus

$$\left.\begin{aligned} (\phi v^r)|_i &= \phi_{,i}v^r+\phi v^r|_i \\ (v^r v_r)|_i &= (v^r v_r)_{,i} = v^r|_i v_r+v^r v_r|_i \\ &= v^r_{,i}v_r+v^r v_{r,i} \\ (A_{ij}+B_{ij})|_r &= A_{ij}|_r+B_{ij}|_r \\ (A_{ij}B^{mn})|_r &= A_{ij}|_r B^{mn}+A_{ij}B^{mn}|_r \end{aligned}\right\}, \tag{1.12.19}$$

remembering that covariant differentiation of a scalar is the same as partial differentiation.

Since the components of the metric tensor are constant in orthogonal cartesian coordinates the corresponding Christoffel symbols are zero, and the covariant derivatives of the metric tensor vanish in these coordinates. Therefore the covariant derivatives of the metric tensor vanish in all coordinate systems. That is†

$$\left.\begin{aligned} g_{ij}|_r = g^{ij}|_r = g^i_j|_r = \delta^i_j|_r = 0 \\ g^{ij}|^r = 0 \end{aligned}\right\}. \tag{1.12.20}$$

This is known as Ricci's lemma. For a similar reason the covariant derivatives of the ϵ-systems are also zero. Thus

$$\epsilon_{rst}|_i = 0, \qquad \epsilon^{rst}|_i = 0. \tag{1.12.21}$$

A direct consequence of these results is that when we are finding the covariant derivative of any combination of tensors, the metric tensor and the ϵ-system can all be treated as constants for the purposes of covariant differentiation.

† These results also follow directly from (1.12.16) and (1.12.5).

We have now seen how to obtain the covariant derivatives of tensors and since these are also tensors we can find their covariant derivatives. The resulting tensors are called the second covariant derivatives of the original tensors, and we can continue in this way to find covariant derivatives of any order. Let us examine the second covariant derivative of the covariant vector v_r. After some calculation we find that

$$v_r|_{st} - v_r|_{ts} = R^p_{\cdot rst}\, v_p, \tag{1.12.22}$$

where

$$R^p_{\cdot rst} = \Gamma^p_{rt,s} - \Gamma^p_{rs,t} + \Gamma^m_{rt}\Gamma^p_{ms} - \Gamma^m_{rs}\Gamma^p_{mt}. \tag{1.12.23}$$

The function $R^p_{\cdot rst}$ is a tensor called the *Riemann–Christoffel tensor* and it will be noticed that it consists only of the components of the metric tensor and their derivatives up to the second order. If we lower the index p we obtain the associated tensor

$$R_{prst} = g_{pm} R^m_{\cdot rst}, \tag{1.12.24}$$

which may be written in the form

$$R_{prst} = \tfrac{1}{2}(g_{pt,rs} + g_{rs,pt} - g_{ps,rt} - g_{rt,ps}) + g^{mn}(\Gamma_{rsm}\Gamma_{ptn} - \Gamma_{rtm}\Gamma_{psn}). \tag{1.12.25}$$

We deduce from (1.12.25) that R_{prst} satisfies the relations

$$R_{prst} = -R_{rpst},$$
$$R_{prst} = -R_{prts},$$
$$R_{prst} = R_{stpr}.$$

The first two of these equations express the fact that R_{prst} is skew-symmetric in p, r and also in s, t. We see that there are only six independent components of R_{prst}, namely

$$R_{3131},\ R_{3232},\ R_{1212},\ R_{3132},\ R_{3212},\ R_{3112}.$$

The Riemann–Christoffel tensor is identically zero in a cartesian coordinate system and therefore it is also zero in all other coordinate systems which can be established in our Euclidean space. Thus

$$R^p_{\cdot rst} = 0. \tag{1.12.26}$$

Formulae for the second covariant derivative of a tensor of order two are

$$\left.\begin{aligned} A_{ij}|_{rs} - A_{ij}|_{sr} &= A_{mj} R^m_{\cdot irs} + A_{im} R^m_{\cdot jrs} \\ A^{ij}|_{rs} - A^{ij}|_{sr} &= -A^{mj} R^i_{\cdot mrs} - A^{im} R^j_{\cdot mrs} \end{aligned}\right\}. \tag{1.12.27}$$

When the space is Euclidean, as in our case, we see from (1.12.22), (1.12.26), and (1.12.27) that

$$\left.\begin{aligned} A_r|_{st} &= A_r|_{ts} \\ A_{ij}|_{rs} &= A_{ij}|_{sr} \\ A^{ij}|_{rs} &= A^{ij}|_{sr} \end{aligned}\right\}. \tag{1.12.28}$$

This means that the order of covariant differentiation is immaterial.

Let us now consider the differential equation

$$A|_r^r = 0 \tag{1.12.29}$$

where A is an invariant. Using (1.12.18) this equation may also be put in the form

$$g^{rs}A|_{rs} = 0. \tag{1.12.30}$$

In cartesian coordinates this reduces to

$$\delta^{rs}A_{,rs} = 0$$

or

$$A_{,rr} = 0, \tag{1.12.31}$$

which is the well-known potential equation written in suffix notation. Functions which satisfy (1.12.29), or its equivalent (1.12.30), are called harmonic or potential functions.

Similarly we find that the differential equation

$$A|_{rs}^{rs} = 0 \tag{1.12.32}$$

reduces to

$$A_{,rrss} = 0 \tag{1.12.33}$$

in cartesian coordinates, and this is the well-known biharmonic equation. Functions which satisfy (1.12.32) are called biharmonic functions.

We conclude this section with the definitions of gradient (grad), divergence (div), and curl and with Green's theorem which relates volume and surface integrals. Thus

$$\left.\begin{aligned} \operatorname{grad}\phi &= \phi_{,r}\,\mathbf{g}^r \\ \operatorname{div}\mathbf{F} &= F^r|_r \\ \operatorname{curl}\mathbf{A} &= \epsilon^{rst}A_s|_r\,\mathbf{g}_t \end{aligned}\right\}, \tag{1.12.34}$$

and we observe that these functions are invariant under general transformations of coordinates. These expressions may be written more compactly with the aid of the operator $\boldsymbol{\nabla}$, where

$$\boldsymbol{\nabla} = \mathbf{g}^r \frac{\partial}{\partial\theta^r}. \tag{1.12.35}$$

The partial derivative $\partial/\partial\theta^r$ acts on any function following the operator ∇. Hence

$$\left.\begin{aligned}\operatorname{grad}\phi &= \nabla\phi = \mathbf{g}^r\frac{\partial\phi}{\partial\theta^r}\\ \operatorname{div}\mathbf{F} &= \nabla.\mathbf{F} = \mathbf{g}^r.\frac{\partial}{\partial\theta^r}(F^s\mathbf{g}_s)\\ &= \mathbf{g}^r.\mathbf{g}_s F^s|_r\\ &= F^r|_r = \frac{1}{\sqrt{g}}(\sqrt{g}F^r)_{,r}\\ \operatorname{curl}\mathbf{A} &= \nabla\times\mathbf{A} = \mathbf{g}^r\frac{\partial}{\partial\theta^r}\times A_s\mathbf{g}^s\\ &= \mathbf{g}^r\times\frac{\partial}{\partial\theta^r}(A_s\mathbf{g}^s)\\ &= \mathbf{g}^r\times\mathbf{g}^s A_s|_r\\ &= \epsilon^{rst}A_s|_r\,\mathbf{g}_t\end{aligned}\right\}. \tag{1.12.36}$$

In these formulae the partial derivative $\partial/\partial\theta^r$ is interchangeable with . and $\times$ but otherwise acts on the complete function which follows the ∇ operator. We also note that

$$\begin{aligned}\operatorname{div}\operatorname{grad}\phi &= \nabla^2\phi = (g^{rs}\phi_{,s})|_r\\ &= g^{rs}\phi|_{rs} = \frac{1}{\sqrt{g}}(\sqrt{g}\,g^{rs}\phi_{,s})_{,r}.\end{aligned} \tag{1.12.37}$$

If $\mathbf{v}$ is a vector defined throughout a volume τ which is bounded by a closed surface S, and if $\mathbf{n}$ is a unit vector normal to the surface S at which it is defined, then the usual form of Green's theorem is

$$\iiint\limits_{\tau}\operatorname{div}\mathbf{v}\,d\tau = \iint\limits_{S}\mathbf{v}.\mathbf{n}\,dS. \tag{1.12.38}$$

Alternatively, in tensor notation this becomes

$$\iiint\limits_{\tau} v^r|_r\,d\tau = \iiint\limits_{\tau}\frac{1}{\sqrt{g}}(v^r\sqrt{g})_{,r}\,d\tau = \iint\limits_{S} v^r n_r\,dS, \tag{1.12.39}$$

where
$$\mathbf{n} = n_r\mathbf{g}^r. \tag{1.12.40}$$

1.13. Geometry of a surface

We now express the position vector of a point in the special form

$$\mathbf{R} = \mathbf{r}+\theta_3\mathbf{a}_3, \tag{1.13.1}$$

where $\mathbf{r}$ is a function of θ_1, θ_2 only and $\mathbf{a}_3$ is a vector of unit magnitude which also depends only on θ_1, θ_2 (see Fig. 1.7). We use the convention that Greek indices range over the values 1, 2 while we still require Latin indices to have the values 1, 2, 3.

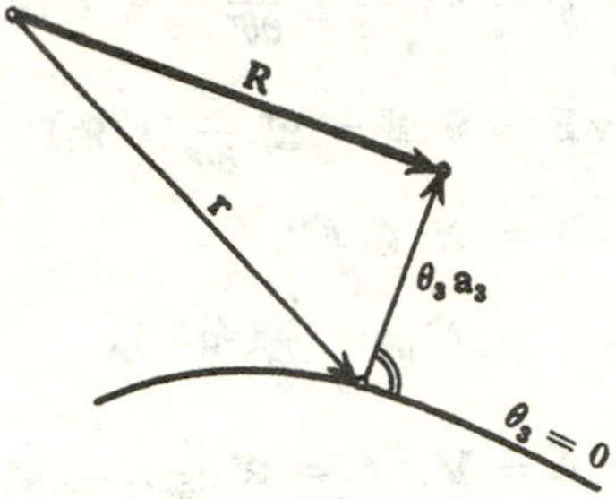

FIG. 1.7. Position vector of a surface.

The equation $\theta_3 = 0$ determines a surface

$$\mathbf{r} = \mathbf{r}(\theta_1, \theta_2). \tag{1.13.2}$$

The vector $\mathbf{a}_3$ is perpendicular to the surface and is called the normal vector. If θ_2 is kept constant then $\mathbf{r}$ describes a curve which lies wholly on the surface and as the constant varies we get a family of curves which we call the θ_1-curves. Similarly we have another family of curves, the θ_2-curves, along which θ_1 is constant. We often refer to θ_1- and θ_2-curves briefly as *coordinate* curves and we call θ_1, θ_2 a system of curvilinear coordinates on the surface.

We may clearly define other sets of curvilinear coordinates on the surface and transformations between two different coordinate systems θ_α, $\bar{\theta}_\alpha$ may be written

$$\left.\begin{aligned} \theta_\alpha &= \theta_\alpha(\bar{\theta}_1, \bar{\theta}_2) \\ \bar{\theta}_\alpha &= \bar{\theta}_\alpha(\theta_1, \theta_2) \end{aligned}\right\}, \tag{1.13.3}$$

where restrictions similar to those in § 1.7 are placed on the functions defining the transformation. A theory of tensors of different orders may be defined for transformations in these variables which is similar to that already developed for the variables θ_r in § 1.3. If we wish to distinguish tensors in Greek indices from those in Latin indices we call them *surface tensors* since they are tensors under a surface transformation of coordinates.

We observe that $d\theta^\alpha$ forms a contravariant surface tensor of order one. Kronecker deltas for the surface are defined by

$$\left.\begin{aligned} \delta^\alpha_\beta &= 0 \quad (\alpha \neq \beta) \\ \delta^\alpha_\beta &= 1 \quad (\alpha = \beta,\ \beta \text{ not summed}) \end{aligned}\right\}, \tag{1.13.4}$$

and since
$$\frac{\partial\bar{\theta}^\alpha}{\partial\theta^\beta}\frac{\partial\theta^\beta}{\partial\bar{\theta}^\gamma} = \delta^\alpha_\gamma, \tag{1.13.5}$$

they form a mixed surface tensor of order two.

Using (1.9.12) we find that the covariant base vectors for the space (1.13.1) are
$$\left.\begin{aligned}\mathbf{g}_\alpha &= \mathbf{a}_\alpha + \theta_3\,\mathbf{a}_{3,\alpha}\\ \mathbf{g}_3 &= \mathbf{a}_3\end{aligned}\right\}, \tag{1.13.6}$$
where
$$\mathbf{a}_\alpha = \mathbf{r}_{,\alpha} \tag{1.13.7}$$

are called the covariant base vectors of the surface. The form (1.13.7) shows that $\mathbf{a}_\alpha$ transform according to the covariant rule for the surface transformation of coordinates. We notice that
$$\mathbf{a}_3 . \mathbf{a}_\alpha = 0, \qquad \mathbf{a}_3 . \mathbf{a}_3 = 1, \qquad \mathbf{a}_3 . \mathbf{a}_{3,\alpha} = 0, \tag{1.13.8}$$
and that when $\theta_3 = 0$, $\mathbf{g}_r = \mathbf{a}_r$. We now denote the values of $\mathbf{g}^r$, g_{rs}, g^{rs}, g by $\mathbf{a}^r$, a_{rs}, a^{rs}, a respectively when $\theta_3 = 0$. It follows from (1.9.7) that
$$\mathbf{a}_\alpha . \mathbf{a}_\beta = a_{\alpha\beta}, \qquad a_{3\alpha} = 0, \qquad a_{33} = 1, \tag{1.13.9}$$
and then, from (1.7.7),
$$a_{\alpha\beta}\,a^{\beta\gamma} = \delta^\gamma_\alpha, \qquad a^{3\alpha} = 0, \qquad a^{33} = 1. \tag{1.13.10}$$
Hence
$$a^{11} = \frac{a_{22}}{a}, \qquad a^{12} = a^{21} = -\frac{a_{12}}{a}, \qquad a^{22} = \frac{a_{11}}{a}, \tag{1.13.11}$$
where
$$a = |a_{\alpha\beta}| = a_{11}\,a_{22} - a_{12}^2. \tag{1.13.12}$$

Again, putting $\theta_3 = 0$ in (1.9.7) we find
$$\left.\begin{aligned}\mathbf{a}^\alpha &= a^{\alpha\beta}\,\mathbf{a}_\beta, & \mathbf{a}_\alpha &= a_{\alpha\beta}\,\mathbf{a}^\beta\\ \mathbf{a}^\alpha . \mathbf{a}^\beta &= a^{\alpha\beta}, & \mathbf{a}^3 &= \mathbf{a}_3\end{aligned}\right\}, \tag{1.13.13}$$
and from (1.9.8)
$$\left.\begin{aligned}|\mathbf{a}_\alpha| &= \sqrt{(\mathbf{a}_\alpha . \mathbf{a}_\alpha)} = \sqrt{a_{\alpha\alpha}}\\ |\mathbf{a}^\alpha| &= \sqrt{(\mathbf{a}^\alpha . \mathbf{a}^\alpha)} = \sqrt{a^{\alpha\alpha}}\end{aligned}\right\} \quad (\alpha \text{ not summed}). \tag{1.13.14}$$

The vectors $\mathbf{a}^\alpha$ are the contravariant base vectors of the surface and $a_{\alpha\beta}$, $a^{\alpha\beta}$ are the symmetric metric surface tensors. All these tensors satisfy the appropriate laws for transformations of surface coordinates. To prove this consider, for example, $a_{\alpha\beta}$. We know that g_{rs} transforms generally according to the rule
$$\bar{g}_{rs} = \frac{\partial\theta^m}{\partial\bar{\theta}^r}\frac{\partial\theta^n}{\partial\bar{\theta}^s}g_{mn}.$$

If we now restrict the transformation to be the surface transformation (1.13.3), together with $\bar{\theta}_3 = \theta_3$, and then put $\theta_3 = 0$, we obtain

$$\bar{a}_{\alpha\beta} = \frac{\partial\theta^\lambda}{\partial\bar{\theta}^\alpha}\frac{\partial\theta^\mu}{\partial\bar{\theta}^\beta}a_{\lambda\mu}, \tag{1.13.15}$$

which is the required form for a covariant surface tensor of order two.

Since
$$d\mathbf{r} = \mathbf{a}_\alpha d\theta^\alpha = \mathbf{a}^\alpha d\theta_\alpha, \tag{1.13.16}$$
where
$$d\theta_\alpha = a_{\alpha\beta}d\theta^\beta, \tag{1.13.17}$$
the line element of the surface takes the form

$$\left.\begin{aligned} ds^2 = d\mathbf{r}.d\mathbf{r} &= a_{\alpha\beta}d\theta^\alpha d\theta^\beta = a^{\alpha\beta}d\theta_\alpha d\theta_\beta \\ &= \delta^\alpha_\beta d\theta_\alpha d\theta^\beta = d\theta_\alpha d\theta^\alpha \end{aligned}\right\}. \tag{1.13.18}$$

Equation (1.13.18) is called the *first fundamental form* of the surface.

The line elements along the coordinate curves may be represented by vectors
$$d\mathbf{s}_\alpha = \mathbf{a}_\alpha d\theta^\alpha \quad (\alpha \text{ not summed}), \tag{1.13.19}$$
with magnitudes
$$ds_\alpha = \sqrt{(a_{\alpha\alpha})}\,d\theta^\alpha, \tag{1.13.20}$$
and the angle ϕ between the coordinate curves is given by

$$\cos\phi = \frac{d\mathbf{s}_1.d\mathbf{s}_2}{ds_1\,ds_2} = \frac{a_{12}}{\sqrt{(a_{11}a_{22})}}. \tag{1.13.21}$$

If the coordinate curves are orthogonal then

$$a^{12} = a_{12} = 0.$$

The surface element dS is obtained in the form

$$dS = \sqrt{(a)}\,d\theta^1 d\theta^2 \tag{1.13.22}$$

by putting $\theta_3 = 0$ in (1.9.18).

The ϵ-systems for surfaces are defined to be the values of $\epsilon_{\alpha\beta 3}$, $\epsilon^{\alpha\beta 3}$ when $\theta_3 = 0$, and are denoted respectively by $\epsilon_{\alpha\beta}$, $\epsilon^{\alpha\beta}$. Hence

$$\left.\begin{aligned} \epsilon_{12} = -\epsilon_{21} = \sqrt{a}, \qquad \epsilon^{12} = -\epsilon^{21} = 1/\sqrt{a} \\ \epsilon_{11} = \epsilon_{22} = \epsilon^{11} = \epsilon^{22} = 0 \end{aligned}\right\}. \tag{1.13.23}$$

Since ϵ_{rst} transforms according to the rule

$$\bar{\epsilon}_{rst} = \frac{\partial\theta^i}{\partial\bar{\theta}^r}\frac{\partial\theta^j}{\partial\bar{\theta}^s}\frac{\partial\theta^k}{\partial\bar{\theta}^t}\epsilon_{ijk}$$

we see that, when the transformation is of the form (1.13.3), together with $\bar{\theta}_3 = \theta_3$, the transformation for $\epsilon_{\alpha\beta}$ is

$$\bar{\epsilon}_{\alpha\beta} = \frac{\partial\theta^\lambda}{\partial\bar{\theta}^\alpha}\frac{\partial\theta^\mu}{\partial\bar{\theta}^\beta}\epsilon_{\lambda\mu},$$

so that $\epsilon_{\alpha\beta}$ is a covariant surface tensor of order two. Similarly, $\epsilon^{\alpha\beta}$ is a contravariant surface tensor of order two. We observe that

$$\left.\begin{aligned} \epsilon_{\alpha\beta 3} &= \sqrt{(g/a)}\epsilon_{\alpha\beta} \\ \epsilon^{\alpha\beta 3} &= \sqrt{(a/g)}\epsilon^{\alpha\beta} \end{aligned}\right\}. \tag{1.13.24}$$

We also have from (1.7.15)

$$\left.\begin{aligned} \epsilon^{\alpha\beta} &= a^{\alpha\lambda}a^{\beta\mu}\epsilon_{\lambda\mu} \\ \epsilon_{\alpha\beta} &= a_{\alpha\lambda}a_{\beta\mu}\epsilon^{\lambda\mu} \end{aligned}\right\}, \tag{1.13.25}$$

and from (1.7.16) and (1.7.17)

$$\left.\begin{aligned} \epsilon^{\alpha\beta}\epsilon^{\lambda\mu}a_{\alpha\lambda}a_{\beta\mu} &= 2 \\ \epsilon_{\alpha\beta}\epsilon_{\lambda\mu}a^{\alpha\lambda}a^{\beta\mu} &= 2 \\ a^{\alpha\beta} &= \epsilon^{\alpha\lambda}\epsilon^{\beta\mu}a_{\lambda\mu} \\ a_{\alpha\beta} &= \epsilon_{\alpha\lambda}\epsilon_{\beta\mu}a^{\lambda\mu} \end{aligned}\right\}. \tag{1.13.26}$$

We denote the product of $\epsilon^{\alpha\beta}$ and $\epsilon_{\lambda\mu}$ by

$$\delta^{\alpha\beta}_{\lambda\mu} = \epsilon^{\alpha\beta}\epsilon_{\lambda\mu} \tag{1.13.27}$$

so that, for example,

$$\delta^{\alpha\beta}_{\lambda\mu}A^{\lambda\mu} = A^{\alpha\beta} - A^{\beta\alpha}. \tag{1.13.28}$$

In particular, from (1.13.25) and (1.13.26),

$$\left.\begin{aligned} \delta^{\alpha\beta}_{\lambda\beta} &= \epsilon^{\alpha\beta}\epsilon_{\lambda\beta} = a_{\lambda\mu}a_{\beta\rho}\epsilon^{\alpha\beta}\epsilon^{\mu\rho} \\ &= a^{\alpha\mu}a_{\lambda\mu} \\ &= \delta^{\alpha}_{\lambda} \\ \delta^{\alpha\beta}_{\alpha\beta} &= 2 \end{aligned}\right\}. \tag{1.13.29}$$

If we now put $\theta_3 = 0$ in (1.9.13) and (1.9.14) we obtain

$$\left.\begin{aligned} \mathbf{a}_\alpha \times \mathbf{a}_\beta &= \epsilon_{\alpha\beta}\mathbf{a}_3 \\ \mathbf{a}^\alpha \times \mathbf{a}^\beta &= \epsilon^{\alpha\beta}\mathbf{a}_3 \\ \mathbf{a}_3 \times \mathbf{a}_\beta &= \epsilon_{\beta\rho}\mathbf{a}^\rho \\ \mathbf{a}_3 \times \mathbf{a}^\beta &= \epsilon^{\beta\rho}\mathbf{a}_\rho \end{aligned}\right\}. \tag{1.13.30}$$

Also, from (1.9.15),

$$[\mathbf{a}_\alpha\mathbf{a}_\beta\mathbf{a}_3] = \epsilon_{\alpha\beta}, \qquad [\mathbf{a}^\alpha\mathbf{a}^\beta\mathbf{a}^3] = \epsilon^{\alpha\beta}. \tag{1.13.31}$$

The scalar product

$$d\mathbf{r}.d\mathbf{a}_3 = -b_{\alpha\beta}\,d\theta^\alpha d\theta^\beta = -b^{\alpha\beta}\,d\theta_\alpha d\theta_\beta = -b^\alpha_\beta\,d\theta_\alpha d\theta^\beta,$$

where

$$b_{\alpha\beta} = -\mathbf{a}_\alpha.\mathbf{a}_{3,\beta} = -\mathbf{a}_\beta.\mathbf{a}_{3,\alpha} = \mathbf{a}_3.\mathbf{a}_{\alpha,\beta} = \mathbf{a}_3.\mathbf{a}_{\beta,\alpha} = [\mathbf{a}_{\alpha,\beta}\mathbf{a}_1\mathbf{a}_2]/\sqrt{a}, \tag{1.13.32}$$

is called the *second fundamental form* of the surface. The coefficients $b_{\alpha\beta}$, $b^{\alpha\beta}$ are symmetric surface tensors of order two which are related by the formulae†

$$\left.\begin{aligned} b^{\alpha}_{\beta} &= a^{\alpha\lambda}b_{\beta\lambda} = a_{\beta\lambda}b^{\alpha\lambda} \\ b^{\alpha\beta} &= a^{\alpha\lambda}b^{\beta}_{\lambda} \\ b_{\alpha\beta} &= a_{\alpha\lambda}b^{\lambda}_{\beta} \end{aligned}\right\}. \tag{1.13.33}$$

The invariant
$$2H = b^{\alpha}_{\alpha} \tag{1.13.34}$$
is called the *mean curvature* of the surface and
$$K = \frac{|b_{\alpha\beta}|}{a} = b^1_1 b^2_2 - b^1_2 b^2_1 \tag{1.13.35}$$
is called the *Gaussian curvature* of the surface.

The Christoffel symbols with respect to the surface $\theta_3 = 0$ are found by putting $\theta_3 = 0$ in (1.12.5) and by using (1.12.7) and (1.13.6). Denoting these symbols by a bar we have

$$\left.\begin{aligned} \bar{\Gamma}_{\beta\gamma\alpha} &= \tfrac{1}{2}(a_{\alpha\beta,\gamma}+a_{\alpha\gamma,\beta}-a_{\beta\gamma,\alpha}) \\ \bar{\Gamma}^{\alpha}_{\beta\gamma} &= a^{\alpha\lambda}\bar{\Gamma}_{\beta\gamma\lambda} = \mathbf{a}^{\alpha}.\mathbf{a}_{\gamma,\beta} = \mathbf{a}^{\alpha}.\mathbf{a}_{\beta,\gamma} = -\mathbf{a}_{\gamma}.\mathbf{a}^{\alpha}{}_{,\beta} \\ \bar{\Gamma}^{\alpha}_{\beta 3} &= \mathbf{a}^{\alpha}.\mathbf{a}_{3,\beta} = -\mathbf{a}_{3}.\mathbf{a}^{\alpha}{}_{,\beta} = -b^{\alpha}_{\beta} \\ \bar{\Gamma}^{3}_{\alpha\beta} &= \mathbf{a}^{3}.\mathbf{a}_{\alpha,\beta} = -\mathbf{a}_{\beta}.\mathbf{a}^{3}{}_{,\alpha} = b_{\alpha\beta} \\ \bar{\Gamma}^{3}_{\alpha 3} &= \mathbf{a}^{3}.\mathbf{a}_{3,\alpha} = 0 \\ \bar{\Gamma}^{3}_{33} &= 0, \quad \bar{\Gamma}^{\alpha}_{33} = 0, \quad \bar{\Gamma}^{\lambda}_{\lambda\alpha} = \frac{1}{\sqrt{a}}\frac{\partial\sqrt{a}}{\partial\theta^{\alpha}} \end{aligned}\right\}. \tag{1.13.36}$$

The values of the Riemann–Christoffel tensor (1.12.23) on the surface $\theta_3 = 0$ can be put in the form

$$\left.\begin{aligned} R^{\lambda}_{\cdot\alpha\beta\gamma} &= \bar{R}^{\lambda}_{\cdot\alpha\beta\gamma} + \bar{\Gamma}^{3}_{\alpha\gamma}\bar{\Gamma}^{\lambda}_{3\beta} - \bar{\Gamma}^{3}_{\alpha\beta}\bar{\Gamma}^{\lambda}_{3\gamma} \\ R^{3}_{\cdot\alpha\beta\gamma} &= \bar{\Gamma}^{3}_{\alpha\gamma,\beta} - \bar{\Gamma}^{3}_{\alpha\beta,\gamma} + \bar{\Gamma}^{m}_{\alpha\lambda}\bar{\Gamma}^{3}_{m\beta} - \bar{\Gamma}^{m}_{\alpha\beta}\bar{\Gamma}^{3}_{m\gamma} \end{aligned}\right\}, \tag{1.13.37}$$

where
$$\bar{R}^{\lambda}_{\cdot\alpha\beta\gamma} = \bar{\Gamma}^{\lambda}_{\alpha\gamma,\beta} - \bar{\Gamma}^{\lambda}_{\alpha\beta,\gamma} + \bar{\Gamma}^{\mu}_{\alpha\gamma}\bar{\Gamma}^{\lambda}_{\mu\beta} - \bar{\Gamma}^{\mu}_{\alpha\beta}\bar{\Gamma}^{\lambda}_{\mu\gamma}, \tag{1.13.38}$$
and is called the Riemann–Christoffel tensor of the surface. Since, from (1.12.26), $R^{\lambda}_{\cdot\alpha\beta\gamma}$ vanishes we have
$$\bar{R}^{\lambda}_{\cdot\alpha\beta\gamma} = \bar{\Gamma}^{3}_{\alpha\beta}\bar{\Gamma}^{\lambda}_{3\gamma} - \bar{\Gamma}^{3}_{\alpha\gamma}\bar{\Gamma}^{\lambda}_{3\beta}. \tag{1.13.39}$$
Lowering the upper index we find that
$$\bar{R}_{\lambda\alpha\beta\gamma} = a_{\lambda\mu}\bar{R}^{\mu}_{\cdot\alpha\beta\gamma} \tag{1.13.40}$$
and, in particular,
$$\left.\begin{aligned} \bar{R}_{\alpha\alpha\beta\gamma} &= \bar{R}_{\alpha\beta\gamma\gamma} = 0 \quad (\alpha,\ \gamma \text{ not summed}) \\ \bar{R}_{1212} &= \bar{R}_{2121} = -\bar{R}_{2112} = -\bar{R}_{1221} \end{aligned}\right\}. \tag{1.13.41}$$

† $b_{\alpha\beta}$, $b^{\alpha\beta}$ are associated surface tensors (see p. 38).

Hence every non-zero component of $\bar{R}_{\alpha\beta\gamma\delta}$ is equal either to $\bar{R}_{1212}$ or to $-\bar{R}_{1212}$. Using (1.13.36), (1.13.39), and (1.13.40) we find that

$$\bar{R}_{1212} = |b_{\alpha\beta}| = b_{11}b_{22}-b_{12}^2, \tag{1.13.42}$$

and hence, from (1.13.35), the *Gaussian curvature* becomes

$$K = \frac{\bar{R}_{1212}}{a}. \tag{1.13.43}$$

We may also put K in the form

$$K = \tfrac{1}{4}\epsilon^{\lambda\alpha}\epsilon^{\beta\gamma}\bar{R}_{\lambda\alpha\beta\gamma}, \tag{1.13.44}$$

which shows that K is an invariant of the surface. This last result is known as the Gauss equation of the surface.

Since $R^3{}_{.\alpha\beta\gamma}$ also vanishes we obtain from† (1.13.37)

$$b_{\alpha1}|_2 = b_{\alpha2}|_1. \tag{1.13.45}$$

These equations are the *Codazzi equations of the surface.*

Equations (1.13.42) and (1.13.45) have been derived by putting to zero three of the six independent components of the Riemann–Christoffel tensor at the surface $\theta_3 = 0$. The other three components, which must be zero, are

$$R_{\alpha3\beta3} = 0,$$

which leads to the relations

$$g_{\alpha\beta,33} = \tfrac{1}{2}g^{\lambda\mu}g_{\alpha\lambda,3}g_{\beta\mu,3}. \tag{1.13.46}$$

Expressions for the derivatives of the base vectors of the surface may be found immediately by putting $\theta_3 = 0$ in (1.12.3) and (1.12.6) and using (1.13.36). These results are known as the formulae of Weingarten and Gauss. Thus

$$\left.\begin{aligned} \mathbf{a}_{\alpha,\beta} &= \bar{\Gamma}^{\lambda}_{\alpha\beta}\mathbf{a}_{\lambda}+b_{\alpha\beta}\mathbf{a}_3 \\ \mathbf{a}^{\alpha}{}_{,\beta} &= -\bar{\Gamma}^{\alpha}_{\beta\lambda}\mathbf{a}^{\lambda}+b^{\alpha}_{\beta}\mathbf{a}_3 \\ \mathbf{a}_{3,\alpha} &= -b^{\lambda}_{\alpha}\mathbf{a}_{\lambda} \end{aligned}\right\}. \tag{1.13.47}$$

From the last equation of (1.13.47) follows

$$\mathbf{a}_{3,\alpha}.\mathbf{a}_{3,\beta} = b_{\alpha\lambda}b^{\lambda}_{\beta}, \tag{1.13.48}$$

and hence

$$d\mathbf{a}_3.d\mathbf{a}_3 = b_{\alpha\beta}b^{\beta}_{\gamma}d\theta^{\alpha}d\theta^{\gamma}, \tag{1.13.49}$$

which is called the *third fundamental form* of the surface.

Let us now consider a vector in the space (1.13.1) which may be put in the form

$$\mathbf{v} = v^{\alpha}\mathbf{a}_{\alpha}+v^3\mathbf{a}_3 = v_{\alpha}\mathbf{a}^{\alpha}+v_3\mathbf{a}^3. \tag{1.13.50}$$

† See (1.13.54) below for the definition of $b_{\alpha1}|_2$ and $b_{\alpha2}|_1$.

Using (1.13.47) the derivatives of v are found to be

$$\left.\begin{aligned}\mathbf{v}_{,\alpha} &= (v_\lambda|_\alpha - b_{\alpha\lambda}v_3)\mathbf{a}^\lambda + (v_{3,\alpha} + b^\lambda_\alpha v_\lambda)\mathbf{a}_3\\ &= (v^\lambda|_\alpha - b^\lambda_\alpha v^3)\mathbf{a}_\lambda + (v^3{}_{,\alpha} + b_{\lambda\alpha}v^\lambda)\mathbf{a}_3\\ \mathbf{v}_{,3} &= v_{\alpha,3}\mathbf{a}^\alpha + v_{3,3}\mathbf{a}_3\\ &= v^\alpha{}_{,3}\mathbf{a}_\alpha + v^3{}_{,3}\mathbf{a}_3\end{aligned}\right\}, \tag{1.13.51}$$

where we have put

$$\left.\begin{aligned}v_\lambda|_\alpha &= v_{\lambda,\alpha} - \bar{\Gamma}^\mu_{\lambda\alpha}v_\mu\\ v^\lambda|_\alpha &= v^\lambda{}_{,\alpha} + \bar{\Gamma}^\lambda_{\mu\alpha}v^\mu\end{aligned}\right\}. \tag{1.13.52}$$

In a general transformation of space coordinates v_r and $\mathbf{g}^r$ transform respectively according to covariant and contravariant types of transformations. It follows that, when we specialize the transformations to the surface transformation (1.13.3), together with $\bar{\theta}_3 = \theta_3$, and then put $\theta_3 = 0$, we find that v_α, $\mathbf{a}^\alpha$ transform respectively according to the covariant and contravariant rules for transformations of surface coordinates. We see, therefore, that $v_\alpha \mathbf{a}^\alpha$ (and also $v^\alpha \mathbf{a}_\alpha$) are invariants under transformations of surface coordinates, and they are called *surface vectors* with covariant and contravariant components v_α, v^α respectively; or, briefly, v^α and v_α are called contravariant and covariant surface vectors. The functions $v_\lambda|_\alpha$, $v^\lambda|_\alpha$ are called the covariant derivatives of the surface vectors and these form surface tensors.

Since

$$v_\alpha \mathbf{a}^\alpha = v^\alpha \mathbf{a}_\alpha$$

and, from (1.13.13),

$$\mathbf{a}^\alpha = a^{\alpha\beta}\mathbf{a}_\beta, \qquad \mathbf{a}_\alpha = a_{\alpha\beta}\mathbf{a}^\beta,$$

it follows that

$$v^\alpha = a^{\alpha\beta}v_\beta, \qquad v_\alpha = a_{\alpha\beta}v^\beta, \tag{1.13.53}$$

which establishes a method for raising or lowering suffixes for surface vectors.

As in § 1.10 for tensors, associated surface tensors are formed with the help of the surface metric components $a_{\alpha\beta}$, $a^{\alpha\beta}$, in a manner similar to that given in (1.13.53) for surface vectors. Examples of associated tensors have already been given in (1.13.33).

As in three dimensions, we may extend the idea of covariant differentiation of surface vectors to surface tensors of any order. For example

$$\left.\begin{aligned}A_{\alpha\beta}|_\gamma &= A_{\alpha\beta,\gamma} - \bar{\Gamma}^\lambda_{\alpha\gamma}A_{\lambda\beta} - \bar{\Gamma}^\lambda_{\beta\gamma}A_{\alpha\lambda}\\ A^{\alpha}_{.\beta}|_\gamma &= A^{\alpha}_{.\beta,\gamma} + \bar{\Gamma}^\alpha_{\gamma\lambda}A^{\lambda}_{.\beta} - \bar{\Gamma}^\lambda_{\beta\gamma}A^{\alpha}_{.\lambda}\\ A^{\alpha\beta}|_\gamma &= A^{\alpha\beta}{}_{,\gamma} + \bar{\Gamma}^\alpha_{\gamma\lambda}A^{\lambda\beta} + \bar{\Gamma}^\beta_{\gamma\lambda}A^{\alpha\lambda}\end{aligned}\right\}. \tag{1.13.54}$$

All the covariant derivatives of g_{rs}, g^{rs}, ϵ_{rst}, ϵ^{rst} are zero in three-dimensional Euclidean space, so that, in particular, their covariant derivatives with respect to θ_α are zero. Hence, putting $\theta_3 = 0$ we see that

$$a_{\alpha\beta}|_\lambda = a^{\alpha\beta}|_\lambda = \epsilon_{\alpha\beta}|_\lambda = \epsilon^{\alpha\beta}|_\lambda = 0. \tag{1.13.55}$$

By analogy with (1.12.22) and (1.12.27) we find that the second covariant derivatives of surface vectors and tensors are given by

$$\left.\begin{aligned} A_\alpha|_{\beta\gamma} - A_\alpha|_{\gamma\beta} &= \bar{R}^\lambda_{.\alpha\beta\gamma} A_\lambda \\ A_{\alpha\beta}|_{\gamma\delta} - A_{\alpha\beta}|_{\delta\gamma} &= \bar{R}^\lambda_{.\alpha\gamma\delta} A_{\lambda\beta} + \bar{R}^\lambda_{.\beta\gamma\delta} A_{\alpha\lambda} \\ A^{\alpha\beta}|_{\gamma\delta} - A^{\alpha\beta}|_{\delta\gamma} &= -\bar{R}^\alpha_{.\lambda\gamma\delta} A^{\lambda\beta} - \bar{R}^\beta_{.\lambda\gamma\delta} A^{\alpha\lambda} \end{aligned}\right\}. \tag{1.13.56}$$

This means that in general the order of covariant differentiation cannot be altered without altering the result. The order is only immaterial for certain classes of surfaces for which the Riemann–Christoffel surface tensor vanishes.

In the special case when the surface (1.13.2) is a *plane* the vector $\mathbf{a}_3$ is a *constant* unit vector perpendicular to the plane surface. In this case

$$\mathbf{a}_{3,\alpha} = \mathbf{0} \tag{1.13.57}$$

and hence, from (1.13.32) and (1.13.42)

$$b_{\alpha\beta} = 0, \qquad \bar{R}_{1212} = 0, \tag{1.13.58}$$

so that, since the Riemann–Christoffel tensor for the surface $\theta_3 = 0$ vanishes, the order of surface covariant differentiation is immaterial. We also record the following special results:

$$\left.\begin{aligned} &\mathbf{g}_\alpha = \mathbf{r}_{,\alpha} = \mathbf{a}_\alpha, \quad \mathbf{g}_3 = \mathbf{a}_3 = \mathbf{a}^3 \\ &g_{\alpha\beta} = \mathbf{a}_\alpha . \mathbf{a}_\beta = a_{\alpha\beta}, \quad g_{\alpha 3} = \mathbf{a}_\alpha . \mathbf{a}_3 = 0, \quad g_{33} = \mathbf{a}_3 . \mathbf{a}_3 = 1 \\ &g^{\alpha\beta} = \mathbf{a}^\alpha . \mathbf{a}^\beta = a^{\alpha\beta}, \quad g^{\alpha 3} = \mathbf{a}^\alpha . \mathbf{a}^3 = 0, \quad g^{33} = \mathbf{a}^3 . \mathbf{a}^3 = 1 \\ &g = |g_{ij}| = |a_{\alpha\beta}| = a \\ &\Gamma^3_{\alpha i} = 0, \quad \Gamma^\alpha_{\beta 3} = 0 \\ &\mathbf{a}_{\alpha,\beta} = \bar{\Gamma}^\lambda_{\alpha\beta} \mathbf{a}_\lambda, \quad \mathbf{a}^\alpha{}_{,\beta} = -\bar{\Gamma}^\alpha_{\beta\lambda} \mathbf{a}^\lambda \end{aligned}\right\}, \tag{1.13.59}$$

and we see that all geometrical quantities (except $\mathbf{R}$) are independent of θ_3.

A corresponding theorem to (1.12.39) exists between an integral around a curve c on a surface and an integral over the surface bounded by c. If $\mathbf{u}$ is the unit outward normal in the surface to the curve c then

$$\mathbf{u} = u_\alpha \mathbf{a}^\alpha, \qquad u_\alpha = \epsilon_{\alpha\beta} \frac{d\theta^\beta}{ds}, \tag{1.13.60}$$

and

$$\iint\limits_S v^\alpha|_\alpha \, dS = \iint\limits_S \frac{1}{\sqrt{a}} (\sqrt{a}\, v^\alpha)_{,\alpha} \, dS = \int\limits_c u_\alpha v^\alpha \, ds = \int\limits_c \epsilon_{\alpha\beta} v^\alpha \frac{d\theta^\beta}{ds} \, ds. \tag{1.13.61}$$

1.14. Smooth curves. Hölder condition

We now summarize some definitions and results in complex variable theory which will be used in the solution of two-dimensional problems. We consider curves lying in the plane defined by a right-handed cartesian system of axes Ox, Oy. Unless stated otherwise curves are always assumed to be simple, that is, they do not intersect.

A *smooth arc* is defined by the parametric equations

$$x = x(s), \qquad y = y(s) \quad (s_a \leqslant s \leqslant s_b),$$

where s_a, s_b are finite constants and $x(s)$, $y(s)$ are functions, continuous in the closed interval $\langle s_a, s_b \rangle$, with the following properties:

1. $x(s)$, $y(s)$ have continuous first derivatives $x'(s)$, $y'(s)$ in $\langle s_a, s_b \rangle$ and these derivatives never vanish simultaneously. The values of $x'(s)$, $y'(s)$ at the ends of the interval are to be interpreted as $x'(s_a+0)$, $y'(s_a+0)$, and $x'(s_b-0)$, $y'(s_b-0)$ respectively.

2. When $s_a \leqslant s_1, s_2 \leqslant s_b$, $s_1 \neq s_2$ then either†

$$x(s_1) \neq x(s_2) \quad \text{or} \quad y(s_1) \neq y(s_2).$$

The points a and b, corresponding to the values of s_a and s_b, are the ends of the arc and belong to the arc. When it is necessary to stress this fact the arc is called closed but when the end points do not belong to the arc it is called open. For each smooth arc considered, a definite positive direction is chosen which corresponds to an increase of the parameter s. The arc with the end points a, b will be denoted by ab, where the order of letters indicates that the positive direction is from a to b.

Curves will be called *smooth contours* if they differ from smooth arcs in that, in condition 2, either†

$$x(s_1) \neq x(s_2) \quad \text{or} \quad y(s_1) \neq y(s_2)$$

whenever $s_a < s_1, s_2 < s_b$, $s_1 \neq s_2$, but

$$x(s_b) = x(s_a), \qquad y(s_b) = y(s_a),$$
$$x'(s_b-0) = x'(s_a+0), \qquad y'(s_b-0) = y'(s_a+0).$$

The union of a finite number of non-intersecting smooth arcs or contours is called a sectionally smooth curve L, which may therefore consist of several disconnected parts. If arcs form a part of L the ends of the arcs are called ends of L. Around each point t on L, not coinciding with an end, a circle may be drawn with radius so small that it is divided by L into two parts lying respectively to the left and right of L when viewed from the positive direction of L, and they will be regarded as

† Alternatively, both inequalities are satisfied.

the left and right neighbourhoods of t. If L consists of contours and is the boundary of some connected region, the positive direction of L is selected so that, if L is described in that direction, that region is always on the left or on the right. The part of the plane which lies on the left will be denoted by S^+ and that to the right by S^-.

Let L be a given sectionally smooth curve, α an arbitrary acute angle ($0 < \alpha < \frac{1}{2}\pi$), and let $R_0 = R_0(\alpha)$ be a positive number depending on α, but not on the position of the point t on L, such that

1. The part of L, lying within a circle Γ with radius $R \leqslant R_0$ with its centre at any point t on L, consists of a single arc ab. If L consists only of a contour the ends a, b of this arc lie always on the circumference of Γ. If an arc is a part of L one or both of its end points may be inside Γ.

2. The non-obtuse angle between the tangents at any two points of the arc ab does not exceed α.

Then, for a given angle α, R_0 is called the standard radius; the circle Γ_0 of radius R_0 the standard circle; the arc ab, cut out of the curve L by the circle Γ_0 drawn at any point on it, the standard arc.

Suppose $\phi(t)$ is a given (complex) function of position on the arc L, where t denotes both the point (x, y) and the corresponding complex number $t = x+iy$. The function $\phi(t)$ is said to satisfy a *Hölder condition* {$H(\mu)$ condition} on L, if for any two points t_1, t_2 on L

$$|\phi(t_2)-\phi(t_1)| \leqslant A\,|t_2-t_1|^{\mu}, \tag{1.14.1}$$

where A and μ are positive real constants.

If L is a sectionally smooth curve then $\phi(t)$ is said to satisfy a Hölder condition on L if (1.14.1) holds for each arc or contour belonging to L.

1.15. Sectionally continuous functions

Let the function $\Phi(z)$ of the point $z = x+iy$ be defined and continuous in the neighbourhood of a sectionally smooth curve L, and perhaps also for points on L. Let t be a point on L not coinciding with an end-point. It will be said that $\Phi(z)$ is continuous at the point t from the left (or from the right) if $\Phi(z)$ tends to a definite limit $\Phi^+(t)$ {or $\Phi^-(t)$} when z approaches t along any path which remains on the left (or on the right) of L.

If the function $\Phi(z)$ is continuous from the left (or right) at any point of some part L' of the curve L, then $\Phi(z)$ is said to be continuous from the left (or right) on L'. In this case the function $\Phi^+(t)$ {or $\Phi^-(t)$} is necessarily continuous on L'.

If t is an end point, $\Phi(z)$ is continuous at the end t if $\Phi(z)$ tends to a definite limit as z tends to t along any path which does not touch L. This limit will sometimes be denoted by $\Phi^{\pm}(t)$. Thus, in the case of end points, no distinction is made between the limits from the left and from the right but it will sometimes be said of a function continuous at an end, that it is continuous at this end from the left and right.

A function $\Phi(z)$, continuous in some neighbourhood of a sectionally smooth curve L and continuous on L from the left and right, including the ends, is said to be *sectionally continuous* in the neighbourhood of L, including its ends. If the continuity does not extend to the ends a corresponding reservation is made. The curve will be called a curve of discontinuity of the sectionally continuous function $\Phi(z)$.

1.16. Sectionally holomorphic functions

Let $\Phi(z)$ be a function which is holomorphic in each finite region not containing points of a sectionally smooth curve L. Further, let $\Phi(z)$ be continuous on L from the left and from the right, with the possible exception of the ends, but in the neighbourhood of the ends $\Phi(z)$ satisfies the condition

$$|\Phi(z)| \leqslant \frac{C}{|z-c|^{\alpha}}, \tag{1.16.1}$$

where c is an end and C is a positive real constant. Also α is real and $0 \leqslant \alpha < 1$. Such functions $\Phi(z)$ are called *sectionally holomorphic functions* with curve of discontinuity L; the curve L will sometimes be called the boundary.

If in the neighbourhood of the point at infinity the expansion

$$\Phi(z) = \sum_{n=-\infty}^{\infty} a_n z^n \tag{1.16.2}$$

of $\Phi(z)$ contains only a finite number of positive powers of z, then $\Phi(z)$ is said to be of finite degree at infinity. If a_k is the last coefficient in the expansion (1.16.2) which is different from zero then the degree of $\Phi(z)$ at infinity will be said to be equal to k. (We exclude the case when all a_n are zero, in which case $\Phi(z) \equiv 0$ in some region which includes the point at infinity.) When $k > 0$ the point at infinity is a pole of order k of the function $\Phi(z)$, while for $k < 0$ it is a zero of order k. For $k = 0$, i.e. when $\Phi(z)$ tends to a definite limit a_0 different from zero as $|z|$ tends to infinity, $\Phi(z)$ may conveniently be said to have a pole or zero of zero-order at infinity. Finally, when $k \leqslant 0$, $\Phi(z)$ is said to be sectionally holomorphic including the point at infinity.

We now recall a well-known property of holomorphic functions. Let S_1 and S_2 be two regions of the plane, bounded by smooth contours, and having no interior points in common but which meet along some sectionally smooth curve L, a common part of their boundaries; the ends of L are not included in L. Let $\Phi_1(z)$ and $\Phi_2(z)$ be functions, holomorphic respectively in S_1 and S_2, continuous on L from S_1 and S_2 respectively and with boundary values along L equal, so that

$$\Phi_1(t) = \Phi_2(t),$$

where t denotes a point on L and $\Phi_1(t)$, $\Phi_2(t)$ are the boundary values of the functions $\Phi_1(z)$ and $\Phi_2(z)$. Then the function

$$\left.\begin{aligned} \Phi(z) &= \Phi_1(z) && (z \text{ in } S_1) \\ \Phi(z) &= \Phi_2(z) && (z \text{ in } S_2) \\ \Phi(z) &= \Phi(t) = \Phi_1(t) = \Phi_2(t) && (z \text{ on } L) \end{aligned}\right\} \tag{1.16.3}$$

is holomorphic in S_1+S_2+L.

The previous result may be modified in the following way. Retaining the above conditions we suppose that $\Phi_1(z)$ and $\Phi_2(z)$ are continuous on L from S_1 and S_2 respectively except possibly at a finite number of points $c_1, c_2,..., c_m$ on L, in the neighbourhood of which

$$|\Phi_1(z)| \leqslant \frac{\text{const}}{|z-c_n|^\alpha}, \qquad |\Phi_2(z)| \leqslant \frac{\text{const}}{|z-c_n|^\alpha} \quad (\alpha < 1,\ n = 1, 2,..., m).$$

Then the function $\Phi(z)$ defined by (1.16.3) is holomorphic in S_1+S_2+L if it is assigned suitable values at the points $c_1, c_2,..., c_m$.

If now $\Phi(z)$ is a sectionally holomorphic function with a curve of discontinuity L and if on any part L' of L, with the possible exception of the end points of L', $\Phi^+(t) = \Phi^-(t)$, then this part of L may be removed and the function $\Phi(z)$ will have a curve of discontinuity $L-L'$.

1.17. Cauchy integrals

We state, without proof, some important properties of the Cauchy integral

$$\Phi(z) = \frac{1}{2\pi i}\int\limits_L \frac{\phi(t)\,dt}{t-z}, \tag{1.17.1}$$

where the integration is along a sectionally smooth curve or contour L lying in a finite region of the plane. If the 'density function' $\phi(t)$ satisfies the $H(\mu)$ condition on L, then the function $\Phi(z)$ is a sectionally holomorphic function, vanishing at infinity, with the curve of discontinuity L. Moreover,

$$\left.\begin{aligned}\Phi^+(t_0)-\Phi^-(t_0) &= \phi(t_0)\\ \Phi^+(t_0)+\Phi^-(t_0) &= \frac{1}{\pi i}\int\limits_L \frac{\phi(t)\,dt}{t-t_0}\end{aligned}\right\}, \tag{1.17.2}$$

where t_0 is a point on L which does not coincide with those end points at which $\phi(t) \neq 0$. If t_0 coincides with an end where $\phi(t_0) = 0$ then $\Phi^+(t_0) = \Phi^-(t_0) = \Phi(t_0)$. The formulae (1.17.2) are known as the Plemelj formulae.

Again, if
$$\Phi(t_0) = \frac{1}{2\pi i}\int\limits_L \frac{\phi(t)\,dt}{t-t_0}, \tag{1.17.3}$$

and if $\phi(t)$ satisfies the $H(\mu)$ condition on L, then $\Phi^+(t)$, $\Phi^-(t)$, and $\Phi(t_0)$ also satisfy everywhere, except in arbitrarily small neighbourhoods of those ends at which $\phi(t) \neq 0$, the $H(\mu)$ condition for $\mu < 1$ and the $H(\mu-\epsilon)$ condition for $\mu = 1$, where ϵ is an arbitrarily small positive quantity.

We observe that the integrals occurring in (1.17.2) and (1.17.3) are interpreted as the principal value of the Cauchy integrals.

When the Cauchy integral (1.17.1) is taken along a curve L extending to infinity the above conditions satisfied by $\phi(t)$ are not sufficient. When, for example, L is an infinite straight line we still assume that $\phi(t)$ satisfies the $H(\mu)$ condition on L and that it takes the definite finite value $\phi(\infty)$ when $t \to \pm\infty$. Also, we suppose that, for sufficiently large $|t|$,

$$|\phi(t)-\phi(\infty)| < \frac{\text{const}}{|t|^\alpha} \quad (\alpha > 0),$$

which takes the part of the H condition at the point at infinity. The integral (1.17.1) then has a definite sense if it is understood to be the limit of the integral, taken over the segment ab, where a and b are points, tending to infinity, on either side and at equal distances from some fixed point. The Plemelj formulae still remain true, but the integral (1.17.1) for $\Phi(z)$ does not tend to zero as $|z| \to \infty$. In fact

$$\Phi(z) \to \pm\tfrac{1}{2}\phi(\infty)$$

when $|z| \to \infty$, remaining in S^+ or S^-, the upper sign being taken for S^+ and the lower for S^-. If $\phi(\infty) = 0$ then $\Phi(z) \to 0$ as $|z| \to \infty$.

It is useful for later work to have a simple bound for the modulus of the derivative of a Cauchy integral near the boundary. As before, let L be a sectionally smooth curve and let $\phi(t)$ satisfy the $H(\mu)$ condition on L. Consider the derivative of the Cauchy integral

$$\Phi'(z) = \frac{1}{2\pi i}\int_L \frac{\phi(t)\,dt}{(t-z)^2}. \tag{1.17.4}$$

Let t_0 be an arbitrary point on L, such that the distance from t_0 to the nearest end of L, if the latter contains arcs, is not less than some arbitrarily fixed number R. Let $R_0 = R_0(\alpha)$ be the standard radius for L, corresponding to some arbitrarily fixed acute angle α, and ρ any positive constant such that

$$\rho < R_0, \qquad \rho < R.$$

When L consists only of contours the condition $\rho < R$ is omitted.

If the distance of the point z from t_0 does not exceed ρ, so that

$$|z-t_0| \leqslant \rho,$$

and if the non-obtuse angle between $t_0 z$ and the tangent to L at t_0 is not less than some fixed quantity $\beta > \alpha$, then

$$\left.\begin{aligned} |\Phi'(z)| &< C|z-t_0|^{\mu-1} \quad &(\mu < 1)\\ |\Phi'(z)| &< C|\log(z-t_0)| \quad &(\mu = 1)\end{aligned}\right\}, \tag{1.17.5}$$

where C is a constant.

If L contains arcs and if the condition that t_0 lies at a finite distance from the ends does not hold and therefore the condition $\rho < R$ is omitted, then, retaining all other conditions on the position of z,

$$|\Phi'(z)| < \frac{C}{|z-t_0|}, \tag{1.17.6}$$

where C is a constant.

The results given in this section may also be used when the density function $\phi(t)$ of the Cauchy integral (1.17.1) has ordinary discontinuities at a finite number of points of L, not necessarily coinciding with the ends. It is sufficient to study the case when $L = ab$ is a simple smooth arc and $\phi(t)$ has only one point of discontinuity at c on ab. Let $\phi(t)$ be some function, given everywhere on the arc ab except possibly at the point c, and let its limits $\phi(c-0)$ and $\phi(c+0)$ exist, when t approaches the point c remaining on ac and cb respectively. Then the function $\phi(t)$ on the closed arcs ac or cb will be given the values $\phi(c-0)$ and $\phi(c+0)$ respectively at c. It is then clear how the statement that the function $\phi(t)$ satisfies the H condition on each of the closed arcs ac and cb separately is to be understood.

Other types of discontinuity in $\phi(t)$ may also be included but the above is sufficient for our purpose.

1.18. Sectionally holomorphic functions with given discontinuities on contours

Let S^+ be a connected region bounded by one or more smooth non-intersecting contours $L_0, L_1, \ldots, L_n$ of which L_0 contains all the others (Fig. 1.8). The contour L_0 may be absent (i.e. it may be at infinity),

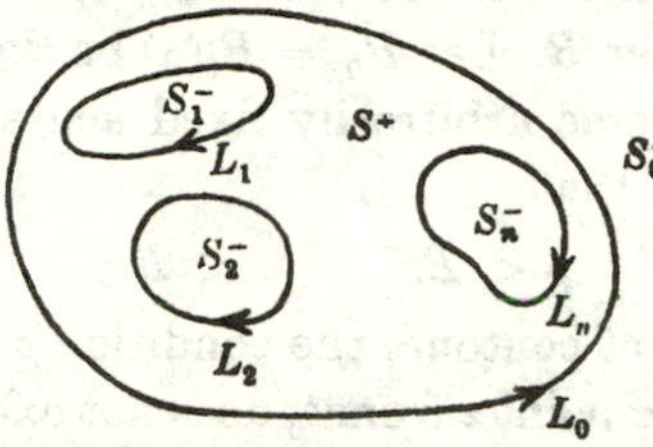

FIG. 1.8. Connected region bounded by smooth non-intersecting contours.

in which case the region S^+ will be an infinite region with certain contours as internal boundaries. The union of the contours $L_0, L_1, \ldots, L_n$ is denoted by L and S^- is the complement of $S^+ + L$. Thus S^- consists of the finite regions $S_1^-, S_2^-, \ldots, S_n^-$ bounded by $L_1, L_2, \ldots, L_n$ and, if L_0 exists, of the infinite region S_0^- bounded by L_0. If $\phi(t)$ satisfies the $H(\mu)$ condition on L we wish to find a function $\Phi(z)$, sectionally holomorphic everywhere, zero at infinity, satisfying the given boundary condition

$$\Phi^+(t_0) - \Phi^-(t_0) = \phi(t_0) \quad \text{on } L. \tag{1.18.1}$$

The problem is immediately solved with the help of Plemelj's formula (1.17.2). Thus

$$\Phi(z) = \frac{1}{2\pi i} \int_L \frac{\phi(t)\, dt}{t - z} \tag{1.18.2}$$

satisfies all the required conditions. The problem has no other solution, because the difference $\Psi(z)$ of two solutions satisfies the condition $\Psi^+(t_0) - \Psi^-(t_0) = 0$. Hence $\Psi(z)$ is holomorphic in the entire plane and, as it vanishes at infinity, $\Psi(z) = 0$.

We may generalize this result by finding a function, sectionally holomorphic everywhere, except possibly at the point at infinity, having finite non-negative degree k there, satisfying the boundary condition (1.18.1) on L. The most general solution is

$$\Phi(z) = \frac{1}{2\pi i} \int_L \frac{\phi(t)\, dt}{t - z} + P_k(z), \tag{1.18.3}$$

where $P_k(z)$ is an arbitrary polynomial of degree not higher than k.

This solution may be shown to be unique by a similar method to that used above.

Similar results hold when $\phi(t)$ satisfies the $H(\mu)$ condition on a smooth curve L which is a union of smooth, non-intersecting arcs $L_1, L_2,..., L_n$ with a definite positive direction, but in this case we consider a more general type of density function $\phi(t)$.

1.19. Functions of classes H and H^*

Let L be a union of smooth arcs $L_1, L_2,..., L_n$ whose ends are denoted by a_r, b_r in such a way that the positive direction of L_r is from a_r to b_r; 'the ends a' and 'the ends b' are distinguished accordingly. When it is of no importance which end is referred to, it is denoted by c_r or c. The plane cut along $L = L_1+L_2+...+L_n$ is denoted by S; the boundary L does not belong to S.

A function $\phi(t)$ of the point t on L is said to belong to the class H on L if it satisfies for some $\mu > 0$ the $H(\mu)$ condition on each of the closed arcs L_r of L (including the ends). If $\phi(t)$ satisfies the $H(\mu)$ condition only in the neighbourhood of some end c of L, including c, then $\phi(t)$ is said to belong to the class H in the neighbourhood of c.

If $\phi(t)$ satisfies the $H(\mu)$ condition on every closed part of L not containing ends, and if near an end c it is of the form

$$\phi(t) = \frac{\phi^*(t)}{(t-c)^\alpha} \quad (0 \leqslant \alpha < 1), \tag{1.19.1}$$

where $\phi^*(t)$ belongs to the class H, then $\phi(t)$ is said to belong to the class H^* on L. If (1.19.1) holds only in the neighbourhood of a given end c then $\phi(t)$ is said to belong to the class H^* in the neighbourhood of c.

1.20. Sectionally holomorphic functions for given discontinuities on arcs

If $\phi(t)$ is a function of the class H^* given on L, then

$$\Phi(z) = \frac{1}{2\pi i}\int\limits_L \frac{\phi(t)\,dt}{t-z} + P_k(z) \tag{1.20.1}$$

is a sectionally holomorphic function, except possibly at infinity, having finite degree $k \geqslant -1$ there, and satisfying the boundary condition

$$\Phi^+(t_0) - \Phi^-(t_0) = \phi(t_0) \quad \text{on } L,\dagger \tag{1.20.2}$$

where $P_k(z)$ is an arbitrary polynomial of degree not greater than k, while $P_{-1}(z) = 0$. The sectionally holomorphic function $\Phi(z)$ satisfying

† Excluding the ends of L.

these conditions is uniquely given by (1.20.1). For, if $\Phi_1(z)$, $\Phi_2(z)$ are two possible solutions then $\Psi(z) = \Phi_1(z) - \Phi_2(z)$, if it is given the value $\Psi^+(t) = \Psi^-(t)$ on L, is holomorphic everywhere in the finite part of the plane, except possibly at the ends a_r and b_r. These points can, however, only be isolated singularities, because near them the degree of infinity of the function $\Psi(z)$ is necessarily less than 1, and hence $\Psi(z)$ is bounded near these points. Thus $\Psi(z)$ can be taken to be holomorphic in the entire plane, and since at infinity it is of finite degree k then $\Psi(z)$ is a polynomial of degree not greater than k.

In particular, for $k = -1$, corresponding to the requirement that $\Phi(z)$ vanishes at infinity, the problem has the unique solution

$$\Phi(z) = \frac{1}{2\pi i} \int\limits_L \frac{\phi(t)\,dt}{t-z}. \tag{1.20.3}$$

1.21. Hilbert problem for arcs

In this section we obtain solutions for two special cases of Hilbert's problem for arcs.

Problem 1. To find the sectionally holomorphic function $\Phi(z)$, having finite degree k at infinity, satisfying the boundary condition

$$\Phi^+(t_0) + \kappa\Phi^-(t_0) = \phi(t_0) \tag{1.21.1}$$

on a sectionally smooth curve L consisting of a union of arcs, where κ is a positive constant and $\phi(t)$ satisfies the $H(\mu)$ condition and is given everywhere on L.

We observe that since $\Phi(z)$ is to be sectionally holomorphic it satisfies the condition (1.16.1) at ends c of L and may therefore become infinite at c with degree less than 1. We can, however, also obtain solutions in which $\Phi(z)$ remains finite at m particular ends of L.

We denote the ends a_r, b_r of the arcs L_r $(r = 1, 2,..., n)$ by c_s $(s = 1, 2,..., 2n)$ in any order and define

$$R(z) = \frac{\prod\limits_{r=1}^{m} (z-c_r)}{\prod\limits_{r=m+1}^{2n} (z-c_r)}, \qquad R_0(z) = \prod_{r=1}^{n} \left(\frac{z-a_r}{z-b_r}\right). \tag{1.21.2}$$

If m is 0 or $2n$ the numerator and denominator respectively of $R(z)$ are replaced by 1. The function $\sqrt{\{R(z)\}}$ refers to that branch which is holomorphic in S, i.e. in the plane cut along L, and which is such that near the point at infinity the expansion in decreasing powers of z has the form

$$\sqrt{\{R(z)\}} = z^{m-n} + A_1 z^{m-n-1} + \dots . \tag{1.21.3}$$

The boundary value of $\sqrt{\{R(z)\}}$ from the left is denoted by

$$[\sqrt{\{R(t)\}}]^+ = \sqrt{\{R(t)\}} \tag{1.21.4}$$

so that

$$[\sqrt{\{R(t)\}}]^- = -\sqrt{\{R(t)\}}. \tag{1.21.5}$$

Also, if we put

$$\ln \kappa = 2\pi\gamma \tag{1.21.6}$$

then $\{R_0(z)\}^{i\gamma}$ denotes that branch which is holomorphic in the plane cut along L and which is such that

$$\{R_0(z)\}^{i\gamma} \to 1 \quad \text{as} \quad |z| \to \infty. \tag{1.21.7}$$

The boundary value of $\{R_0(z)\}^{i\gamma}$ on L, from the left, is denoted by

$$[\{R_0(t)\}^{i\gamma}]^+ = \{R_0(t)\}^{i\gamma} \tag{1.21.8}$$

and therefore

$$[\{R_0(t)\}^{i\gamma}]^- = \frac{\{R_0(t)\}^{i\gamma}}{\kappa}. \tag{1.21.9}$$

If, now,

$$X_m(z) = \{R_0(z)\}^{i\gamma}\sqrt{\{R(z)\}}, \tag{1.21.10}$$

then $X_m(z)$ is holomorphic in the plane cut along L, and at infinity

$$X_m(z) = O(z^{m-n}). \tag{1.21.11}$$

Also

$$X_m^+(t) = \{R_0(t)\}^{i\gamma}\sqrt{\{R(t)\}}, \qquad X_m^-(t) = -\frac{X_m^+(t)}{\kappa}, \tag{1.21.12}$$

and $X_m(z)$ is zero at m ends c_r $(r = 1,..., m)$ whilst at $2n-m$ ends c_s $(s = m+1,..., 2n)$ it is infinite of order $\frac{1}{2}$.

We now find a solution of problem 1 which is such that $\Phi(z)$ is finite at m ends c_r $(r = 1, 2,..., m)$, but may be infinite at the remaining ends with degree less than 1. Consider the function $\Phi(z)/X_m(z)$, which is holomorphic in the plane cut along L, and which is of finite degree $k+n-m$ at infinity. Since $\Phi(z)$ remains bounded near those ends $c_1,..., c_m$ at which $X_m(z)$ becomes zero, then near these ends

$$\left|\frac{\Phi(z)}{X_m(z)}\right| < \frac{\text{const}}{|z-c_r|^\alpha} \quad (\alpha < 1),$$

whilst near the remaining ends $\Phi(z)/X_m(z)$ satisfies a similar condition since $\Phi(z)$ may be infinite of degree less than 1 and $X_m(z)$ is infinite of degree $\frac{1}{2}$. Hence $\Phi(z)/X_m(z)$ is sectionally holomorphic in the entire plane, of finite degree $k+n-m$ at infinity, and, from (1.21.1) and (1.21.12), satisfying the boundary condition

$$\left[\frac{\Phi(t_0)}{X_m(t_0)}\right]^+ - \left[\frac{\Phi(t_0)}{X_m(t_0)}\right]^- = \frac{\phi(t_0)}{X_m^+(t_0)} \tag{1.21.13}$$

on L excluding its ends. Here $\phi(t)/X_m^+(t)$ is a function of the class H^* on L, and from (1.20.1) the required solution is

$$\frac{\Phi(z)}{X_m(z)} = \frac{1}{2\pi i}\int_L \frac{\phi(t)\,dt}{(t-z)X_m^+(t)} + P_{k+n-m}(z) \quad (k+n-m \geqslant -1),$$

or

$$\Phi(z) = \frac{\{R_0(z)\}^{i\gamma}\sqrt{\{R(z)\}}}{2\pi i}\int_L \frac{\phi(t)\,dt}{(t-z)\{R_0(t)\}^{i\gamma}\sqrt{\{R(t)\}}}+ \\ +\{R_0(z)\}^{i\gamma}\sqrt{\{R(z)\}}P_{k+n-m}(z) \quad (1.21.14)$$

where $P_{k+n-m}(z)$ is an arbitrary polynomial of degree not greater than $k+n-m$ and $P_{-1}(z) = 0$. The solution is also valid for $k+n-m < -1$ if $P_{k+n-m}(z) = 0$ and $\phi(t)$ satisfies further conditions.

PROBLEM 2. The functions $\Phi(z)$ and $\phi(t)$ satisfy the same conditions as before but (1.21.1) is replaced by

$$\Phi^+(t_0)-e^{2\pi i(\frac{1}{2}+\beta)}\Phi^-(t_0) = \phi(t_0) \quad (0 \leqslant \beta < \tfrac{1}{2}) \qquad (1.21.15)$$

on L. Also here we shall allow $\Phi(z)$ to be infinite with degree less than 1 at all end points.

To solve this problem we put

$$S(z) = \prod_{r=1}^{n} (z-a_r)^{\frac{1}{2}+\beta}(z-b_r)^{\frac{1}{2}-\beta}, \qquad (1.21.16)$$

where $S(z)$ is holomorphic in the plane cut along L, and

$$S(z) \to z^n \quad \text{as} \quad |z| \to \infty. \qquad (1.21.17)$$

As before we put

$$S^+(t) = S(t) \qquad (1.21.18)$$

and hence

$$S^-(t) = e^{2\pi i(\frac{1}{2}+\beta)}S(t). \qquad (1.21.19)$$

The boundary condition (1.21.15) now becomes

$$\Phi^+(t_0)S^+(t_0)-\Phi^-(t_0)S^-(t_0) = \phi(t_0)S(t_0), \qquad (1.21.20)$$

and $\phi(t)S(t)$ satisfies a Hölder condition on L. Also $\Phi(z)S(z)$ is sectionally holomorphic everywhere, except at infinity, where it is of degree $k+n$, and therefore

$$\Phi(z)S(z) = \frac{1}{2\pi i}\int_L \frac{S(t)\phi(t)}{t-z}\,dt+Q_{k+n}(z),$$

where $Q_{k+n}(z)$ is an arbitrary polynomial of degree not greater than $k+n$. Also $k+n \geqslant -1$ and $Q_{-1}(z) = 0$. Hence

$$\Phi(z) = \frac{1}{2\pi i S(z)}\int_L \frac{S(t)\phi(t)}{t-z}\,dt+\frac{Q_{k+n}(z)}{S(z)}. \qquad (1.21.21)$$

1.22. Fourier integrals

In this section we state, without proof, some properties of Fourier integrals which will be needed in later parts of the book.

Let $f(t)$ be of bounded variation in the neighbourhood of the point $t = x$ and let the integral $\int_{-\infty}^{\infty} f(t)\,dt$ be absolutely convergent. Then

$$\tfrac{1}{2}\{f(x+0)+f(x-0)\} = \frac{1}{\pi}\int_0^{\infty} du \int_{-\infty}^{\infty} f(t)\cos u(x-t)\,dt$$

$$= \frac{1}{2\pi}\int_{-\infty}^{\infty} e^{-ixu}\,du \int_{-\infty}^{\infty} f(t)e^{iut}\,dt. \tag{1.22.1}$$

If, in addition, $f(t)$ is an *odd* function of t,

$$\tfrac{1}{2}\{f(x+0)+f(x-0)\} = \frac{2}{\pi}\int_0^{\infty} \sin ux\,du \int_0^{\infty} f(t)\sin ut\,dt, \tag{1.22.2}$$

and, if $f(t)$ is an *even* function of t,

$$\tfrac{1}{2}\{f(x+0)+f(x-0)\} = \frac{2}{\pi}\int_0^{\infty} \cos ux\,du \int_0^{\infty} f(t)\cos ut\,dt. \tag{1.22.3}$$

When $f(t)$ is continuous at the point $t = x$ it is more convenient to state the above results in the following forms:

1. Fourier sine and cosine transforms:

$$\left.\begin{aligned} F(u) &= \int_0^{\infty} f(t)\begin{matrix}\sin ut\\ \cos ut\end{matrix}\,dt \\ f(x) &= \frac{2}{\pi}\int_0^{\infty} F(u)\begin{matrix}\sin ux\\ \cos ux\end{matrix}\,du \end{aligned}\right\}. \tag{1.22.4}$$

2. Complex Fourier transforms:

$$\left.\begin{aligned} F(u) &= \int_{-\infty}^{\infty} f(t)e^{iut}\,dt \\ f(x) &= \frac{1}{2\pi}\int_{-\infty}^{\infty} F(u)e^{-ixu}\,du \end{aligned}\right\}. \tag{1.22.5}$$

For many purposes a form of Fourier's theorem for holomorphic functions is useful. Let $f(z)$ be a holomorphic function in the region $a < y < b$. In any strip interior to $a < y < b$ let

$$f(z) = \begin{cases} O(e^{(\lambda+\epsilon)x}) & (x \to \infty), \\ O(e^{(\mu-\epsilon)x}) & (x \to -\infty), \end{cases} \tag{1.22.6}$$

for every positive ϵ. Then if

$$F(w) = \frac{1}{\sqrt{(2\pi)}} \int_{-\infty}^{\infty} f(z) e^{izw}\, dz, \tag{1.22.7}$$

$F(w)$ is a function holomorphic in the region $\lambda < v < \mu$ where $w = u+iv$, and in any strip interior to this region

$$F(w) = \begin{cases} O(e^{(-b+\epsilon)u}) & (u \to \infty), \\ O(e^{-(a+\epsilon)u}) & (u \to -\infty). \end{cases} \tag{1.22.8}$$

Moreover,

$$f(z) = \frac{1}{\sqrt{(2\pi)}} \int_{ic-\infty}^{ic+\infty} F(w) e^{-iwz}\, dw, \tag{1.22.9}$$

for every z in the strip $a < y < b$, where c is a real constant and $\lambda < c < \mu$.

2 General Theory

In this chapter we develop a general theory of elasticity for finite displacements using the tensor and vector notations that were summarized in Chapter 1. A comprehensive historical and critical account of the mechanical foundations of elasticity and fluid dynamics up to 1952 has been given by C. Truesdell.† The reader is referred to this paper for a comparison between the various presentations of non-linear elasticity theory and for an excellent list of references to pioneer contributions by a number of different writers. A monograph by Truesdell and Toupin‡ is wider in its scope and contains references up to 1960.

The basic theory is presented in general coordinates in §§ 2.1–2.4 and these sections are sufficient for understanding the remaining chapters in the book. In § 2.5 onwards we return to rectangular cartesian coordinates and obtain the fundamental equations for a dynamical theory of any continuum using only an energy equation, invariance principles under superposed rigid-body motions, and an entropy production inequality. From these equations explicit results for stresses are deduced for an elastic continuum, allowing for temperature changes, thus justifying, from the point of view of thermodynamics, the equations obtained earlier in this chapter.

2.1. Geometrical relations

Let every point of a continuous three-dimensional body, called briefly the body B_0, be at rest, at time $t = t_0$, relative to a fixed rectangular cartesian system of axes x_i. The position vector of a typical point P_0 of the body B_0 referred to the origin is

$$\mathbf{r} = x_k \mathbf{i}_k, \tag{2.1.1}$$

where $\mathbf{i}_k$ are unit vectors along the fixed axes.

† *J. rat. Mech. Analysis* **1** (1952) 125. Reference should also be made to F. D. Murnaghan, *Finite Deformation of an Elastic Solid* (New York, 1951).

‡ C. Truesdell and R. A. Toupin, 'The Classical Field Theories', *Handbuch der Physik* (Berlin–Göttingen–Heidelberg, 1960).

We suppose that the body B_0 is deformed so that at time t a typical point P_0 has moved to P. The position vector of P referred to the same origin is

$$\mathbf{R} = y_k \mathbf{i}_k. \tag{2.1.2}$$

The position vector of the point P relative to P_0 is denoted by $\mathbf{v}$ and is called the displacement vector. Thus

$$\mathbf{v} = \mathbf{R} - \mathbf{r} = (y_k - x_k)\mathbf{i}_k. \tag{2.1.3}$$

We assume that each point P, at time t, is related to its original position P_0 at time $t = t_0$ by the equations

$$y_i = y_i(x_1, x_2, x_3, t), \tag{2.1.4}$$

$$x_i = x_i(y_1, y_2, y_3, t), \tag{2.1.5}$$

dependence on t_0 being understood, where y_i, x_i are single-valued and continuously differentiable with respect to each of their variables as many times as may be required, except possibly at singular points, curves, and surfaces. If this deformation is to be possible in a real material then

$$\left|\frac{\partial y_i}{\partial x_j}\right| > 0. \tag{2.1.6}$$

Referred to our fixed system of axes the components of velocity at the point P at time t are w_i, where

$$w_i = \dot{y}_i, \tag{2.1.7}$$

a dot denoting differentiation with respect to t holding x_j fixed in (2.1.4). Thus

$$\dot{\mathbf{v}} = w_k \mathbf{i}_k. \tag{2.1.8}$$

The acceleration vector $\ddot{\mathbf{v}}$ is given by

$$\ddot{\mathbf{v}} = \dot{w}_k \mathbf{i}_k \tag{2.1.9}$$

and we may write $\dot{w}_k$ in the alternative form

$$\dot{w}_k = \frac{\partial w_k}{\partial t} + w_m \frac{\partial w_k}{\partial y_m}, \tag{2.1.10}$$

if w_k is regarded as a function of y_m and t. In (2.1.10), $\partial/\partial t$ denotes partial differentiation with respect to t holding y_m fixed.

In view of (2.1.6) we may write the displacement gradients $\partial y_i/\partial x_j$ in the form

$$\frac{\partial y_i}{\partial x_j} = R_{ik} M_{kj}, \tag{2.1.11}$$

where M_{kj} is a positive definite symmetric tensor and R_{ik} is a rotation tensor, so that

$$R_{ik}R_{jk} = R_{ki}R_{kj} = \delta_{ij}, \quad |R_{ij}| = 1. \tag{2.1.12}$$

We define a rate of deformation tensor d_{ij} and vorticity tensor ω_{ij} at time t by the formulae

$$2d_{ij} = \frac{\partial w_i}{\partial y_j} + \frac{\partial w_j}{\partial y_i}, \quad 2\omega_{ij} = \frac{\partial w_i}{\partial y_j} - \frac{\partial w_j}{\partial y_i}. \tag{2.1.13}$$

The tensor d_{ij} is symmetric and ω_{ij} is skew-symmetric.

The original body B_0 may also be described by a general curvilinear set of coordinates θ_i so that

$$x_i = x_i(\theta_1, \theta_2, \theta_3), \tag{2.1.14}$$

where x_i is single-valued and continuously differentiable as many times as required except possibly at singular points, curves, or surfaces. We may imagine that this curvilinear coordinate system moves continuously† with the body as we pass from the original state B_0 at time t_0 to the state B at time t. It will therefore form a curvilinear system in B so that, from (2.1.4) and (2.1.14),

$$y_i = y_i(\theta_1, \theta_2, \theta_3, t). \tag{2.1.15}$$

By using § 1.7 we may define, in B_0, a contravariant vector $d\theta^i$, from the relation (2.1.14). Thus

$$d\theta^i = \frac{\partial \theta^i}{\partial x^j}dx^j, \qquad dx^i = \frac{\partial x^i}{\partial \theta^j}d\theta^j. \tag{2.1.16}$$

From (2.1.4) and (2.1.16) we have

$$dy^i = \frac{\partial y^i}{\partial x^j}dx^j = \frac{\partial y^i}{\partial \theta^j}d\theta^j, \qquad d\theta^i = \frac{\partial \theta^i}{\partial y^j}dy^j, \tag{2.1.17}$$

at a fixed time, so that $d\theta^i$ remains a contravariant vector under all transformations including transformations to coordinate systems in B.

Using (2.1.14) and (2.1.15) the position vectors (2.1.1) and (2.1.2) of the points P_0 and P respectively take the forms

$$\mathbf{r} = \mathbf{r}(\theta_1, \theta_2, \theta_3), \quad \mathbf{R} = \mathbf{R}(\theta_1, \theta_2, \theta_3, t), \tag{2.1.18}$$

and the displacement vector (2.1.3) becomes

$$\mathbf{v} = \mathbf{v}(\theta_1, \theta_2, \theta_3, t). \tag{2.1.19}$$

† The values of θ_i which define a point P_0 of B_0 remain fixed with P_0 as it moves from its position in B_0 to its position P in B. The functional forms (2.1.15) and (2.1.4) are, in general, different.

Base vectors $\mathbf{g}_i$, $\mathbf{g}^i$, and metric tensors g_{ij}, g^{ij} may be defined for the curvilinear system θ_i in the body B_0, so that

$$\left.\begin{aligned} \mathbf{g}_i &= \mathbf{r}_{,i}, \quad \mathbf{g}^i.\mathbf{g}_j = \delta^i_j \\ g_{ij} &= \mathbf{g}_i.\mathbf{g}_j = \frac{\partial x^r}{\partial \theta^i}\frac{\partial x^r}{\partial \theta^j} \\ g^{ij} &= \mathbf{g}^i.\mathbf{g}^j = \frac{\partial \theta^i}{\partial x^r}\frac{\partial \theta^j}{\partial x^r} \end{aligned}\right\}, \tag{2.1.20}$$

where a comma denotes partial differentiation with respect to θ_i.

Similarly, base vectors $\mathbf{G}_i$, $\mathbf{G}^i$ and metric tensors G_{ij}, G^{ij} may be defined for the curvilinear system in the body B at time t. Thus

$$\left.\begin{aligned} \mathbf{G}_i &= \mathbf{R}_{,i}, \quad \mathbf{G}^i.\mathbf{G}_j = \delta^i_j \\ G_{ij} &= \mathbf{G}_i.\mathbf{G}_j = \frac{\partial y^r}{\partial \theta^i}\frac{\partial y^r}{\partial \theta^j} \\ G^{ij} &= \mathbf{G}^i.\mathbf{G}^j = \frac{\partial \theta^i}{\partial y^r}\frac{\partial \theta^j}{\partial y^r} \end{aligned}\right\}. \tag{2.1.21}$$

We define a symmetric tensor† γ_{ij}, subsequently called a strain tensor, by the equation

$$\gamma_{ij} = \tfrac{1}{2}(G_{ij} - g_{ij}). \tag{2.1.22}$$

We may interpret γ_{ij} as a tensor which measures the difference of the squares of the corresponding line elements at points θ_i in the bodies B_0 and B. For the line elements corresponding to the vectors $\mathbf{r}$ and $\mathbf{R}$ are ds_0 and ds respectively, where

$$ds_0^2 = g_{ij}\,d\theta^i d\theta^j, \quad ds^2 = G_{ij}\,d\theta^i d\theta^j, \tag{2.1.23}$$

and

$$ds^2 - ds_0^2 = 2\gamma_{ij}\,d\theta^i d\theta^j. \tag{2.1.24}$$

When the deformation from the body B_0 to the body B is rigid then $\gamma_{ij} = 0$.

The strain tensor γ_{ij} may be expressed in terms of the displacement vector $\mathbf{v}$, or its components with respect to base vectors $\mathbf{g}_i$ or $\mathbf{G}_i$. From (2.1.3) we see that

$$\mathbf{G}_i = \mathbf{R}_{,i} = \mathbf{r}_{,i} + \mathbf{v}_{,i} = \mathbf{g}_i + \mathbf{v}_{,i}. \tag{2.1.25}$$

Hence, using (2.1.20), (2.1.21), and (2.1.22),

$$\gamma_{ij} = \tfrac{1}{2}(\mathbf{g}_i.\mathbf{v}_{,j} + \mathbf{g}_j.\mathbf{v}_{,i} + \mathbf{v}_{,i}.\mathbf{v}_{,j}) = \tfrac{1}{2}(\mathbf{G}_i.\mathbf{v}_{,j} + \mathbf{G}_j.\mathbf{v}_{,i} - \mathbf{v}_{,i}.\mathbf{v}_{,j}). \tag{2.1.26}$$

The displacement vector $\mathbf{v}$ may be expressed in terms of the base vectors of B_0 or of B. Thus

$$\mathbf{v} = v_m\,\mathbf{g}^m, \quad \mathbf{v}_{,i} = v_m|_i\,\mathbf{g}^m, \tag{2.1.27}$$

or

$$\mathbf{v} = V_m\,\mathbf{G}^m, \quad \mathbf{v}_{,i} = V_m\|_i\,\mathbf{G}^m, \tag{2.1.28}$$

† Tensors are now tensors under changes of curvilinear coordinates θ_i.

where, according to (1.12.15), covariant differentiation means

$$v_m|_i = v_{m,i} - {}_0\Gamma^r_{mi} v_r, \tag{2.1.29}$$

$$V_m\|_i = V_{m,i} - \Gamma^r_{mi} V_r. \tag{2.1.30}$$

The Christoffel symbols in (2.1.29) are to be calculated for the body B_0 from the metric tensors g_{ij}, g^{ij}, the vertical line denoting covariant differentiation with respect to B_0. The double vertical line in (2.1.30) denotes covariant differentiation with respect to the body B using Christoffel symbols calculated from G_{ij}, G^{ij}. Introducing (2.1.27) and (2.1.28) into (2.1.26), and using (2.1.20) and (2.1.21), the strain tensor γ_{ij} becomes

$$\gamma_{ij} = \tfrac{1}{2}(v_i|_j + v_j|_i + v^r|_i v_r|_j) = \tfrac{1}{2}(V_i\|_j + V_j\|_i - V^r\|_i V_r\|_j). \tag{2.1.31}$$

Mixed strain tensors can be formed in two ways according to the choice of metric. For our purpose we define

$$\gamma^i_j = g^{ik}\gamma_{kj} = \tfrac{1}{2}(g^{ik}G_{kj} - \delta^i_j). \tag{2.1.32}$$

Three invariants may be formed from a symmetric second order tensor. In particular, strain invariants may be obtained from the mixed strain tensor γ^i_j and are the coefficients of powers of λ in the expansion of the determinant

$$\begin{aligned} |\lambda\delta^r_s + \delta^r_s + 2\gamma^r_s| &= |\lambda\delta^r_s + g^{rm}G_{ms}| \\ &= (G/g)|\lambda G^{rm}g_{ms} + \delta^r_s| = \lambda^3 + I_1\lambda^2 + I_2\lambda + I_3. \end{aligned} \tag{2.1.33}$$

Hence

$$\left.\begin{aligned} I_1 &= 3 + 2\gamma^r_r = g^{rs}G_{rs} \\ I_2 &= 3 + 4\gamma^r_r + 2(\gamma^r_r\gamma^s_s - \gamma^r_s\gamma^s_r) \\ &= \tfrac{1}{2}(I_1^2 - g^{rm}g^{sn}G_{rs}G_{mn}) \\ &= G^{rs}g_{rs}I_3 \\ I_3 &= |\delta^r_s + 2\gamma^r_s| = |g^{rm}G_{ms}| = G/g \end{aligned}\right\}, \tag{2.1.34}$$

where

$$G = |G_{ij}|, \qquad g = |g_{ij}|. \tag{2.1.35}$$

If the continuum is incompressible, volume elements are conserved during deformation so that $G = g$ and

$$I_3 = 1. \tag{2.1.36}$$

2.2. Stress. Equations of motion

We assume that the body B_0 is moved to its strained position B by the action of body forces **F** per unit mass and surface forces **P** per unit area of the boundary of B.† We shall usually assume that **F** and **P**

† Most of the work of this section, however, depends only on the existence of a body B acted on by body and surface forces.

vary with time and are continuous and sectionally continuous functions respectively with respect to their space variables.

We consider an element ΔS of a surface which is situated in the strained body B and which contains within its area a point O. Let $\mathbf{n}$ be a unit normal to ΔS at O measured from one side of ΔS which we call negative to the other side which we call positive. We may think of the portion of the body which is on the positive side exerting force on the negative side across this surface. We suppose that the force, which is thus exerted across the particular area ΔS is statically equivalent to a force $\Delta \mathbf{T}$ at O and a couple $\Delta \mathbf{G}$. We now imagine that the area ΔS tends to zero in any manner, keeping the point O always within it. We assume that the vector $\Delta \mathbf{T}/\Delta S$ tends to a definite limit, and that the vector $\Delta \mathbf{G}/\Delta S$ tends to zero.† The vector

$$\mathbf{t} = \lim_{\Delta S \to 0} \frac{\Delta \mathbf{T}}{\Delta S} \tag{2.2.1}$$

is called the *stress vector* belonging to an element whose unit normal is $\mathbf{n}$ and represents the force, per unit area of surface in the deformed body, exerted by material lying on the positive side of the area, on material lying on the negative side. We assume that $\mathbf{t}$ satisfies sufficient continuity and differentiability conditions for the validity of subsequent analysis.

We assume that the force exerted by the negative side of the surface on the positive side is $-\mathbf{t}$ per unit area. When two bodies are in contact we assume that the nature of the action between them, over the surfaces in contact, is the same as the nature of the action between two portions of the same body.

The acceleration of each point of the body B is denoted by a vector $\mathbf{f}$ which is assumed to be continuous. Consider an arbitrary volume τ in the body B bounded by a closed surface S. We postulate Cauchy's equation of motion of the form‡

$$\int_S \mathbf{t}\, dS + \int_\tau \rho(\mathbf{F}-\mathbf{f})\, d\tau = 0, \tag{2.2.2}$$

for the volume τ, and Cauchy's equation of moments (with respect to fixed axes)

$$\int_S \mathbf{R}\times\mathbf{t}\, dS + \int_\tau \rho\mathbf{R}\times(\mathbf{F}-\mathbf{f})\, d\tau = 0, \tag{2.2.3}$$

where ρ is the density of the body B.

† Throughout this book we restrict attention to the stress vector, but satisfactory theories have been developed when stress couples, and higher order stresses, are present.

‡ In § 2.7 we show that (2.2.2) and (2.2.3) may be deduced from the equation of energy.

At any point P of the strained body B we construct a parallelepiped which is bounded by the faces $\theta_i =$ constant, $\theta_i + d\theta^i =$ constant, and a tetrahedron whose edges are formed by the coordinate curves PP_i of lengths ds_i, and the curves P_1P_2, P_2P_3, P_3P_1 (see Fig. 2.1), which in the limit are defined by the three vectors

$$d\mathbf{s}_j - d\mathbf{s}_i = \mathbf{G}_j\,d\theta^j - \mathbf{G}_i\,d\theta^i \quad (i, j \text{ not summed, } i \neq j). \tag{2.2.4}$$

The surfaces $\theta_i =$ constant of the tetrahedron have areas $\frac{1}{2}dS_i$ and may be represented vectorially by

$$\frac{\mathbf{G}^i\,dS_i}{2\sqrt{G^{ii}}}. \tag{2.2.5}$$

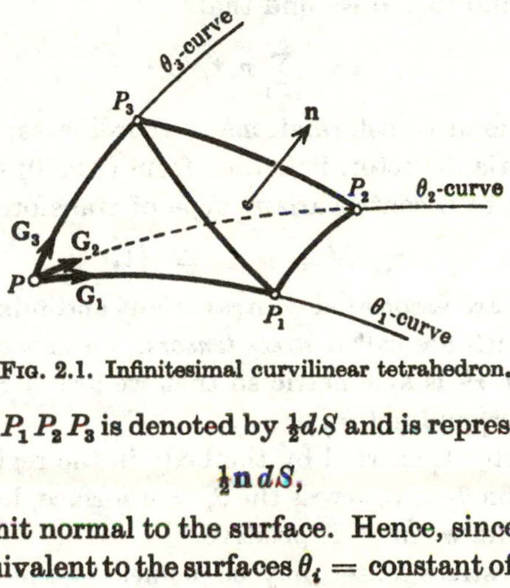

FIG. 2.1. Infinitesimal curvilinear tetrahedron.

Also, the area of $P_1P_2P_3$ is denoted by $\frac{1}{2}dS$ and is represented vectorially by

$$\tfrac{1}{2}\mathbf{n}\,dS, \tag{2.2.6}$$

where $\mathbf{n}$ is the unit normal to the surface. Hence, since the area $P_1P_2P_3$ is vectorially equivalent to the surfaces $\theta_i =$ constant of the tetrahedron, we have

$$\mathbf{n}\,dS = \sum_{i=1}^{3} \frac{\mathbf{G}^i dS_i}{\sqrt{G^{ii}}}, \tag{2.2.7}$$

so that if n_i are the covariant components of $\mathbf{n}$ with respect to base vectors $\mathbf{G}^i$, it follows that

$$n_i\sqrt{(G^{ii})}\,dS = dS_i. \tag{2.2.8}$$

At P in the strained body B a stress vector $\mathbf{t}$, as defined above, is associated with elements of area which are normal to the unit vector $\mathbf{n}$. If the coordinates θ_i are changed to a new set of coordinates $\bar{\theta}_i$, whilst the unit vector $\mathbf{n}$ remains unchanged, then the stress vector $\mathbf{t}$ is an invariant. Stress vectors $-\mathbf{t}_i$ are also associated with elements of area at P in the surfaces $\theta_i =$ constant. They are not invariant under a change of coordinates from θ_i to $\bar{\theta}_j$ if, after such a transformation, we wish to consider stress vectors $-\bar{\mathbf{t}}_j$ associated with elements of areas at P in the surfaces $\bar{\theta}_j =$ constant.† The use of the term 'vector' for these stresses is justified, however, since each $-\mathbf{t}_i$, associated with elements of area at P in *fixed* surfaces coinciding with $\theta_i =$ constant, is an invariant.

By applying the equation of motion (2.2.2) to the infinitesimal tetrahedron $PP_1P_2P_3$ we have, in the limit, keeping the direction of $\mathbf{n}$ fixed,

$$\mathbf{t}\,dS = \mathbf{t}_i\,dS_i. \tag{2.2.9}$$

Volume forces and mass-accelerations acting on the tetrahedron do not appear in this equation since they are of higher order of smallness than the surface forces.

From (2.2.8) and (2.2.9) we find that

$$\mathbf{t} = \sum_{i=1}^{3} n_i\,\mathbf{t}_i\sqrt{G^{ii}}. \tag{2.2.10}$$

Since, under general transformations of coordinates, $\mathbf{t}$ is an invariant and n_i is a covariant vector, it follows from (2.2.10) that $\mathbf{t}_i\sqrt{G^{ii}}$ transforms according to a contravariant type of transformation. We may therefore write

$$\mathbf{t}_i\sqrt{G^{ii}} = \tau^{ij}\mathbf{G}_j = \tau^i_j\,\mathbf{G}^j, \tag{2.2.11}$$

where τ^{ij} and τ^i_j are associated contravariant and mixed tensors of the second order which are called *stress tensors*. Later we shall prove that the stress tensor τ^{ij} is symmetric so that we are allowed to write the mixed tensor as τ^i_j and not $\tau^i_{.j}$ or $\tau_j^{\,.i}$.

The stress vector $\mathbf{t}_i$ exerted by the body in the region $\theta_i > 0$ on the body in the region $\theta_i < 0$, across the $\theta_i = 0$ surface, has positive values for its components when τ^{ij} is positive.

The covariant stress tensor may be defined by the equations

$$\tau_{ij} = G_{ir}\,\tau^r_j. \tag{2.2.12}$$

From (2.2.11) we see that the contravariant stress tensor τ^{ij} and the mixed stress tensor τ^i_j are related to the vectors $\mathbf{t}_i\sqrt{G^{ii}}$. The covariant stress tensor τ_{ij} cannot be related to the vectors $\mathbf{t}_i\sqrt{G^{ii}}$ in a simple way and is therefore, in general, of less importance.

From (2.2.10) and (2.2.11) we obtain

$$\mathbf{t} = \frac{n_i\,\mathbf{T}_i}{\sqrt{G}} = \tau^{ij}n_i\,\mathbf{G}_j = \tau^i_j\,n_i\,\mathbf{G}^j, \tag{2.2.13}$$

where
$$\mathbf{T}_i = \mathbf{t}_i\sqrt{(GG^{ii})} = \sqrt{(G)}\tau^{ij}\mathbf{G}_j = \sqrt{(G)}\tau^i_j\,\mathbf{G}^j, \tag{2.2.14}$$

and is introduced for later convenience. The quantity $\mathbf{T}_i$ has a simple

† The suffixes do *not* indicate that $\mathbf{t}_i$ transform into $\bar{\mathbf{t}}_j$ according to the covariant type of transformation.

interpretation. An element of area at a point θ_r of the body B in the θ_i-surface is, from (1.9.18),

$$\sqrt{(GG^{ii})}\,d\theta^j d\theta^k \quad (i \text{ not summed, } i, j, k \text{ unequal}),$$

and the force across this element is

$$\mathbf{t}_i\sqrt{(GG^{ii})}\,d\theta^j d\theta^k = \mathbf{T}_i\,d\theta^j d\theta^k. \tag{2.2.15}$$

The three stress vectors $\mathbf{t}_i$ in (2.2.11) may be put in the form

$$\mathbf{t}_i = \sum_{j=1}^{3} \frac{\sigma_{(ij)}\mathbf{G}_j}{\sqrt{G_{jj}}}, \tag{2.2.16}$$

where $\mathbf{G}_j/\sqrt{G_{jj}}$ are unit vectors along the coordinate curves and

$$\sigma_{(ij)} = \sqrt{(G_{jj}/G^{ii})}\tau^{ij}. \tag{2.2.17}$$

The *physical* components of the stress tensor referred to oblique axes along $\mathbf{G}_j$ are therefore $\sigma_{(ij)}$. The functions $\sigma_{(ij)}$ are not, of course, the components of a tensor.

With the help of (2.2.13) and (1.12.39) equation (2.2.2) becomes

$$\int_\tau \{\mathbf{T}_{i,i}+\rho(\mathbf{F}-\mathbf{f})\sqrt{G}\}\frac{d\tau}{\sqrt{G}} = 0 \tag{2.2.18}$$

for all arbitrary volumes τ in B. Hence, provided the integrand is a continuous function of θ_i,

$$\mathbf{T}_{i,i}+\rho\mathbf{F}\sqrt{G} = \rho\mathbf{f}\sqrt{G}. \tag{2.2.19}$$

Again, using (2.2.13), (1.12.39), and (2.2.19), equation (2.2.3) reduces to

$$\int_\tau \mathbf{G}_i\times\mathbf{T}_i\frac{d\tau}{\sqrt{G}} = 0$$

for all arbitrary volumes τ in B. If the integrand is a continuous function of θ_i

$$\mathbf{G}_i\times\mathbf{T}_i = 0. \tag{2.2.20}$$

It follows from (2.2.14) and (2.2.20) that

$$\tau^{ij} = \tau^{ji}. \tag{2.2.21}$$

We may express the equations of motion (2.2.19) in a number of alternative forms, and we give the most useful of these below. If

$$\left.\begin{aligned} \mathbf{F} &= F^i\mathbf{G}_i = F_i\mathbf{G}^i \\ \mathbf{f} &= f^i\mathbf{G}_i = f_i\mathbf{G}^i \end{aligned}\right\}, \tag{2.2.22}$$

then, from (2.2.14) and (2.2.19), we have

$$\left.\begin{aligned} \tau^{ij}\|_i+\rho F^j &= \rho f^j \\ \tau^i_j\|_i+\rho F_j &= \rho f_j \end{aligned}\right\}, \tag{2.2.23}$$

where the double line again denotes covariant differentiation with respect to the strained body B.

The conditions at the boundary surface of the body at which the surface forces are prescribed require that

$$\mathbf{t} = \mathbf{P}. \tag{2.2.24}$$

If we express the surface forces $\mathbf{P}$ in terms of their components,

$$\mathbf{P} = P^j\mathbf{G}_j = P_j\mathbf{G}^j, \tag{2.2.25}$$

and use (2.2.13), the boundary conditions have the alternative forms

$$\tau^{ij}n_i = P^j, \tag{2.2.26}$$

or

$$\tau_j^i n_i = P_j. \tag{2.2.27}$$

2.3. Elasticity

Consider an arbitrary volume τ in the strained body B bounded by a closed surface S. The rate of work of the surface forces over S plus the rate of work of body forces throughout τ minus the rate of increase of kinetic energy of the mass in S is†

$$R = \int_S \mathbf{t}.\dot{\mathbf{v}}\,dS + \int_\tau \rho\mathbf{F}.\dot{\mathbf{v}}\,d\tau - \int_\tau \rho\mathbf{f}.\dot{\mathbf{v}}\,d\tau. \tag{2.3.1}$$

Using (2.2.13) this becomes

$$R = \int_S (\mathbf{T}_i.\dot{\mathbf{v}})\frac{n_i\,dS}{\sqrt{G}} + \int_\tau \rho(\mathbf{F}-\mathbf{f}).\dot{\mathbf{v}}\,d\tau.$$

Transforming the surface integral to a volume integral with the help of (1.12.39), and using (2.2.19), we have

$$R = \int_\tau \mathbf{T}_i.\dot{\mathbf{v}}_{,i}\frac{d\tau}{\sqrt{G}},$$

or

$$R = \int_\tau \tau^{ij}\mathbf{G}_j.\dot{\mathbf{v}}_{,i}\,d\tau, \tag{2.3.2}$$

if we use (2.2.14). Since τ^{ij} is symmetric

$$R = \tfrac{1}{2}\int_\tau \tau^{ij}(\mathbf{G}_i.\dot{\mathbf{v}}_{,j}+\mathbf{G}_j.\dot{\mathbf{v}}_{,i})\,d\tau. \tag{2.3.3}$$

Now, from (2.1.22), we have

$$\dot{\gamma}_{ij} = \tfrac{1}{2}\dot{G}_{ij} = \tfrac{1}{2}(\mathbf{G}_i.\dot{\mathbf{v}}_{,j}+\mathbf{G}_j.\dot{\mathbf{v}}_{,i}), \tag{2.3.4}$$

so that

$$R = \int_\tau \tau^{ij}\dot{\gamma}_{ij}\,d\tau. \tag{2.3.5}$$

† A dot denotes differentiation with respect to time holding θ_i (or x_i) fixed.

We confine our attention to a body B which is such that

$$\tau^{ij}\dot{\gamma}_{ij} = \rho\dot{E} \tag{2.3.6}$$

and

$$R = \int_\tau \rho\dot{E}\,d\tau, \tag{2.3.7}$$

where τ^{ij} and E are single-valued functions depending only on the state of strain of B at time t, the initial metric tensor g_{ij}, and tensors which represent the physical properties of the body B_0. Thus we write

$$E = E(\gamma_{ij}), \tag{2.3.8}$$

dependence of E on g_{ij} and the physical tensors being understood, and therefore

$$\dot{E} = \frac{1}{2}\left(\frac{\partial E}{\partial \gamma_{ij}} + \frac{\partial E}{\partial \gamma_{ji}}\right)\dot{\gamma}_{ij}. \tag{2.3.9}$$

When these conditions are satisfied the body B_0 is said to be *elastic.*† The values of $\dot{\gamma}_{ij}$ are arbitrary, subject to $\dot{\gamma}_{ij} = \dot{\gamma}_{ji}$, so it follows that

$$\tau^{ij} = \tfrac{1}{2}\rho\left(\frac{\partial E}{\partial \gamma_{ij}} + \frac{\partial E}{\partial \gamma_{ji}}\right). \tag{2.3.10}$$

In (2.3.9) and (2.3.10) differentiation with respect to γ_{ij} is understood to be performed holding all other components of the strain tensor constant, including γ_{ji} if $i \neq j$.

The function E is called the *elastic potential* of the body (per unit mass). It will be shown later in this chapter that E exists in at least two physical cases, either when the changes which take place are reversible and isothermal, or when they are reversible and isentropic, the corresponding functions E being, in general, different functions. These two possibilities cover many important practical problems.

It is convenient to express (2.3.10) in terms of an elastic potential W measured per unit volume of the unstrained body B_0. Since mass elements are conserved under deformation

$$\rho_0\sqrt{g} = \rho\sqrt{G}, \tag{2.3.11}$$

where ρ_0 is the density of B_0. Hence

$$\tau^{ij} = \frac{1}{2\sqrt{I_3}}\left(\frac{\partial W}{\partial \gamma_{ij}} + \frac{\partial W}{\partial \gamma_{ji}}\right), \tag{2.3.12}$$

where $W = \rho_0 E$ and $\qquad W = W(\gamma_{ij}).$ (2.3.13)

The relation between stress and strain in the theory which has been developed so far, is of such a form that an elastic potential exists with

† A more complete definition of elasticity will be given in § 2.9 which deals with thermodynamical equations.

the property (2.3.12). We may assume values for W and derive stress-strain relations from these equations. In the remainder of this section we consider bodies which are of constant density in the unstrained state, and which are *homogeneous* and *isotropic*. A body is said to be *elastically isotropic* or, more simply, *isotropic*, if the elastic potential W depends only on the three strain invariants I_1, I_2, I_3 and on scalar functions of the coordinates θ_i. If these scalar functions are constants the body is said to be *elastically homogeneous*. If therefore, the body is *homogeneous* and *isotropic*†

$$W = W(I_1, I_2, I_3). \tag{2.3.14}$$

From (2.3.12) and (2.3.14) we obtain the stress tensor in the form

$$\begin{aligned} 2\tau^{ij}\sqrt{I_3} &= \frac{\partial W}{\partial \gamma_{ij}} + \frac{\partial W}{\partial \gamma_{ji}} \\ &= \frac{\partial W}{\partial I_1}\left(\frac{\partial I_1}{\partial \gamma_{ij}} + \frac{\partial I_1}{\partial \gamma_{ji}}\right) + \frac{\partial W}{\partial I_2}\left(\frac{\partial I_2}{\partial \gamma_{ij}} + \frac{\partial I_2}{\partial \gamma_{ji}}\right) + \frac{\partial W}{\partial I_3}\left(\frac{\partial I_3}{\partial \gamma_{ij}} + \frac{\partial I_3}{\partial \gamma_{ji}}\right). \end{aligned} \tag{2.3.15}$$

With the help of (2.1.32) and (2.1.34) we have

$$\left.\begin{aligned} \frac{\partial I_1}{\partial \gamma_{ij}} + \frac{\partial I_1}{\partial \gamma_{ji}} &= 4g^{ij} \\ \frac{\partial I_2}{\partial \gamma_{ij}} + \frac{\partial I_2}{\partial \gamma_{ji}} &= 4(g^{ij}g^{rs} - g^{ir}g^{js})G_{rs} \\ \frac{\partial I_3}{\partial \gamma_{ij}} + \frac{\partial I_3}{\partial \gamma_{ji}} &= 4I_3\, G^{ij} \end{aligned}\right\}, \tag{2.3.16}$$

and therefore (2.3.15) becomes

$$\tau^{ij} = \Phi g^{ij} + \Psi B^{ij} + pG^{ij}, \tag{2.3.17}$$

where

$$\left.\begin{aligned} \Phi = \frac{2}{\sqrt{I_3}}\frac{\partial W}{\partial I_1}, \quad \Psi = \frac{2}{\sqrt{I_3}}\frac{\partial W}{\partial I_2}, \quad p = 2\sqrt{I_3}\frac{\partial W}{\partial I_3} \\ B^{ij} = I_1 g^{ij} - g^{ir}g^{js}G_{rs} = e^{irm}e^{jsn}g_{rs}\,G_{mn}/g \end{aligned}\right\}. \tag{2.3.18}$$

The three functions Φ, Ψ, and p depend only on I_1, I_2, and I_3 and are therefore scalar invariant functions.

If, in addition to being homogeneous and isotropic, the body B_0 is incompressible, the third invariant of strain I_3 is unity and we only know the value of the strain energy function W in terms of I_1 and I_2 when $I_3 = 1$. Thus

$$I_3 = 1, \quad W = W(I_1, I_2). \tag{2.3.19}$$

† A more complete discussion of the elastic potential for isotropic bodies, and for all the crystal classes, is given in the book by A. E. Green and J. Adkins, *Large Elastic Deformations* (Oxford, 1960).

The stress–strain equations (2.3.17) retain the same form but now

$$\Phi = 2\frac{\partial W}{\partial I_1}, \qquad \Psi = 2\frac{\partial W}{\partial I_2}, \tag{2.3.20}$$

and p cannot be evaluated from (2.3.18) since it is the value of the derivative $2(\partial W/\partial I_3)$ at $I_3 = 1$. The function p is an unknown scalar function† which represents a hydrostatic pressure, and p will be found from the equations of equilibrium and the boundary conditions.

We shall not be very much concerned here with particular forms for W but we mention that for certain incompressible rubber-like bodies an elastic potential of the form

$$W = C_1(I_1-3)+C_2(I_2-3), \tag{2.3.21}$$

where C_1, C_2 are constants, postulated by Mooney,‡ appears to be very suitable. The form (2.3.21) for W represents the first two terms in an expansion of W in a double power series of I_1-3, I_2-3 and the necessary and sufficient§ condition that $W \geqslant 0$ is that $C_1 \geqslant 0$, $C_2 \geqslant 0$. When $C_2 = 0$ the expression for W reduces to a form which Rivlin|| uses for a so-called neo-Hookean solid, and this form has also been obtained theoretically by Wall, Flory, Treloar, and others. The reader is referred to a book by Treloar¶ for more detailed discussion of the properties of rubber-like materials.

2.4. Alternative form for basic equations

For some purposes it is convenient to express the results of the previous two sections in alternative forms. The stress vector $\mathbf{t}$ is referred to a surface S at time t in the body B and is measured per unit area of S, while the stress tensor τ^{ij} is referred to θ_i-coordinates in B and is measured per unit area of these coordinate surfaces. The forces acting on an element of area in the θ_i-surface at a point θ_r in B are given by (2.2.15) and this may be expressed as

$$\mathbf{T}_i\, d\theta^j d\theta^k = {}_0\mathbf{t}_i \sqrt{(gg^{ii})}\, d\theta^j d\theta^k, \tag{2.4.1}$$

where, using (2.2.14), we have

$${}_0\mathbf{t}_i \sqrt{(gg^{ii})} = \mathbf{T}_i = \sqrt{(g)} s^{ij}\mathbf{G}_j, \quad s^{ij} = \tau^{ij}\sqrt{I_3}. \tag{2.4.2}$$

It follows from (2.4.1) that ${}_0\mathbf{t}_i$ is a stress vector acting across the θ_i-surface in B, but measured per unit area of the corresponding θ_i-surface

† See also § 2.9. ‡ M. Mooney, *J. appl. Phys.* **11** (1940) 582.

§ Further investigation is, however, needed before we can insist on the condition $W \geqslant 0$.

|| See references at beginning of Chapter 3.

¶ *The Physics of Rubber Elasticity*, 2nd ed. (Oxford, 1958). See also A. E. Green and J. E. Adkins, loc. cit., p. 64.

in B_0. Also, from (2.4.2), s^{ij} is a stress tensor referred to θ_i-coordinates in B but measured per unit area of the corresponding θ_i-surfaces in B_0. If S_0 is a surface in B_0 which becomes the surface S in B after deformation, and if ${}_0\mathbf{n}$ is a unit normal to S_0 such that

$$ {}_0\mathbf{n} = {}_0n_i\,\mathbf{g}^i = {}_0n^i\mathbf{g}_i, \tag{2.4.3}$$

then

$$ {}_0\mathbf{t} = {}_0n_i\,s^{ij}\mathbf{G}_j = {}_0n_i\,s_j^i\,\mathbf{G}^j = \frac{{}_0n_i\,\mathbf{T}_i}{\sqrt{g}}, \tag{2.4.4}$$

where ${}_0\mathbf{t}$ is the stress vector across the surface S measured per unit area of S_0.

From (2.1.25) and (2.1.27) we see that

$$ \mathbf{G}_i = \mathbf{g}_i + \mathbf{v}_{,i} = (\delta_i^r + v^r|_i)\mathbf{g}_r. \tag{2.4.5}$$

Hence equations (2.2.14) and (2.4.2) may be re-written as

$$ \mathbf{T}_i = \sqrt{(G)}\pi^{ij}\mathbf{g}_j = \sqrt{(g)}t^{ij}\mathbf{g}_j, \tag{2.4.6}$$

where

$$ \pi^{ij} = \tau^{ir}(\delta_r^j + v^j|_r), \qquad t^{ij} = s^{ir}(\delta_r^j + v^j|_r). \tag{2.4.7}$$

The stresses π^{ij} are measured per unit area of θ_i-surfaces in B but are referred to base vectors in B_0; the stresses t^{ij} are measured per unit area of θ_i-surfaces in B_0 and are referred to base vectors in B_0. These stresses are not symmetric. Using (2.2.14) and (2.4.2) we see that

$$ \pi^{ij}\mathbf{g}_j = \tau^{ij}\mathbf{G}_j, \qquad t^{ij}\mathbf{g}_j = s^{ij}\mathbf{G}_j, \tag{2.4.8}$$

so that, as τ^{ij} and s^{ij} are symmetric, the stress tensors π^{ij} and t^{ij} satisfy the conditions

$$ \left.\begin{aligned} \pi^{ir}\mathbf{g}_r\,.\,\mathbf{G}^j &= \pi^{jr}\mathbf{g}_r\,.\,\mathbf{G}^i \\ t^{ir}\mathbf{g}_r\,.\,\mathbf{G}^j &= t^{jr}\mathbf{g}_r\,.\,\mathbf{G}^i \end{aligned}\right\}. \tag{2.4.9}$$

Writing the body force $\mathbf{F}$ and acceleration vector $\mathbf{f}$ in terms of base vectors in B_0 in the forms

$$ \mathbf{F} = {}_0F_i\,\mathbf{g}^i = {}_0F^i\mathbf{g}_i, \qquad \mathbf{f} = {}_0f_i\,\mathbf{g}^i = {}_0f^i\mathbf{g}_i, \tag{2.4.10}$$

we see from (2.2.19) and (2.4.6) that

$$ t^{ij}|_i + \rho_0\,{}_0F^j = \rho_0\,{}_0f^j. \tag{2.4.11}$$

Equations of motion may also be found in terms of stresses π^{ij} and s^{ij} but are omitted.

From (2.3.12) and (2.4.2) we have

$$ \dot{W} = s^{ij}\dot{\gamma}_{ij}. \tag{2.4.12}$$

Also, using (2.4.2) and remembering that s^{ij} is symmetric

$$ \begin{aligned} s^{ij}\dot{\gamma}_{ij} &= \tfrac{1}{2}s^{ij}(\mathbf{G}_i\,.\,\dot{\mathbf{G}}_j + \dot{\mathbf{G}}_i\,.\,\mathbf{G}_j) \\ &= s^{ij}\mathbf{G}_j\,.\,\dot{\mathbf{G}}_i \\ &= s^{ij}\mathbf{G}_j\,.\,\dot{\mathbf{v}}_{,i} = \mathbf{T}_i\,.\,\dot{\mathbf{v}}_{,i}/\sqrt{g}. \end{aligned} \tag{2.4.13}$$

But $$\mathbf{v}_{,i} = v_j|_i \mathbf{g}^j, \qquad \dot{\mathbf{v}}_{,i} = \dot{v}_j|_i \mathbf{g}^j, \tag{2.4.14}$$
so that, with the help of (2.4.6), equation (2.4.12) becomes
$$\dot{W} = t^{ij}\dot{v}_j|_i. \tag{2.4.15}$$
Hence $$\pi^{ij}\sqrt{I_3} = t^{ij} = \frac{\partial W}{\partial v_j|_i}. \tag{2.4.16}$$

Conditions at the boundary surface of the body B at which the surface forces are prescribed have been given in (2.2.25)–(2.2.27). If the applied surface force is measured per unit area of the boundary of B_0 which corresponds to the boundary of B, then recalling (2.4.2) and (2.4.4), we have
$$_0\mathbf{t} = {}_0\mathbf{P}, \qquad {}_0n_i t^{ij} = {}_0P^j, \tag{2.4.17}$$
if $$_0\mathbf{P} = {}_0P^i \mathbf{g}_i = {}_0P_i \mathbf{g}^i. \tag{2.4.18}$$

In the rest of this chapter we consider the fundamental thermodynamical equations for a continuum and we use these to derive the basic constitutive relations for elasticity.

2.5. Further kinematics

We now return to the notation introduced at the beginning of § 2.1 and refer the motion of the body B to a fixed system of rectangular cartesian axes. We consider motions of the body B which differ from those given by (2.1.4) only by superposed rigid-body motions, at different times. Thus
$$y_i^* = c_i^* + Q_{ij}(y_j - c_j), \tag{2.5.1}$$
where c_i, c_i^* are vector functions of t and t^* $(= t+a)$ respectively, a is an arbitrary constant, and Q_{ij} is a proper orthogonal tensor which depends on t. In § 2.1 vectors and tensors are defined in terms of the motion (2.1.4) and we denote corresponding quantities defined from (2.5.1) by the same letters to which we add an asterisk. From (2.5.1) we have
$$w_i^* = \dot{c}_i^* + Q_{ij}(w_j - \dot{c}_j) + \Omega_{ir}(y_r^* - c_r^*), \tag{2.5.2}$$
where a dot denotes differentiation with respect to t or t^* holding x_j fixed and
$$\dot{Q}_{ij} = \Omega_{ir} Q_{rj}, \qquad \Omega_{ij} = -\Omega_{ji}. \tag{2.5.3}$$
From (2.5.2) we have
$$\frac{\partial w_i^*}{\partial y_j^*} = Q_{ir} Q_{js} \frac{\partial w_r}{\partial y_s} + \Omega_{ij}. \tag{2.5.4}$$
Hence
$$\left.\begin{aligned} d_{ij}^* &= Q_{ir} Q_{js} d_{rs} \\ \omega_{ij}^* &= Q_{ir} Q_{js} \omega_{rs} + \Omega_{ij} \end{aligned}\right\}. \tag{2.5.5}$$

Let ϕ be a function of t and y_k and let ϕ^* be its value regarded as a function of t^* and y_j^* when the motion of the body B is changed by superposed rigid-body motions. If

$$\phi(t, y_j) = \phi^*(t^*, y_j^*), \tag{2.5.6}$$

then
$$\dot{\phi} = \frac{\partial \phi}{\partial t} + w_i \frac{\partial \phi}{\partial y_i} = \frac{\partial \phi^*}{\partial t^*} + w_i^* \frac{\partial \phi^*}{\partial y_i^*} = \dot{\phi}^*. \tag{2.5.7}$$

2.6. Body and surface forces. Stress

Let w_i be an arbitrary velocity field of the continuum at time t. If F_i is a vector such that the scalar

$$F_i w_i \tag{2.6.1}$$

is a rate of work per unit mass at time t then F_i is called a body force per unit mass. The total rate of work of F_i distributed throughout a volume τ at time t is

$$\int_\tau \rho F_i w_i \, d\tau, \tag{2.6.2}$$

where ρ is density at time t.

Suppose A is a surface whose unit normal at a point y_i in a specified direction is n_k. If w_i is an arbitrary velocity field at time t and if t_i is a vector such that the scalar

$$t_i w_i \tag{2.6.3}$$

is a rate of work per unit area of A at time t, then t_i is called a surface force per unit area. The total rate of work of t_i distributed over a surface A at time t is

$$\int_A t_i w_i \, dA. \tag{2.6.4}$$

When the surface A coincides with the y_k-coordinate plane at the point y_i we denote the value of the surface force by σ_{ki} measured† per unit area of the y_k-plane, and we call σ_{ki} components of stress. The index k is not necessarily a tensor index at present but just denotes the plane over which the force acts.

2.7. The rate of work equation

We consider an arbitrary *fixed* surface A enclosing a volume τ of the continuum and we postulate a rate of change of energy and rate of work

† The physical components of stress $\sigma_{(ki)}$ in curvilinear coordinates, defined in (2.2.17), do not occur again and need not be confused with σ_{ki} used in the rest of this chapter.

balance at time t in the form

$$\frac{\partial}{\partial t}\int_{\tau}(\tfrac{1}{2}w_i w_i+U)\rho\, d\tau$$
$$= -\int_{A}(\tfrac{1}{2}w_i w_i+U)\rho n_k w_k\, dA+\int_{\tau}\rho(r+F_i w_i)\, d\tau+\int_{A}(t_i w_i-h)\, dA, \tag{2.7.1}$$

where the scalar U is internal energy per unit mass, r is a heat supply function per unit mass per unit time (due to heat sources and radiation from external sources), and the scalar h is the flux of heat across A measured per unit area of A per unit time. Equation (2.7.1) states that the rate of increase of kinetic energy and internal energy inside the surface A is equal to the flux of kinetic and internal energy across A plus the rate of work of body and surface forces and the contributions due to a volume distribution of heat supply and the flux of heat across A. Equation (2.7.1) may be replaced by

$$\int_{\tau}\{\rho(w_i \dot{w}_i+\dot{U})+(\tfrac{1}{2}w_i w_i+U)(\dot{\rho}+\rho d_{kk})\}\, d\tau$$
$$= \int_{\tau}\rho(r+F_i w_i)\, d\tau+\int_{A}(t_i w_i-h)\, dA. \tag{2.7.2}$$

We suppose that the continuum has arrived at the given state at time t through some prescribed motion. We consider a second motion which differs from the given motion only by a *constant* superposed rigid-body translational velocity.† We assume that ρ, U, h, r, t_i, and $F_i-\dot{w}_i$ are unaltered by such superposed rigid-body velocity. In view of (2.5.7) it follows that $\dot{\rho}$ and $\dot{U}$ are unaltered and, from (2.1.10), $\dot{w}_i$ is unaltered. The velocity w_i, however, is changed to w_i+a_i where a_i is an arbitrary constant. Equation (2.7.2) then holds when w_i is replaced by w_i+a_i, every other term being unaltered, so that, by subtraction

$$\left[\int_{\tau}\{\rho(F_i-\dot{w}_i)-w_i(\dot{\rho}+\rho d_{kk})\}\, d\tau+\int_{A}t_i\, dA\right]a_i-$$
$$-\tfrac{1}{2}a_i a_i\int_{\tau}(\dot{\rho}+\rho d_{kk})\, d\tau = 0, \tag{2.7.3}$$

for all a_i, the other quantities in (2.7.3) being independent of a_i. Replacing a_i by αa_i where α is an arbitrary scalar we see that

$$\int_{\tau}(\dot{\rho}+\rho d_{kk})\, d\tau = 0, \tag{2.7.4}$$

$$\int_{\tau}\{\rho(F_i-\dot{w}_i)-w_i(\dot{\rho}+\rho d_{kk})\}\, d\tau+\int_{A}t_i\, dA = 0, \tag{2.7.5}$$

† The independent thermodynamical variable, such as entropy S per unit mass, is unaltered.

for all arbitrary volumes τ. Hence, provided the integrand in (2.7.4) is a continuous function, we have

$$\dot{\rho}+\rho d_{kk} = 0. \tag{2.7.6}$$

Equation (2.7.6) is the equation of continuity and is equivalent to the statement that mass elements are conserved. Using (2.7.6) we see that (2.7.5) reduces to

$$\int\limits_{\tau} \rho(F_i-\dot{w}_i)\, d\tau+\int\limits_{A} t_i\, dA = 0 \tag{2.7.7}$$

for all arbitrary volumes τ. Equations (2.7.5), or equations (2.7.7) and (2.7.6) are equivalent to the equations

$$\frac{\partial}{\partial t}\int\limits_{\tau} \rho w_i\, d\tau = -\int\limits_{A} \rho w_i n_k w_k\, dA+\int\limits_{\tau} \rho F_i\, d\tau+\int\limits_{A} t_i\, dA, \tag{2.7.8}$$

which state that the rate of increase of momentum inside a fixed surface A is equal to the flux of momentum into A across its surface plus the total force acting on the material inside A. These are the usual Cauchy equations of motion and are equivalent to those postulated in (2.2.2).

We consider a tetrahedron element bounded by the coordinate planes at the point y_i and a plane whose unit normal is n_k measured from inside to outside of the tetrahedron. If we apply (2.7.7) to this tetrahedron and take the limit as the tetrahdron tends to zero with n_k being unaltered we have

$$t_i = n_k \sigma_{ki}, \tag{2.7.9}$$

provided the contributions from the volume integrals may be neglected compared with the surface integrals, in the limit. From (2.7.9) we see that σ_{ki} is a tensor with respect to both indices k, i under changes of rectangular cartesian axes, where the stresses in each coordinate system are associated with the three coordinate planes in that system. With the help of (2.7.9) equations (2.7.7) reduce to

$$\int\limits_{\tau} \left\{\rho(F_i-\dot{w}_i)+\frac{\partial \sigma_{ki}}{\partial y_k}\right\} d\tau = 0$$

for all arbitrary volumes τ. Provided the integrand is continuous we conclude that

$$\frac{\partial \sigma_{ki}}{\partial y_k}+\rho F_i = \rho \dot{w}_i, \tag{2.7.10}$$

the classical equations of motion.

With the help of (2.7.6), (2.7.9), and (2.7.10) the energy equation (2.7.2) reduces to

$$\int\limits_{\tau} \rho \dot{U}\, d\tau = \int\limits_{\tau} \left(\rho r+\sigma_{ki}\frac{\partial w_i}{\partial y_k}\right) d\tau-\int\limits_{A} h\, dA. \tag{2.7.11}$$

We apply this equation to a tetrahedron element bounded by the coordinate planes at the point y_i and a plane surface whose unit normal is n_k. Provided the contributions from the volume integrals may be neglected compared with the surface integrals in the limit as the tetrahedron tends to zero, we have

$$h = n_k Q_k, \tag{2.7.12}$$

where Q_k is the flux of heat across the y_k-plane at y_i, per unit area of this plane and per unit time. It follows from (2.7.12) that Q_k is a vector under changes of rectangular cartesian axes, where the heat flux in each coordinate system is associated with the coordinate planes in that system. If we substitute h from (2.7.12) into (2.7.11) we see that

$$\int_\tau \left(\rho\dot{U}-\rho r-\sigma_{ki}\frac{\partial w_i}{\partial y_k}+\frac{\partial Q_k}{\partial y_k}\right) d\tau = 0$$

for all arbitrary volumes τ. Provided that the integrand is continuous it follows that

$$\rho\dot{U}-\rho r-\sigma_{ki}\frac{\partial w_i}{\partial y_k}+\frac{\partial Q_k}{\partial y_k} = 0. \tag{2.7.13}$$

We consider a motion of the continuum which is such that the velocities differ from the velocities of the given motion only by a superposed uniform rigid-body angular velocity, the body having the same orientation in space at time t, and we assume that ρ, U, r, Q_k, and σ_{ki} are unaltered by such a rigid-body motion. From (2.5.4) we see that $\partial w_i/\partial y_k$ is replaced by $\partial w_i/\partial y_k+\Omega_{ik}$, where Ω_{ik} is an arbitrary constant skew-symmetric tensor. Hence, by subtraction,

$$\sigma_{ki}\Omega_{ik} = 0,$$

and therefore

$$\sigma_{ki} = \sigma_{ik} \tag{2.7.14}$$

since σ_{ik} is independent of Ω_{ik}. With the help of (2.7.14) equation (2.7.13) becomes

$$\rho\dot{U}-\rho r+\frac{\partial Q_k}{\partial y_k}-\sigma_{ki}d_{ik} = 0. \tag{2.7.15}$$

An examination of (2.7.15) suggests that under a general superposed rigid-body motion, in which (2.1.4) is replaced by (2.5.1), we assume that

$$\left.\begin{aligned} \rho^* = \rho, \quad U^* = U, \quad Q_k^* = Q_{kr}Q_r \\ \sigma_{ij}^* = Q_{ir}Q_{js}\sigma_{rs} \end{aligned}\right\}, \tag{2.7.16}$$

where an asterisk denotes the values of the functions corresponding to the motion (2.5.1). The equation (2.7.15) is then invariant under superposed rigid-body motions. The invariance conditions (2.7.16) under superposed rigid-body motions include the special conditions already imposed on ρ, U, Q_k, and σ_{ij}.

When we have a particular material for our continuum constitutive equations must be postulated for U, Q_k, and σ_{ij} and these can be reduced to canonical forms with the help of the invariance conditions (2.7.16). Further restrictions may be imposed by the entropy production inequality when used with (2.7.15) and we consider this inequality in the next section.

2.8. Entropy production inequality

We postulate an entropy production inequality

$$\frac{\partial}{\partial t}\int_\tau \rho S\,d\tau+\int_A \rho S n_k w_k\,dA-\int_\tau \rho\frac{r}{T}\,d\tau+\int_A \frac{h}{T}\,dA \geqslant 0 \tag{2.8.1}$$

for an arbitrary fixed surface A, where the scalar S is entropy per unit mass and the scalar T (>0) is local temperature. With the help of (2.7.6) and (2.7.12) we may obtain the equation

$$\rho T\dot{S}-\rho r+\frac{\partial Q_k}{\partial y_k}-\frac{Q_k}{T}\frac{\partial T}{\partial y_k} \geqslant 0 \tag{2.8.2}$$

from (2.8.1), provided the left-hand side of (2.8.2) is continuous.

If we substitute for ρr from (2.7.15) into (2.8.2) we have

$$\rho(T\dot{S}-\dot{U})+\sigma_{ki}d_{ik}-\frac{Q_k}{T}\frac{\partial T}{\partial y_k} \geqslant 0. \tag{2.8.3}$$

For some purposes it is convenient to use the Helmholtz free-energy function

$$A = U-TS \tag{2.8.4}$$

so that (2.8.3) becomes

$$-\rho(S\dot{T}+\dot{A})+\sigma_{ki}d_{ik}-\frac{Q_k}{T}\frac{\partial T}{\partial y_k} \geqslant 0. \tag{2.8.5}$$

2.9. Thermo-elasticity

We define an elastic continuum to be one for which a reference state x_i exists such that the following constitutive equations† hold at each point of the continuum and for all time t:

$$A = A\left(T, \frac{\partial y_i}{\partial x_j}\right), \tag{2.9.1}$$

$$S = S\left(T, \frac{\partial y_i}{\partial x_j}\right), \tag{2.9.2}$$

† If we adopt Truesdell's principle of equipresence then we assume that A, S, and σ_{ij} also depend on $\partial T/\partial y_i$. However, use of the energy equation and entropy production inequality shows that these functions then reduce to the forms assumed in (2.9.1)–(2.9.3). The initial density of the reference state should also be included as an independent variable in (2.9.1)–(2.9.4) but is omitted here for convenience. Throughout the book this density is taken to be constant.

$$\sigma_{ij} = \sigma_{ij}\left(T, \frac{\partial y_i}{\partial x_j}\right), \tag{2.9.3}$$

$$Q_k = Q_k\left(T, \frac{\partial T}{\partial y_r}, \frac{\partial y_p}{\partial x_q}\right). \tag{2.9.4}$$

For convenience we shall call the reference state of the continuum a body B_0 and the state at time t a body B. We observe that $y_i - x_i$ cannot occur explicitly in the constitutive equations, since these equations must be unaltered by superposed rigid-body translations. The functions A and S must satisfy the same invariance condition as U, being unaltered by superposed rigid-body motions. Hence

$$A\left(T, \frac{\partial y_i}{\partial x_j}\right) = A\left(T, Q_{ir}\frac{\partial y_r}{\partial x_j}\right) \tag{2.9.5}$$

for all proper orthogonal tensors Q_{ij}. Recalling (2.1.11) we choose the special value R_{ki} for Q_{ik} so that

$$A\left(T, \frac{\partial y_i}{\partial x_j}\right) = A(T, M_{ij}).$$

Since M_{ij} is a positive definite symmetric tensor we may replace A by the different function

$$A = A(T, E_{ij}), \tag{2.9.6}$$

where

$$E_{ij} = M_{ik} M_{kj} = \frac{\partial y_r}{\partial x_i}\frac{\partial y_r}{\partial x_j}. \tag{2.9.7}$$

The function (2.9.6) satisfies a condition of the form (2.9.5) for all proper orthogonal tensors Q_{ij}. We replace (2.9.6) by

$$A = A(T, e_{ij}), \tag{2.9.8}$$

where

$$2e_{ij} = E_{ij} - \delta_{ij}. \tag{2.9.9}$$

Also, using (2.1.7) and (2.1.13), we have

$$\dot{e}_{ij} = \frac{\partial y_r}{\partial x_i}\frac{\partial y_s}{\partial x_j} d_{rs}. \tag{2.9.10}$$

The tensor e_{ij} is usually called a strain tensor and is a measure of the difference of the squares of the line elements at corresponding points of the bodies B and B_0.

With the help of (2.9.2)–(2.9.4), (2.9.8), and (2.9.10), the inequality (2.8.5) becomes

$$\left\{\sigma_{ki} - \frac{\rho}{2}\frac{\partial y_i}{\partial x_r}\frac{\partial y_k}{\partial x_s}\left(\frac{\partial A}{\partial e_{rs}} + \frac{\partial A}{\partial e_{sr}}\right)\right\} d_{ik} - \rho\left(S + \frac{\partial A}{\partial T}\right)\dot{T} - \frac{Q_k}{T}\frac{\partial T}{\partial y_k} \geqslant 0, \tag{2.9.11}$$

if we recall that e_{rs}, d_{ik}, and σ_{ik} are symmetric. For a given deformation and temperature distribution (2.9.11) is valid for all arbitrary values of $\dot{T}$ and d_{rs}, subject to $d_{rs} = d_{sr}$. Both $\dot{T}$ and d_{rs} may be chosen arbitrarily in view of the presence of the heat radiation term r in the energy equation (2.7.15).† We take an arbitrary homogeneous temperature distribution so that (2.9.11) reduces to

$$\left\{\sigma_{ki} - \frac{\rho}{2}\frac{\partial y_i}{\partial x_r}\frac{\partial y_k}{\partial x_s}\left(\frac{\partial A}{\partial e_{rs}} + \frac{\partial A}{\partial e_{sr}}\right)\right\}d_{ik} - \rho\left(S + \frac{\partial A}{\partial T}\right)\dot{T} \geqslant 0, \tag{2.9.12}$$

for arbitrary $\dot{T}$ and d_{ik}. Then

$$S = -\frac{\partial A}{\partial T}, \tag{2.9.13}$$

$$\sigma_{ki} = \frac{\rho}{2}\frac{\partial y_i}{\partial x_r}\frac{\partial y_k}{\partial x_s}\left(\frac{\partial A}{\partial e_{rs}} + \frac{\partial A}{\partial e_{sr}}\right). \tag{2.9.14}$$

Also, (2.9.11) now reduces to

$$-Q_k\frac{\partial T}{\partial y_k} \geqslant 0. \tag{2.9.15}$$

Alternatively, from (2.9.13) and (2.8.4), we may, in general, express U in the form

$$U = U(S, e_{rs}), \tag{2.9.16}$$

with S and e_{rs} as independent variables, so that

$$T = \frac{\partial U}{\partial S}, \tag{2.9.17}$$

$$\sigma_{ki} = \frac{\rho}{2}\frac{\partial y_i}{\partial x_r}\frac{\partial y_k}{\partial x_s}\left(\frac{\partial U}{\partial e_{rs}} + \frac{\partial U}{\partial e_{sr}}\right). \tag{2.9.18}$$

With the help of these results the energy equation (2.7.15) reduces to

$$\rho T\dot{S} - \rho r + \frac{\partial Q_k}{\partial y_k} = 0. \tag{2.9.19}$$

If the continuum is incompressible then

$$\left|\frac{\partial y_i}{\partial x_j}\right| = 1 \tag{2.9.20}$$

and

$$\rho = \rho_0 \tag{2.9.21}$$

where ρ_0 is density of the body B_0. Then $\dot{\rho}$ is zero, and from (2.1.13) and (2.7.6) we have

$$d_{ii} = 0. \tag{2.9.22}$$

The inequality (2.9.12) is now subject to (2.9.22) so that, introducing

† When d_{rs}, $\dot{T}$, and the deformation and temperature are given, r is determined from (2.7.15) and body forces are determined from (2.7.10), if we use equations (2.9.1)–(2.9.4). See B. D. Coleman and W. Noll, *Archs ration. Mech. Analysis* **13** (1963) 167.

the scalar function p as a Lagrangian multiplier, we see that (2.9.14) and (2.9.18) are replaced by

$$\sigma_{ki} = p\delta_{ki}+\frac{\rho_0}{2}\frac{\partial y_i}{\partial x_r}\frac{\partial y_k}{\partial x_s}\left(\frac{\partial A}{\partial e_{rs}}+\frac{\partial A}{\partial e_{sr}}\right)$$

$$= p\delta_{ki}+\frac{\rho_0}{2}\frac{\partial y_i}{\partial x_r}\frac{\partial y_k}{\partial x_s}\left(\frac{\partial U}{\partial e_{rs}}+\frac{\partial U}{\partial e_{sr}}\right). \tag{2.9.23}$$

For a discussion of the heat conduction vector the reader is referred to the book by Green and Adkins.†

From (2.1.22), (2.9.7), and (2.9.9) we have

$$\gamma_{ij} = \frac{\partial x^r}{\partial \theta^i}\frac{\partial x^s}{\partial \theta^j}e_{rs}. \tag{2.9.24}$$

Since
$$\tau^{ij} = \frac{\partial \theta^i}{\partial y^r}\frac{\partial \theta^j}{\partial y^s}\sigma_{rs}, \tag{2.9.25}$$

it follows from (2.9.14) and (2.9.24) that

$$\tau^{ij} = \frac{1}{2\sqrt{I_3}}\left(\frac{\partial W}{\partial \gamma_{ij}}+\frac{\partial W}{\partial \gamma_{ji}}\right), \qquad W = \rho_0 A. \tag{2.9.26}$$

The result (2.3.12) is a special case of this formula when temperature T is constant. Alternatively, using (2.9.18) instead of (2.9.14), we obtain the first formula in (2.9.26) where W now stands for $\rho_0 U$. In this case we recover the result (2.3.12) when entropy S is constant. In view of (2.9.25) we may at once write the equation of motion (2.7.10) in the general form (2.2.23). Alternative forms for equations of motion and stress–strain relations can now be obtained as in §§ 2.2–2.4.

When the elastic body is initially homogeneous and isotropic we may obtain expressions for cartesian components of stress directly from (2.9.14) or from (2.3.17) and (2.9.25). Thus

$$\sigma_{ij} = p\delta_{ij}+(\Phi+I_1\Psi)C_{ij}-\Psi C_{ik}C_{kj}, \tag{2.9.27}$$

where
$$C_{ij} = \frac{\partial y_i}{\partial x_m}\frac{\partial y_j}{\partial x_m} \tag{2.9.28}$$

and p, Φ, Ψ are given by (2.3.18) if we restrict attention either to the case of constant entropy or constant temperature. Then, from (2.9.27), or using (2.9.23) directly, we see that (2.9.27) still holds when the body is incompressible, where Φ and Ψ are given by (2.3.20), and p is an arbitrary scalar function.

† loc. cit., p. 64.

Before proceeding to Chapter 3, in which we solve special problems for bodies which are homogeneous, isotropic, and incompressible, we close the present chapter by adding a summary of the more important notations and formulae.

2.10. Summary of notations and formulae

θ_i	Coordinate system which moves with the body.
g_{ij}, g^{ij}	Metric tensors of the unstrained body.
g	Determinant of the components g_{ij}.
$B^i\|_j$, $B^i_r\|_j$	This denotes covariant differentiation of any functions B^i,..., with respect to θ_j and g_{mn}.
G_{ij}, G^{ij}	Metric tensors of the strained body.
G	Determinant of the components G_{ij}.
$B^i\|\|_j$, $B^i_r\|\|_j$	This denotes covariant differentiation of any functions B^i,..., with respect to θ_j and G_{mn}.
ρ_0	Density of unstrained body.
ρ	Density of strained body.
$\mathbf{g}_i$, $\mathbf{g}^i$	Covariant and contravariant base vectors of the unstrained body.
$\mathbf{G}_i$, $\mathbf{G}^i$	Covariant and contravariant base vectors of the strained body.
$\mathbf{v}$	Displacement vector.
v_i, v^i	Covariant and contravariant components of $\mathbf{v}$ referred to $\mathbf{g}^i$, $\mathbf{g}_i$ respectively.
V_i, V^i	Covariant and contravariant components of $\mathbf{v}$ referred to $\mathbf{G}^i$, $\mathbf{G}_i$ respectively.
γ_{ij}	Strain tensor.
$\mathbf{t}$	Stress vector, per unit area of deformed body.
${}_0\mathbf{t}$	Stress vector, per unit area of undeformed body.
$\mathbf{t}_i$	Stress vectors for each coordinate surface, per unit area of deformed body.
${}_0\mathbf{t}_i$	Stress vectors for each coordinate surface, per unit area of undeformed body.
$\mathbf{n}$	Unit normal to surface over which $\mathbf{t}$ acts.
${}_0\mathbf{n}$	Unit normal to surface in undeformed body which corresponds to surface over which ${}_0\mathbf{t}$ acts.
n_i, n^i	Covariant and contravariant components of $\mathbf{n}$ referred to $\mathbf{G}^i$, $\mathbf{G}_i$ respectively.
${}_0n_i$, ${}_0n^i$	Covariant and contravariant components of ${}_0\mathbf{n}$ referred to $\mathbf{g}^i$, $\mathbf{g}_i$ respectively.

P	Surface force vector, per unit area of deformed body.
$_0\mathbf{P}$	Surface force vector, per unit area of undeformed body.
P_i, P^i	Covariant and contravariant components of **P** referred to $\mathbf{G}^i$, $\mathbf{G}_i$ respectively.
$_0P_i$, $_0P^i$	Covariant and contravariant components of $_0\mathbf{P}$ referred to $\mathbf{g}^i$, $\mathbf{g}_i$ respectively.
F	Body force vector per unit mass.
f	Acceleration vector.
F_i, F^i; f_i, f^i	Covariant and contravariant components of **F** and **f** referred to $\mathbf{G}^i$, $\mathbf{G}_i$ respectively.
$_0F_i$, $_0F^i$; $_0f_i$, $_0f^i$	Covariant and contravariant components of **F** and **f** referred to $\mathbf{g}^i$, $\mathbf{g}_i$ respectively.
E	Elastic potential per unit mass.
W	Elastic potential per unit volume of the unstrained body.
U	Internal energy per unit mass.
A	Helmholtz free energy per unit mass.

A list of the more important strain and stress relations follows, assuming that the deformed body is either at constant temperature or at constant entropy.

$$\gamma_{ij} = \tfrac{1}{2}(G_{ij}-g_{ij}). \tag{2.10.1}$$

$$\mathbf{t}_i\sqrt{G^{ii}} = \tau^{ij}\mathbf{G}_j = \pi^{ij}\mathbf{g}_j, \quad \tau^{ij} = \tau^{ji}. \tag{2.10.2}$$

$$_0\mathbf{t}_i\sqrt{g^{ii}} = s^{ij}\mathbf{G}_j = t^{ij}\mathbf{g}_j, \quad s^{ij} = s^{ji}. \tag{2.10.3}$$

$$\mathbf{T}_i = \sqrt{(G)}\tau^{ij}\mathbf{G}_j = \sqrt{(g)}s^{ij}\mathbf{G}_j = \sqrt{(G)}\pi^{ij}\mathbf{g}_j = \sqrt{(g)}t^{ij}\mathbf{g}_j. \tag{2.10.4}$$

$$\mathbf{t} = \tau^{ij}n_i\mathbf{G}_j = \pi^{ij}n_i\mathbf{g}_j = \frac{n_i\mathbf{T}_i}{\sqrt{G}}. \tag{2.10.5}$$

$$_0\mathbf{t} = s^{ij}{}_0n_i\mathbf{G}_j = t^{ij}{}_0n_i\mathbf{g}_j = \frac{_0n_i\mathbf{T}_i}{\sqrt{g}}. \tag{2.10.6}$$

$$\tau^{ij}\sqrt{I_3} = s^{ij} = \frac{1}{2}\left(\frac{\partial W}{\partial\gamma_{ij}}+\frac{\partial W}{\partial\gamma_{ji}}\right). \tag{2.10.7}$$

$$\pi^{ij}\sqrt{I_3} = t^{ij} = \frac{\partial W}{\partial v_j|_i}. \tag{2.10.8}$$

$$t^{ij} = s^{ir}(\delta_r^j+v^j|_r), \qquad \pi^{ij} = \tau^{ir}(\delta_r^j+v^j|_r). \tag{2.10.9}$$

$$\mathbf{T}_{i,i}+\rho\mathbf{F}\sqrt{G} = \rho\mathbf{f}\sqrt{G}. \tag{2.10.10}$$

$$\tau^{ij}\|_i+\rho F^j = \rho f^j. \tag{2.10.11}$$

$$t^{ij}|_i+\rho_0\,{}_0F^j = \rho_0\,{}_0f^j. \tag{2.10.12}$$

$$\mathbf{t} = \mathbf{P}, \qquad \tau^{ij}n_i = P^j. \tag{2.10.13}$$

$$_0\mathbf{t} = {}_0\mathbf{P}, \qquad {}_0n_it^{ij} = {}_0P^j. \tag{2.10.14}$$

For homogeneous isotropic bodies

$$W = W(I_1, I_2, I_3), \tag{2.10.15}$$

and

$$\tau^{ij} = \Phi g^{ij} + \Psi B^{ij} + pG^{ij}, \tag{2.10.16}$$

where

$$\left.\begin{aligned} \Phi = \frac{2}{\sqrt{I_3}}\frac{\partial W}{\partial I_1}, \quad \Psi = \frac{2}{\sqrt{I_3}}\frac{\partial W}{\partial I_2}, \quad p = 2\sqrt{I_3}\frac{\partial W}{\partial I_3} \\ B^{ij} = I_1 g^{ij} - g^{ir}g^{js}G_{rs} = \frac{1}{g}e^{irm}e^{jsn}g_{rs}G_{mn} \end{aligned}\right\}, \tag{2.10.17}$$

and

$$I_1 = g^{rs}G_{rs}, \qquad I_2 = g_{rs}G^{rs}I_3, \qquad I_3 = G/g. \tag{2.10.18}$$

When the body is incompressible

$$I_3 = 1, \qquad W = W(I_1, I_2). \tag{2.10.19}$$

The stress–strain relations (2.10.16) still have the same form but now

$$\Phi = 2\frac{\partial W}{\partial I_1}, \qquad \Psi = 2\frac{\partial W}{\partial I_2}, \tag{2.10.20}$$

and p is an arbitrary scalar function.

3 Finite Deformation: Solution of Special Problems

In this chapter we consider the solution of certain problems for bodies which are homogeneous and isotropic and we are mostly concerned with incompressible bodies.† The method used throughout is to assume that the strained body B is obtained from the unstrained body B_0 by suitable displacements and then to find the system of forces which is necessary to maintain the equilibrium of B. This means that we shall find in each problem a state of equilibrium under a certain system of forces, but the possibility of alternative states of equilibrium under the same system of forces should be borne in mind. Readers may refer to the summary at the end of Chapter 2 for the various formulae and equations which are used.

3.1. Uniform extensions

In the first part of this section we are concerned with a compressible material and we take fixed cartesian axes (x, y, z) to define the unstrained body B_0, and identify the θ_i-coordinates with (x,y,z) so that

$$\theta_1 = x_1 = x, \qquad \theta_2 = x_2 = y, \qquad \theta_3 = x_3 = z. \tag{3.1.1}$$

We now consider a strain in which a unit cube of B_0 whose sides are parallel to the axes is deformed into a cube of dimensions λ_1, λ_2, λ_3 parallel to the x-, y-, z-axes respectively. The coordinates of the strained body B may be referred to a fixed cartesian set of axes y_i, which coincide with the axes (x,y,z), so that

$$y_1 = \lambda_1 x, \qquad y_2 = \lambda_2 y, \qquad y_3 = \lambda_3 z. \tag{3.1.2}$$

† Most of the problems in this chapter have been discussed by R. S. Rivlin, *Phil. Trans. R. Soc.* **A240** (1948) 459, 491, 509; **A241** (1948) 379; **A242** (1949) 173; *Proc. R. Soc.* **A195** (1949) 463; *Proc. Camb. phil. Soc. math. phys. Sci.* **45** (1949) 485. The method of solution which is given here, however, differs from that used by Rivlin. See also A. E. Green and R. T. Shield, *Proc. R. Soc.* **A202** (1950) 407. Further developments are given by A. E. Green and J. E. Adkins, loc. cit., p. 64.

The components of the metric tensor for the unstrained and strained bodies are respectively

$$g_{ik} = g^{ik} = \delta_{ik}, \qquad g = 1, \tag{3.1.3}$$

$$G_{ik} = \begin{bmatrix} \lambda_1^2 & 0 & 0 \\ 0 & \lambda_2^2 & 0 \\ 0 & 0 & \lambda_3^2 \end{bmatrix}, \qquad G^{ik} = \begin{bmatrix} \lambda_1^{-2} & 0 & 0 \\ 0 & \lambda_2^{-2} & 0 \\ 0 & 0 & \lambda_3^{-2} \end{bmatrix}, \qquad G = \lambda_1^2\lambda_2^2\lambda_3^2, \tag{3.1.4}$$

and the strain invariants (2.10.18) are

$$I_1 = \lambda_1^2+\lambda_2^2+\lambda_3^2, \qquad I_2 = \lambda_1^2\lambda_2^2+\lambda_2^2\lambda_3^2+\lambda_3^2\lambda_1^2, \qquad I_3 = \lambda_1^2\lambda_2^2\lambda_3^2. \tag{3.1.5}$$

It follows from (2.10.17) that Φ, Ψ, p are constants and that

$$B^{ik} = \begin{bmatrix} \lambda_2^2+\lambda_3^2 & 0 & 0 \\ 0 & \lambda_3^2+\lambda_1^2 & 0 \\ 0 & 0 & \lambda_1^2+\lambda_2^2 \end{bmatrix}. \tag{3.1.6}$$

Hence the stress components τ^{ik} which are given by (2.10.16) are constants, and the stress equations of equilibrium (2.10.11) are satisfied when body forces are absent. Using (2.10.16) the physical stress components (2.2.17) are found to be

$$\left.\begin{aligned} \sigma_{11} &= \lambda_1^2\tau^{11} = \lambda_1^2\Phi+\lambda_1^2(\lambda_2^2+\lambda_3^2)\Psi+p \\ \sigma_{22} &= \lambda_2^2\tau^{22} = \lambda_2^2\Phi+\lambda_2^2(\lambda_3^2+\lambda_1^2)\Psi+p \\ \sigma_{33} &= \lambda_3^2\tau^{33} = \lambda_3^2\Phi+\lambda_3^2(\lambda_1^2+\lambda_2^2)\Psi+p \\ \sigma_{12} &= \sigma_{23} = \sigma_{31} = 0 \end{aligned}\right\}. \tag{3.1.7}$$

With the help of (3.1.5) these may be rewritten in the form

$$\left.\begin{aligned} \sigma_{11} &= \lambda_1^2\Phi-\lambda_2^2\lambda_3^2\Psi+p' \\ \sigma_{22} &= \lambda_2^2\Phi-\lambda_3^2\lambda_1^2\Psi+p' \\ \sigma_{33} &= \lambda_3^2\Phi-\lambda_1^2\lambda_2^2\Psi+p' \end{aligned}\right\}, \tag{3.1.8}$$

where

$$p' = p+\Psi I_2. \tag{3.1.9}$$

When the body is incompressible the stresses (3.1.8) may be further reduced to

$$\sigma_{11} = \lambda_1^2\Phi-\Psi/\lambda_1^2+p', \tag{3.1.10}$$

etc., since

$$\lambda_1^2\lambda_2^2\lambda_3^2 = 1. \tag{3.1.11}$$

For the particular case of simple extension under a force parallel to the x-axis, $\lambda_2 = \lambda_3$, $\sigma_{22} = \sigma_{33} = 0$ and, hence, for a compressible body, from (3.1.8),

$$p' = -\lambda_2^2\Phi+\lambda_1^2\lambda_2^2\Psi, \tag{3.1.12}$$

and

$$\sigma_{11} = (\lambda_1^2-\lambda_2^2)(\Phi+\lambda_2^2\Psi). \tag{3.1.13}$$

For an incompressible body $\lambda_2^2 = 1/\lambda_1$ and therefore

$$\sigma_{11} = (\lambda_1^2-1/\lambda_1)(\Phi+\Psi/\lambda_1). \tag{3.1.14}$$

3.2. Simple shear

Consider a body B_0 which, in the undeformed state, is a cuboid whose faces are

$$x = \pm a, \qquad y = \pm b, \qquad z = \pm c, \tag{3.2.1}$$

where a, b, c are constants and where the coordinate system in B_0 is the same as that given by (3.1.1). If B_0 is subject to a simple shearing deformation, in which each point of the material moves parallel to the x-axis by an amount which is proportional to its y-coordinate, then the final coordinates of the point which is initially at (x, y, z), referred to cartesian axes y_i which coincide with the axes (x, y, z), are

$$y_1 = x+Ky, \qquad y_2 = y, \qquad y_3 = z, \tag{3.2.2}$$

where K is a constant. A section by the plane z = constant of the unstrained and strained bodies is shown in Fig. 3.1.

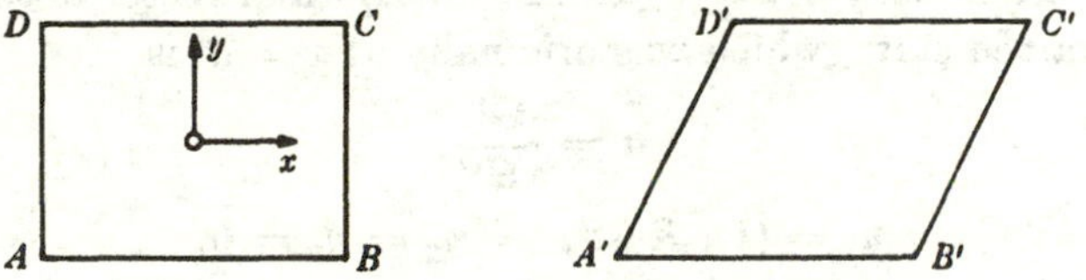

FIG. 3.1. Simple shear of cuboid.

The components of the metric tensor of B_0 are again given by (3.1.3), and the components of the metric tensor of B are now

$$G_{ik} = \begin{bmatrix} 1 & K & 0 \\ K & 1+K^2 & 0 \\ 0 & 0 & 1 \end{bmatrix}, \qquad G^{ik} = \begin{bmatrix} 1+K^2 & -K & 0 \\ -K & 1 & 0 \\ 0 & 0 & 1 \end{bmatrix}, \qquad G = 1, \tag{3.2.3}$$

and the strain invariants (2.10.18) are

$$I_1 = I_2 = 3+K^2, \qquad I_3 = 1. \tag{3.2.4}$$

Also

$$B^{ik} = \begin{bmatrix} 2+K^2 & -K & 0 \\ -K & 2 & 0 \\ 0 & 0 & 2+K^2 \end{bmatrix}. \tag{3.2.5}$$

Once again it follows that Φ, Ψ, and p are constants and, if the body forces are zero, the equations of equilibrium are satisfied.

From (2.10.16) the components of stress are found to be

$$\left.\begin{aligned} \tau^{11} &= \Phi+(2+K^2)\Psi+(1+K^2)p \\ \tau^{22} &= \Phi+2\Psi+p \\ \tau^{33} &= \Phi+(2+K^2)\Psi+p \\ \tau^{12} &= -K(\Psi+p) \\ \tau^{23} &= \tau^{31} = 0 \end{aligned}\right\}. \tag{3.2.6}$$

It should be remembered that these contravariant components of the stress tensor refer to surfaces $\theta_i =$ constant in B, i.e. to surfaces which in the unstrained state are parallel to the coordinate planes x, y, z.

Since $\tau^{23} = \tau^{31} = 0$ there is no stress across any plane $z =$ constant provided $\tau^{33} = 0$, that is, for an incompressible body,

$$p = -\Phi-(2+K^2)\Psi, \tag{3.2.7}$$

and in this case the non-zero stress components (3.2.6) reduce to

$$\left.\begin{aligned} \tau^{11} &= -K^2\{\Phi+(2+K^2)\Psi\} \\ \tau^{22} &= -K^2\Psi \\ \tau^{12} &= K\{\Phi+(1+K^2)\Psi\} \end{aligned}\right\}. \tag{3.2.8}$$

The surface forces which must be applied to maintain this state of stress may be found from (2.10.13) and (3.2.8). Consider first the surfaces which were originally at $x = \pm a$. The unit normal vector to the surface in the strained state, which was originally at $x = a$, is

$$\mathbf{n} = \frac{\mathbf{G}^1}{\sqrt{G^{11}}}, \tag{3.2.9}$$

so that $$n_1 = (1+K^2)^{-\frac{1}{2}}, \qquad n_2 = n_3 = 0. \tag{3.2.10}$$

The surface force acting over the surface $x = a$ is given from (2.10.13) and (3.2.10) by the vector

$$\mathbf{P} = \frac{\tau^{1k}\mathbf{G}_k}{(1+K^2)^{\frac{1}{2}}}, \tag{3.2.11}$$

so that the tangential component of surface force (along $B'C'$) is

$$\frac{\mathbf{P}.\mathbf{G}_2}{\sqrt{G_{22}}} = \frac{K\tau^{11}+(1+K^2)\tau^{12}}{1+K^2} = \frac{K(\Phi+\Psi)}{1+K^2}, \tag{3.2.12}$$

and the normal component of surface force is†

$$\frac{\mathbf{P}.\mathbf{G}^1}{\sqrt{G^{11}}} = \frac{\tau^{11}}{1+K^2} = -\frac{K^2\{\Phi+(2+K^2)\Psi\}}{1+K^2}. \tag{3.2.13}$$

The unit normal vector to the surface in the strained state which was originally at $y = b$ is defined by

$$\mathbf{n} = \frac{\mathbf{G}^2}{\sqrt{G^{22}}}, \qquad n_1 = 0, \qquad n_2 = 1, \qquad n_3 = 0, \tag{3.2.14}$$

and the surface forces are defined by the vector

$$\mathbf{P} = \tau^{2k}\mathbf{G}_k.$$

Hence the tangential component of force (along $D'C'$) is

$$\frac{\mathbf{P}.\mathbf{G}_1}{\sqrt{G_{11}}} = \tau^{21}+K\tau^{22} = K(\Phi+\Psi), \tag{3.2.15}$$

† The tangential and normal components are here the resolved parts of **P** in these directions.

and the normal component is†

$$\frac{\mathbf{P}\cdot\mathbf{G}^2}{\sqrt{G^{22}}} = \tau^{22} = -K^2\Psi. \tag{3.2.16}$$

We see that the state of shear described by equation (3.2.2) cannot be maintained by tangential tractions applied to the surfaces $y = \pm b$, $x = \pm a$, as in the case of small deformations of infinitesimal elasticity. It should be remembered that the components of surface traction (3.2.12), (3.2.13) and (3.2.15), (3.2.16) refer to unit area of the surface measured in the deformed state of the body.

The problem of biaxial shear in the (x, y) and (x, z) planes can be solved in a similar manner.

We have now considered the simplest problems of finite deformation for which rectangular coordinates are suitable and it will be observed that we have specified the state of deformation by reference to coordinates of points of the initial or unstrained body B_0. It would be just as easy, however, in these examples, to consider coordinates of points in the strained body B as reference points, since we know the form of B. In the next sections we consider problems for which cylindrical polar or spherical polar coordinates are more suitable and in these problems it is slightly more convenient to take points in B as reference points, since the stress equations of equilibrium have their simplest form in these coordinates. This is only possible because we can guess the form of B. When the form of B is not known *ab initio* (as is the case in general problems) we then usually need reference points in B_0. Unless stated otherwise attention in the following sections is confined to incompressible bodies so that $I_3 = 1$.

3.3. Pure torsion of a right circular cylinder

We consider a cylinder which in the undeformed state has a length l and radius a, and we consider a torsional deformation of the cylinder in which planes normal to the axis remain plane and suffer only a pure rotation proportional to their distance from one end of the cylinder. The body remains cylindrical in form and in the strained state will have a length l and radius a in view of incompressibility.

The curvilinear coordinate system θ_i in the strained state of the cylinder is taken to be a system of cylindrical polar coordinates (r, θ, z) so that

$$\left.\begin{aligned} \theta_1 = r, \quad \theta_2 = \theta, \quad \theta_3 = z \\ y_1 = r\cos\theta, \quad y_2 = r\sin\theta, \quad y_3 = z \end{aligned}\right\}, \tag{3.3.1}$$

† The tangential and normal components are here the resolved parts of **P** in these directions.

and the cylinder occupies the region $r \leqslant a$, $0 \leqslant z \leqslant l$ in both the strained and unstrained states. The y_i-axes are taken to coincide with the x_i-axes and, with the above assumptions about the deformation, the point (r, θ, z) in the deformed state was originally at $(r, \theta-\psi z, z)$, where ψ is a constant. Hence

$$x_1 = r\cos(\theta-\psi z), \quad x_2 = r\sin(\theta-\psi z), \quad x_3 = z. \tag{3.3.2}$$

From (3.3.1) and (3.3.2) the components of the metric tensors are found to be

$$G_{ik} = \begin{bmatrix} 1 & 0 & 0 \\ 0 & r^2 & 0 \\ 0 & 0 & 1 \end{bmatrix}, \qquad G^{ik} = \begin{bmatrix} 1 & 0 & 0 \\ 0 & \frac{1}{r^2} & 0 \\ 0 & 0 & 1 \end{bmatrix}, \tag{3.3.3}$$

$$g_{ik} = \begin{bmatrix} 1 & 0 & 0 \\ 0 & r^2 & -\psi r^2 \\ 0 & -\psi r^2 & 1+\psi^2 r^2 \end{bmatrix}, \qquad g^{ik} = \begin{bmatrix} 1 & 0 & 0 \\ 0 & \psi^2+\frac{1}{r^2} & \psi \\ 0 & \psi & 1 \end{bmatrix}. \tag{3.3.4}$$

Since
$$G = g = r^2 \tag{3.3.5}$$
the incompressibility condition $I_3 = 1$ is satisfied and the strain invariants I_1, I_2 are, from (2.10.18),

$$I_1 = I_2 = 3+\psi^2 r^2, \tag{3.3.6}$$

and therefore Φ and Ψ are functions of r only. Also

$$B^{ik} = \begin{bmatrix} 2+\psi^2 r^2 & 0 & 0 \\ 0 & \psi^2+\frac{2}{r^2} & \psi \\ 0 & \psi & 2 \end{bmatrix}. \tag{3.3.7}$$

Hence, from (2.10.16) and (3.3.3) to (3.3.7),

$$\left.\begin{aligned} \tau^{11} &= \Phi+(2+\psi^2 r^2)\Psi+p \\ r^2\tau^{22} &= (1+\psi^2 r^2)\Phi+(2+\psi^2 r^2)\Psi+p \\ \tau^{33} &= \Phi+2\Psi+p \\ \tau^{23} &= \psi(\Phi+\Psi), \qquad \tau^{31} = \tau^{12} = 0 \end{aligned}\right\}. \tag{3.3.8}$$

When the strained cylinder is in equilibrium, and the body forces are zero, the equations of equilibrium (2.10.11) become

$$\tau^{ik}\|_i = 0,$$

or
$$\tau^{ik}{}_{,i}+\Gamma^i_{ir}\tau^{rk}+\Gamma^k_{ir}\tau^{ir} = 0, \tag{3.3.9}$$

and the only non-zero Christoffel symbols, derived from the metric tensor of the strained body B, are

$$\Gamma^1_{22} = -r, \qquad \Gamma^2_{12} = \Gamma^2_{21} = \frac{1}{r}. \tag{3.3.10}$$

From (3.3.8)–(3.3.10) we obtain

$$\left.\begin{aligned} \frac{\partial}{\partial r}\{\Phi+(2+\psi^2 r^2)\Psi+p\}-\psi^2 r\Phi = 0 \\ \frac{\partial p}{\partial \theta} = \frac{\partial p}{\partial z} = 0 \end{aligned}\right\}. \tag{3.3.11}$$

Hence p depends only on r and is determined, apart from an arbitrary constant, by the equation

$$p = -\Phi-(2+\psi^2 r^2)\Psi+\psi^2 \int^r r\Phi\, dr. \tag{3.3.12}$$

The unit normal to the curved surface of the cylinder is directed along the vector $\mathbf{G}^1$ so that

$$\mathbf{n} = \frac{\mathbf{G}^1}{\sqrt{G^{11}}} = \mathbf{G}^1, \qquad n_1 = 1, \qquad n_2 = n_3 = 0. \tag{3.3.13}$$

Hence, if the curved surface $r = a$ of the cylinder is free from applied stress the boundary condition (2.10.13) becomes, on using (3.3.13),

$$\tau^{1k} = 0 \quad (r = a;\ k = 1,2,3). \tag{3.3.14}$$

This condition is satisfied identically when $k = 2$, 3, and, when $k = 1$, it follows from (3.3.8) that

$$p = -\Phi-(2+\psi^2 r^2)\Psi \quad (r = a).$$

Hence, (3.3.12) may now be written

$$p = -\Phi-(2+\psi^2 r^2)\Psi+\psi^2 \int_a^r r\Phi\, dr. \tag{3.3.15}$$

At the end $z = l$ of the cylinder the unit normal is parallel to $\mathbf{G}^3$ and therefore

$$\mathbf{n} = \frac{\mathbf{G}^3}{\sqrt{G^{33}}} = \mathbf{G}^3, \qquad n_3 = 1, \qquad n_1 = n_2 = 0. \tag{3.3.16}$$

The forces which must be applied at $z = l$ are therefore, from (2.10.13),

$$\left.\begin{aligned} P^1 = 0, \qquad P^2 = \tau^{23} = \psi(\Phi+\Psi) \\ P^3 = \tau^{33} = \psi^2\left(\int_a^r r\Phi\, dr - r^2\Psi\right) \end{aligned}\right\}. \tag{3.3.17}$$

Similar forces, only in opposite directions, must be applied over the end $z = 0$.

The physical components of stress at any point may now be written down from (2.2.17), (3.3.8), and (3.3.15). Using an obvious notation

$$\left.\begin{aligned}\sigma_{11} &= \tau^{11} = \psi^2 \int_a^r r\Phi\, dr\\ \sigma_{22} &= r^2\tau^{22} = \psi^2\left(\int_a^r r\Phi\, dr + r^2\Phi\right)\\ \sigma_{33} &= \tau^{33} = \psi^2\left(\int_a^r r\Phi\, dr - r^2\Psi\right)\\ \sigma_{23} &= r\tau^{23} = \psi r(\Phi+\Psi), \quad \sigma_{31} = \sigma_{12} = 0\end{aligned}\right\}. \tag{3.3.18}$$

The distributed surface tractions have physical components $(0, rP^2, P^3)$ over an end of the cylinder, and are statically equivalent to a couple M about the axis of the cylinder and a force N parallel to the axis, where

$$\left.\begin{aligned}M &= 2\pi \int_0^a r^3P^2\, dr = 2\pi\psi \int_0^a r^3(\Phi+\Psi)\, dr\\ N &= 2\pi \int_0^a rP^3\, dr = 2\pi\psi^2 \int_0^a r\, dr\left(\int_a^r x\Phi\, dx - r^2\Psi\right)\end{aligned}\right\}. \tag{3.3.19}$$

These formulae may be written in alternative forms. We see from (3.3.6) that I_1, I_2 depend only on r and therefore W is a function of r. Hence, writing $R = r^2$, we have

$$I_1 = I_2 = 3+\psi^2R,$$

$$2\frac{dW}{dR} = 2\psi^2\left(\frac{\partial W}{\partial I_1}+\frac{\partial W}{\partial I_2}\right) = \psi^2(\Phi+\Psi),$$

and the formulae (3.3.19) for the resultant force and couple become

$$M = \frac{2\pi}{\psi}\int_0^{a^2} R\frac{dW}{dR}\, dR = \frac{2\pi}{\psi}\left(a^2(W)_{r=a} - \int_0^{a^2} W\, dR\right), \tag{3.3.20}$$

$$\begin{aligned}N &= -\tfrac{1}{2}\pi\psi^2 \int_0^{a^2} R(\Phi+2\Psi)\, dR\\ &= -\tfrac{1}{2}\psi M - \tfrac{1}{2}\pi\psi^2 \int_0^{a^2} R\Psi\, dR\\ &= -\psi M + \tfrac{1}{2}\pi\psi^2 \int_0^{a^2} R\Phi\, dR.\end{aligned} \tag{3.3.21}$$

This completes one of the simplest problems associated with a circular cylinder. We now go on to consider more general types of displacements which can be sustained by circular cylindrical tubes. These displacements may conveniently be divided into two groups which will be examined in §§ 3.4 to 3.6 and §§ 3.7 to 3.8.

3.4. Simultaneous extension, inflation, and torsion of a cylindrical tube

We consider a circular cylindrical tube which in the unstrained state is of length l, and has external and internal radii a_1, a_2 respectively. We shall assume that the tube is strained by the following successive displacements:

(i) a uniform simple extension of extension ratio λ, parallel to the axis of the tube;

(ii) a uniform inflation of the tube in which its length remains constant and its external and internal radii change to $r_1 = \mu_1 a_1$ and $r_2 = \mu_2 a_2$ respectively;

(iii) a uniform simple torsion in which planes perpendicular to the axis of the tube are rotated in their own plane through an angle proportional to the distance of the plane considered from one end, the constant of proportionality being ψ.

We take as reference frame cylindrical polar coordinates (r, θ, z) in the deformed body which are defined in (3.3.1). Then the point (r, θ, z) was initially at the point $(\rho, \theta-\psi z, z/\lambda)$, where ρ is a function of r only, and the points of the undeformed body are given by

$$x_1 = rQ\cos(\theta-\psi z), \qquad x_2 = rQ\sin(\theta-\psi z), \qquad x_3 = z/\lambda, \tag{3.4.1}$$

where, for convenience, we have put

$$\rho(r) = rQ(r). \tag{3.4.2}$$

The components G_{ik}, G^{ik} of the metric tensor of the strained body are again given by (3.3.3) and the components g_{ik} for the unstrained body, obtained from (3.4.1), are

$$g_{ik} = \begin{bmatrix} (Q+rQ_r)^2 & 0 & 0 \\ 0 & r^2Q^2 & -\psi r^2Q^2 \\ 0 & -\psi r^2Q^2 & \psi^2r^2Q^2+1/\lambda^2 \end{bmatrix}, \tag{3.4.3}$$

where $Q_r = dQ/dr$. The incompressibility condition $I_3 = 1$ gives

$$g = G = r^2,$$

or

$$(Q+rQ_r)rQ = \lambda r, \tag{3.4.4}$$

and, integrating, this becomes

$$\rho = rQ = \{\lambda(r^2+K)\}^{\frac{1}{2}}, \tag{3.4.5}$$

where K is a constant. Since $\rho = a_1$ when $r = \mu_1 a_1$, and $\rho = a_2$ when $r = \mu_2 a_2$,

$$\lambda K = a_1^2(1-\lambda\mu_1^2) = a_2^2(1-\lambda\mu_2^2). \tag{3.4.6}$$

If λ is positive K may be either positive or negative according as $\lambda\mu_1^2$ is less or greater than unity. If λ is negative, then K is negative and $\mu_2^2 a_2^2 > \mu_1^2 a_1^2$ or $r_2 > r_1$. This means that the tube is turned inside out.

If λ is positive, then r^2+K is positive for all values of r between r_2 and r_1, and if λ is negative, r^2+K is negative over this range of r.

Using (3.4.3) and (3.4.5) we find that

$$g_{ik} = \begin{bmatrix} \lambda^2/Q^2 & 0 & 0 \\ 0 & r^2Q^2 & -\psi r^2Q^2 \\ 0 & -\psi r^2Q^2 & \psi^2r^2Q^2+1/\lambda^2 \end{bmatrix},$$

$$g^{ik} = \begin{bmatrix} Q^2/\lambda^2 & 0 & 0 \\ 0 & \psi^2\lambda^2+\dfrac{1}{r^2Q^2} & \psi\lambda^2 \\ 0 & \psi\lambda^2 & \lambda^2 \end{bmatrix}, \tag{3.4.7}$$

and from (2.10.18), (3.3.3), and (3.4.7) the strain invariants are

$$I_1 = \frac{Q^2}{\lambda^2}+\frac{1}{Q^2}+\lambda^2+\psi^2\lambda^2r^2, \qquad I_2 = \frac{\lambda^2}{Q^2}+Q^2+\frac{1}{\lambda^2}+\psi^2r^2Q^2. \tag{3.4.8}$$

Also, from (2.10.17),

$$B^{ik} = \begin{bmatrix} \dfrac{1}{\lambda^2}+Q^2+\psi^2r^2Q^2 & 0 & 0 \\ 0 & \dfrac{1}{\lambda^2r^2}+\dfrac{\lambda^2}{r^2Q^2}+\psi^2Q^2 & \psi Q^2 \\ 0 & \psi Q^2 & Q^2+\dfrac{\lambda^2}{Q^2} \end{bmatrix}. \tag{3.4.9}$$

The components of the stress tensor τ^{ik} are now obtained from (2.10.16), (3.3.3), (3.4.7), and (3.4.9) in the form

$$\left.\begin{aligned} \tau^{11} &= \frac{Q^2\Phi}{\lambda^2}+\left(\frac{1}{\lambda^2}+Q^2+\psi^2r^2Q^2\right)\Psi+p \\ \tau^{22} &= \left(\psi^2\lambda^2+\frac{1}{r^2Q^2}\right)\Phi+\left(\frac{1}{\lambda^2r^2}+\frac{\lambda^2}{r^2Q^2}+\psi^2Q^2\right)\Psi+\frac{p}{r^2} \\ \tau^{33} &= \lambda^2\Phi+\left(Q^2+\frac{\lambda^2}{Q^2}\right)\Psi+p \\ \tau^{23} &= \psi\lambda^2\Phi+\psi Q^2\Psi, \qquad \tau^{31} = \tau^{12} = 0 \end{aligned}\right\}. \tag{3.4.10}$$

The equations of equilibrium again reduce to (3.3.9) where the Christoffel symbols are given by (3.3.10), and hence, from these equations and (3.4.10),

$$\left.\begin{aligned}\frac{\partial}{\partial r}\left\{p+\frac{Q^2\Phi}{\lambda^2}+\left(\frac{1}{\lambda^2}+Q^2+\psi^2r^2Q^2\right)\Psi\right\}+\\ +\left(\frac{Q^2}{\lambda^2}-\frac{1}{Q^2}-\psi^2\lambda^2r^2\right)\frac{\Phi}{r}+\left(Q^2-\frac{\lambda^2}{Q^2}\right)\frac{\Psi}{r}=0\\ \frac{\partial p}{\partial\theta}=\frac{\partial p}{\partial z}=0\end{aligned}\right\}. \tag{3.4.11}$$

The pressure p, therefore, depends only on r and is given by

$$p=-\frac{Q^2\Phi}{\lambda^2}-\left(\frac{1}{\lambda^2}+Q^2+\psi^2r^2Q^2\right)\Psi-L(r)+H, \tag{3.4.12}$$

where H is a constant and

$$L(r)=\int_{r_1}^{r}\left\{\left(\frac{Q^2}{\lambda^2}-\frac{1}{Q^2}-\psi^2\lambda^2r^2\right)\Phi+\left(Q^2-\frac{\lambda^2}{Q^2}\right)\Psi\right\}\frac{dr}{r}. \tag{3.4.13}$$

From (3.4.8),

$$\left.\begin{aligned}\frac{dI_1}{dr}&=\frac{2}{r}\left(\frac{1}{\lambda}-\frac{Q^2}{\lambda^2}-\frac{\lambda}{Q^4}+\frac{1}{Q^2}+\psi^2\lambda^2r^2\right)\\ \frac{dI_2}{dr}&=\frac{2}{r}\left(\lambda-Q^2-\frac{\lambda^3}{Q^4}+\frac{\lambda^2}{Q^2}+\psi^2\lambda r^2\right)\end{aligned}\right\}, \tag{3.4.14}$$

and hence (3.4.13) may be expressed in the alternative form

$$\begin{aligned}L(r)&=\frac{1}{2}\int_{r_1}^{r}\left\{\left(-\frac{dI_1}{dr}+\frac{2}{\lambda r}-\frac{2\lambda}{rQ^4}\right)\Phi+\left(-\frac{dI_2}{dr}+\frac{2\lambda}{r}-\frac{2\lambda^3}{rQ^4}+2\lambda\psi^2r\right)\Psi\right\}dr\\ &=W(r_1)-W(r)+\int_{r_1}^{r}\left\{\left(\frac{1}{\lambda}-\frac{\lambda}{Q^4}\right)\Phi+\left(\lambda-\frac{\lambda^3}{Q^4}+\lambda\psi^2r^2\right)\Psi\right\}\frac{dr}{r}.\end{aligned} \tag{3.4.15}$$

Using the value of p given by (3.4.12), the non-zero components of the stress tensor in (3.4.10) become

$$\left.\begin{aligned}\tau^{11}&=H-L(r)\\ r^2\tau^{22}&=H-L(r)+\left(\frac{1}{Q^2}-\frac{Q^2}{\lambda^2}+\psi^2\lambda^2r^2\right)\Phi+\left(\frac{\lambda^2}{Q^2}-Q^2\right)\Psi\\ \tau^{33}&=H-L(r)+\left(\lambda^2-\frac{Q^2}{\lambda^2}\right)\Phi+\left(\frac{\lambda^2}{Q^2}-\frac{1}{\lambda^2}-\psi^2r^2Q^2\right)\Psi\\ \tau^{23}&=\psi\lambda^2\Phi+\psi Q^2\Psi\end{aligned}\right\}. \tag{3.4.16}$$

The unit normal to the surface $r = r_1$ of the deformed cylinder is again given by (3.3.13) so that the components of the stress vector acting over this cylinder, namely,

$$P^k = \tau^{ik}n_i$$

reduce to

$$P^1 = (\tau^{11})_{r=r_1} = H - L(r_1) = H, \qquad P^2 = P^3 = 0.$$

If there is no traction on this surface

$$H = 0. \tag{3.4.17}$$

The unit normal to the surface $r = r_2$ of the deformed cylinder has the opposite sign to that given in (3.3.13), so that, over this surface

$$-P^1 = (\tau^{11})_{r=r_2} = H - L(r_2), \qquad P^2 = P^3 = 0. \tag{3.4.18}$$

The surface traction on the inner surface $r = r_2$ therefore reduces to a (physical) radial component R_{ν_2}, measured positively along the outward drawn normal to this surface, where

$$R_{\nu_2} = -L(r_2) \tag{3.4.19}$$

if we use the condition (3.4.17). If the traction over this surface is zero

$$L(r_2) = 0. \tag{3.4.20}$$

On the plane end of the tube, the normal to which is $\mathbf{G}^3$, we have

$$\left.\begin{aligned} P^1 &= \tau^{31} = 0. \qquad P^2 = \tau^{32} = \psi\lambda^2\Phi + \psi Q^2\Psi \\ P^3 &= \tau^{33} = H - L(r) + \left(\lambda^2 - \frac{Q^2}{\lambda^2}\right)\Phi + \left(\frac{\lambda^2}{Q^2} - \frac{1}{\lambda^2} - \psi^2 r^2 Q^2\right)\Psi \end{aligned}\right\}. \tag{3.4.21}$$

The physical components of stress are $(0, rP^2, P^3)$ so that the resultant couple M about the axis of the tube is

$$M = 2\pi \int_{r_2}^{r_1} (rP^2)r^2\,dr = 2\pi\psi \int_{r_2}^{r_1} r^3(\lambda^2\Phi + Q^2\Psi)\,dr, \tag{3.4.22}$$

and the resultant longitudinal force is

$$\begin{aligned} N &= 2\pi \int_{r_2}^{r_1} P^3 r\,dr \\ &= -2\pi \int_{r_2}^{r_1} L(r) r\,dr + 2\pi \int_{r_2}^{r_1} \left\{\left(\lambda^2 - \frac{Q^2}{\lambda^2}\right)\Phi + \left(\frac{\lambda^2}{Q^2} - \frac{1}{\lambda^2} - \psi^2 r^2 Q^2\right)\Psi\right\} r\,dr \end{aligned}$$

assuming that $H = 0$. An integration by parts gives

$$-2\int_{r_2}^{r_1} L(r)r\,dr = r_2^2 L(r_2) + \int_{r_2}^{r_1} r^2 \frac{dL(r)}{dr}\,dr$$

$$= r_2^2 L(r_2) + \int_{r_2}^{r_1} \left\{\left(\frac{Q^2}{\lambda^2} - \frac{1}{Q^2} - \psi^2\lambda^2 r^2\right)\Phi + \left(Q^2 - \frac{\lambda^2}{Q^2}\right)\Psi\right\} r\,dr;$$

hence, when $R_{\nu_2} = 0$,

$$N = \pi \int_{r_2}^{r_1} \left\{\left(2\lambda^2 - \frac{Q^2}{\lambda^2} - \frac{1}{Q^2} - \psi^2\lambda^2 r^2\right)\Phi + \left(\frac{\lambda^2}{Q^2} - \frac{2}{\lambda^2} + Q^2 - 2\psi^2 r^2 Q^2\right)\Psi\right\} r\,dr. \tag{3.4.23}$$

In the more general case when $R_{\nu_2} \neq 0$,

$$N = \pi \int_{r_2}^{r_1} \left\{\left(2\lambda^2 - \frac{Q^2}{\lambda^2} - \frac{1}{Q^2} - \psi^2\lambda^2 r^2\right)\Phi + \left(\frac{\lambda^2}{Q^2} - \frac{2}{\lambda^2} + Q^2 - 2\psi^2 r^2 Q^2\right)\Psi\right\} r\,dr -$$

$$-\pi r_2^2 \int_{r_2}^{r_1} \left\{\left(\frac{Q^2}{\lambda^2} - \frac{1}{Q^2} - \psi^2\lambda^2 r^2\right)\Phi + \left(Q^2 - \frac{\lambda^2}{Q^2}\right)\Psi\right\} \frac{dr}{r}. \tag{3.4.24}$$

3.5. The deformation of a cylindrical rod

If we take $a_2 = 0$ in the previous section we obtain the solution of the problem of simultaneous extension and torsion of a circular cylinder. Since ρ must then vanish when $r = 0$ we see from (3.4.5) and (3.4.6) that

$$K = 0, \qquad Q = \sqrt{\lambda}, \qquad \mu_1 = 1/\sqrt{\lambda}. \tag{3.5.1}$$

Equations (3.4.22), (3.4.24) yield

$$\left.\begin{aligned} M &= 2\pi\psi\lambda \int_0^{r_1} (\lambda\Phi + \Psi) r^3\,dr \\ N &= 2\pi\left(\lambda - \frac{1}{\lambda^2}\right) \int_0^{r_1} (\lambda\Phi + \Psi) r\,dr - \pi\lambda\psi^2 \int_0^{r_1} (\lambda\Phi + 2\Psi) r^3\,dr \end{aligned}\right\}. \tag{3.5.2}$$

From (3.4.8),

$$\left.\begin{aligned} I_1 &= \lambda^2 + \frac{2}{\lambda} + \psi^2\lambda^2 r^2, \qquad I_2 = 2\lambda + \frac{1}{\lambda^2} + \psi^2\lambda r^2 \\ \frac{dI_1}{dR} &= \psi^2\lambda^2, \qquad \frac{dI_2}{dR} = \psi^2\lambda, \qquad R = r^2 \end{aligned}\right\}, \tag{3.5.3}$$

and therefore

$$2\frac{dW}{dR} = \Phi\frac{dI_1}{dR}+\Psi\frac{dI_2}{dR} = \psi^2\lambda(\lambda\Phi+\Psi). \tag{3.5.4}$$

Thus equations (3.5.2) become

$$M = \frac{2\pi}{\psi}\int_0^{r_1^2}\frac{dW}{dR}R\,dR = \frac{2\pi}{\psi}\left(r_1^2(W)_{r=r_1}-\int_0^{r_1^2}W(R)\,dR\right), \tag{3.5.5}$$

$$\begin{aligned}N &= \frac{2\pi}{\psi^2}\left(1-\frac{1}{\lambda^3}\right)\int_0^{r_1^2}\frac{dW}{dR}\,dR-\pi\int_0^{r_1^2}\frac{dW}{dR}R\,dR-\frac{\pi\lambda\psi^2}{2}\int_0^{r_1^2}\Psi R\,dR\\ &= \frac{2\pi}{\psi^2}\left(1-\frac{1}{\lambda^3}\right)\{(W)_{r=r_1}-(W)_{r=0}\}-\pi\left(r_1^2(W)_{r=r_1}-\int_0^{r_1^2}W\,dR\right)-\\ &\qquad -\frac{\pi\lambda\psi^2}{2}\int_0^{r_1^2}\Psi R\,dR\\ &= -\tfrac{1}{2}\psi M+\frac{2\pi}{\psi^2}\left(1-\frac{1}{\lambda^3}\right)\{(W)_{r=r_1}-(W)_{r=0}\}-\frac{\pi\lambda\psi^2}{2}\int_0^{r_1^2}\Psi R\,dR.\end{aligned} \tag{3.5.6}$$

When $\lambda = 1$, i.e. the torsion is unaccompanied by extension, equations (3.5.5) and (3.5.6) reduce to (3.3.20) and (3.3.21) respectively.

If $\psi = 0$, i.e. the torsion is zero, then $M = 0$ and, from the second of equations (3.5.2),

$$N = \pi a_1^2\left(\lambda-\frac{1}{\lambda^2}\right)\left(\Phi+\frac{\Psi}{\lambda}\right)_{\psi=0}, \tag{3.5.7}$$

since Φ and Ψ are constants when $\psi = 0$. Also, from the first of equations (3.5.2),

$$\left(\frac{M}{\psi}\right)_{\psi=0} = \tfrac{1}{2}\pi a_1^4\left(\Phi+\frac{\Psi}{\lambda}\right)_{\psi=0}. \tag{3.5.8}$$

Equations (3.5.7) and (3.5.8) yield

$$\left(\frac{Na_1^2}{M/\psi}\right)_{\psi=0} = 2\left(\lambda-\frac{1}{\lambda^2}\right). \tag{3.5.9}$$

This law relates the force necessary to produce a large simple extension, with the torsional modulus for a small torsion superposed on that simple extension, and is independent of the particular form of the strain energy function.

3.6. Tube turned inside out

When $\psi = 0$ and there is no traction on the curved surfaces of the tube the condition (3.4.20) gives, on using (3.4.13),

$$\int_{r_1}^{r_2} \left(\frac{Q^2}{\lambda^2}-\frac{1}{Q^2}\right)(\Phi+\lambda^2\Psi)\,\frac{dr}{r} = 0. \tag{3.6.1}$$

This equation may be written in a different form. From (3.4.14), when $\psi = 0$,

$$\frac{dI_2}{dr} = \lambda^2\frac{dI_1}{dr} = \frac{2\lambda^2}{r}\left(\frac{Q^2}{\lambda^2}-\frac{1}{Q^2}\right)\left(\frac{\lambda}{Q^2}-1\right),$$

and, from (3.4.5),

$$\frac{\lambda}{Q^2}-1 = -\frac{K}{r^2+K} = -\frac{K\lambda}{\rho^2}.$$

Hence, (3.6.1) reduces to

$$\int_{r_1}^{r_2} \rho^2\frac{dW}{dr}\,dr = 0 \quad \text{or} \quad \int_{a_2}^{a_1} \rho^2\frac{dW}{d\rho}\,d\rho = 0, \tag{3.6.2}$$

and integrating by parts this becomes

$$a_1^2(W)_{\rho=a_1}-a_2^2(W)_{\rho=a_2} = \int_{a_2^2}^{a_1^2} W\,dR, \tag{3.6.3}$$

where $R = \rho^2$.

If, in addition, the total force on the ends of the cylinder is zero, $N = 0$, and hence, from (3.4.23),

$$\int_{r_2}^{r_1} \left\{\left(2\lambda^2-\frac{Q^2}{\lambda^2}-\frac{1}{Q^2}\right)\Phi+\left(\frac{\lambda^2}{Q^2}-\frac{2}{\lambda^2}+Q^2\right)\Psi\right\}r\,dr = 0. \tag{3.6.4}$$

Equation (3.6.4), together with one of the equations (3.6.1), (3.6.2), or (3.6.3), are the equations for the determination of the extension ratio and changes in the radii of the tube when it is turned inside out and rests in a deformed state under the action of no forces over the curved surfaces and no resultant force over the ends of the cylinder.

For further discussion of this problem when the strain energy function takes the special form (2.3.21) the reader is referred to the original paper by Rivlin.† In this paper Rivlin also discusses the simultaneous torsion and extension of a tube using the special strain energy function (2.3.21).

† *Phil. Trans. R. Soc.* **A242** (1949) 173.

3.7. Simultaneous extension, inflation, and shear of a cylindrical tube

We now suppose that the body described in § 3.4 is subjected to the following successive deformations:

(i) a uniform simple extension of extension ratio λ;

(ii) a uniform inflation of the tube in which its length remains constant and its external and internal radii change to $r_1 = \mu_1 a_1$ and $r_2 = \mu_2 a_2$ respectively;

(iii) a simple shear of the tube about its axis, in which each point moves about the axis through an angle ϕ which is dependent only on the radial position of the point;

(iv) a simple shear of the tube, in which each point moves parallel to the axis of the tube through a distance w which depends only upon the radial position of the point.

We again take cylindrical polar coordinates (r, θ, z) in the deformed body as defined in (3.3.1), with corresponding components of the metric tensor given by (3.3.3). The initial coordinates are then (ρ, ψ, ζ) where

$$\rho = \rho(r) = rQ(r), \quad \theta = \psi + \phi(r), \quad z = \lambda\zeta + w(r), \tag{3.7.1}$$

and hence

$$x_1 = rQ(r)\cos(\theta - \phi), \quad x_2 = rQ(r)\sin(\theta - \phi), \quad x_3 = \{z - w(r)\}/\lambda. \tag{3.7.2}$$

From (3.7.2) we have

$$g_{ik} = \begin{bmatrix} (Q+rQ_r)^2 + r^2Q^2\phi_r^2 + w_r^2/\lambda^2 & -r^2Q^2\phi_r & -w_r/\lambda^2 \\ -r^2Q^2\phi_r & r^2Q^2 & 0 \\ -w_r/\lambda^2 & 0 & 1/\lambda^2 \end{bmatrix},$$

and the incompressibility condition $g = G = r^2$ gives

$$Q(Q + rQ_r) = \lambda. \tag{3.7.3}$$

As in § 3.4 this leads to

$$\rho = rQ = \{\lambda(r^2 + K)\}^{\frac{1}{2}}, \tag{3.7.4}$$

where K is given by (3.4.6). Using (3.7.3) and (3.7.4) we find that the components g_{ik}, g^{ik} of the metric tensor of the unstrained body become

$$g_{ik} = \begin{bmatrix} \dfrac{\lambda^2}{Q^2} + r^2Q^2\phi_r^2 + \dfrac{w_r^2}{\lambda^2} & -r^2Q^2\phi_r & -\dfrac{w_r}{\lambda^2} \\ -r^2Q^2\phi_r & r^2Q^2 & 0 \\ -\dfrac{w_r}{\lambda^2} & 0 & \dfrac{1}{\lambda^2} \end{bmatrix}, \tag{3.7.5}$$

$$g^{ik} = \begin{bmatrix} \dfrac{Q^2}{\lambda^2} & \dfrac{Q^2\phi_r}{\lambda^2} & \dfrac{Q^2 w_r}{\lambda^2} \\ \dfrac{Q^2\phi_r}{\lambda^2} & \dfrac{1}{r^2Q^2}+\dfrac{Q^2\phi_r^2}{\lambda^2} & \dfrac{Q^2\phi_r w_r}{\lambda^2} \\ \dfrac{Q^2 w_r}{\lambda^2} & \dfrac{Q^2\phi_r w_r}{\lambda^2} & \lambda^2+\dfrac{Q^2 w_r^2}{\lambda^2} \end{bmatrix}. \tag{3.7.6}$$

Also, from (2.10.18), (3.3.3), (3.7.5), and (3.7.6) we have

$$\left.\begin{aligned} I_1 &= \frac{1}{Q^2}+\lambda^2+\frac{Q^2}{\lambda^2}+\frac{r^2Q^2\phi_r^2}{\lambda^2}+\frac{Q^2w_r^2}{\lambda^2} \\ &= \frac{1}{Q^2}+\lambda^2+\frac{Q^2}{\lambda^2}+r^2\phi_\rho^2+w_\rho^2 \\ I_2 &= Q^2+\frac{1}{\lambda^2}+\frac{\lambda^2}{Q^2}+r^2Q^2\phi_r^2+\frac{w_r^2}{\lambda^2} \\ &= Q^2+\frac{1}{\lambda^2}+\frac{\lambda^2}{Q^2}+\lambda^2r^2\phi_\rho^2+\frac{w_\rho^2}{Q^2} \end{aligned}\right\}, \tag{3.7.7}$$

where the alternative forms for I_1, I_2 are obtained by using, from (3.7.4), the results

$$\phi_r = \frac{\lambda\phi_\rho}{Q}, \qquad w_r = \frac{\lambda w_\rho}{Q}. \tag{3.7.8}$$

The components of the contravariant tensor B^{ik} are found, from (2.10.17), to be

$$\left.\begin{aligned} &B^{11} = \frac{1}{\lambda^2}+Q^2, \qquad B^{12} = Q^2\phi_r, \qquad B^{13} = \frac{w_r}{\lambda^2} \\ &B^{22} = \frac{\lambda^2}{r^2Q^2}+\frac{1}{\lambda^2r^2}+Q^2\phi_r^2+\frac{w_r^2}{\lambda^2r^2} \\ &B^{33} = Q^2+\frac{\lambda^2}{Q^2}+r^2Q^2\phi_r^2+\frac{w_r^2}{\lambda^2}, \qquad B^{23} = 0 \end{aligned}\right\}. \tag{3.7.9}$$

Hence, from the stress–strain relations (2.10.16) we obtain

$$\left.\begin{aligned} \tau^{11} &= \frac{Q^2\Phi}{\lambda^2}+\left(\frac{1}{\lambda^2}+Q^2\right)\Psi+p \\ r^2\tau^{22} &= \left(\frac{1}{Q^2}+\frac{r^2Q^2\phi_r^2}{\lambda^2}\right)\Phi+\left(\frac{\lambda^2}{Q^2}+\frac{1}{\lambda^2}+r^2Q^2\phi_r^2+\frac{w_r^2}{\lambda^2}\right)\Psi+p \\ \tau^{33} &= \left(\lambda^2+\frac{Q^2w_r^2}{\lambda^2}\right)\Phi+\left(Q^2+\frac{\lambda^2}{Q^2}+r^2Q^2\phi_r^2+\frac{w_r^2}{\lambda^2}\right)\Psi+p \\ \tau^{12} &= \frac{Q^2\phi_r\Phi}{\lambda^2}+Q^2\phi_r\Psi \\ \tau^{31} &= \frac{Q^2w_r\Phi}{\lambda^2}+\frac{w_r\Psi}{\lambda^2}, \qquad \tau^{23} = \frac{Q^2\phi_r w_r\Phi}{\lambda^2} \end{aligned}\right\}, \tag{3.7.10}$$

if p is a function of r only.

In the absence of body forces the equations of equilibrium are equations (3.3.9) and, using (3.3.10) and noting that the components of the stress tensor are functions of r only, we find that

$$\left.\begin{aligned} \frac{d\tau^{11}}{dr}+\frac{\tau^{11}-r^2\tau^{22}}{r} &= 0 \\ \frac{d\tau^{12}}{dr}+\frac{3\tau^{12}}{r} &= 0 \\ \frac{d\tau^{13}}{dr}+\frac{\tau^{13}}{r} &= 0 \end{aligned}\right\}. \tag{3.7.11}$$

The last two equations give

$$r^3\tau^{12} = \frac{2B}{\lambda^2}, \qquad r\tau^{13} = \frac{2D}{\lambda^2}, \tag{3.7.12}$$

where B, D are constants. Equations (3.7.10) and (3.7.12) yield

$$\left.\begin{aligned} \phi_r &= \frac{2B}{r^3Q^2(\Phi+\lambda^2\Psi)} \\ w_r &= \frac{2D}{r(Q^2\Phi+\Psi)} \end{aligned}\right\}. \tag{3.7.13}$$

It is seen from (3.7.7) that I_1, I_2 depend on ϕ_r, w_r and since, in general, Φ and Ψ are functions of I_1, I_2, equations (3.7.13) are non-linear differential equations for the determination of ϕ and w as functions of r. We must therefore assume definite forms for the strain energy function W in order to make further progress. If W is given by (2.3.21) then Φ and Ψ are constants and are given by

$$\Phi = 2C_1, \qquad \Psi = 2C_2.$$

From (3.7.4),

$$\frac{dr}{dQ} = \frac{rQ}{\lambda-Q^2} = -\frac{r^3Q}{\lambda K}, \tag{3.7.14}$$

and so (3.7.13) becomes

$$\frac{d\phi}{dQ} = -\frac{B}{\lambda K(C_1+\lambda^2C_2)Q}, \qquad \frac{dw}{dr} = \frac{Dr}{\lambda(r^2+K)C_1+C_2r^2},$$

and therefore

$$\left.\begin{aligned} \phi &= -\frac{B}{\lambda K(C_1+\lambda^2C_2)}\ln\frac{Q}{Q_2}+B' \\ w &= \frac{D}{2(\lambda C_1+C_2)}\ln\left\{\frac{(C_1Q^2+C_2)r^2}{(C_1Q_2^2+C_2)r_2^2}\right\}+D' \end{aligned}\right\}, \tag{3.7.15}$$

where B' and D' are constants and where

$$a_2 = Q_2r_2. \tag{3.7.16}$$

If we now suppose that w and ϕ are zero at $r = r_2$ and that $w = w_0$, $\phi = \phi_0$ when $r = r_1$ we obtain

$$\left.\begin{aligned} B' = 0, \qquad D' = 0, \\ B = -\frac{\lambda K(C_1+\lambda^2 C_2)\phi_0}{\ln(Q_1/Q_2)}, \qquad D = \frac{2(\lambda C_1+C_2)w_0}{\ln\left\{\dfrac{(C_1 Q_1^2+C_2)r_1^2}{(C_1 Q_2^2+C_2)r_2^2}\right\}}\end{aligned}\right\}, \tag{3.7.17}$$

where
$$a_1 = Q_1 r_1. \tag{3.7.18}$$

From (3.7.15) and (3.7.17) we have

$$\phi = \phi_0' \ln\frac{Q}{Q_2}, \qquad w = w_0' \ln\left\{\frac{(Q^2C_1+C_2)r^2}{(Q_2^2 C_1+C_2)r_2^2}\right\}, \tag{3.7.19}$$

where we have put

$$\phi_0' = \frac{\phi_0}{\ln(Q_1/Q_2)}, \qquad w_0' = \frac{w_0}{\ln\left\{\dfrac{(C_1 Q_1^2+C_2)r_1^2}{(C_1 Q_2^2+C_2)r_2^2}\right\}}. \tag{3.7.20}$$

Substituting from (3.7.17) and (3.7.20) in (3.7.13), we obtain

$$\phi_r = -\frac{\lambda K\phi_0'}{r^3 Q^2}, \qquad w_r = \frac{2(\lambda C_1+C_2)w_0'}{r(C_1 Q^2+C_2)}. \tag{3.7.21}$$

The first of equations (3.7.11) yields

$$\frac{dp}{dr} = -\frac{d}{dr}(\tau^{11}-p)+\frac{r^2\tau^{22}-\tau^{11}}{r}$$

and with (3.7.10) this becomes

$$p = -2\left\{\frac{C_1 Q^2}{\lambda^2}+C_2\left(\frac{1}{\lambda^2}+Q^2\right)\right\}+$$
$$+2\int^r \left\{\left(\frac{1}{Q^2}+\frac{r^2Q^2\phi_r^2}{\lambda^2}-\frac{Q^2}{\lambda^2}\right)(C_1+C_2\lambda^2)+\frac{C_2 w_r^2}{\lambda^2}\right\}\frac{dr}{r}. \tag{3.7.22}$$

Introducing the expressions (3.7.21) into (3.7.22) and performing the integrations with the help of (3.7.14), we obtain

$$p = \left(\frac{C_1+C_2\lambda^2}{\lambda^2}\right)\{2\lambda\ln Q+Q^2-\phi_0'^2(Q^2-2\lambda\ln Q)\}-$$
$$-\frac{4C_2(\lambda C_1+C_2)w_0'^2}{\lambda^2 r^2(C_1 Q^2+C_2)}-2\left\{\frac{C_1 Q^2}{\lambda^2}+C_2\left(\frac{1}{\lambda^2}+Q^2\right)\right\}+F, \tag{3.7.23}$$

where F is a constant of integration.

We now consider the surface tractions. For the curved surface $r = r_1$ the unit normal is given by (3.3.13) and therefore, from (2.10.13),

$$P^k_{(1)} = (\tau^{1k})_{r=r_1},$$

the suffix (1) referring to this outer surface in the deformed state. Thus

$$\left.\begin{aligned} P^1_{(1)} &= 2\left\{\frac{C_1 Q_1^2}{\lambda^2}+C_2\left(\frac{1}{\lambda^2}+Q_1^2\right)\right\}+p_1 \\ P^2_{(1)} &= -\frac{2K(C_1+C_2\lambda^2)\phi_0'}{\lambda r_1^3} \\ P^3_{(1)} &= \frac{4(C_1\lambda+C_2)w_0'}{\lambda^2 r_1} \end{aligned}\right\}, \tag{3.7.24}$$

where

$$p_1 = p(r_1).$$

For the curved surface $r = r_2$

$$\mathbf{n} = -\mathbf{G}^1 = -\mathbf{G}_1,$$

and

$$P^k_{(2)} = -(\tau^{1k})_{r=r_2}.$$

The initially plane ends of the cylinder deform into the surfaces

$$z = \pm\lambda l+w(r)$$

and the unit normal to the first of these surfaces is

$$\mathbf{n} = \frac{\left(-\dfrac{dz}{dr}\mathbf{G}^1+\mathbf{G}^3\right)}{\left\{1+\left(\dfrac{dz}{dr}\right)^2\right\}^{\frac{1}{2}}} = \frac{-w_r\,\mathbf{G}^1+\mathbf{G}^3}{(1+w_r^2)^{\frac{1}{2}}}. \tag{3.7.25}$$

Hence, using an obvious notation,

$$\left.\begin{aligned} (1+w_r^2)^{\frac{1}{2}}P^1_{(3)} &= -w_r\,\tau^{11}+\tau^{13} \\ &= -w_r(2C_2\,Q^2+p) \\ (1+w_r^2)^{\frac{1}{2}}P^2_{(3)} &= -w_r\,\tau^{12}+\tau^{23} \\ &= -2C_2\,Q^2\phi_r\,w_r \\ (1+w_r^2)^{\frac{1}{2}}P^3_{(3)} &= -w_r\,\tau^{13}+\tau^{33} \\ &= 2C_1\lambda^2+2C_2\left(Q^2+\frac{\lambda^2}{Q^2}+r^2Q^2\phi_r^2\right)+p \end{aligned}\right\}, \tag{3.7.26}$$

The forces which must be applied to the curved surfaces of the tube are:

(i) normal surface tractions R_1 and R_2 per unit length of the tube measured in the deformed state, acting respectively on the surfaces $r = r_1$ and $r = r_2$, in the directions of the outward normals to these surfaces,

(ii) an axial couple M per unit length of the deformed tube, acting on each curved surface,

(iii) a longitudinal force L, per unit length of the deformed tube, acting on each curved surface.

These are given, from (3.7.24), by

$$\left.\begin{aligned} R_1 &= 2\pi r_1 P^1_{(1)} = 2\pi a_1\left\{\frac{2C_1 Q_1}{\lambda^2}+2C_2\left(\frac{1}{\lambda^2 Q_1}+Q_1\right)+\frac{p_1}{Q_1}\right\} \\ R_2 &= -2\pi r_2 P^1_{(2)} = 2\pi a_2\left\{\frac{2C_1 Q_2}{\lambda^2}+2C_2\left(\frac{1}{\lambda^2 Q_2}+Q_2\right)+\frac{p_2}{Q_2}\right\} \\ M &= 2\pi r_1^2(r_1 P^2_{(1)}) = -4\pi K(C_1+C_2\lambda^2)\phi_0'/\lambda \\ L &= 2\pi r_1 P^3_{(1)} = 8\pi(C_1\lambda+C_2)w_0'/\lambda^2 \end{aligned}\right\}. \tag{3.7.27}$$

The surface tractions (3.7.26) must also be applied over the ends of the cylinder in order to produce the deformation considered. It should be noticed throughout this analysis that the hydrostatic pressure—and therefore the forces R_1 and R_2, given by (3.7.27)—are undetermined to the extent of an arbitrary constant. The value of the constant can be determined if the conditions at the ends of the cylinder are investigated in detail. Its value does not, however, affect the couple M and the tangential force L given by (3.7.27).

3.8. Shear of a cylindrical tube

We now consider a special case of the previous section in which the tube is not extended and not inflated so that $\mu_1 = \mu_2 = \lambda = 1$. Hence $Q = Q_1 = Q_2 = 1$ and, from (3.7.4), $K = 0$. Equations (3.7.13) become, on using (2.3.21),

$$\phi_r = \frac{B}{(C_1+C_2)r^3}, \qquad w_r = \frac{D}{(C_1+C_2)r}. \tag{3.8.1}$$

Integrating these equations, we obtain

$$\phi = -\frac{B}{2(C_1+C_2)r^2}+B', \qquad w = \frac{D\ln(r/a_2)}{C_1+C_2}+D'. \tag{3.8.2}$$

Taking $w = 0$ and $\phi = 0$ when $r = a_2$, and $w = w_0$, $\phi = \phi_0$ when $r = a_1$, as before we obtain

$$\left.\begin{aligned} D' &= 0, & B' &= \frac{B}{2(C_1+C_2)a_2^2} \\ D &= \frac{(C_1+C_2)w_0}{\ln(a_1/a_2)}, & B &= \frac{2(C_1+C_2)a_1^2 a_2^2\phi_0}{a_1^2-a_2^2} \end{aligned}\right\}. \tag{3.8.3}$$

Hence
$$\phi = \frac{\phi_0 a_1^2}{a_1^2-a_2^2}\left(1-\frac{a_2^2}{r^2}\right), \qquad w = \frac{w_0\ln(r/a_2)}{\ln(a_1/a_2)}. \tag{3.8.4}$$

If we put $Q = \lambda = 1$ in the equation (3.7.22) for p, we have

$$p = 2\int^{r} \{(C_1+C_2)r^2\phi_r^2+C_2 w_r^2\}\frac{dr}{r}$$

$$= 2\int^{r} \left\{\frac{B^2}{(C_1+C_2)r^4}+\frac{C_2 D^2}{(C_1+C_2)^2 r^2}\right\}\frac{dr}{r}$$

$$= -\frac{B^2}{2(C_1+C_2)r^4}-\frac{C_2 D^2}{(C_1+C_2)^2 r^2}+F, \tag{3.8.5}$$

where F is again an integration constant. Also

$$\left.\begin{aligned} P^1_{(1)} &= 2(C_1+2C_2)+p_1, \quad P^2_{(1)} = \frac{2B}{a_1^3}, \quad P^3_{(1)} = \frac{2D}{a_1} \\ R_{(1)} &= 2\pi a_1(2C_1+4C_2+p_1), \quad M = 4\pi B, \quad L = 4\pi D \end{aligned}\right\}. \tag{3.8.6}$$

The results (3.8.6) are found by the same process as that which led to (3.7.24) and (3.7.27). The tractions over the ends of the cylinder are again given by (3.7.26), where ϕ, w are now given by (3.8.4).

3.9. Rotation of a right circular cylinder about its axis

As in previous sections we use cylindrical polar coordinates (r, θ, z) to define the points of the cylinder in the strained state, and the cylinder of incompressible isotropic material has its axis along the z-axis. The state of stress of the cylinder when it is rotated about its axis with constant angular velocity ω is identical with the stress system which would be produced in the cylinder if it were stationary under the action of a body force of magnitude $r\omega^2$ directed along the r-axis, provided that the boundary conditions remain unaltered, and we shall consider this equivalent problem to avoid the use of rotating axes.

Assuming that in the strained state there is a uniform extension λ along the z-axis so that the point (r, θ, z) was initially at the point $(r', \theta, z/\lambda)$, the incompressibility condition implies that $r' = r\sqrt{\lambda}$, since we must have $r^2 z = r'^2 z/\lambda$. The initial coordinates of the point (r, θ, z) are therefore

$$x_1 = r\sqrt{(\lambda)}\cos\theta, \quad x_2 = r\sqrt{(\lambda)}\sin\theta, \quad x_3 = z/\lambda, \tag{3.9.1}$$

since the x_i- and y_i-axes are taken to coincide. From (3.9.1) we find that

$$g^{ik} = \begin{bmatrix} \dfrac{1}{\lambda} & 0 & 0 \\ 0 & \dfrac{1}{\lambda r^2} & 0 \\ 0 & 0 & \lambda^2 \end{bmatrix}, \tag{3.9.2}$$

while the metric tensor G^{ik} of the strained body is given by (3.3.3). A little calculation suffices to show that (2.10.17) and (2.10.18) now give

$$B^{ik} = \begin{bmatrix} \lambda+\dfrac{1}{\lambda^2} & 0 & 0 \\ 0 & \dfrac{\lambda}{r^2}+\dfrac{1}{\lambda^2 r^2} & 0 \\ 0 & 0 & 2\lambda \end{bmatrix}, \tag{3.9.3}$$

$$I_1 = \lambda^2+\frac{2}{\lambda}, \qquad I_2 = 2\lambda+\frac{1}{\lambda^2}. \tag{3.9.4}$$

From (3.3.3), (3.9.2), and (3.9.3) we see that

$$g^{ik}-G^{ik}/\lambda, \qquad B^{ik}-\left(\lambda+\frac{1}{\lambda^2}\right)G^{ik}$$

both reduce to tensors which have a constant component corresponding to $i = 3$, $k = 3$, all other components being zero. Hence

$$g^{ik}\|_i = 0, \qquad B^{ik}\|_i = 0. \tag{3.9.5}$$

Since I_1, I_2 are constants it follows that Φ and Ψ are constants, and hence, from (2.10.16) and (3.9.5),

$$\tau^{ik}\|_i = G^{ik}p_{,i}. \tag{3.9.6}$$

If we now put $F^k = (r\omega^2, 0, 0)$ and $f^k = 0$ in (2.10.11), and use (3.9.6), we find that the equations of equilibrium reduce to

$$\frac{\partial p}{\partial r} = -\rho r\omega^2, \qquad \frac{\partial p}{\partial \theta} = \frac{\partial p}{\partial z} = 0, \tag{3.9.7}$$

and it follows that

$$p = -\tfrac{1}{2}\rho r^2\omega^2+B, \tag{3.9.8}$$

where B is a constant to be determined. With the help of (3.9.2), (3.9.3), and (3.9.8) the stress tensor (2.10.16) becomes

$$\left.\begin{aligned} \tau^{11} &= r^2\tau^{22} = \frac{\Phi}{\lambda}+\left(\lambda+\frac{1}{\lambda^2}\right)\Psi-\tfrac{1}{2}\rho r^2\omega^2+B \\ \tau^{33} &= \lambda^2\Phi+2\lambda\Psi-\tfrac{1}{2}\rho r^2\omega^2+B \\ \tau^{23} &= \tau^{31} = \tau^{12} = 0 \end{aligned}\right\}. \tag{3.9.9}$$

On the curved surface of the cylinder, which we suppose to be free from traction, $\mathbf{n} = \mathbf{G}^1 = \mathbf{G}_1$ so that the boundary conditions (2.10.13) reduce to

$$\tau^{11} = 0 \quad (r = a/\sqrt{\lambda}), \tag{3.9.10}$$

where a is the initial radius of the cylinder. Hence

$$B = \frac{\rho a^2\omega^2}{2\lambda}-\frac{\Phi}{\lambda}-\left(\lambda+\frac{1}{\lambda^2}\right)\Psi, \tag{3.9.11}$$

and, using (2.2.17), (3.9.9), and (3.9.11) we obtain the physical components of stress in the form

$$\left.\begin{aligned} \sigma_{11} &= \sigma_{22} = \tau^{11} = r^2\tau^{22} = \frac{\rho\omega^2}{2}\left(\frac{a^2}{\lambda}-r^2\right) \\ \sigma_{33} &= \tau^{33} = \frac{\rho\omega^2}{2}\left(\frac{a^2}{\lambda}-r^2\right)+\left(\lambda^2-\frac{1}{\lambda}\right)\Phi+\left(\lambda-\frac{1}{\lambda^2}\right)\Psi \\ \sigma_{23} &= \sigma_{31} = \sigma_{12} = 0 \end{aligned}\right\}. \tag{3.9.12}$$

We see from (3.9.12), or alternatively from the boundary conditions (2.10.13), that the surface tractions on the plane ends of the cylinder consist of a normal tension of magnitude

$$N = \frac{\rho\omega^2}{2}\left(\frac{a^2}{\lambda}-r^2\right)+\left(\lambda^2-\frac{1}{\lambda}\right)\Phi+\left(\lambda-\frac{1}{\lambda^2}\right)\Psi. \tag{3.9.13}$$

If the traction on each end of the cylinder has no resultant then

$$\int\limits_0^{a/\sqrt{\lambda}} Nr\,dr = 0$$

or, using (3.9.13),

$$\left(\lambda^2-\frac{1}{\lambda}\right)\Phi+\left(\lambda-\frac{1}{\lambda^2}\right)\Psi+\frac{\rho a^2\omega^2}{4\lambda} = 0. \tag{3.9.14}$$

Unless explicit expressions are assumed for Φ, Ψ it is difficult to discuss the nature of the roots of this equation. In the case of the Mooney material, however, when the strain energy is given by (2.3.21), equation (3.9.14) becomes

$$\lambda^4+k\lambda^3+(n^2-1)\lambda-k = 0, \tag{3.9.15}$$

where $n^2 = \rho\omega^2a^2/(8C_1)$, $k = C_2/C_1$, and $k \geqslant 0$. When $k = 0$ equation (3.9.15) has one positive real root provided $n^2 < 1$, i.e. provided $\rho\omega^2a^2 < 8C_1$. When $k > 0$ equation (3.9.15) always has only one real positive root. We shall assume therefore that equation (3.9.14) has a unique positive real root $\lambda = \lambda_1$ in the general case.

Substituting $\lambda = \lambda_1$ in (3.9.12) we find that the stresses are

$$\left.\begin{aligned} \sigma_{11} &= \sigma_{22} = \frac{\rho\omega^2}{2}\left(\frac{a^2}{\lambda_1}-r^2\right) \\ \sigma_{33} &= \frac{\rho\omega^2}{4}\left(\frac{a^2}{\lambda_1}-2r^2\right) \end{aligned}\right\}. \tag{3.9.16}$$

When the strain is infinitesimal λ_1 is slightly less than unity and, to the order of approximation of the infinitesimal theory, we may put $\lambda_1 = 1$ in (3.9.16), obtaining the classical results for the special case when the material is incompressible.

3.10. Symmetrical expansion of a thick spherical shell

We identify the curvilinear coordinate system θ_i in the strained body with a system of spherical polar coordinates (R, θ, ϕ) which has its origin at the centre of the shell, so that

$$\left.\begin{aligned} \theta_1 = R, \quad \theta_2 = \theta, \quad \theta_3 = \phi \\ y_1 = R\sin\theta\cos\phi, \quad y_2 = R\sin\theta\sin\phi, \quad y_3 = R\cos\theta \end{aligned}\right\}, \tag{3.10.1}$$

and
$$G^{ik} = \begin{bmatrix} 1 & 0 & 0 \\ 0 & \dfrac{1}{R^2} & 0 \\ 0 & 0 & \dfrac{1}{R^2\sin^2\theta} \end{bmatrix}, \qquad G = R^4\sin^2\theta. \tag{3.10.2}$$

We assume that the displacement of the unstrained body possesses spherical symmetry so that the point (R, θ, ϕ) was originally at the point (r, θ, ϕ) and we write
$$Q(R) = r/R \tag{3.10.3}$$
for convenience in the algebra. The internal and external radii of the shell in the unstrained and strained states are denoted by r_1, r_2 and R_1, R_2 respectively. For incompressibility we must have
$$r^3 - R^3 = r_1^3 - R_1^3 = r_2^3 - R_2^3 \tag{3.10.4}$$
and therefore
$$Q(R) = \left(1 + \frac{r_1^3 - R_1^3}{R^3}\right)^{\frac{1}{3}}. \tag{3.10.5}$$

Taking the x_i-axes to coincide with the y_i-axes we have, using (3.10.3),
$$x_1 = RQ\sin\theta\cos\phi, \quad x_2 = RQ\sin\theta\sin\phi, \quad x_3 = RQ\cos\theta. \tag{3.10.6}$$
From (3.10.5) we deduce that
$$\frac{dQ}{dR} = \frac{1}{R}\left(\frac{1}{Q^2} - Q\right), \tag{3.10.7}$$
and using this result and (3.10.6) we find that
$$g^{ik} = \begin{bmatrix} Q^4 & 0 & 0 \\ 0 & \dfrac{1}{R^2Q^2} & 0 \\ 0 & 0 & \dfrac{1}{R^2Q^2\sin^2\theta} \end{bmatrix}, \qquad g = R^4\sin^2\theta. \tag{3.10.8}$$

We also have, from (2.10.17) and (2.10.18),

$$B^{ik} = \begin{bmatrix} 2Q^2 & 0 & 0 \\ 0 & \dfrac{Q^2}{R^2}+\dfrac{1}{Q^4R^2} & 0 \\ 0 & 0 & \dfrac{Q^6+1}{Q^4R^2\sin^2\theta} \end{bmatrix}, \tag{3.10.9}$$

$$I_1 = Q^4+\frac{2}{Q^2}, \qquad I_2 = 2Q^2+\frac{1}{Q^4}. \tag{3.10.10}$$

The stress–strain relations (2.10.16) give, with the help of (3.10.2), (3.10.8), and (3.10.9),

$$\left.\begin{aligned} \tau^{11} &= Q^4\Phi+2Q^2\Psi+p \\ \tau^{22} = \tau^{33}\sin^2\theta &= \frac{\Phi}{R^2Q^2}+\left(Q^2+\frac{1}{Q^4}\right)\frac{\Psi}{R^2}+\frac{p}{R^2} \\ \tau^{23} = \tau^{31} &= \tau^{12} = 0 \end{aligned}\right\}. \tag{3.10.11}$$

In the absence of body forces the equations of equilibrium of the strained spherical shell are (3.3.9), where the non-zero Christoffel symbols of the second kind are now

$$\left.\begin{aligned} \Gamma^1_{22} = -R, \qquad \Gamma^1_{33} = -R\sin^2\theta, \qquad \Gamma^2_{33} = -\sin\theta\cos\theta \\ \Gamma^2_{12} = \Gamma^3_{13} = \frac{1}{R}, \qquad \Gamma^3_{23} = \frac{\cos\theta}{\sin\theta} \end{aligned}\right\}. \tag{3.10.12}$$

The second and third equations in (3.3.9) show immediately that p is a function of r only, and the first equation of equilibrium reduces, with the help of (3.10.12), (3.10.7), and (3.10.11), to

$$\frac{dp}{dQ}+Q^4\frac{d\Phi}{dQ}+2Q^2\frac{d\Psi}{dQ}+2(Q^3-1)\Phi+2\left(Q-\frac{1}{Q^2}\right)\Psi = 0. \tag{3.10.13}$$

Integrating and using the rule of integration by parts we obtain

$$p = -Q^2(Q^2\Phi+2\Psi)+2\int^Q \left\{(Q^3+1)\Phi+\left(Q+\frac{1}{Q^2}\right)\Psi\right\} dQ \tag{3.10.14}$$

and the function p is determined apart from an arbitrary constant.

If Π_1 is the normal pressure on the inner surface of the shell, the unit normal to which is $\mathbf{n} = -\mathbf{G}_1 = -\mathbf{G}^1$, the boundary conditions (2.10.13) on this surface show that

$$P^1 = \Pi_1 = -\tau^{11}, \qquad P^2 = P^3 = 0$$

at $R = R_1$, and using the first of these equations together with (3.10.11) and (3.10.14) we find that

$$p = -Q^2(Q^2\Phi+2\Psi)+K(R)-\Pi_1, \tag{3.10.15}$$

where

$$K(R) = 2\int\limits_{Q_1}^{Q}\left\{(Q^3+1)\Phi+\left(Q+\frac{1}{Q^2}\right)\Psi\right\}dQ, \qquad Q_1 = \frac{r_1}{R_1}. \tag{3.10.16}$$

The physical components of stress may now be determined from (2.2.17) and (3.10.11), and we obtain

$$\left.\begin{aligned} \sigma_{11} = \tau^{11} &= K(R)-\Pi_1 \\ \sigma_{22} = \sigma_{33} = R^2\tau^{22} &= R^2\tau^{33}\sin^2\theta \\ &= \left(\frac{1}{Q^2}-Q^4\right)\Phi+\left(\frac{1}{Q^4}-Q^2\right)\Psi+K(R)-\Pi_1 \\ \sigma_{23} = \sigma_{31} &= \sigma_{12} = 0 \end{aligned}\right\}. \tag{3.10.17}$$

The boundary conditions (2.10.13) and the equations (3.10.17) show that if Π_2 is the normal pressure on the external surface of the shell then

$$\Pi_2 = \Pi_1-K(R_2). \tag{3.10.18}$$

Since we can express R_2 in terms of R_1 by the incompressibility condition (3.10.4), this equation serves to determine R_1 if the difference in pressure $\Pi_1-\Pi_2$ and the form of the elastic potential W are known; alternatively, it gives the value of $\Pi_1-\Pi_2$ when R_1 is known.

It may happen that the outer radius of the shell is infinite and the body is then an infinite solid containing a spherical cavity. In this case a possible form of the boundary condition at infinity is that the stresses vanish there, and noting that $Q(R) \to 1$ as $R \to \infty$ we see from (3.10.17) that this condition will be satisfied if

$$\Pi_1 = K(\infty),$$

provided Φ and Ψ are finite at infinity.

The expression (3.10.16) for $K(R)$ can be put into a different form. From (3.10.10) we have

$$\frac{dI_1}{dQ} = \frac{4(Q^3-1)(Q^3+1)}{Q^3}, \qquad \frac{dI_2}{dQ} = \frac{4(Q^3-1)}{Q^3}\left(Q+\frac{1}{Q^2}\right),$$

and therefore

$$\frac{dW}{dQ} = \frac{2(Q^3-1)}{Q^3}\left\{(Q^3+1)\Phi+\left(Q+\frac{1}{Q^2}\right)\Psi\right\}.$$

Hence, using (3.10.5), (3.10.16) can be written

$$K(R) = \int_{Q_1}^{Q} \frac{Q^3}{Q^3-1}\frac{dW}{dQ}\,dQ = \int_{R_1}^{R} \frac{(R^3+r_1^3-R_1^3)}{r_1^3-R_1^3}\frac{dW}{dR}\,dR. \tag{3.10.19}$$

Putting $R = R_2$ in (3.10.19) and remembering (3.10.4) we find that

$$K(R_2) = \frac{1}{r_1^3-R_1^3}\left\{r_2^3W(R_2)-r_1^3W(R_1)-\frac{3U}{4\pi}\right\}, \tag{3.10.20}$$

where $U = \int_{R_1}^{R_2} 4\pi R^2 W\,dR$ is the total energy stored elastically by the body. Using (3.10.4) again, (3.10.18) and (3.10.20) give

$$\tfrac{4}{3}\pi(R_1^3-r_1^3)\Pi_1-\tfrac{4}{3}\pi(R_2^3-r_2^3)\Pi_2 = U+\tfrac{4}{3}\pi r_1^3W(R_1)-\tfrac{4}{3}\pi r_2^3W(R_2). \tag{3.10.21}$$

For a Mooney material when W is given by (2.3.21) equation (3.10.16) for $K(R)$ can be integrated to give

$$K(R) = C_1(Q^4+4Q)+2C_2\left(Q^2-\frac{2}{Q}\right)-C_1\left(\frac{1}{\lambda_1^4}+\frac{4}{\lambda_1}\right)-2C_2\left(\frac{1}{\lambda_1^2}-2\lambda_1\right), \tag{3.10.22}$$

where

$$Q_1 = \frac{r_1}{R_1} = \frac{1}{\lambda_1}, \qquad Q = \{1+(1-\lambda_1^3)r_1^3/R^3\}^{\frac{1}{3}}. \tag{3.10.23}$$

The non-zero physical components of stresses in (3.10.17) then become

$$\left.\begin{aligned}\sigma_{11} &= C_1(Q^4+4Q)+2C_2\left(Q^2-\frac{2}{Q}\right)-C_1\left(\frac{1}{\lambda_1^4}+\frac{4}{\lambda_1}\right)-\\ &\qquad -2C_2\left(\frac{1}{\lambda_1^2}-2\lambda_1\right)-\Pi_1\\ \sigma_{22} = \sigma_{33} &= C_1\left(\frac{2}{Q^2}+4Q-Q^4\right)+2C_2\left(\frac{1}{Q^4}-\frac{2}{Q}\right)-\\ &\qquad -C_1\left(\frac{1}{\lambda_1^4}+\frac{4}{\lambda_1}\right)-2C_2\left(\frac{1}{\lambda_1^2}-2\lambda_1\right)-\Pi_1\end{aligned}\right\}. \tag{3.10.24}$$

The radial stress σ_{11} is a monotonic increasing or a monotonic decreasing function of R according as $\lambda_1 > 1$ or $\lambda_1 < 1$. The stress σ_{22} (or σ_{33}) has a turning value when

$$\frac{d\sigma_{22}}{dQ} = 4C_1\left(-\frac{1}{Q^3}+1-Q^3\right)+4C_2\left(\frac{1}{Q^2}-\frac{2}{Q^5}\right) = 0.$$

If $C_2 = 0$ this equation has no real root so that for a neo-Hookean material σ_{22} (and σ_{33}) attain their maximum and minimum values on the surfaces of the shell. When $C_2 \neq 0$ we have to consider the roots of the equation

$$Q^2\{(Q^3-1)^2+Q^3\}+k(2-Q^3) = 0,$$

where $k = C_2/C_1 > 0$. If this equation has a positive root $Q = Q'$ then $Q'^3 > 2$ and

$$k = Q'^2\left(Q'^3+1+\frac{3}{Q'^3-2}\right) > 3.$$

For rubber-like materials k is usually taken to be less than unity and in this case we see that maximum and minimum values of σ_{22} and σ_{33} are attained at the surfaces of the shell.

Equations (3.10.17) and (3.10.18) can be used to derive the approximate formulae for the stresses when the strain is infinitesimal and it is found that their values agree with those obtained from the classical theory.

We close this section by using the above results to obtain the corresponding formulae for a *thin* spherical shell. In this case we put $r_2 = r_1(1+\epsilon)$, where ϵ is small enough to allow us to neglect second degree and higher terms in ϵ. Denoting the extension of the shell by λ, so that $R_1 = \lambda r_1$, we have from (3.10.4)

$$R_2 = \lambda r_1(1+\epsilon/\lambda^3)$$

approximately, λ being considered large compared with ϵ. It follows that

$$Q_1 = \frac{r_1}{R_1} = \frac{1}{\lambda} \qquad Q_2 = \frac{r_2}{R_2} = \frac{1}{\lambda}\{1+\epsilon(1-1/\lambda^3)\}, \tag{3.10.25}$$

and we see that Q_2-Q_1 is of the same order as ϵ. Hence, to the first order in ϵ, we have from (3.10.16)

$$K(R_2) = 2(Q_2-Q_1)\left\{(Q^3+1)\Phi+\left(Q+\frac{1}{Q^2}\right)\Psi\right\}_{R=R_1}. \tag{3.10.26}$$

Introducing the values (3.10.25) of Q_1 and Q_2 into this equation and using (3.10.18) we find that, for a general value of W,

$$\Pi_1-\Pi_2 = 2\epsilon\left\{\left(\frac{1}{\lambda}-\frac{1}{\lambda^7}\right)(\Phi)_{R=R_1}+\left(\lambda-\frac{1}{\lambda^5}\right)(\Psi)_{R=R_1}\right\}. \tag{3.10.27}$$

3.11. Flexure of a cuboid

Suppose that in the undeformed state of the body it is a cuboid bounded by the planes $x_1 = a_1$, $x_1 = a_2$; $x_2 = \pm b$; $x_3 = \pm c$ and let $a_1-a_2 = 2a$. The cuboid is now deformed symmetrically with respect to the x_1-axis so that:

I. Each plane initially normal to the x_1-axis becomes, in the deformed state, a portion of a curved surface of a cylinder whose axis is the x_3-axis.

II. Planes initially normal to the x_2-axis become in the deformed state planes containing the x_3-axis.

III. There is a uniform extension λ in the direction of the x_3-axis.

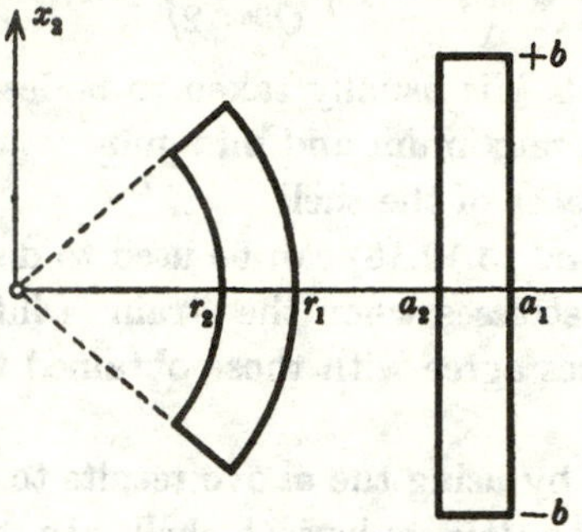

FIG. 3.2. Flexure of cuboid.

We take cylindrical polar coordinates (r, θ, z) to define the strained body and we identify the curvilinear system θ_i with these coordinates so that, if the y_i-axes coincide with x_i-axes,

$$\left.\begin{array}{lll} \theta_1 = r, & \theta_2 = \theta, & \theta_3 = z \\ y_1 = r\cos\theta, & y_2 = r\sin\theta, & y_3 = z \end{array}\right\}. \tag{3.11.1}$$

The assumptions I to III imply that if the point (r, θ, z) was originally at (x_1, x_2, x_3) then

$$x_1 = f(r), \qquad x_2 = g(\theta), \qquad x_3 = z/\lambda. \tag{3.11.2}$$

From (3.11.1) and (3.11.2) we obtain

$$G_{ik} = \begin{bmatrix} 1 & 0 & 0 \\ 0 & r^2 & 0 \\ 0 & 0 & 1 \end{bmatrix}, \qquad G^{ik} = \begin{bmatrix} 1 & 0 & 0 \\ 0 & \dfrac{1}{r^2} & 0 \\ 0 & 0 & 1 \end{bmatrix}, \qquad G = r^2, \tag{3.11.3}$$

$$g_{ik} = \begin{bmatrix} f_r^2 & 0 & 0 \\ 0 & g_\theta^2 & 0 \\ 0 & 0 & \dfrac{1}{\lambda^2} \end{bmatrix},$$

and the incompressibility condition gives

$$f_r^2\, g_\theta^2 = \lambda^2 r^2,$$

so that

$$\frac{f_r(r)}{r} = \frac{\lambda}{g_\theta(\theta)} = A,$$

where A is a constant and suffixes denote differentiation. Hence, since the flexure is symmetric with respect to the x_1-axis,

$$x_1 = f(r) = \tfrac{1}{2}Ar^2+B, \qquad x_2 = g(\theta) = \frac{\lambda\theta}{A}. \tag{3.11.4}$$

If r_1 and r_2 ($r_1 > r_2$) are the radii of the curved surfaces of the deformed body, which are initially the planes $x_1 = a_1$ and $x_1 = a_2$ respectively, then

$$A = \frac{4a}{r_1^2-r_2^2}, \qquad B = \frac{a_2 r_1^2-a_1 r_2^2}{r_1^2-r_2^2}. \tag{3.11.5}$$

From (3.11.3) and (3.11.4) we now find

$$g_{ik} = \begin{bmatrix} A^2r^2 & 0 & 0 \\ 0 & \dfrac{\lambda^2}{A^2} & 0 \\ 0 & 0 & \dfrac{1}{\lambda^2} \end{bmatrix}, \qquad g^{ik} = \begin{bmatrix} \dfrac{1}{A^2r^2} & 0 & 0 \\ 0 & \dfrac{A^2}{\lambda^2} & 0 \\ 0 & 0 & \lambda^2 \end{bmatrix}, \tag{3.11.6}$$

and, from (2.10.17), (2.10.18), (3.11.3), and (3.11.6),

$$\left.\begin{aligned} I_1 &= \frac{1}{A^2r^2}+\frac{A^2r^2}{\lambda^2}+\lambda^2 \\ I_2 &= A^2r^2+\frac{\lambda^2}{A^2r^2}+\frac{1}{\lambda^2} \end{aligned}\right\}, \tag{3.11.7}$$

$$B^{ik} = \begin{bmatrix} \dfrac{\lambda^2}{A^2r^2}+\dfrac{1}{\lambda^2} & 0 & 0 \\ 0 & A^2+\dfrac{1}{\lambda^2r^2} & 0 \\ 0 & 0 & \dfrac{\lambda^2}{A^2r^2}+A^2r^2 \end{bmatrix}. \tag{3.11.8}$$

With the help of (3.11.3), (3.11.6), and (3.11.8) the stress–strain relations (2.10.16) give

$$\left.\begin{aligned} \tau^{11} &= \frac{\Phi}{A^2r^2}+\left(\frac{\lambda^2}{A^2r^2}+\frac{1}{\lambda^2}\right)\Psi+p \\ \tau^{22} &= \frac{A^2\Phi}{\lambda^2}+\left(A^2+\frac{1}{\lambda^2r^2}\right)\Psi+\frac{p}{r^2} \\ \tau^{33} &= \lambda^2\Phi+\left(\frac{\lambda^2}{A^2r^2}+A^2r^2\right)\Psi+p \\ \tau^{23} &= \tau^{31} = \tau^{12} = 0 \end{aligned}\right\}. \tag{3.11.9}$$

When body forces are absent the equations of equilibrium again reduce to (3.3.9), with Christoffel symbols given by (3.3.10). From these equations it follows that p is a function of r only and that

$$\frac{d\tau^{11}}{dr}-\left(\frac{rA^2}{\lambda^2}-\frac{1}{A^2r^3}\right)(\Phi+\lambda^2\Psi) = 0,$$

or, after an integration,

$$p = -\frac{\Phi}{A^2r^2}-\left(\frac{\lambda^2}{A^2r^2}+\frac{1}{\lambda^2}\right)\Psi+\int^r\left(\frac{rA^2}{\lambda^2}-\frac{1}{A^2r^3}\right)(\Phi+\lambda^2\Psi)\,dr. \tag{3.11.10}$$

Now, from (3.11.7),

$$\frac{dI_1}{dr} = \frac{1}{\lambda^2}\frac{dI_2}{dr} = 2\left(\frac{rA^2}{\lambda^2}-\frac{1}{A^2r^3}\right); \tag{3.11.11}$$

hence

$$\begin{aligned} p &= -\frac{\Phi}{A^2r^2}-\left(\frac{\lambda^2}{A^2r^2}+\frac{1}{\lambda^2}\right)\Psi+\int^r\left(\frac{\partial W}{\partial I_1}\frac{dI_1}{dr}+\frac{\partial W}{\partial I_2}\frac{dI_2}{dr}\right)dr \\ &= -\frac{\Phi}{A^2r^2}-\left(\frac{\lambda^2}{A^2r^2}+\frac{1}{\lambda^2}\right)\Psi+W(r)+K, \end{aligned} \tag{3.11.12}$$

where K is a constant. The non-zero stresses in (3.11.9) now become

$$\left.\begin{aligned} \tau^{11} &= W(r)+K \\ r^2\tau^{22} &= \left(\frac{A^2r^2}{\lambda^2}-\frac{1}{A^2r^2}\right)(\Phi+\lambda^2\Psi)+W+K \\ &= r\frac{dW}{dr}+W+K \\ \tau^{33} &= \left(\lambda^2-\frac{1}{A^2r^2}\right)\left(\Phi+\frac{A^2r^2\Psi}{\lambda^2}\right)+W+K \end{aligned}\right\}. \tag{3.11.13}$$

The unit normals to the surfaces in the deformed state which were originally normal to the x_1, x_2, x_3 axes are $\pm\mathbf{G}_1 = \pm\mathbf{G}^1$, $\pm\mathbf{G}_2/r = \pm r\mathbf{G}^2$, and $\pm\mathbf{G}_3 = \pm\mathbf{G}^3$ respectively. It follows from (2.10.13) and (3.11.9) that the surface tractions are normal to the surfaces on which they act.

On $r = r_1$, $$P^1 = W(r_1)+K. \tag{3.11.14}$$

On the plane end which was initially the plane $x_2 = b$,

$$P^2 = \frac{dW}{dr}+\frac{W+K}{r}. \tag{3.11.15}$$

On the surface $z = \lambda c$,

$$P^3 = \left(\lambda^2-\frac{1}{A^2r^2}\right)\left(\Phi+\frac{A^2r^2\Psi}{\lambda^2}\right)+W+K. \tag{3.11.16}$$

If we denote the normal tensions applied to the surfaces $r = r_1$; $r = r_2$ by R_1 and R_2 respectively; to the surfaces which are planes through the z-axis by Θ; and to the surfaces normal to the z-axis by Z, we have

$$\left.\begin{aligned} R_1 = W(r_1)+K, \qquad R_2 = W(r_2)+K \\ \Theta = rP^2 = r\frac{dW}{dr}+W+K \\ Z = P^3 = \left(\lambda^2-\frac{1}{A^2r^2}\right)\left(\Phi+\frac{A^2r^2\Psi}{\lambda^2}\right)+W+K \end{aligned}\right\}. \tag{3.11.17}$$

If $R_1 = R_2$, $\qquad W(r_1) = W(r_2) = W_0 \quad \text{(say)}, \qquad$ (3.11.18)

and if, further, $R_1 = R_2 = 0$, so that the curved surfaces are free from applied traction, we have

$$K = -W_0.$$

The resultant force F acting normally on a surface which is initially at $x_2 = \pm b$ is given by

$$F = 2\lambda c\int_{r_2}^{r_1}\Theta\,dr = 2\lambda c[r(W+K)]_{r_2}^{r_1}. \tag{3.11.19}$$

If the curved surfaces are free from applied force so that (3.11.18) is satisfied and $K = -W_0$, we have, from (3.11.19), $F = 0$; i.e. the forces acting on each of the surfaces initially at $x_2 = \pm b$ are statically equivalent to a couple, as is evident from considerations of symmetry. The couple M acting on each of the surfaces is given, with $K = -W_0$, by

$$M = 2\lambda c\int_{r_2}^{r_1} r\Theta\,dr = \lambda c\left\{(r_1^2-r_2^2)W_0-2\int_{r_2}^{r_1} rW\,dr\right\}. \tag{3.11.20}$$

Since W is a function of I_1, I_2, it is seen from (3.11.7) that the relation (3.11.18) can be satisfied if

$$A^2 = \frac{\lambda}{r_1 r_2}. \tag{3.11.21}$$

If the radius of one of the curved surfaces is given, this equation completes the definition of the deformation.

In the special case when there is no extension in the x_3-direction, $\lambda = 1$ and

$$I_1 = I_2 = 1+A^2r^2+\frac{1}{A^2r^2} = I. \tag{3.11.22}$$

Then W is a function of I and, provided that it is a monotonic function, it follows from (3.11.18) that

$$(I)_{r=r_1} = (I)_{r=r_2},$$

which gives $A^2 = 1/(r_1 r_2)$. If A^2 has this value, the line which is not changed in length in the flexure has in the final strained state a radius r_0, where

$$r_0 = (r_1 r_2)^{\frac{1}{2}}. \tag{3.11.23}$$

3.12. Experiments on rubber

Some of the general solutions of special problems which have been given in this chapter have been used† to interpret the load deformation curves obtained for certain simple types of deformation of vulcanized rubber test-pieces in terms of a strain energy function W. The experiments showed that $\partial W/\partial I_1$ was approximately independent of both I_1 and I_2 and that $\partial W/\partial I_2$ is independent of I_1 and falls with increase of I_2. The reader is referred to Rivlin's paper for a full discussion.

On referring to equation (3.5.9) we see that the ratio of the force required to produce a large simple extension in a circular cylindrical rod, to the torsional modulus for a small torsion superposed on the simple extension, is independent of the particular form assumed for the strain energy function. This remarkable result has been confirmed by experimental tests on circular cylindrical rods made from vulcanized rubber.† A generalization of this formula is given in (4.3.31) for cylinders with arbitrary constant cross-sections.

Theoretical calculations concerned with thin plane sheets‡ and thin shells§ have been compared with experiment by Rivlin and Thomas, and by Adkins and Rivlin. Further experiments on the mechanics of rubber are reported by Gent and Rivlin.||

† R. S. Rivlin and D. W. Saunders, *Phil. Trans. R. Soc.* **A243** (1951) 251.
‡ R. S. Rivlin and A. G. Thomas, ibid. 289.
§ J. E. Adkins and R. S. Rivlin, ibid. **244** (1952) 505.
|| A. N. Gent and R. S. Rivlin, *Proc. phys. Soc.* **B65** (1952) 118, 487, 645.

4 Small Deformations Superposed on Finite Deformation

MOST of the problems which were considered in Chapter 3 were solved exactly without any assumptions being made beyond those of homogeneity, isotropy, and incompressibility, and some solutions were valid for compressible bodies. It was only possible to solve these problems exactly because of some symmetry in the deformation which reduced the equations of equilibrium to ordinary differential equations in one variable. In this chapter we consider unsymmetrical deformations and we obtain certain types of approximations to the exact theory. We begin by presenting a general theory of small deformations which are superposed on a known finite deformation of a compressible homogeneous isotropic body.†

4.1. General theory

Adopting the notations and formulae which are summarized at the end of Chapter 2 we consider a deformation of the unstrained body B_0 which is such that the state of strain and stress at any time differs only slightly from the state in a known finite deformation of B_0. We suppose that the points P_0 of the body B_0 are displaced to P' and form, at time t, a strained body B' which may be obtained by an infinitesimal displacement of the points P of a strained body B, where the deformation of the body B_0 to the body B is assumed to be completely determined. The state of the body B is described as in Chapter 2 so that the displacement vector $\overrightarrow{P_0 P'}$ may be written

$$\mathbf{v}(\theta_1, \theta_2, \theta_3, t) + \epsilon \mathbf{w}(\theta_1, \theta_2, \theta_3, t), \tag{4.1.1}$$

where $\mathbf{v}$ is the displacement vector $\overrightarrow{P_0 P}$ and ϵ is a constant which is small so that squares and higher powers of ϵ may be neglected compared with ϵ.

† See A. E. Green, R. S. Rivlin, R. T. Shield, *Proc. R. Soc.* A211 (1952) 128.

The covariant base vectors of the coordinate system θ_i at points P' of the body B' are denoted by $\mathbf{G}_i+\epsilon\mathbf{G}'_i = \mathbf{r}_{,i}+\mathbf{v}_{,i}+\epsilon\mathbf{w}_{,i}$ so that

$$\mathbf{G}'_i = \mathbf{w}_{,i}. \tag{4.1.2}$$

The displacement vector $\mathbf{w}$ may be expressed in terms of components in various ways, the most convenient for our purpose being

$$\mathbf{w} = w_m \mathbf{G}^m = w^m \mathbf{G}_m, \tag{4.1.3}$$

so that w_m, w^m are the components of $\mathbf{w}$ referred to base vectors at points P of the body B. Hence

$$\mathbf{G}'_i = w_m\|_i \mathbf{G}^m = w^m\|_i \mathbf{G}_m. \tag{4.1.4}$$

The covariant metric tensor of the body B', evaluated at a given time t, is

$$G_{ij}+\epsilon G'_{ij} = (\mathbf{G}_i+\epsilon\mathbf{G}'_i).(\mathbf{G}_j+\epsilon\mathbf{G}'_j),$$

so that, to the first order in ϵ,

$$G'_{ij} = \mathbf{G}_i.\mathbf{G}'_j+\mathbf{G}_j.\mathbf{G}'_i = w_i\|_j+w_j\|_i \tag{4.1.5}$$

if we use (4.1.4). The contravariant metric tensor of B' is $G^{ij}+\epsilon G'^{ij}$, and since we have

$$(G_{ij}+\epsilon G'_{ij})(G^{jk}+\epsilon G'^{jk}) = \delta_i^k,$$

it follows that

$$G'_{ij}\, G^{jk}+G_{ij}\, G'^{jk} = 0.$$

Alternatively, solving for G'^{ij} we obtain

$$G'^{ij} = -G^{ir}G^{js}G'_{rs}. \tag{4.1.6}$$

The contravariant base vectors at P' in the body B' are

$$\mathbf{G}^i+\epsilon\mathbf{G}'^i = (G^{ij}+\epsilon G'^{ij})(\mathbf{G}_j+\epsilon\mathbf{G}'_j),$$

and therefore

$$\mathbf{G}'^i = G^{ij}\mathbf{G}'_j+G'^{ij}\mathbf{G}_j. \tag{4.1.7}$$

If the determinant of the metric tensor components $G_{ij}+\epsilon G'_{ij}$ is denoted by $G+\epsilon G'$ then

$$|G^{jr}(G_{ij}+\epsilon G'_{ij})| = |\delta_i^r+\epsilon G^{jr}G'_{ij}| = |G^{jr}|(G+\epsilon G') = 1+\epsilon\frac{G'}{G},$$

and, therefore, to the first order in ϵ,

$$G' = GG^{ij}G'_{ij}. \tag{4.1.8}$$

The strain invariants associated with the body B' are $I_1+\epsilon I'_1$, $I_2+\epsilon I'_2$, $I_3+\epsilon I'_3$, where, from (2.10.18),

$$\left.\begin{aligned} I_1+\epsilon I'_1 &= g^{rs}(G_{rs}+\epsilon G'_{rs}) \\ I_2+\epsilon I'_2 &= g_{rs}(G^{rs}+\epsilon G'^{rs})(I_3+\epsilon I'_3) \\ I_3+\epsilon I'_3 &= (G+\epsilon G')/g \end{aligned}\right\}, \tag{4.1.9}$$

and, to the first order in ϵ, this gives

$$\left.\begin{aligned} I'_1 &= g^{rs}G'_{rs} \\ I'_2 &= g_{rs}(G'^{rs}I_3+G^{rs}I'_3) \\ I'_3 &= G'/g = I_3\,G^{ij}G'_{ij} \end{aligned}\right\}. \tag{4.1.10}$$

For homogeneous isotropic materials the elastic potential for the body B has the form (2.10.15). For the strained body B' the potential becomes

$$W(I_1+\epsilon I'_1,\ I_2+\epsilon I'_2,\ I_3+\epsilon I'_3).$$

The scalar invariants Φ, Ψ, p which, for the body B, are functions of I_1, I_2, I_3 become functions of $I_1+\epsilon I'_1$, $I_2+\epsilon I'_2$, $I_3+\epsilon I'_3$ for the strained body B', and may be denoted by $\Phi+\epsilon\Phi'$, $\Psi+\epsilon\Psi'$, $p+\epsilon p'$. By Taylor's expansion, to the first order in ϵ, we obtain

$$\Phi' = I'_1\frac{\partial\Phi}{\partial I_1}+I'_2\frac{\partial\Phi}{\partial I_2}+I'_3\frac{\partial\Phi}{\partial I_3},$$

$$\Psi' = I'_1\frac{\partial\Psi}{\partial I_1}+I'_2\frac{\partial\Psi}{\partial I_2}+I'_3\frac{\partial\Psi}{\partial I_3},$$

$$p' = I'_1\frac{\partial p}{\partial I_1}+I'_2\frac{\partial p}{\partial I_2}+I'_3\frac{\partial p}{\partial I_3},$$

and, using (2.10.17), these may be written

$$\left.\begin{aligned} \Phi' &= AI'_1+FI'_2+EI'_3-\frac{\Phi}{2I_3}\,I'_3 \\ \Psi' &= FI'_1+BI'_2+DI'_3-\frac{\Psi}{2I_3}\,I'_3 \\ p' &= I_3(EI'_1+DI'_2+CI'_3)+\frac{p}{2I_3}\,I'_3 \end{aligned}\right\}, \tag{4.1.11}$$

where

$$\left.\begin{aligned} A &= \frac{2}{\sqrt{I_3}}\frac{\partial^2W}{\partial I_1^2}, & B &= \frac{2}{\sqrt{I_3}}\frac{\partial^2W}{\partial I_2^2}, & C &= \frac{2}{\sqrt{I_3}}\frac{\partial^2W}{\partial I_3^2} \\ D &= \frac{2}{\sqrt{I_3}}\frac{\partial^2W}{\partial I_2\,\partial I_3}, & E &= \frac{2}{\sqrt{I_3}}\frac{\partial^2W}{\partial I_3\,\partial I_1}, & F &= \frac{2}{\sqrt{I_3}}\frac{\partial^2W}{\partial I_1\,\partial I_2} \end{aligned}\right\}. \tag{4.1.12}$$

The elastic potential W which appears in (4.1.12) is given by (2.10.15) and depends only on I_1, I_2, I_3. The functions A, B,..., F are therefore invariants which depend only on I_1, I_2, I_3.

For an incompressible body the potential is a function of I_1 and I_2 only, and also $I'_3 = 0$. The invariants A, B, and F can still be found from equations (4.1.12), where I_3 is put equal to 1, and from (4.1.11) we have

$$\Phi' = AI'_1+FI'_2, \qquad \Psi' = FI'_1+BI'_2, \tag{4.1.13}$$

but p' cannot be found from the elastic potential and is a scalar invariant function of the coordinates for each value of the time t.

The tensor B^{ij} in (2.10.17) becomes $B^{ij}+\epsilon B'^{ij}$ for the strained body B', where

$$B'^{ij} = (g^{ij}g^{rs}-g^{ir}g^{js})G'_{rs} = e^{irm}e^{jsn}g_{rs}\,G'_{mn}/g. \tag{4.1.14}$$

The symmetric stress for the strained body B' is $\tau^{ij}+\epsilon\tau'^{ij}$ and $\mathbf{T}_i$ becomes $\mathbf{T}_i+\epsilon\mathbf{T}'_i$, so that, from (2.10.4), (2.10.16), (4.1.4)–(4.1.14),

$$\mathbf{T}'_i = \sqrt{(G)}\lambda^{ij}\mathbf{G}_j, \qquad \lambda^{ij} = \tau'^{ij}+\tau^{im}w^j\|_m+\tau^{ij}w^m\|_m, \tag{4.1.15}$$

$$\tau'^{ij} = g^{ij}\Phi'+B^{ij}\Psi'+B'^{ij}\Psi+G'^{ij}p+G^{ij}p'. \tag{4.1.16}$$

If the body force and acceleration vectors for B' are respectively $\mathbf{F}+\epsilon\mathbf{F}'$, $\mathbf{f}+\epsilon\mathbf{f}'$ then the equations of motion for B and B' are (2.10.10) and

$$\mathbf{T}'_{i,i}+\rho\mathbf{F}'\sqrt{G} = \rho\mathbf{f}'\sqrt{G}. \tag{4.1.17}$$

Alternatively, using (4.1.15), equations (4.1.17) can be put in the form

$$\lambda^{ij}\|_i+\rho F'^j = \rho f'^j, \tag{4.1.18}$$

where

$$\mathbf{F}' = F'^j\mathbf{G}_j, \qquad \mathbf{f}' = f'^j\mathbf{G}_j. \tag{4.1.19}$$

The stress vector $\mathbf{t}+\epsilon\mathbf{t}'$, associated with a surface in B' whose unit normal in its position in B is $\mathbf{n}$, and measured per unit area of B, is

$$\mathbf{t}+\epsilon\mathbf{t}' = n_i(\mathbf{T}_i+\epsilon\mathbf{T}'_i)/\sqrt{G} = n_i(\tau^{ij}+\epsilon\lambda^{ij})\mathbf{G}_j. \tag{4.1.20}$$

If $\mathbf{P}+\epsilon\mathbf{P}'$ is an applied surface force vector at the boundary surface of B', measured per unit area of the corresponding surface in B, then the surface condition is

$$\mathbf{t} = \mathbf{P}, \qquad \mathbf{t}' = \mathbf{P}', \tag{4.1.21}$$

or, using (4.1.20) and putting

$$\mathbf{P} = P^j\mathbf{G}_j, \qquad \mathbf{P}' = P'^j\mathbf{G}_j, \tag{4.1.22}$$

we have

$$n_i\tau^{ij} = P^j, \qquad n_i\lambda^{ij} = P'^j. \tag{4.1.23}$$

4.2. Small deformation superposed on finite uniform extensions

We now assume that the body B_0 is deformed into the body B by uniform finite extensions along three perpendicular directions and that B is maintained in equilibrium without the aid of body forces. The strained body B' is then obtained from B by superposing a small deformation of the type described in § 4.1. We take our moving coordinates θ_i to coincide with a fixed rectangular system of coordinates (x, y, z) in the strained body B, the axes being directed along the three perpendicular directions. Thus we have

$$\theta_1 = x, \qquad \theta_2 = y, \qquad \theta_3 = z. \tag{4.2.1}$$

If the rectangular axes x_i which define points P_0 of the unstrained body B_0 are taken to coincide with the axes (x, y, z) then

$$x_1 = \frac{x}{\lambda_1}, \qquad x_2 = \frac{y}{\lambda_2}, \qquad x_3 = \frac{z}{\lambda_3}, \tag{4.2.2}$$

where λ_1, λ_2, λ_3 are the constant extension ratios. It follows from Chapter 2 that

$$g_{ij} = \begin{bmatrix} \frac{1}{\lambda_1^2} & 0 & 0 \\ 0 & \frac{1}{\lambda_2^2} & 0 \\ 0 & 0 & \frac{1}{\lambda_3^2} \end{bmatrix}, \qquad g^{ij} = \begin{bmatrix} \lambda_1^2 & 0 & 0 \\ 0 & \lambda_2^2 & 0 \\ 0 & 0 & \lambda_3^2 \end{bmatrix}, \qquad g = \frac{1}{\lambda_1^2\lambda_2^2\lambda_3^2}, \tag{4.2.3}$$

and $\qquad G_{ij} = G^{ij} = \delta_{ij}, \quad G = 1, \quad \mathbf{G}^m = \mathbf{G}_m, \quad \Gamma^r_{ij} = 0. \qquad$ (4.2.4)

Also, from (2.10.18),

$$I_1 = \lambda_1^2+\lambda_2^2+\lambda_3^2, \qquad I_2 = \lambda_2^2\lambda_3^2+\lambda_3^2\lambda_1^2+\lambda_1^2\lambda_2^2, \qquad I_3 = \lambda_1^2\lambda_2^2\lambda_3^2, \tag{4.2.5}$$

and from (2.10.17) we see that Φ, Ψ, and p are independent of x, y, z and

$$\left.\begin{aligned} B^{11} = \lambda_1^2(\lambda_2^2+\lambda_3^2), \quad B^{22} = \lambda_2^2(\lambda_3^2+\lambda_1^2), \quad B^{33} = \lambda_3^2(\lambda_1^2+\lambda_2^2) \\ B^{12} = B^{23} = B^{31} = 0 \end{aligned}\right\}. \tag{4.2.6}$$

The stress components τ^{ij} for the body B are given by (2.10.16) and are

$$\left.\begin{aligned} \tau^{11} &= \Phi\lambda_1^2+\Psi\lambda_1^2(\lambda_2^2+\lambda_3^2)+p \\ \tau^{22} &= \Phi\lambda_2^2+\Psi\lambda_2^2(\lambda_3^2+\lambda_1^2)+p \\ \tau^{33} &= \Phi\lambda_3^2+\Psi\lambda_3^2(\lambda_1^2+\lambda_2^2)+p \\ \tau^{12} &= \tau^{23} = \tau^{31} = 0 \end{aligned}\right\}, \tag{4.2.7}$$

so that the stress equations of equilibrium are satisfied when body forces are zero.

From (4.2.4) and (4.1.3) we see that $w_m = w^m$, and from (4.1.5) and (4.1.6),

$$-G'^{ij} = G'_{ij} = \frac{\partial w_i}{\partial \theta^j}+\frac{\partial w_j}{\partial \theta^i}. \tag{4.2.8}$$

Equations (4.1.10) now reduce to

$$\left.\begin{aligned} I'_1 &= 2\left(\lambda_1^2\frac{\partial u}{\partial x}+\lambda_2^2\frac{\partial v}{\partial y}+\lambda_3^2\frac{\partial w}{\partial z}\right) \\ I'_2 &= 2\lambda_1^2(\lambda_2^2+\lambda_3^2)\frac{\partial u}{\partial x}+2\lambda_2^2(\lambda_3^2+\lambda_1^2)\frac{\partial v}{\partial y}+2\lambda_3^2(\lambda_1^2+\lambda_2^2)\frac{\partial w}{\partial z} \\ I'_3 &= 2\lambda_1^2\lambda_2^2\lambda_3^2\left(\frac{\partial u}{\partial x}+\frac{\partial v}{\partial y}+\frac{\partial w}{\partial z}\right) \end{aligned}\right\}, \tag{4.2.9}$$

where we have written

$$w_1 = u, \qquad w_2 = v, \qquad w_3 = w. \tag{4.2.10}$$

The coefficients $A, B,..., F$ in (4.1.12) are constants and the tensor B'^{ij}, which is given by (4.1.14), has components

$$\left.\begin{aligned} B'^{11} &= 2\lambda_1^2\left(\lambda_2^2\frac{\partial v}{\partial y}+\lambda_3^2\frac{\partial w}{\partial z}\right), & B'^{22} &= 2\lambda_2^2\left(\lambda_3^2\frac{\partial w}{\partial z}+\lambda_1^2\frac{\partial u}{\partial x}\right)\\ B'^{33} &= 2\lambda_3^2\left(\lambda_1^2\frac{\partial u}{\partial x}+\lambda_2^2\frac{\partial v}{\partial y}\right), & B'^{12} &= -\lambda_1^2\lambda_2^2\left(\frac{\partial u}{\partial y}+\frac{\partial v}{\partial x}\right)\\ B'^{23} &= -\lambda_2^2\lambda_3^2\left(\frac{\partial v}{\partial z}+\frac{\partial w}{\partial y}\right), & B'^{31} &= -\lambda_3^2\lambda_1^2\left(\frac{\partial w}{\partial x}+\frac{\partial u}{\partial z}\right) \end{aligned}\right\}, \tag{4.2.11}$$

and substitution in (4.1.16) from (4.1.11), (4.2.3)–(4.2.11) shows that the components of the stress tensor τ'^{ij} are given by

$$\left.\begin{aligned} \tau'^{11} &= c_{11}\frac{\partial u}{\partial x}+c_{12}\frac{\partial v}{\partial y}+c_{13}\frac{\partial w}{\partial z}\\ \tau'^{22} &= c_{21}\frac{\partial u}{\partial x}+c_{22}\frac{\partial v}{\partial y}+c_{23}\frac{\partial w}{\partial z}\\ \tau'^{33} &= c_{31}\frac{\partial u}{\partial x}+c_{32}\frac{\partial v}{\partial y}+c_{33}\frac{\partial w}{\partial z} \end{aligned}\right\}, \tag{4.2.12}$$

$$\left.\begin{aligned} \tau'^{23} &= c_{44}\left(\frac{\partial v}{\partial z}+\frac{\partial w}{\partial y}\right)\\ \tau'^{31} &= c_{55}\left(\frac{\partial w}{\partial x}+\frac{\partial u}{\partial z}\right)\\ \tau'^{12} &= c_{66}\left(\frac{\partial u}{\partial y}+\frac{\partial v}{\partial x}\right) \end{aligned}\right\}, \tag{4.2.13}$$

where

$$\begin{aligned} c_{11} &= -\tau^{11}+2A\lambda_1^4+2B\lambda_1^4(\lambda_2^2+\lambda_3^2)^2+2C\lambda_1^4\lambda_2^4\lambda_3^4+4D\lambda_1^4\lambda_2^2\lambda_3^2(\lambda_2^2+\lambda_3^2)+\\ &\qquad +4E\lambda_1^4\lambda_2^4\lambda_3^4+4F\lambda_1^4(\lambda_2^2+\lambda_3^2),\\ c_{12} &= -\Phi\lambda_1^2+\Psi\lambda_1^2(\lambda_2^2-\lambda_3^2)+p+2A\lambda_1^2\lambda_2^2+2B\lambda_1^2\lambda_2^2(\lambda_2^2+\lambda_3^2)(\lambda_3^2+\lambda_1^2)+\\ &\qquad +2C\lambda_1^4\lambda_2^4\lambda_3^4+2D\lambda_1^2\lambda_2^2\lambda_3^2(\lambda_2^2\lambda_3^2+\lambda_3^2\lambda_1^2+2\lambda_1^2\lambda_2^2)+\\ &\qquad +2E\lambda_1^2\lambda_2^2\lambda_3^2(\lambda_1^2+\lambda_2^2)+2F\lambda_1^2\lambda_2^2(\lambda_1^2+\lambda_2^2+2\lambda_3^2), \end{aligned} \tag{4.2.14}$$

c_{22}, c_{33} being obtained from c_{11} by cyclic permutation of λ_1, λ_2, λ_3 and τ^{11}, τ^{22}, τ^{33}, and c_{23}, c_{31} being obtained from c_{12} by cyclic permutation of λ_1, λ_2, λ_3. Also

$$\left.\begin{aligned} c_{21}-c_{12} &= \tau^{11}-\tau^{22}\\ c_{32}-c_{23} &= \tau^{22}-\tau^{33}\\ c_{13}-c_{31} &= \tau^{33}-\tau^{11} \end{aligned}\right\}, \tag{4.2.15}$$

$$\left.\begin{aligned} c_{44} &= -\Psi\lambda_2^2\lambda_3^2-p \\ c_{55} &= -\Psi\lambda_3^2\lambda_1^2-p \\ c_{66} &= -\Psi\lambda_1^2\lambda_2^2-p \end{aligned}\right\}. \tag{4.2.16}$$

When the body is incompressible $I_3 = 1$ and $I_3' = 0$ so that we have

$$\lambda_1^2\lambda_2^2\lambda_3^2 = 1, \qquad \frac{\partial u}{\partial x}+\frac{\partial v}{\partial y}+\frac{\partial w}{\partial z} = 0, \tag{4.2.17}$$

and the stress tensor components are obtained from (4.1.16), (4.1.13), (4.2.3) to (4.2.11) in the form

$$\left.\begin{aligned} \tau'^{11} &= p'+a_{11}\frac{\partial u}{\partial x}+a_{12}\frac{\partial v}{\partial y}+a_{13}\frac{\partial w}{\partial z} \\ \tau'^{22} &= p'+a_{12}\frac{\partial u}{\partial x}+a_{22}\frac{\partial v}{\partial y}+a_{23}\frac{\partial w}{\partial z} \\ \tau'^{33} &= p'+a_{13}\frac{\partial u}{\partial x}+a_{23}\frac{\partial v}{\partial y}+a_{33}\frac{\partial w}{\partial z} \end{aligned}\right\}, \tag{4.2.18}$$

where

$$\left.\begin{aligned} a_{11} &= -2p+2\lambda_1^4\{A+B(\lambda_2^2+\lambda_3^2)^2+2F(\lambda_2^2+\lambda_3^2)\} \\ a_{22} &= -2p+2\lambda_2^4\{A+B(\lambda_3^2+\lambda_1^2)^2+2F(\lambda_3^2+\lambda_1^2)\} \\ a_{33} &= -2p+2\lambda_3^4\{A+B(\lambda_1^2+\lambda_2^2)^2+2F(\lambda_1^2+\lambda_2^2)\} \\ a_{12} &= 2\lambda_1^2\lambda_2^2\{\Psi+A+B(\lambda_2^2+\lambda_3^2)(\lambda_3^2+\lambda_1^2)+F(\lambda_1^2+\lambda_2^2+2\lambda_3^2)\} \\ a_{23} &= 2\lambda_2^2\lambda_3^2\{\Psi+A+B(\lambda_3^2+\lambda_1^2)(\lambda_1^2+\lambda_2^2)+F(\lambda_2^2+\lambda_3^2+2\lambda_1^2)\} \\ a_{13} &= 2\lambda_3^2\lambda_1^2\{\Psi+A+B(\lambda_1^2+\lambda_2^2)(\lambda_2^2+\lambda_3^2)+F(\lambda_3^2+\lambda_1^2+2\lambda_2^2)\} \end{aligned}\right\}. \tag{4.2.19}$$

The remaining stress components are given by (4.2.13) and (4.2.16).

When $F'^j = 0$ the equations of motion (4.1.18) reduce to

$$\frac{\partial\tau'^{ij}}{\partial\theta^i}+\tau^{im}\frac{\partial^2 w_j}{\partial\theta^i\partial\theta^m}+\tau^{ij}\frac{\partial^2 w_m}{\partial\theta^i\partial\theta^m} = \rho f'^j. \tag{4.2.20}$$

If we interchange the summation indices in the last term and remember that the stress components τ^{ij} are constants this equation may be written in the form

$$\frac{\partial}{\partial\theta^i}\left(\tau'^{ij}+\tau^{ir}\frac{\partial w_j}{\partial\theta^r}+\tau^{rj}\frac{\partial w_i}{\partial\theta^r}\right) = \rho f'^j. \tag{4.2.21}$$

For some purposes it is more convenient to have the stress components in the strained body B' referred to rectangular cartesian coordinates y_i which coincide with the axes (x, y, z). Thus

$$y_i = \theta_i+\epsilon w_i, \tag{4.2.22}$$

and we denote the components of the stress tensor referred to y_i-axes by $t^{rs}+\epsilon t'^{rs}$,† these being related to the stress components $\tau^{ij}+\epsilon\tau'^{ij}$ by the tensor transformation

$$t^{rs}+\epsilon t'^{rs} = \frac{\partial y_r}{\partial\theta^m}\frac{\partial y_s}{\partial\theta^n}(\tau^{mn}+\epsilon\tau'^{mn})$$
$$= \left(\delta^r_m+\epsilon\frac{\partial w_r}{\partial\theta^m}\right)\left(\delta^s_n+\epsilon\frac{\partial w_s}{\partial\theta^n}\right)(\tau^{mn}+\epsilon\tau'^{mn}). \qquad (4.2.23)$$

Thus, to the first order in ϵ,

$$t^{rs} = \tau^{rs}, \qquad t'^{rs} = \tau'^{rs}+\tau^{ms}\frac{\partial w_r}{\partial\theta^m}+\tau^{rm}\frac{\partial w_s}{\partial\theta^m}, \qquad (4.2.24)$$

and we see that the equations of motion (4.2.21) take the alternative form

$$\frac{\partial t'^{rs}}{\partial\theta^s} = \rho f'^r, \qquad (4.2.25)$$

where, since $f^r = 0$, $\epsilon f'^r$ are also the acceleration components along the y_r-axes. This result may also be obtained by observing that the equations of motion of the body B', referred to y_i-axes, are

$$\frac{\partial}{\partial y_s}(t^{rs}+\epsilon t'^{rs}) = \epsilon\rho f'^r,$$

and, using (4.2.22) and remembering that the stresses t^{rs} are constant, this reduces to (4.2.25) when we retain terms up to the order ϵ.

We now examine conditions to be satisfied at a boundary surface of B' at which surface forces are prescribed. Suppose that the corresponding surface in the body B is given in the parametric form

$$F(\theta_1,\theta_2,\theta_3) = F(x,y,z) = 0, \qquad (4.2.26)$$

and let $\mathbf{n}$ be the unit normal to this surface so that

$$\mathbf{n} = n_r\mathbf{G}^r, \qquad (4.2.27)$$

where

$$n_r = k\frac{\partial F}{\partial\theta^r}, \qquad (4.2.28)$$

and k is a constant. Since $\mathbf{n}$ is a unit vector

$$k\left(\frac{\partial F}{\partial\theta^r}\frac{\partial F}{\partial\theta^r}\right)^{\frac{1}{2}} = 1. \qquad (4.2.29)$$

The applied surface tractions at the surface of B', measured per unit area of the corresponding surface in B, are then given by (4.1.20)–(4.1.23) in the forms

$$\mathbf{P}+\epsilon\mathbf{P}' = k\frac{\partial F}{\partial\theta^i}(\tau^{ij}+\epsilon\lambda^{ij})\mathbf{G}_j, \qquad (4.2.30)$$

or

$$P^j = k\tau^{ij}\frac{\partial F}{\partial\theta^i}, \qquad P'^j = k\lambda^{ij}\frac{\partial F}{\partial\theta^i}, \qquad (4.2.31)$$

† These are also physical components of stress referred to y_i-axes.

where, from (4.1.15),

$$\lambda^{ij} = \tau'^{ij} + \tau^{im}\frac{\partial w_j}{\partial \theta^m} + \tau^{ij}\frac{\partial w_m}{\partial \theta^m}. \tag{4.2.32}$$

4.3. Small twist of a cylinder superposed on finite extension

We now apply the general theory of §§ 4.1, 4.2 to the solution of a particular problem.† We suppose that the unstrained body B_0 is a cylinder of constant cross-section R_0 whose generators are parallel to the x_3-axis, the plane ends of the cylinder being $x_3 = 0$, $x_3 = l_0$. The cylinder is first deformed into another cylinder B by a uniform finite extension of extension ratio $\lambda_3 = \lambda$ along the x_3-axis and by equal finite extensions of extension ratio $\lambda_1 = \lambda_2 = \mu$ along the other axes. With the notation of § 4.2,

$$x_1 = \frac{x}{\mu}, \qquad x_2 = \frac{y}{\mu}, \qquad x_3 = \frac{z}{\lambda}. \tag{4.3.1}$$

and the cylinder B has length $l = \lambda l_0$. We shall only consider the case where the cylinder B is maintained in equilibrium by a force parallel to its axis so that, from (4.2.7),

$$\left.\begin{aligned} \tau^{11} = \tau^{22} &= \Phi\mu^2 + \Psi\mu^2(\lambda^2+\mu^2) + p = 0 \\ \tau^{33} &= \Phi\lambda^2 + 2\Psi\lambda^2\mu^2 + p \end{aligned}\right\}, \tag{4.3.2}$$

or, eliminating p,

$$\tau^{33} = (\lambda^2 - \mu^2)(\Phi + \Psi\mu^2) = H \quad \text{(say)}. \tag{4.3.3}$$

The cylinder B now receives a small twist about the z-axis so that we assume the displacements w_r to have the form

$$u = -yz, \qquad v = zx, \qquad w = \phi(x, y), \tag{4.3.4}$$

and we take the parameter ϵ to represent the amount of twist per unit length of the cylinder B. It follows from (4.3.4) and the appropriate formulae in §§ 4.1, 4.2 that

$$-G'^{ij} = G'_{ij} = \begin{bmatrix} 0 & 0 & \phi_x - y \\ 0 & 0 & \phi_y + x \\ \phi_x - y & \phi_y + x & 0 \end{bmatrix}, \tag{4.3.5}$$

$$G' = I'_1 = I'_2 = I'_3 = \Phi' = \Psi' = p' = 0, \tag{4.3.6}$$

$$B'^{ij} = -\lambda^2\mu^2 G'_{ij}, \tag{4.3.7}$$

$$\left.\begin{aligned} \lambda^{23} = \tau'^{23} &= -(\Psi\lambda^2\mu^2 + p)(\phi_y + x) = \mu^2(\Phi + \Psi\mu^2)(\phi_y + x) \\ \lambda^{13} = \tau'^{13} &= -(\Psi\lambda^2\mu^2 + p)(\phi_x - y) = \mu^2(\Phi + \Psi\mu^2)(\phi_x - y) \\ \lambda^{11} = \lambda^{12} = \lambda^{21} &= \lambda^{22} = \tau'^{11} = \tau'^{22} = \tau'^{33} = \tau'^{12} = 0 \end{aligned}\right\}, \tag{4.3.8}$$

† This problem was originally solved without the aid of the present general theory by A. E. Green and R. T. Shield, *Phil. Trans. R. Soc.* **A244** (1951) 47.

and

$$\left.\begin{aligned} \lambda^{31} &= t'^{31} = (\Phi+\Psi\mu^2)(\mu^2\phi_x-\lambda^2 y) \\ \lambda^{32} &= t'^{32} = (\Phi+\Psi\mu^2)(\mu^2\phi_y+\lambda^2 x) \\ t^{33} &= \tau^{33} = H \\ \lambda^{33} &= t^{11} = t^{22} = t'^{11} = t'^{22} = t'^{33} = t'^{12} = 0 \end{aligned}\right\}. \tag{4.3.9}$$

The equations of motion (4.2.25), with $f'^r = 0$, are therefore satisfied provided

$$\nabla_1^2\phi = \phi_{xx}+\phi_{yy} = 0. \tag{4.3.10}$$

We now consider the boundary conditions. We suppose that the curved surface of the cylinder B is the surface

$$F(x,y) = 0, \tag{4.3.11}$$

where

$$F(\mu x_1, \mu x_2) = 0 \tag{4.3.12}$$

was the curved surface of the cylinder in the unstrained state B_0. If the bounding surface (4.3.11) of the final strained cylinder is to be free from applied traction then $P^j = P'^j = 0$, and from (4.2.31) and (4.3.11)

$$\tau^{1j}F_x+\tau^{2j}F_y = 0, \qquad \lambda^{1j}F_x+\lambda^{2j}F_y = 0, \tag{4.3.13}$$

where suffixes denote partial differentiation with respect to x and y. In view of (4.3.8) and (4.3.9) we see that the only surviving boundary condition corresponds to $j = 3$ in the second equation of (4.3.13), and this reduces to

$$(\phi_x-y)F_x+(\phi_y+x)F_y = 0 \quad \text{on } F(x,y) = 0. \tag{4.3.14}$$

The displacement function $\phi(x,y)$ is a plane harmonic function in any cross-section of the strained cylinder B, subject to the boundary condition (4.3.14) on the boundary of the cross-section of B. Any cross-section of the original unstrained cylinder B_0 is bounded by the curve (4.3.12), and the equation (4.3.10) and the boundary condition (4.3.14) may be written in the forms

$$\frac{\partial^2\phi}{\partial x_1^2}+\frac{\partial^2\phi}{\partial x_2^2} = 0,$$

$$\left(\frac{\partial\phi}{\partial x_1}-\mu^2 x_2\right)\frac{\partial F}{\partial x_1}+\left(\frac{\partial\phi}{\partial x_2}+\mu^2 x_1\right)\frac{\partial F}{\partial x_2} = 0 \quad \text{on } F(\mu x_1, \mu x_2) = 0.$$

We may therefore write

$$\phi(x,y) = \mu^2 w'(x_1,x_2), \tag{4.3.15}$$

where $w'(x_1,x_2)$ is the classical torsion function for a small torsion of the unstrained cylinder B_0. The function w' therefore satisfies the equation

$$\frac{\partial^2 w'}{\partial x_1^2}+\frac{\partial^2 w'}{\partial x_2^2} = 0, \tag{4.3.16}$$

and also the boundary condition

$$\left(\frac{\partial w'}{\partial x_1}-x_2\right)\frac{\partial F}{\partial x_1}+\left(\frac{\partial w'}{\partial x_2}+x_1\right)\frac{\partial F}{\partial x_2}=0 \tag{4.3.17}$$

on the curved surface of B_0.

Since the surface tractions on the curved surface of the cylinder are zero the cylinder is maintained in equilibrium by forces over its ends. The end $x_3 = l_0$ of the unstrained cylinder becomes the surface $z = l = \lambda l_0$ in the strained state B with a corresponding surface in the final strained state still being defined parametrically by $z = l$. The applied surface traction over this surface, measured per unit area of the surface $z = l$ in B, is from (4.2.30), (4.3.8), and (4.3.9),

$$\mathbf{P}+\epsilon\mathbf{P}' = \epsilon\lambda^{31}\mathbf{G}_1+\epsilon\lambda^{32}\mathbf{G}_2+H\mathbf{G}_3. \tag{4.3.18}$$

The element of area of the surface $z = l$ in the strained state B is

$$d\theta^1 d\theta^2 = dxdy. \tag{4.3.19}$$

Denoting the components of resultant force over the end of the cylinder by Y^i, referred to (x, y, z) axes, we have from (4.3.18), (4.3.19), and (4.3.9),

$$Y^1 = \epsilon \iint\limits_R (\Phi+\Psi\mu^2)(\mu^2\phi_x-\lambda^2 y)\, dxdy,$$

$$Y^2 = \epsilon \iint\limits_R (\Phi+\Psi\mu^2)(\mu^2\phi_y+\lambda^2 x)\, dxdy,$$

$$Y^3 = \iint\limits_R H\, dxdy.$$

These expressions for Y^i may be simplified. Since ϕ is a plane harmonic function of x and y,

$$\iint (\phi_x-y)\, dxdy = \iint \left[\frac{\partial}{\partial x}\{x(\phi_x-y)\}+\frac{\partial}{\partial y}\{x(\phi_y+x)\}\right] dxdy$$

$$= \int x\{(\phi_x-y)F_x+(\phi_y+x)F_y\}\, ds = 0,$$

where the line integral, which is evaluated over the boundary of the region R, vanishes in view of the boundary condition (4.3.14). Similarly

$$\iint (\phi_y+x)\, dxdy = \int y\{(\phi_x-y)F_x+(\phi_y+x)F_y\}\, ds = 0.$$

Hence, recalling (4.3.3),

$$Y^1 = -\epsilon HA_x, \qquad Y^2 = \epsilon HA_y, \qquad Y^3 = HA = H\mu^2 A_0, \tag{4.3.20}$$

where A, A_0 are the areas of the cross-section of the strained cylinder B

and the unstrained cylinder B_0 respectively, and

$$A_x = \iint\limits_R y\,dxdy, \qquad A_y = \iint\limits_R x\,dxdy. \tag{4.3.21}$$

The moments M^1, M^2 of the total traction on the end $z = l$ of the final deformed cylinder about axes through the point $(0, 0, l)$ and parallel to the x-axis and y-axis respectively, are given by

$$M^1 = \iint\limits_R \{Hy_2 - \epsilon\lambda^{32}(y_3 - l)\}_{z=l}\,dxdy,$$

$$M^2 = \iint\limits_R \{\epsilon(y_3 - l)\lambda^{31} - y_1 H\}_{z=l}\,dxdy,$$

and neglecting second-order terms in ϵ and using (4.3.9), these become

$$\left.\begin{aligned} M^1 &= H\iint\limits_R (y+\epsilon lx)\,dxdy = HA_x + \epsilon lHA_y \\ M^2 &= -H\iint\limits_R (x - \epsilon ly)\,dxdy = -HA_y + \epsilon lHA_x \end{aligned}\right\}. \tag{4.3.22}$$

The moment M^3 of the traction on $z = l$ about the z-axis is given by

$$M^3 = \epsilon \iint\limits_R (y_1\lambda^{32} - y_2\lambda^{31})\,dxdy$$

or, to first order in ϵ, using also (4.3.9),

$$\begin{aligned} M^3 &= \epsilon(\Phi + \Psi\mu^2)\iint\limits_R \{\mu^2(x\phi_y - y\phi_x) + \lambda^2(x^2+y^2)\}\,dxdy \\ &= \epsilon HI + \epsilon\mu^2(\Phi+\Psi\mu^2)S, \end{aligned} \tag{4.3.23}$$

where

$$\left.\begin{aligned} I &= \iint\limits_R (x^2+y^2)\,dxdy \\ S &= \iint\limits_R (x^2+y^2+x\phi_y - y\phi_x)\,dxdy \end{aligned}\right\}. \tag{4.3.24}$$

Using (4.3.1) and (4.3.15) we see that

$$S = \mu^4 S_0, \qquad I = \mu^4 I_0, \tag{4.3.25}$$

where

$$\left.\begin{aligned} I_0 &= \iint\limits_{R_0} (x_1^2 + x_2^2)\,dx_1\,dx_2 \\ S_0 &= \iint\limits_{R_0} \left(x_1^2 + x_2^2 + x_1\frac{\partial w'}{\partial x_2} - x_2\frac{\partial w'}{\partial x_1}\right)dx_1\,dx_2 \end{aligned}\right\}, \tag{4.3.26}$$

so that I_0 is the moment of inertia of the unstrained cross-section R_0 about the x_3-axis and S_0 is the geometrical torsional rigidity of the unextended cylinder when subject to a small twist. The function S_0 is known for a variety of cross-sections from classical theory and can be evaluated when the classical boundary value problem defined by equations (4.3.16) and (4.3.17) has been solved.† It is well known that the geometrical torsional rigidity S_0 is independent of the axis of torsion, but the twisting couple M^3 in (4.3.23) depends on the axis of torsion, which is here taken to be the x_3-axis, because of the term containing I. If H, which is given by (4.3.3), is positive, and $\lambda > \mu$, then M^3 is least when I (or I_0) is least, that is when the axis of torsion coincides with the line of centroids of the cross-sections of the unstrained cylinder B_0 (or of the strained cylinder B). In this case, A_x and A_y are zero and, from (4.3.20) and (4.3.22), we see that Y^1, Y^2, M^1, M^2 all vanish.

From (4.3.3), (4.3.23), (4.3.25), and (4.3.26),

$$M^3 = \epsilon\mu^4(\Phi+\Psi\mu^2)\{\lambda^2 I_0-\mu^2(I_0-S_0)\}. \tag{4.3.27}$$

It is known‡ that $S_0 \leqslant I_0$, and the equality sign holds only when the cross-section is a circle or a circular ring bounded by two concentric circles. Hence we see from (4.3.27) that when the cylinder is not a circular cylinder or circular cylindrical tube, the twisting couple will be zero, to the first order in ϵ at least, when

$$\frac{\lambda^2}{\mu^2} = \frac{I_0-S_0}{I_0} < 1, \tag{4.3.28}$$

provided $\mu^4(\Phi+\Psi\mu^2)$ is finite for this value of λ.

The value of μ is given in terms of λ by (4.3.2) and cannot be determined explicitly unless the particular form for the strain energy is known. For most bodies, however, it is probable that μ will be greater than unity when λ is less than unity and the cylinder is compressed in the direction of its length, and in this case a value of λ may exist for which (4.3.28) is true.

When the body is incompressible $\mu = 1/\surd\lambda$ and equation (4.3.28) becomes

$$\lambda^3 = \frac{I_0-S_0}{I_0}, \tag{4.3.29}$$

and the value of λ determined by this equation is independent of the particular form of the strain energy function which is used.

From (4.3.20) and (4.3.27),

$$\frac{Y^3}{M^3/\epsilon} = \frac{(1/\mu^2-1/\lambda^2)A_0}{I_0-(I_0-S_0)\mu^2/\lambda^2}, \tag{4.3.30}$$

and this relates the force necessary to produce a large simple extension

† See, e.g., I. S. Sokolnikoff, *Mathematical Theory of Elasticity* (2nd ed., New York, 1956).
‡ J. B. Diaz and A. Weinstein, *Am. J. Math.* 70 (1948) 107.

with the torsional modulus for a small twist superposed on that simple extension. When the material is incompressible (4.3.30) becomes

$$\frac{Y^3}{M^3/\epsilon} = \frac{(\lambda - 1/\lambda^2)A_0}{I_0 - (I_0 - S_0)/\lambda^3}, \tag{4.3.31}$$

and, in this case, the relation is independent of the particular form of the strain energy function which applies to the body.

When the cylinder is circular $I_0 = S_0 = \frac{1}{2}\pi a^4$, $A_0 = \pi a^2$, where a is the radius in the undeformed state, so that (4.3.31) reduces to

$$\frac{a^2 Y^3}{M^3/\epsilon} = 2(\lambda - 1/\lambda^2), \tag{4.3.32}$$

which was obtained previously in (3.5.9).

4.4. Solution in terms of potential functions: compressible case

We now return to the general theory of § 4.2 and develop a general method of solution for problems in which the finite extensions are maintained by equal forces in the x_1- and x_2-directions so that

$$\lambda_1 = \lambda_2 = \mu, \qquad \lambda_3 = \lambda, \tag{4.4.1}$$

and, from (4.2.7),

$$\left.\begin{aligned} \tau^{11} = \tau^{22} &= \Phi\mu^2 + \Psi\mu^2(\lambda^2+\mu^2) + p \\ \tau^{33} &= \Phi\lambda^2 + 2\Psi\lambda^2\mu^2 + p \end{aligned}\right\}. \tag{4.4.2}$$

From (4.2.12)–(4.2.16) and (4.4.1) we have

$$\left.\begin{aligned} \tau'^{11} &= c_{11}\frac{\partial u}{\partial x} + c_{12}\frac{\partial v}{\partial y} + c_{13}\frac{\partial w}{\partial z} \\ \tau'^{22} &= c_{12}\frac{\partial u}{\partial x} + c_{11}\frac{\partial v}{\partial y} + c_{13}\frac{\partial w}{\partial z} \\ \tau'^{33} &= c_{31}\left(\frac{\partial u}{\partial x} + \frac{\partial v}{\partial y}\right) + c_{33}\frac{\partial w}{\partial z} \end{aligned}\right\}, \tag{4.4.3}$$

$$\left.\begin{aligned} \tau'^{12} &= \tfrac{1}{2}(c_{11} - c_{12})\left(\frac{\partial u}{\partial y} + \frac{\partial v}{\partial x}\right) \\ \tau'^{23} &= c_{44}\left(\frac{\partial v}{\partial z} + \frac{\partial w}{\partial y}\right) \\ \tau'^{31} &= c_{44}\left(\frac{\partial w}{\partial x} + \frac{\partial u}{\partial z}\right) \end{aligned}\right\}, \tag{4.4.4}$$

where the coefficients $c_{11},...$ now take the values

$$\left.\begin{aligned}
c_{11} &= -\tau^{11}+2\mu^4\{A+B(\lambda^2+\mu^2)^2+C\lambda^4\mu^4+\\
&\qquad +2D\lambda^2\mu^2(\lambda^2+\mu^2)+2E\lambda^2\mu^2+2F(\lambda^2+\mu^2)\}\\
c_{12} &= -\Phi\mu^2+\Psi\mu^2(\mu^2-\lambda^2)+p+2\mu^4\{A+B(\lambda^2+\mu^2)^2+\\
&\qquad +C\lambda^4\mu^4+2D\lambda^2\mu^2(\lambda^2+\mu^2)+2E\lambda^2\mu^2+2F(\lambda^2+\mu^2)\}\\
c_{13} &= -\Phi\mu^2+\Psi\mu^2(\lambda^2-\mu^2)+p+\\
&\qquad +2\lambda^2\mu^2\{A+2B\mu^2(\lambda^2+\mu^2)+C\lambda^2\mu^6+\\
&\qquad +D\mu^4(3\lambda^2+\mu^2)+E\mu^2(\lambda^2+\mu^2)+F(\lambda^2+3\mu^2)\}\\
c_{33} &= -\tau^{33}+2\lambda^4\{A+4B\mu^4+C\mu^8+4D\mu^6+\\
&\qquad +2E\mu^4+4F\mu^2\}\\
c_{31}-c_{13} &= (\mu^2-\lambda^2)(\Phi+\Psi\mu^2) = \tau^{11}-\tau^{33}\\
c_{11}-c_{12} &= -2(\Psi\mu^4+p)\\
c_{44} &= -\Psi\lambda^2\mu^2-p
\end{aligned}\right\}. \tag{4.4.5}$$

Also, from (4.2.24),

$$t^{11} = t^{22} = \tau^{11}, \qquad t^{33} = \tau^{33}, \qquad t^{12} = t^{23} = t^{31} = 0, \tag{4.4.6}$$

$$\left.\begin{aligned}
t'^{11} &= \tau'^{11}+2\tau^{11}\frac{\partial u}{\partial x}, & t'^{12} &= \tau'^{12}+\tau^{11}\left(\frac{\partial u}{\partial y}+\frac{\partial v}{\partial x}\right)\\
t'^{22} &= \tau'^{22}+2\tau^{11}\frac{\partial v}{\partial y}, & t'^{23} &= \tau'^{23}+\tau^{11}\frac{\partial w}{\partial y}+\tau^{33}\frac{\partial v}{\partial z}\\
t'^{33} &= \tau'^{33}+2\tau^{33}\frac{\partial w}{\partial z}, & t'^{31} &= \tau'^{31}+\tau^{11}\frac{\partial w}{\partial x}+\tau^{33}\frac{\partial u}{\partial z}
\end{aligned}\right\}, \tag{4.4.7}$$

and with the help of (4.4.3) and (4.4.4), equations (4.4.7) become

$$\left.\begin{aligned}
t'^{11} &= d_{11}\frac{\partial u}{\partial x}+d_{12}\frac{\partial v}{\partial y}+d_{13}\frac{\partial w}{\partial z}\\
t'^{22} &= d_{12}\frac{\partial u}{\partial x}+d_{11}\frac{\partial v}{\partial y}+d_{13}\frac{\partial w}{\partial z}\\
t'^{33} &= d_{31}\left(\frac{\partial u}{\partial x}+\frac{\partial v}{\partial y}\right)+d_{33}\frac{\partial w}{\partial z}
\end{aligned}\right\}, \tag{4.4.8}$$

$$\left.\begin{aligned}
t'^{12} &= \tfrac{1}{2}(d_{11}-d_{12})\left(\frac{\partial u}{\partial y}+\frac{\partial v}{\partial x}\right)\\
t'^{23} &= d_{44}\frac{\partial v}{\partial z}+d_{55}\frac{\partial w}{\partial y}\\
t'^{31} &= d_{44}\frac{\partial u}{\partial z}+d_{55}\frac{\partial w}{\partial x}
\end{aligned}\right\}, \tag{4.4.9}$$

where

$$\left.\begin{aligned} d_{11} &= c_{11}+2\tau^{11}, \qquad d_{33} = c_{33}+2\tau^{33} \\ d_{12} &= c_{12}, \qquad d_{13} = c_{13}, \qquad d_{31} = c_{31} \\ d_{11}-d_{12} &= c_{11}-c_{12}+2\tau^{11} = 2\mu^2(\Phi+\Psi\lambda^2) \\ d_{44} &= c_{44}+\tau^{33} = \lambda^2(\Phi+\Psi\mu^2) \\ d_{55} &= c_{44}+\tau^{11} = \mu^2(\Phi+\Psi\mu^2) \\ d_{55}-d_{44} &= c_{31}-c_{13} = d_{31}-d_{13} = \tau^{11}-\tau^{33} \end{aligned}\right\}. \tag{4.4.10}$$

Equations (4.4.8) and (4.4.9) are in some respects similar to the corresponding equations in infinitesimal elasticity when the material is transversely isotropic.†

We put
$$u = \frac{\partial\phi_1}{\partial x}+\frac{\partial\phi_3}{\partial y}, \qquad v = \frac{\partial\phi_1}{\partial y}-\frac{\partial\phi_3}{\partial x}, \qquad w = \frac{\partial\phi_2}{\partial z}. \tag{4.4.11}$$

If u, v, w are given functions of x, y, z, then

$$\phi_1 = \phi_1'+\frac{\partial\psi_1}{\partial y}, \qquad \phi_3 = \phi_3'-\frac{\partial\psi_1}{\partial x}, \qquad \phi_2 = \phi_2'+\psi_2, \tag{4.4.12}$$

where ϕ_1', ϕ_2', ϕ_3' are any particular integrals of the equations

$$\nabla_1^2\phi_1' = \frac{\partial u}{\partial x}+\frac{\partial v}{\partial y}, \qquad \nabla_1^2\phi_3' = \frac{\partial u}{\partial y}-\frac{\partial v}{\partial x}, \qquad \frac{\partial\phi_2'}{\partial z} = w. \tag{4.4.13}$$

Also ψ_1 is an arbitrary function of x, y, z subject to the condition

$$\nabla_1^2\psi_1 = 0, \tag{4.4.14}$$

where
$$\nabla_1^2 = \frac{\partial^2}{\partial x^2}+\frac{\partial^2}{\partial y^2}, \tag{4.4.15}$$

and ψ_2 is an arbitrary function of x, y. We shall not examine here the conditions under which ϕ_1', ϕ_2', ϕ_3' can be found. The functions ψ_1, ψ_2 do not contribute to the displacements and may be omitted from ϕ_1, ϕ_2, ϕ_3.

It follows from (4.4.8), (4.4.9), and (4.4.11) that

$$\left.\begin{aligned} t'^{11} &= d_{11}\frac{\partial^2\phi_1}{\partial x^2}+d_{12}\frac{\partial^2\phi_1}{\partial y^2}+(d_{11}-d_{12})\frac{\partial^2\phi_3}{\partial x\partial y}+d_{13}\frac{\partial^2\phi_2}{\partial z^2} \\ t'^{22} &= d_{12}\frac{\partial^2\phi_1}{\partial x^2}+d_{11}\frac{\partial^2\phi_1}{\partial y^2}-(d_{11}-d_{12})\frac{\partial^2\phi_3}{\partial x\partial y}+d_{13}\frac{\partial^2\phi_2}{\partial z^2} \\ t'^{33} &= d_{31}\nabla_1^2\phi_1+d_{33}\frac{\partial^2\phi_2}{\partial z^2} \end{aligned}\right\}, \tag{4.4.16}$$

† See § 5.12.

and

$$\left.\begin{aligned} t'^{12} &= \tfrac{1}{2}(d_{11}-d_{12})\left(2\frac{\partial^2\phi_1}{\partial x\partial y}-\frac{\partial^2\phi_3}{\partial x^2}+\frac{\partial^2\phi_3}{\partial y^2}\right) \\ t'^{23} &= d_{44}\left(\frac{\partial^2\phi_1}{\partial y\partial z}-\frac{\partial^2\phi_3}{\partial x\partial z}\right)+d_{55}\frac{\partial^2\phi_2}{\partial y\partial z} \\ t'^{31} &= d_{44}\left(\frac{\partial^2\phi_1}{\partial x\partial z}+\frac{\partial^2\phi_3}{\partial y\partial z}\right)+d_{55}\frac{\partial^2\phi_2}{\partial x\partial z} \end{aligned}\right\}. \tag{4.4.17}$$

If the strained body is in equilibrium the stress equations of equilibrium (4.2.25) with $f'^r = 0$ are satisfied provided

$$\frac{\partial}{\partial x}\left\{d_{11}\nabla_1^2\phi_1+d_{44}\frac{\partial^2\phi_1}{\partial z^2}+(d_{13}+d_{55})\frac{\partial^2\phi_2}{\partial z^2}\right\}+ \\ +\frac{\partial}{\partial y}\left\{\tfrac{1}{2}(d_{11}-d_{12})\nabla_1^2\phi_3+d_{44}\frac{\partial^2\phi_3}{\partial z^2}\right\}=0,$$

$$\frac{\partial}{\partial y}\left\{d_{11}\nabla_1^2\phi_1+d_{44}\frac{\partial^2\phi_1}{\partial z^2}+(d_{13}+d_{55})\frac{\partial^2\phi_2}{\partial z^2}\right\}- \\ -\frac{\partial}{\partial x}\left\{\tfrac{1}{2}(d_{11}-d_{12})\nabla_1^2\phi_3+d_{44}\frac{\partial^2\phi_3}{\partial z^2}\right\}=0,$$

$$\frac{\partial}{\partial z}\left\{(d_{31}+d_{44})\nabla_1^2\phi_1+d_{55}\nabla_1^2\phi_2+d_{33}\frac{\partial^2\phi_2}{\partial z^2}\right\}=0.$$

Hence

$$\left.\begin{aligned} \tfrac{1}{2}(d_{11}-d_{12})\nabla_1^2\phi_3+d_{44}\frac{\partial^2\phi_3}{\partial z^2} &= -d_{44}\frac{\partial f_1}{\partial x} \\ d_{11}\nabla_1^2\phi_1+d_{44}\frac{\partial^2\phi_1}{\partial z^2}+(d_{13}+d_{55})\frac{\partial^2\phi_2}{\partial z^2} &= d_{44}\frac{\partial f_1}{\partial y} \\ (d_{31}+d_{44})\nabla_1^2\phi_1+d_{55}\nabla_1^2\phi_2+d_{33}\frac{\partial^2\phi_2}{\partial z^2} &= d_{55}f_2 \end{aligned}\right\}, \tag{4.4.18}$$

where f_2 is an arbitrary function of x, y and where f_1 is an arbitrary function of x, y, z satisfying the equation

$$\nabla_1^2 f_1 = 0. \tag{4.4.19}$$

Let $g(x,y,z)$ be any particular integral of the equation

$$\frac{\partial^2 g}{\partial z^2} = f_1(x,y,z). \tag{4.4.20}$$

Then

$$\frac{\partial^2}{\partial z^2}\nabla_1^2 g = \nabla_1^2\frac{\partial^2 g}{\partial z^2} = \nabla_1^2 f_1 = 0, \tag{4.4.21}$$

and hence

$$\nabla_1^2 g = \alpha(x,y)z+\beta(x,y), \tag{4.4.22}$$

where α, β are arbitrary functions of x, y. Next define

$$g_1(x,y,z) = g(x,y,z)+A(x,y)z+B(x,y), \tag{4.4.23}$$

where
$$\nabla_1^2 A = -\alpha, \qquad \nabla_1^2 B = -\beta. \tag{4.4.24}$$

It follows that
$$\frac{\partial^2 g_1}{\partial z^2} = \frac{\partial^2 g}{\partial z^2} = f_1, \tag{4.4.25}$$

and
$$\nabla_1^2 g_1 = 0. \tag{4.4.26}$$

We can now take a particular integral of the equations (4.4.18) in the form
$$\phi_1 = \frac{\partial g_1}{\partial y}, \qquad \phi_3 = -\frac{\partial g_1}{\partial x}, \qquad \phi_2 = g_2(x,y), \tag{4.4.27}$$

where g_1 satisfies (4.4.26) and g_2 is a particular integral of the equation
$$\nabla_1^2 g_2 = f_2(x,y). \tag{4.4.28}$$

The particular integrals (4.4.27), however, do not contribute to the displacements and may be ignored, so that without loss of generality we may put $f_1 = f_2 \equiv 0$ in the right-hand side of (4.4.18).

In order to complete the solution we put
$$\frac{d_{11}\nu - d_{44}}{d_{13}+d_{55}} = \frac{(d_{31}+d_{44})\nu}{d_{33}-d_{55}\nu} = k, \tag{4.4.29}$$

so that
$$d_{11}d_{55}\nu^2 + \{(d_{13}+d_{55})^2 - d_{11}d_{33} - d_{44}d_{55}\}\nu + d_{33}d_{44} = 0. \tag{4.4.30}$$

At present we assume that the quadratic equation (4.4.30) has two distinct roots ν_1 and ν_2 so that the corresponding values k_1, k_2 of k, given by (4.4.29), are distinct. Without loss of generality we may now put
$$\phi_1 = \chi_1 + \chi_2, \qquad \phi_2 = k_1\chi_1 + k_2\chi_2, \tag{4.4.31}$$

and substitute into the second and third of equations (4.4.18). Thus, using (4.4.29), we find that
$$\left(\nabla_1^2 + \nu_1\frac{\partial^2}{\partial z^2}\right)\chi_1 + \left(\nabla_1^2 + \nu_2\frac{\partial^2}{\partial z^2}\right)\chi_2 = 0,$$
$$\frac{k_1}{\nu_1}\left(\nabla_1^2 + \nu_1\frac{\partial^2}{\partial z^2}\right)\chi_1 + \frac{k_2}{\nu_2}\left(\nabla_1^2 + \nu_2\frac{\partial^2}{\partial z^2}\right)\chi_2 = 0.$$

If $k_1/\nu_1 \neq k_2/\nu_2$ it follows that
$$\left(\nabla_1^2 + \nu_\alpha\frac{\partial^2}{\partial z^2}\right)\chi_\alpha = 0 \quad (\alpha = 1, 2;\ \alpha \text{ not summed}). \tag{4.4.32}$$

To complete the notation we put
$$\phi_3 = \chi_3, \tag{4.4.33}$$

where, from (4.4.18),

$$\left(\nabla_1^2+\nu_3\frac{\partial^2}{\partial z^2}\right)\chi_3 = 0, \qquad \nu_3 = \frac{2d_{44}}{d_{11}-d_{12}}. \tag{4.4.34}$$

In its present form the theory is valid for a body which is homogeneous and isotropic in its unstrained state and we may proceed to write down the appropriate values of the stresses and displacements in terms of χ_α and χ_3. We shall not, however, continue this general discussion here but will consider the special case of an incompressible body in more detail in the next section. We observe that the above solution is degenerate in certain circumstances, e.g. when $\nu_1 = \nu_2$. In most problems the solution in any degenerate case can be found by a limiting process from the general solution.

4.5. Solution in terms of potential functions: incompressible case

When the body is incompressible we have, from (4.2.17) and (4.4.1),

$$\lambda\mu^2 = 1, \qquad \frac{\partial u}{\partial x}+\frac{\partial v}{\partial y}+\frac{\partial w}{\partial z} = 0 \tag{4.5.1}$$

and the scalar invariant p is found in terms of stresses from (4.4.2). Also, from (4.2.18), (4.2.19), and (4.5.1), we find that

$$\left.\begin{aligned} \tau'^{11} &= p'+\alpha\frac{\partial u}{\partial x}+\beta\frac{\partial v}{\partial y} \\ \tau'^{22} &= p'+\beta\frac{\partial u}{\partial x}+\alpha\frac{\partial v}{\partial y} \\ \tau'^{33} &= p'+\gamma\frac{\partial w}{\partial z} \end{aligned}\right\}, \tag{4.5.2}$$

where

$$\left.\begin{aligned} \alpha &= -2\Psi\lambda^2\mu^2-2p+2\mu^2(\mu^2-\lambda^2)\{A+B\mu^2(\lambda^2+\mu^2)+F(\lambda^2+2\mu^2)\} \\ \beta &= 2\mu^2(\mu^2-\lambda^2)\{\Psi+A+B\mu^2(\mu^2+\lambda^2)+F(2\mu^2+\lambda^2)\} \\ \gamma &= -2\Psi\lambda^2\mu^2-2p+2\lambda^2(\lambda^2-\mu^2)(A+2B\mu^4+3F\mu^2) \end{aligned}\right\}. \tag{4.5.3}$$

The remaining components of stress τ'^{12}, τ'^{23}, τ'^{31} are still given by (4.4.4), and we note that

$$c_{11}-c_{12} = \alpha-\beta, \qquad c_{44} = -\Psi\lambda^2\mu^2-p. \tag{4.5.4}$$

From (4.4.7) and (4.5.2),

$$\left.\begin{aligned} t'^{11} &= p'+a\frac{\partial u}{\partial x}+b\frac{\partial v}{\partial y} \\ t'^{22} &= p'+b\frac{\partial u}{\partial x}+a\frac{\partial v}{\partial y} \\ t'^{33} &= p'+c\frac{\partial w}{\partial z} \end{aligned}\right\}, \tag{4.5.5}$$

where $\quad a = \alpha+2\tau^{11}, \quad b = \beta, \quad c = \gamma+2\tau^{33},$ (4.5.6)

the components t'^{12}, t'^{23}, t'^{31} being given by (4.4.9), and

$$d_{11}-d_{12} = a-b, \quad d_{44} = \lambda^2(\Phi+\Psi\mu^2), \quad d_{55} = \mu^2(\Phi+\Psi\mu^2). \tag{4.5.7}$$

The method of solution is now similar to that used in § 4.4 for the general compressible material. Assuming the displacements to be of the form (4.4.11) we have, because of the incompressibility condition (4.5.1),

$$\nabla_1^2\phi_1+\frac{\partial^2\phi_2}{\partial z^2} = 0. \tag{4.5.8}$$

Also, from (4.4.9), (4.5.5), (4.4.11), and (4.5.7),

$$\left.\begin{aligned} t'^{11} &= p'+a\frac{\partial^2\phi_1}{\partial x^2}+b\frac{\partial^2\phi_1}{\partial y^2}+(a-b)\frac{\partial^2\phi_3}{\partial x\partial y} \\ t'^{22} &= p'+b\frac{\partial^2\phi_1}{\partial x^2}+a\frac{\partial^2\phi_1}{\partial y^2}-(a-b)\frac{\partial^2\phi_3}{\partial x\partial y} \\ t'^{33} &= p'+c\frac{\partial^2\phi_2}{\partial z^2} \end{aligned}\right\}, \tag{4.5.9}$$

$$\left.\begin{aligned} t'^{12} &= \tfrac{1}{2}(a-b)\left(2\frac{\partial^2\phi_1}{\partial x\partial y}-\frac{\partial^2\phi_3}{\partial x^2}+\frac{\partial^2\phi_3}{\partial y^2}\right) \\ t'^{23} &= d_{44}\left(\frac{\partial^2\phi_1}{\partial y\partial z}-\frac{\partial^2\phi_3}{\partial x\partial z}\right)+d_{55}\frac{\partial^2\phi_2}{\partial y\partial z} \\ t'^{31} &= d_{44}\left(\frac{\partial^2\phi_1}{\partial x\partial z}+\frac{\partial^2\phi_3}{\partial y\partial z}\right)+d_{55}\frac{\partial^2\phi_2}{\partial x\partial z} \end{aligned}\right\}. \tag{4.5.10}$$

By an analysis similar to that given in § 4.4 we may show that the equations of equilibrium (4.2.25) (with $f'^r = 0$) can be integrated to give

$$\left.\begin{aligned} p'+(a-d_{55})\nabla_1^2\phi_1+d_{44}\frac{\partial^2\phi_1}{\partial z^2} &= 0 \\ p'+d_{55}\nabla_1^2\phi_2+(c-d_{44})\frac{\partial^2\phi_2}{\partial z^2} &= 0 \\ \tfrac{1}{2}(a-b)\nabla_1^2\phi_3+d_{44}\frac{\partial^2\phi_3}{\partial z^2} &= 0 \end{aligned}\right\}. \tag{4.5.11}$$

Without loss of generality we now put

$$\phi_1 = \chi_1+\chi_2, \quad \phi_2 = k_1\chi_1+k_2\chi_2, \quad \phi_3 = \chi_3, \tag{4.5.12}$$

where k_1, k_2 are assumed to be distinct roots of the quadratic equation

$$k^2 d_{55}+k(d_{44}+d_{55}-a-c)+d_{44} = 0. \tag{4.5.13}$$

It follows from (4.5.8) and (4.5.11) that

$$\left(\nabla_1^2+k_\alpha\frac{\partial^2}{\partial z^2}\right)\chi_\alpha = 0 \quad (\alpha = 1,2,3;\ \alpha \text{ not summed}), \tag{4.5.14}$$

where
$$k_3 = \frac{2d_{44}}{a-b},$$

and

$$\left.\begin{aligned} p' &= (k_1 a-k_1 d_{55}-d_{44})\frac{\partial^2\chi_1}{\partial z^2}+(k_2 a-k_2 d_{55}-d_{44})\frac{\partial^2\chi_2}{\partial z^2} \\ &= k_1(k_1 d_{55}+d_{44}-c)\frac{\partial^2\chi_1}{\partial z^2}+k_2(k_2 d_{55}+d_{44}-c)\frac{\partial^2\chi_2}{\partial z^2} \end{aligned}\right\}. \tag{4.5.15}$$

The displacements (4.4.11) are therefore

$$u = \frac{\partial}{\partial x}(\chi_1+\chi_2)+\frac{\partial\chi_3}{\partial y}, \qquad v = \frac{\partial}{\partial y}(\chi_1+\chi_2)-\frac{\partial\chi_3}{\partial x},$$
$$w = k_1\frac{\partial\chi_1}{\partial z}+k_2\frac{\partial\chi_2}{\partial z}. \tag{4.5.16}$$

The components of stress τ'^{rs} can now be found in terms of the functions χ_α, by substituting in (4.4.4) and (4.5.2) from (4.5.15) and (4.5.16), and we have

$$\left.\begin{aligned} \tau'^{11} &= \left(d_{55}+\frac{d_{44}}{k_1}\right)\nabla_1^2\chi_1+\left(d_{55}+\frac{d_{44}}{k_2}\right)\nabla_1^2\chi_2+(b-a)\frac{\partial^2}{\partial y^2}(\chi_1+\chi_2)- \\ &\qquad -2\tau^{11}\left(\frac{\partial^2\chi_1}{\partial x^2}+\frac{\partial^2\chi_2}{\partial x^2}+\frac{\partial^2\chi_3}{\partial x\partial y}\right)+(a-b)\frac{\partial^2\chi_3}{\partial x\partial y} \\ \tau'^{22} &= \left(d_{55}+\frac{d_{44}}{k_1}\right)\nabla_1^2\chi_1+\left(d_{55}+\frac{d_{44}}{k_2}\right)\nabla_1^2\chi_2+(b-a)\frac{\partial^2}{\partial x^2}(\chi_1+\chi_2)- \\ &\qquad -2\tau^{11}\left(\frac{\partial^2\chi_1}{\partial y^2}+\frac{\partial^2\chi_2}{\partial y^2}-\frac{\partial^2\chi_3}{\partial x\partial y}\right)-(a-b)\frac{\partial^2\chi_3}{\partial x\partial y} \\ \tau'^{33} &= k_1(k_1 d_{55}+d_{44}-2\tau^{33})\frac{\partial^2\phi_1}{\partial z^2}+k_2(k_2 d_{55}+d_{44}-2\tau^{33})\frac{\partial^2\phi_2}{\partial z^2} \end{aligned}\right\},$$

(4.5.17

$$\left.\begin{aligned}\tau'^{12} &= (\alpha-\beta)\left(\frac{\partial^2\chi_1}{\partial x\partial y}+\frac{\partial^2\chi_2}{\partial x\partial y}+\frac{1}{2}\frac{\partial^2\chi_3}{\partial y^2}-\frac{1}{2}\frac{\partial^2\chi_3}{\partial x^2}\right)\\ \tau'^{23} &= c_{44}\left\{(1+k_1)\frac{\partial^2\chi_1}{\partial y\partial z}+(1+k_2)\frac{\partial^2\chi_2}{\partial y\partial z}-\frac{\partial^2\chi_3}{\partial x\partial z}\right\}\\ \tau'^{31} &= c_{44}\left\{(1+k_1)\frac{\partial^2\chi_1}{\partial z\partial x}+(1+k_2)\frac{\partial^2\chi_2}{\partial z\partial x}+\frac{\partial^2\chi_3}{\partial y\partial z}\right\}\end{aligned}\right\}. \tag{4.5.18}$$

Also, from (4.5.9) and (4.5.12)–(4.5.15), we find that the components t'^{rs} of stress are

$$\left.\begin{aligned}t'^{11} &= \left(d_{55}+\frac{d_{44}}{k_1}\right)\nabla_1^2\chi_1+\left(d_{55}+\frac{d_{44}}{k_2}\right)\nabla_1^2\chi_2+\\ &\qquad +(b-a)\left(\frac{\partial^2\chi_1}{\partial y^2}+\frac{\partial^2\chi_2}{\partial y^2}-\frac{\partial^2\chi_3}{\partial x\partial y}\right)\\ t'^{22} &= \left(d_{55}+\frac{d_{44}}{k_1}\right)\nabla_1^2\chi_1+\left(d_{55}+\frac{d_{44}}{k_2}\right)\nabla_1^2\chi_2+\\ &\qquad +(b-a)\left(\frac{\partial^2\chi_1}{\partial x^2}+\frac{\partial^2\chi_2}{\partial x^2}+\frac{\partial^2\chi_3}{\partial x\partial y}\right)\\ t'^{33} &= k_1(k_1 d_{55}+d_{44})\frac{\partial^2\chi_1}{\partial z^2}+k_2(k_2 d_{55}+d_{44})\frac{\partial^2\chi_2}{\partial z^2}\end{aligned}\right\}, \tag{4.5.19}$$

and

$$\left.\begin{aligned}t'^{12} &= (a-b)\left(\frac{\partial^2\chi_1}{\partial x\partial y}+\frac{\partial^2\chi_2}{\partial x\partial y}+\frac{1}{2}\frac{\partial^2\chi_3}{\partial y^2}-\frac{1}{2}\frac{\partial^2\chi_3}{\partial x^2}\right)\\ t'^{23} &= (k_1 d_{55}+d_{44})\frac{\partial^2\chi_1}{\partial y\partial z}+(k_2 d_{55}+d_{44})\frac{\partial^2\chi_2}{\partial y\partial z}-d_{44}\frac{\partial^2\chi_3}{\partial x\partial z}\\ t'^{31} &= (k_1 d_{55}+d_{44})\frac{\partial^2\chi_1}{\partial x\partial z}+(k_2 d_{55}+d_{44})\frac{\partial^2\chi_2}{\partial x\partial z}+d_{44}\frac{\partial^2\chi_3}{\partial y\partial z}\end{aligned}\right\}. \tag{4.5.20}$$

The above results become comparatively simple when we take the Mooney form (2.3.21) for the strain energy function. For this case

$$\left.\begin{aligned}\Phi = 2C_1, \qquad \Psi = 2C_2\\ A = B = F = 0\end{aligned}\right\} \tag{4.5.21}$$

and hence, from (4.5.1), (4.4.2), (4.4.10), (4.5.3), (4.5.6), and (4.5.7),

$$\left.\begin{aligned}a &= 4\mu^2(C_1+C_2\mu^2), & c &= 4\lambda^2(C_1+C_2\mu^2)\\ d_{44} &= 2\lambda^2(C_1+C_2\mu^2), & d_{55} &= 2\mu^2(C_1+C_2\mu^2)\\ \alpha &= a-2\tau^{11}, \quad \gamma = c-2\tau^{33}, & b &= \beta = 4C_2\mu^2(\mu^2-\lambda^2)\end{aligned}\right\}. \tag{4.5.22}$$

The quadratic equation (4.5.13) for k becomes

$$\mu^2k^2-(\lambda^2+\mu^2)k+\lambda^2 = 0,$$

and therefore k has the values

$$k_1 = 1, \qquad k_2 = \frac{\lambda^2}{\mu^2} = \frac{1}{\mu^6} = \lambda^3. \tag{4.5.23}$$

4.6. Indentation problems

The theory of the preceding section will now be used to solve problems in which we take the component of stress τ^{33} to be zero. From (4.4.2), since the body is incompressible, this implies that the scalar function p is given by

$$p = -\lambda^2(\Phi+2\Psi\mu^2), \tag{4.6.1}$$

and therefore

$$\tau^{11} = \tau^{22} = (\mu^2-\lambda^2)(\Phi+\Psi\mu^2). \tag{4.6.2}$$

We suppose that the strained body B, which is obtained from the unstrained body B_0 by the finite deformation of § 4.4, occupies the semi-infinite region $z \geqslant 0$, and the final strained body B' is obtained from B by the small indentation of the plane surface by a rigid punch. The plane boundary is the surface

$$F(x,y,z) \equiv z = 0,$$

and $\partial F/\partial\theta^m = \delta^3_m$. It follows from (4.2.30) that

$$\mathbf{P}+\epsilon\mathbf{P}' = \epsilon\lambda^{3j}\mathbf{G}_j,$$

and with the help of (4.2.32) this reduces to

$$\mathbf{P}+\epsilon\mathbf{P}' = \epsilon\tau'^{31}\mathbf{G}_1+\epsilon\tau'^{32}\mathbf{G}_2+\epsilon\tau'^{33}\mathbf{G}_3 \quad (z = 0). \tag{4.6.3}$$

The components of traction tangential to the surface $z = 0$ in the final strained body B' are

$$(\mathbf{P}+\epsilon\mathbf{P}').(\mathbf{G}_\alpha+\epsilon\mathbf{G}'_\alpha) = \epsilon\tau'^{3\alpha} \quad (\alpha = 1, 2)$$

to the first order in ϵ.

If we assume that the surfaces of the punch and body B' are ideally smooth so that the frictional forces are zero, and if the remaining part of the boundary $z = 0$ of B' is free from applied force, then along the whole boundary

$$\tau'^{31} = \tau'^{32} = 0 \quad (z = 0). \tag{4.6.4}$$

From (4.5.18) we see that (4.6.4) implies that

$$(1+k_1)\frac{\partial\chi_1}{\partial z}+(1+k_2)\frac{\partial\chi_2}{\partial z} = 0, \qquad \frac{\partial\chi_3}{\partial z} = 0 \qquad (z = 0). \tag{4.6.5}$$

These conditions can be satisfied if we put†

$$\left.\begin{aligned} \chi_1 = \frac{\sqrt{k_1}}{1+k_1}\chi(x,y,z_1), \qquad \chi_2 = -\frac{\sqrt{k_2}}{1+k_2}\chi(x,y,z_2), \qquad \chi_3 = \chi_3(x,y,z_3) \\ z_1 = \frac{z}{\sqrt{k_1}}, \qquad z_2 = \frac{z}{\sqrt{k_2}}, \qquad z_3 = \frac{z}{\sqrt{k_3}} \end{aligned}\right\}, \tag{4.6.6}$$

† See § 5.12 for a similar method of solution for transversely isotropic bodies.

where, because of (4.5.14),

$$\nabla^2\chi(x,y,z) = 0, \qquad \nabla^2\chi_3(x,y,z) = 0. \tag{4.6.7}$$

We impose the further condition that all first derivatives of χ and χ_3 vanish at infinity so that the displacements vanish there. In view of the boundary condition (4.6.5) on χ_3, and the condition on its first derivatives at infinity, we see that the solution of $(4.6.7)_2$ is

$$\chi_3 = \text{constant}.$$

If we restrict our attention to rigid punches which have an axis of symmetry in the direction of the z-axis, and which are moved parallel to this direction, then the remaining boundary conditions on $z = 0$ require

$$\left.\begin{aligned} w &= \sigma\frac{\partial\chi(x,y,0)}{\partial z} = f(r) \quad (r^2 = x^2+y^2 \leqslant r_0^2) \\ \tau'^{33} &= \kappa\frac{\partial^2\chi(x,y,0)}{\partial z^2} = 0 \quad (r^2 = x^2+y^2 > r_0^2) \end{aligned}\right\} \tag{4.6.8}$$

where

$$\left.\begin{aligned} \sigma &= \frac{k_1}{1+k_1}-\frac{k_2}{1+k_2} \\ \kappa &= \frac{(k_1 d_{55}+d_{44})\sqrt{k_1}}{1+k_1}-\frac{(k_2 d_{55}+d_{44})\sqrt{k_2}}{1+k_2} \end{aligned}\right\}, \tag{4.6.9}$$

and where r_0 is the radius of the circle of contact between the punch and the surface $z = 0$ in the strained body B'. The known function $\epsilon f(r)$ defines the shape of the punch. We will not examine the problem in detail here but will consider only the distribution of stress under the punch. For this we need the value of $\partial^2\chi/\partial z^2$ at $z = 0$ which is given by†

$$\left.\begin{aligned} \sigma\frac{\partial^2\chi(x,y,0)}{\partial z^2} &= \frac{1}{r}\frac{d}{dr}\int\limits_r^{r_0}\frac{tS(t)}{\sqrt{(t^2-r^2)}}\,dt \quad (r \leqslant r_0) \\ S(r) &= \frac{2}{\pi}\frac{d}{dr}\int\limits_0^r\frac{tf(t)}{\sqrt{(r^2-t^2)}}\,dt \end{aligned}\right\}, \tag{4.6.10}$$

if $f(r)$ is continuously differentiable in $0 \leqslant r \leqslant r_0$.

As an example we consider the small indentation by a rigid sphere of radius R so that

$$f(r) = b+1-r^2/r_0^2, \tag{4.6.11}$$

† Reference should be made to § 5.8 where these results are derived.

where b is a constant to be determined. The actual displacement of the surface under the punch is

$$\epsilon w_{z=0} = \epsilon b + \epsilon(1 - r^2/r_0^2), \tag{4.6.12}$$

and hence, approximately,

$$r_0^2 = 2R\epsilon. \tag{4.6.13}$$

From (4.6.10) and (4.6.11) a simple calculation gives the value of $\partial^2\chi(x,y,0)/\partial z^2$ for $r \leqslant r_0$, and if we remember that the stresses must be finite at $r = r_0$, we have $b = 1$ and

$$\sigma\frac{\partial^2\chi(x,y,0)}{\partial z^2} = -\frac{8\sqrt{(r_0^2 - r^2)}}{\pi r_0^2} \quad (r \leqslant r_0). \tag{4.6.14}$$

The total pressure N exerted by the punch is

$$N = -\iint \epsilon\tau'^{33}\,dxdy = \frac{16\epsilon\kappa r_0}{3\sigma} = \frac{8\kappa}{3\sigma}\frac{r_0^3}{R} = \frac{8\kappa}{3\sigma}d^{\frac{3}{2}}R^{\frac{1}{2}}, \tag{4.6.15}$$

where $d = 2\epsilon = r_0^2/R$ is the maximum depth of penetration by the punch.

The substitution of (4.5.22), (4.5.23), and (4.6.9) in this expression shows that for the special case of the Mooney material the total pressure is

$$N = \frac{16(C_1 + C_2\mu^2)(\mu^9 + \mu^6 + 3\mu^3 - 1)d^{\frac{3}{2}}R^{\frac{1}{2}}}{3\mu^4(\mu^3 + 1)}. \tag{4.6.16}$$

If we put $\mu = 1$ in this expression and remember that $6(C_1 + C_2)$ is then the value of Young's modulus E for this material for small strains, we find that

$$N = \tfrac{16}{9}Ed^{\frac{3}{2}}R^{\frac{1}{2}},$$

which agrees with the classical value† for an incompressible material in which Poisson's ratio is 0·5.

The value of N given by (4.6.16) is zero when μ is approximately $\frac{2}{3}$ and is negative for values less than $\frac{2}{3}$. This result indicates that, when a body is bounded by a plane surface and is acted on by an all-round compressive force in planes parallel to the bounding surface, the equilibrium becomes unstable at certain critical values of the compressive force.

4.7. Small bending of a stretched plate

In this and the next section we develop approximate theories for the small stretching and bending of a plate which has first been extended finitely by uniform stresses τ^{11}, τ^{22} in its plane. In order to avoid undue complications in the analysis we restrict our attention to incompressible bodies although the more general compressible body

† J. W. Harding and I. N. Sneddon, *Proc. Camb. phil. Soc. math. phys. Sci.* **41** (1945) 16.

may be discussed in a similar manner. Using the notation of § 4.2 we suppose that the unstrained body B_0 is bounded by the parallel planes $x_3 = \pm h_0$, where h_0 is a constant, and that it is first extended finitely by uniform forces in the x_1- and x_2-directions only, so putting $\tau^{33} = 0$ in (4.2.7) and eliminating p,

$$\left.\begin{aligned} \tau^{11} = (\lambda_1^2-\lambda_3^2)(\Phi+\Psi\lambda_2^2), \qquad \tau^{22} = (\lambda_2^2-\lambda_3^2)(\Phi+\Psi\lambda_1^2) \\ p = -\lambda_3^2\{\Phi+\Psi(\lambda_1^2+\lambda_2^2)\} \end{aligned}\right\}. \tag{4.7.1}$$

We also repeat here the incompressibility condition (4.2.17),

$$\lambda_1\lambda_2\lambda_3 = 1, \qquad \frac{\partial u}{\partial x}+\frac{\partial v}{\partial y}+\frac{\partial w}{\partial z} = 0. \tag{4.7.2}$$

The stress components τ'^{rs} are given by (4.2.13), (4.2.16), (4.2.18), and (4.2.19), the value of p being given by (4.7.1), and if we eliminate p' and use (4.7.2) we find that

$$\left.\begin{aligned} &\tau'^{11}-\tau'^{33} \\ &= 2\{2(\Phi+\Psi\lambda_2^2)\lambda_3^2+(\lambda_1^2-\lambda_3^2)^2(A+2F\lambda_2^2+B\lambda_2^4)\}\frac{\partial u}{\partial x}+ \\ &\quad +2[\Phi\lambda_3^2+\Psi\lambda_1^2\lambda_2^2+(\lambda_1^2-\lambda_3^2)(\lambda_2^2-\lambda_3^2)\{A+F(\lambda_1^2+\lambda_2^2)+B\lambda_1^2\lambda_2^2\}]\frac{\partial v}{\partial y} \\ &\tau'^{22}-\tau'^{33} \\ &= 2[\Phi\lambda_3^2+\Psi\lambda_1^2\lambda_2^2+(\lambda_1^2-\lambda_3^2)(\lambda_2^2-\lambda_3^2)\{A+F(\lambda_1^2+\lambda_2^2)+B\lambda_1^2\lambda_2^2\}]\frac{\partial u}{\partial x}+ \\ &\quad +2\{2(\Phi+\Psi\lambda_1^2)\lambda_3^2+(\lambda_2^2-\lambda_3^2)^2(A+2F\lambda_1^2+B\lambda_1^4)\}\frac{\partial v}{\partial y} \end{aligned}\right\}, \tag{4.7.3}$$

$$\left.\begin{aligned} \tau'^{12} &= -(\Psi\lambda_1^2\lambda_2^2+p)\left(\frac{\partial u}{\partial y}+\frac{\partial v}{\partial x}\right) \\ \tau'^{23} &= \lambda_3^2(\Phi+\Psi\lambda_1^2)\left(\frac{\partial v}{\partial z}+\frac{\partial w}{\partial y}\right) \\ \tau'^{31} &= \lambda_3^2(\Phi+\Psi\lambda_2^2)\left(\frac{\partial w}{\partial x}+\frac{\partial u}{\partial z}\right) \end{aligned}\right\}. \tag{4.7.4}$$

Remembering that τ^{11}, τ^{22} are the only non-zero components of τ^{rs}, we find from (4.2.24)

$$t^{11} = \tau^{11}, \qquad t^{22} = \tau^{22}, \qquad t^{rs} = 0 \quad \text{otherwise},$$

$$\left.\begin{aligned} t'^{11} &= \tau'^{11}+2\tau^{11}\frac{\partial u}{\partial x}, & t'^{13} &= \tau'^{13}+\tau^{11}\frac{\partial w}{\partial x} \\ t'^{22} &= \tau'^{22}+2\tau^{22}\frac{\partial v}{\partial y}, & t'^{23} &= \tau'^{23}+\tau^{22}\frac{\partial w}{\partial y} \\ t'^{33} &= \tau'^{33}, & t'^{12} &= \tau'^{12}+\tau^{11}\frac{\partial v}{\partial x}+\tau^{22}\frac{\partial u}{\partial y} \end{aligned}\right\}, \tag{4.7.5}$$

so that, using (4.7.3),

$$\left.\begin{aligned} t'^{11}-t'^{33} &= c_{11}\frac{\partial u}{\partial x}+c_{12}\frac{\partial v}{\partial y} \\ t'^{22}-t'^{33} &= c_{12}\frac{\partial u}{\partial x}+c_{22}\frac{\partial v}{\partial y} \\ t'^{12} &= c_{66}\left(\lambda_2^2\frac{\partial u}{\partial y}+\lambda_1^2\frac{\partial v}{\partial x}\right) \end{aligned}\right\} \tag{4.7.6}$$

where

$$\left.\begin{aligned} c_{11} &= 2(\lambda_1^2+\lambda_3^2)(\Phi+\Psi\lambda_2^2)+2(\lambda_1^2-\lambda_3^2)^2(A+2F\lambda_2^2+B\lambda_2^4) \\ c_{12} &= 2(\Phi\lambda_3^2+\Psi\lambda_1^2\lambda_2^2)+2(\lambda_1^2-\lambda_3^2)(\lambda_2^2-\lambda_3^2)\{A+F(\lambda_1^2+\lambda_2^2)+B\lambda_1^2\lambda_2^2\} \\ c_{22} &= 2(\lambda_2^2+\lambda_3^2)(\Phi+\Psi\lambda_1^2)+2(\lambda_2^2-\lambda_3^2)^2(A+2F\lambda_1^2+B\lambda_1^4) \\ c_{66} &= \Phi+\Psi\lambda_3^2 \end{aligned}\right\}, \tag{4.7.7}$$

these coefficients being different from those defined in previous sections. It is convenient to repeat here the equations

$$\left.\begin{aligned} \frac{\partial t'^{11}}{\partial x}+\frac{\partial t'^{12}}{\partial y}+\frac{\partial t'^{13}}{\partial z} &= 0 \\ \frac{\partial t'^{12}}{\partial x}+\frac{\partial t'^{22}}{\partial y}+\frac{\partial t'^{23}}{\partial z} &= 0 \\ \frac{\partial t'^{13}}{\partial x}+\frac{\partial t'^{23}}{\partial y}+\frac{\partial t'^{33}}{\partial z} &= 0 \end{aligned}\right\}, \tag{4.7.8}$$

which must be satisfied if the final strained body B' is in equilibrium.

We now suppose that after the finite extension of the plate by stresses τ^{11}, τ^{22}, the plate is bent by small transverse forces which are antisymmetrical about the plane $z = 0$. The faces of the deformed plate are the surfaces $z = \pm h$, where

$$h = \lambda_3 h_0 = h_0/(\lambda_1\lambda_2),$$

and if we impose the boundary conditions

$$\tau'^{13} = \tau'^{23} = 0, \qquad \tau'^{33} = t'^{33} = \mp\tfrac{1}{2}q \quad (z = \pm h), \tag{4.7.9}$$

where q is a function of x and y, we see, from (4.2.30), that

$$\mathbf{P}+\epsilon\mathbf{P}' = \epsilon\lambda^{3j}\mathbf{G}_j = \epsilon\tau'^{3j}\mathbf{G}_j = \mp\tfrac{1}{2}\epsilon q\mathbf{G}_3 \quad (z = \pm h). \tag{4.7.10}$$

Stress resultants and stress couples may be defined in a variety of ways. We consider the stress couple about the y-direction defined by

$$G_x+\epsilon G'_x = \int (t^{11}+\epsilon t'^{11})y_3\,dy_3, \tag{4.7.11}$$

where the integration is through the thickness of the deformed plate, taken with respect to y_3 keeping y_1 and y_2 constant. Since t^{11} is constant

$$\int t^{11} y_3\, dy_3 = \tfrac{1}{2}\tau^{11}[y_3^2]$$

and, from (4.2.22), $y_3 = z+\epsilon w(x,y,z)$. Owing to the factor ϵ outside the term $w(x,y,z)$ in y_3 we may consider x and y to be constants in this expression instead of y_1, y_2 and therefore, to the first order in ϵ,

$$\int t^{11} y_3\, dy_3 = \epsilon\tau^{11}[zw(x,y,z)]_{z=-h}^{z=+h}.$$

Also, to the first order in ϵ, the term

$$\epsilon \int t'^{11} y_3\, dy_3$$

can be evaluated by replacing y_3 by z and integrating with respect to z from $-h$ to h. Thus, finally,

$$G_x = 0, \qquad G'_x = \tau^{11}[zw]+\int_{-h}^{h} t'^{11} z\, dz, \tag{4.7.12}$$

where $$[zw] = hw(x,y,h)+hw(x,y,-h).$$

Similarly, we may define other stress couples $G_y+\epsilon G'_y$, $H_{xy}+\epsilon H'_{xy}$, and we find that

$$\left.\begin{aligned} G_y = 0, \qquad G'_y &= \tau^{22}[zw]+\int_{-h}^{h} t'^{22} z\, dz \\ H_{xy} = 0, \qquad H'_{xy} &= \int_{-h}^{h} t'^{12} z\, dz \end{aligned}\right\}. \tag{4.7.13}$$

We define stress resultants $N_x+\epsilon N'_x$, $N_y+\epsilon N'_y$, by

$$N_x+\epsilon N'_x = \int (t^{13}+\epsilon t'^{13})\, dy_3,$$

$$N_y+\epsilon N'_y = \int (t^{23}+\epsilon t'^{23})\, dy_3,$$

so that, to the first order in ϵ,

$$\left.\begin{aligned} N_x = 0, \qquad N'_x &= \int_{-h}^{h} t'^{13}\, dz \\ N_y = 0, \qquad N'_y &= \int_{-h}^{h} t'^{23}\, dz \end{aligned}\right\}. \tag{4.7.14}$$

Other stress resultants will be defined in the next section when we consider the further stretching of the plate instead of bending.

The first two equations of (4.7.8) are now multiplied by z and integrated with respect to z from $-h$ to h, and the last equation in (4.7.8) is integrated directly with respect to z from $-h$ to h. Then, using (4.7.5), (4.7.12)–(4.7.14), and the boundary conditions (4.7.9) we find that

$$\left.\begin{aligned} \frac{\partial G'_x}{\partial x}+\frac{\partial H'_{xy}}{\partial y}-N'_x &= 0 \\ \frac{\partial H'_{xy}}{\partial x}+\frac{\partial G'_y}{\partial y}-N'_y &= 0 \\ \frac{\partial N'_x}{\partial x}+\frac{\partial N'_y}{\partial y}-q &= 0 \end{aligned}\right\}. \tag{4.7.15}$$

These equations are of the same form as the classical equations for the transverse bending of a plate, but the stress couples have different forms. If we eliminate N'_x, N'_y, we have

$$\frac{\partial^2 G'_x}{\partial x^2}+2\frac{\partial^2 H'_{xy}}{\partial x \partial y}+\frac{\partial^2 G'_y}{\partial y^2}-q = 0. \tag{4.7.16}$$

From (4.7.6), (4.7.12), and (4.7.13),

$$\left.\begin{aligned} G'_x &= \int_{-h}^{h} t'^{33}z\,dz+\frac{2h^3}{3}\left(c_{11}\frac{\partial \bar{u}}{\partial x}+c_{12}\frac{\partial \bar{v}}{\partial y}\right)+\tau^{11}[zw] \\ G'_y &= \int_{-h}^{h} t'^{33}z\,dz+\frac{2h^3}{3}\left(c_{12}\frac{\partial \bar{u}}{\partial x}+c_{22}\frac{\partial \bar{v}}{\partial y}\right)+\tau^{22}[zw] \\ H'_{xy} &= \frac{2h^3}{3}c_{66}\left(\lambda_2^2\frac{\partial \bar{u}}{\partial y}+\lambda_1^2\frac{\partial \bar{v}}{\partial x}\right) \end{aligned}\right\}, \tag{4.7.17}$$

where we have put

$$\bar{u} = \frac{3}{2h^3}\int_{-h}^{h} uz\,dz, \qquad \bar{v} = \frac{3}{2h^3}\int_{-h}^{h} vz\,dz. \tag{4.7.18}$$

The numerical factor $2h^3/3$ in (4.7.18) is chosen so that if u, v are proportional to z then $\bar{u}$, $\bar{v}$ are respectively equal to the constants of proportionality.

If we multiply the second and third equations in (4.7.4) by $(1-z^2/h^2)$ and use the rule of integration by parts we have

$$\left.\begin{aligned} \frac{3}{4h}\int_{-h}^{h}\left(1-\frac{z^2}{h^2}\right)\tau'^{23}\,dz &= \lambda_3^2(\Phi+\Psi\lambda_1^2)\left(\frac{\partial \bar{w}}{\partial y}+\bar{v}\right) \\ \frac{3}{4h}\int_{-h}^{h}\left(1-\frac{z^2}{h^2}\right)\tau'^{31}\,dz &= \lambda_3^2(\Phi+\Psi\lambda_2^2)\left(\frac{\partial \bar{w}}{\partial x}+\bar{u}\right) \end{aligned}\right\}, \tag{4.7.19}$$

where $\bar{u}$, $\bar{v}$ are given by (4.7.18) and where

$$\bar{w} = \frac{3}{4h}\int_{-h}^{h}\left(1-\frac{z^2}{h^2}\right)w\,dz. \tag{4.7.20}$$

This process avoids the introduction of unknown values of u, v at $z = \pm h$, and the definition of $\bar{w}$ is such that if w is independent of z then $\bar{w} = w$. From (4.7.20),

$$2h\bar{w} = \frac{3}{2}\left[\left(z-\frac{z^3}{3h^2}\right)w\right]_{z=-h}^{z=+h} - \frac{3}{2}\int_{-h}^{h}\left(z-\frac{z^3}{3h^2}\right)\frac{\partial w}{\partial z}\,dz$$

$$= [zw]+\frac{3}{2}\int_{-h}^{h}\left(z-\frac{z^3}{3h^2}\right)\left(\frac{\partial u}{\partial x}+\frac{\partial v}{\partial y}\right)dz$$

if we use (4.7.2), and therefore

$$[zw] = 2h\bar{w}-\frac{3}{2}\int_{-h}^{h}\left(z-\frac{z^3}{3h^2}\right)\left(\frac{\partial u}{\partial x}+\frac{\partial v}{\partial y}\right)dz.$$

The integral in this formula is of the order $h^3(\partial\bar{u}/\partial x+\partial\bar{v}/\partial y)$ so that we evaluate it approximately, as far as terms of this order are concerned, by assuming that u and v are proportional to z. This gives

$$[zw] = 2h\bar{w}-\frac{4h^3}{5}\left(\frac{\partial\bar{u}}{\partial x}+\frac{\partial\bar{v}}{\partial y}\right). \tag{4.7.21}$$

If we substitute this value of $[zw]$ into the expression for G'_x in (4.7.17) and use (4.7.19) to eliminate $\bar{u}$ and $\bar{v}$ we obtain

$$G'_x = 2h\tau^{11}\bar{w}-\frac{2h^3}{3}\left(c_{11}\frac{\partial^2\bar{w}}{\partial x^2}+c_{12}\frac{\partial^2\bar{w}}{\partial y^2}\right)+\frac{4h^3}{5}\tau^{11}\nabla_1^2\bar{w}+\int_{-h}^{h}t'^{33}z\,dz+$$

$$+\frac{h^2}{10\lambda_3^2}\int_{-h}^{h}\left\{\frac{(5c_{11}-6\tau^{11})}{(\Phi+\Psi\lambda_2^2)}\frac{\partial\tau'^{13}}{\partial x}+\frac{(5c_{12}-6\tau^{11})}{(\Phi+\Psi\lambda_1^2)}\frac{\partial\tau'^{23}}{\partial y}\right\}\left(1-\frac{z^2}{h^2}\right)dz.$$

As in the corresponding classical theory (see Chapter 7) we now neglect the last two terms in this expression for G'_x and obtain approximately

$$G'_x = 2h\tau^{11}\bar{w}-\frac{2h^3}{3}\left(c_{11}\frac{\partial^2\bar{w}}{\partial x^2}+c_{12}\frac{\partial^2\bar{w}}{\partial y^2}\right)+\frac{4h^3}{5}\tau^{11}\nabla_1^2\bar{w}. \tag{4.7.22}$$

In a similar way we have approximately

$$\left.\begin{aligned} G'_y &= 2h\tau^{22}\bar{w} - \frac{2h^3}{3}\left(c_{12}\frac{\partial^2\bar{w}}{\partial x^2} + c_{22}\frac{\partial^2\bar{w}}{\partial y^2}\right) + \frac{4h^3}{5}\tau^{22}\nabla_1^2\bar{w} \\ H'_{xy} &= -\frac{2h^3}{3}c_{66}(\lambda_1^2+\lambda_2^2)\frac{\partial^2\bar{w}}{\partial x\partial y} \end{aligned}\right\}. \tag{4.7.23}$$

With these values (4.7.22), (4.7.23) of G'_x, G'_y, H'_{xy}, equation (4.7.16) can be written

$$a\frac{\partial^4\bar{w}}{\partial x^4} + 2b\frac{\partial^4\bar{w}}{\partial x^2\partial y^2} + c\frac{\partial^4\bar{w}}{\partial y^4} + q = T_1\frac{\partial^2\bar{w}}{\partial x^2} + T_2\frac{\partial^2\bar{w}}{\partial y^2}, \tag{4.7.24}$$

where

$$\left.\begin{aligned} a &= \frac{2h^3}{15}(5c_{11} - 6\tau^{11}), \qquad c = \frac{2h^3}{15}(5c_{22} - 6\tau^{22}) \\ b &= \frac{2h^3}{15}\{5c_{12} + 5c_{66}(\lambda_1^2+\lambda_2^2) - 3(\tau^{11}+\tau^{22})\} \\ T_1 &= 2h\tau^{11}, \qquad T_2 = 2h\tau^{22} \end{aligned}\right\}. \tag{4.7.25}$$

When the two finite extensions λ_1, λ_2 are equal (and therefore, by (4.7.2), both equal to $1/\sqrt{\lambda_3}$) then (4.7.24) takes the simple form

$$a\nabla_1^4\bar{w} - T\nabla_1^2\bar{w} + q = 0, \tag{4.7.26}$$

where

$$\left.\begin{aligned} a &= \frac{2h^3}{15}(5c_{11} - 6\tau^{11}), \qquad \tau^{11} = \left(\lambda_1^2 - \frac{1}{\lambda_1^4}\right)(\Phi + \Psi\lambda_1^2) \\ c_{11} &= 2\left(\lambda_1^2 + \frac{1}{\lambda_1^4}\right)(\Phi + \Psi\lambda_1^2) + 2\left(\lambda_1^2 - \frac{1}{\lambda_1^4}\right)^2(A + 2F\lambda_1^2 + B\lambda_1^4) \\ T &= T_1 = T_2 = 2h\tau^{11} \end{aligned}\right\}, \tag{4.7.27}$$

and

$$\nabla_1^2 = \frac{\partial^2}{\partial x^2} + \frac{\partial^2}{\partial y^2}, \qquad \nabla_1^4 = \frac{\partial^4}{\partial x^4} + 2\frac{\partial^4}{\partial x^2\partial y^2} + \frac{\partial^4}{\partial y^4}.$$

As an example we solve (4.7.26) for the case of a circular sheet, stretched finitely by a uniform all-round tension T in its plane so that it becomes a sheet of radius r_0. The sheet is then clamped at its edge and is subject to a constant small transverse pressure uniformly distributed over a face of the sheet. If the resultant transverse force is denoted by W then

$$q = \frac{W}{\pi r_0^2} \tag{4.7.28}$$

and, changing the sign of $\bar{w}$ so that it is measured positively in the direction of W,

$$a\nabla_1^4\bar{w}-T\nabla_1^2\bar{w} = \frac{W}{\pi r_0^2}, \tag{4.7.29}$$

together with the boundary conditions

$$\bar{w} = 0, \qquad \frac{\partial\bar{w}}{\partial r} = 0 \qquad (r = r_0), \tag{4.7.30}$$

where $r = \sqrt{(x^2+y^2)}$. The solution of (4.7.29) subject to the boundary conditions (4.7.30) may be found by elementary analysis. If we suppose that $T > 0$ and put $n^2 = Tr_0^2/a$, then

$$\frac{\pi\bar{w}T}{W} = \tfrac{1}{4}(1-\rho^2)+\frac{I_0(n\rho)-I_0(n)}{2nI_1(n)}, \tag{4.7.31}$$

where $\rho = r/r_0$ and $I_0(x)$, $I_1(x)$ are modified Bessel functions of the first kind. We notice that when $T = 0$, $n = 0$ and the limiting form of (4.7.31) gives

$$\bar{w} = \frac{W(r^2-r_0^2)^2}{512\pi r_0^2 E}, \tag{4.7.32}$$

where E is Young's modulus for small strains, and this is the classical result for an incompressible material. On the other hand, if we consider the limiting form of (4.7.31) as $n \to \infty$ (which corresponds to $h \to 0$) we have

$$\frac{\pi\bar{w}T}{W} = \tfrac{1}{4}(1-\rho^2), \tag{4.7.33}$$

and this is the result for the small transverse displacement of a uniformly loaded membrane which is stretched to tension T.

If $T < 0$ the result corresponding to (4.7.31) may be expressed in terms of Bessel functions $J_0(x)$, $J_1(x)$ of the first kind.

4.8. Generalized plane stress superposed on finite extension

We now consider the deformation produced by small forces in the plane of the plate superposed upon the finite extension of the plate described in § 4.7, the faces of the plate being free from surface traction.

We define stress resultants by

$$\left.\begin{aligned} T_x+\epsilon T'_x &= \int (t^{11}+\epsilon t'^{11})\,dy_3 \\ T_y+\epsilon T'_y &= \int (t^{22}+\epsilon t'^{22})\,dy_3 \\ S_{xy}+\epsilon S'_{xy} &= \int (t^{12}+\epsilon t'^{12})\,dy_3 \end{aligned}\right\}, \tag{4.8.1}$$

where the integration is through the thickness of the deformed plate keeping y_1 and y_2 constant. To our order of approximation we have

$$T_x+\epsilon T'_x = t^{11}[z+\epsilon w]_{z=-h}^{z=+h}+\epsilon\int_{-h}^{h} t'^{11}\,dz$$

$$= 2h\tau^{11}+\epsilon\tau^{11}[w]+\epsilon\int_{-h}^{h} t'^{11}\,dz,$$

where $$[w] = w(x,y,h)-w(x,y,-h),$$

so that $$T_x = 2h\tau^{11},\qquad T'_x = \int_{-h}^{h} t'^{11}\,dz+\tau^{11}[w]. \tag{4.8.2}$$

In the same way we obtain

$$\left.\begin{aligned} T_y &= 2h\tau^{22}, & T'_y &= \int_{-h}^{h} t'^{22}\,dz+\tau^{22}[w] \\ S_{xy} &= 0, & S'_{xy} &= \int_{-h}^{h} t'^{12}\,dz \end{aligned}\right\}. \tag{4.8.3}$$

Since the faces of the plate are free from applied stress

$$\tau'^{13} = \tau'^{23} = 0 \quad (z = \pm h),$$

and if we integrate the first two equations of equilibrium in (4.7.8) with respect to z from $-h$ to h and use this condition we obtain

$$\left.\begin{aligned} \frac{\partial T'_x}{\partial x}+\frac{\partial S'_{xy}}{\partial y} &= 0 \\ \frac{\partial S'_{xy}}{\partial x}+\frac{\partial T'_y}{\partial y} &= 0 \end{aligned}\right\}. \tag{4.8.4}$$

The form of these equations is the same as the form of the classical generalized plane stress equations.

We now write $$U = \int_{-h}^{h} u\,dz,\qquad V = \int_{-h}^{h} v\,dz, \tag{4.8.5}$$

and if we integrate the incompressibility condition (4.7.2) with respect to z from $-h$ to h we get

$$[w] = -\frac{\partial U}{\partial x}-\frac{\partial V}{\partial y}. \tag{4.8.6}$$

The stress component $t'^{33} = \tau'^{33}$ is zero on the faces of the plate, and as in the classical generalized plane stress theory we neglect $\int_{-h}^{h} t'^{33}\,dz$

compared with T'_x, T'_y, S'_{xy}. With this assumption, equations (4.7.6), (4.8.2), (4.8.3), and (4.8.6) give

$$\left.\begin{aligned} T'_x &= (c_{11}-\tau^{11})\frac{\partial U}{\partial x}+(c_{12}-\tau^{11})\frac{\partial V}{\partial y} \\ T'_y &= (c_{12}-\tau^{22})\frac{\partial U}{\partial x}+(c_{22}-\tau^{22})\frac{\partial V}{\partial y} \\ S'_{xy} &= c_{66}\left(\lambda_2^2\frac{\partial U}{\partial y}+\lambda_1^2\frac{\partial V}{\partial x}\right) \end{aligned}\right\}. \tag{4.8.7}$$

These equations can be rearranged to give

$$\left.\begin{aligned} \frac{\partial U}{\partial x} &= s_{11}T'_x+s_{12}T'_y \\ \frac{\partial V}{\partial y} &= s_{21}T'_x+s_{22}T'_y \\ \lambda_2^2\frac{\partial U}{\partial y}+\lambda_1^2\frac{\partial V}{\partial x} &= s_{66}S'_{xy} \end{aligned}\right\}, \tag{4.8.8}$$

where

$$\left.\begin{aligned} s_{11} &= \frac{c_{22}-\tau^{22}}{\Delta}, \qquad s_{22} = \frac{c_{11}-\tau^{11}}{\Delta} \\ s_{12} &= \frac{\tau^{11}-c_{12}}{\Delta}, \qquad s_{21} = \frac{\tau^{22}-c_{12}}{\Delta} \\ s_{66} &= \frac{1}{c_{66}}, \qquad \Delta = (c_{11}-\tau^{11})(c_{22}-\tau^{22})-(c_{12}-\tau^{11})(c_{12}-\tau^{22}) \end{aligned}\right\}. \tag{4.8.9}$$

The stress equations of equilibrium (4.8.4) will be satisfied if we put

$$T'_x = \frac{\partial^2\phi}{\partial y^2}, \qquad T'_y = \frac{\partial^2\phi}{\partial x^2}, \qquad S'_{xy} = -\frac{\partial^2\phi}{\partial x\partial y}, \tag{4.8.10}$$

where the stress function ϕ is a function of x and y. If we also write

$$\zeta = x+iy, \qquad \bar{\zeta} = x-iy,$$

then equations (4.8.10) are equivalent to

$$T'_x+T'_y = 4\frac{\partial^2\phi}{\partial\zeta\partial\bar{\zeta}}, \qquad T'_x-T'_y+2iS'_{xy} = -4\frac{\partial^2\phi}{\partial\bar{\zeta}^2}. \tag{4.8.11}$$

The elimination of U, V from the relations (4.8.8) gives us

$$\lambda_2^2\frac{\partial^2}{\partial y^2}(s_{11}T'_x+s_{12}T'_y)+\lambda_1^2\frac{\partial^2}{\partial x^2}(s_{21}T'_x+s_{22}T'_y)-s_{66}\frac{\partial^2 S'_{xy}}{\partial x\partial y} = 0,$$

that is, if we use the expressions (4.8.10) for T'_x, T'_y, S'_{xy},

$$\lambda_1^2 s_{22}\frac{\partial^4\phi}{\partial x^4}+(\lambda_1^2 s_{21}+\lambda_2^2 s_{12}+s_{66})\frac{\partial^4\phi}{\partial x^2\partial y^2}+\lambda_2^2 s_{11}\frac{\partial^4\phi}{\partial y^4}=0. \quad (4.8.12)$$

This equation is analogous to the corresponding equation for generalized plane stress in an orthotropic material, and we may employ complex variable techniques to solve a variety of problems. We leave the discussion, however, at this point, and refer the reader to Chapters 6 and 9.

5 Infinitesimal Theory

THE classical or infinitesimal theory of elasticity has been established for many years and has been used for the solution of a variety of problems.† Here we regard the classical theory as a limiting case of the general theory of Chapter 2 and we present‡ the results in tensor notation for a general curvilinear system of coordinates θ_i. Solutions of the fundamental equations for an isotropic body are given in terms of harmonic functions and these functions are used to solve a few three-dimensional problems. Solutions of the equations for transversely isotropic bodies are also given.

5.1. Strain

We replace the displacement vector $\mathbf{v}$ by $\epsilon\mathbf{v}$, where ϵ is a small non-dimensional parameter, and we retain only first powers of ϵ in expressions for strain, stress, etc. In certain quantities, such as density ρ, we neglect all powers of ϵ. Then, without loss of generality, we replace ϵ by unity at the end of the approximation process. This is equivalent to assuming that the components of $\mathbf{v}$ with respect to base vectors $\mathbf{g}_m$, $\mathbf{g}^m$, and their derivatives with respect to θ_i and time t, are small so that we may neglect squares and products of these quantities compared with their first powers. With this approximation the covariant strain tensor (2.1.26) and (2.1.31) becomes

$$\begin{aligned}\gamma_{ij} &= \tfrac{1}{2}(\mathbf{g}_i\cdot\mathbf{v}_{,j}+\mathbf{g}_j\cdot\mathbf{v}_{,i}) \\ &= \tfrac{1}{2}(v_i|_j+v_j|_i).\end{aligned} \tag{5.1.1}$$

From (2.1.27), (2.1.28), and (2.1.25),

$$\begin{aligned}V_m &= v_n\,\mathbf{g}^n\cdot(\mathbf{g}_m+\mathbf{v}_{,m}) \\ &= v_m+v_n v^n|_m,\end{aligned}$$

† Reference may be made, e.g. to A. E. H. Love, *A Treatise on the Mathematical Theory of Elasticity*, 4th ed. (Cambridge, 1927); I. S. Sokolnikoff, *Mathematical Theory of Elasticity*, 2nd ed. (New York, 1956); S. Timoshenko and J. N. Goodier, *Theory of Elasticity*, 2nd ed. (New York, 1951).

‡ We assume that either the temperature or the entropy is constant. The infinitesimal thermo-elastic equations are discussed by Green and Adkins, loc. cit., p. 64.

and

$$V_m\|_i = v_n|_i \mathbf{g}^n.(\mathbf{g}_m+\mathbf{v}_{,m})$$
$$= v_m|_i+v_n|_i v^n|_m,$$

and hence, approximately,

$$V_m = v_m, \qquad V_m\|_i = v_m|_i. \tag{5.1.2}$$

It follows that the second form in (2.1.31) for γ_{ij} also reduces to (5.1.1). The components of the strain tensor are small quantities of the same order as displacements or their derivatives.

The mixed strain tensor γ^i_j is still defined by (2.1.32) so that

$$\gamma^i_j = g^{im}\gamma_{mj}, \tag{5.1.3}$$

and we may now define a contravariant strain tensor

$$\gamma^{ij} = g^{im}g^{jn}\gamma_{mn} = g^{im}\gamma^j_m. \tag{5.1.4}$$

The strains γ_{ij} as defined by (5.1.1) are not independent functions. By eliminating the displacement v_i it can be shown that

$$\gamma_{im}|_{jk}+\gamma_{jk}|_{im}-\gamma_{ik}|_{jm}-\gamma_{jm}|_{ik} = 0 \tag{5.1.5}$$

which are usually called the equations of (strain) compatibility.

The change of volume element during deformation is characterized by the expression

$$\frac{\sqrt{G}-\sqrt{g}}{\sqrt{g}}$$

which is called the cubical dilatation or simply the dilatation. Using (2.1.34) and retaining only the major term, we find that

$$\frac{\sqrt{G}-\sqrt{g}}{\sqrt{g}} = \gamma^i_i = g^{im}\gamma_{mi}.$$

With the help of (5.1.3) and (5.1.1) this reduces to

$$\frac{\sqrt{G}-\sqrt{g}}{\sqrt{g}} = \gamma^i_i = v^r|_r. \tag{5.1.6}$$

Moreover, since

$$\rho\sqrt{G} = \rho_0\sqrt{g},$$

we see that

$$\rho = \rho_0, \tag{5.1.7}$$

approximately.

5.2. Stress. Equations of motion

Retaining only the major terms it follows from (2.10.7)–(2.10.9) that

$$\tau^{ij} = s^{ij} = \pi^{ij} = t^{ij}, \tag{5.2.1}$$

and therefore all forms of the stress tensors are symmetric. We assume that τ^{ij} is $O(\epsilon)$. Also, from (2.10.2)–(2.10.6), we have

$${}_0\mathbf{t}_i = \mathbf{t}_i, \qquad {}_0\mathbf{t} = \mathbf{t}, \tag{5.2.2}$$

approximately, and all forms of stress tensors or stress vectors are measured per unit area of the undeformed body. We shall omit the prefix $_0$ from subsequent formulae and use τ^{ij} for our (symmetric) stress tensor, per unit area of the undeformed body. Referring to equations (2.10.2)–(2.10.14), we see that for infinitesimal theory we have

$$\mathbf{t}_i\sqrt{g^{ii}} = \tau^{ij}\mathbf{g}_j = \tau^i_j\,\mathbf{g}^j, \quad \tau^{ij} = \tau^{ji}, \tag{5.2.3}$$

$$\mathbf{t} = \tau^{ij}n_i\,\mathbf{g}_j = \tau^i_j n_i\,\mathbf{g}^j = \frac{n_i\,\mathbf{T}_i}{\sqrt{g}}, \tag{5.2.4}$$

$$\mathbf{T}_i = \sqrt{(gg^{ii})}\mathbf{t}_i = \sqrt{(g)}\tau^{ij}\mathbf{g}_j = \sqrt{(g)}\tau^i_j\,\mathbf{g}^j, \tag{5.2.5}$$

$$\mathbf{T}_{i,i}+\rho\mathbf{F}\sqrt{g} = \rho\mathbf{f}\sqrt{g}, \tag{5.2.6}$$

$$\tau^{ij}|_i+\rho F^j = \rho f^j, \tag{5.2.7}$$

$$\tau^i_j|_i+\rho F_j = \rho f_j, \tag{5.2.8}$$

where

$$\mathbf{n} = n_i\,\mathbf{g}^i = n^i\mathbf{g}_i, \tag{5.2.9}$$

$$\mathbf{F} = F_i\,\mathbf{g}^i = F^i\mathbf{g}_i, \tag{5.2.10}$$

$$\mathbf{f} = f_i\,\mathbf{g}^i = f^i\mathbf{g}_i, \tag{5.2.11}$$

and $\mathbf{n}$ is a unit normal to a surface in the undeformed body. Indices on tensors are raised and lowered with the help of the metric tensors g_{ij}, g^{ij}. In particular

$$\tau^i_j = g_{jr}\,\tau^{ir}, \quad \tau_{ij} = g_{ir}\,\tau^r_j = g_{ir}\,g_{js}\,\tau^{rs}. \tag{5.2.12}$$

Finally, we re-state conditions (2.10.14) which must be satisfied at the boundary surface of the body when surface forces are prescribed, in a form appropriate for infinitesimal theory. Thus, omitting the prefix $_0$,

$$\mathbf{t} = \mathbf{P}, \tag{5.2.13}$$

and if

$$\mathbf{P} = P^i\mathbf{g}_i = P_i\,\mathbf{g}^i, \tag{5.2.14}$$

we have the alternative forms

$$\tau^{ij}n_i = P^j, \qquad \tau^i_j n_i = P_j. \tag{5.2.15}$$

5.3. Stress–strain relations

When the strain is infinitesimal the relation (2.10.7) between the stress tensor τ^{ij} and the elastic potential W becomes

$$\tau^{ij} = \frac{1}{2}\left(\frac{\partial W}{\partial \gamma_{ij}}+\frac{\partial W}{\partial \gamma_{ji}}\right). \tag{5.3.1}$$

In Chapter 2 we restricted further discussion of the stress–strain relations to those bodies which were homogeneous and isotropic. In this chapter we are considering only infinitesimal theory, but we wish to include in our discussion bodies which are not necessarily isotropic.

The elastic potential W is invariant under all transformations of coordinates, and a suitable form to assume for W, which is consistent with approximations already made about infinitesimal strain, is

$$W = \text{constant} + E^{ij}\gamma_{ij} + \tfrac{1}{2}E^{ijrs}\gamma_{ij}\gamma_{rs}, \tag{5.3.2}$$

where E^{ij} is $O(\epsilon)$ and, without loss of generality, we may assume that E^{ij} and E^{ijrs} have the symmetric properties

$$E^{ij} = E^{ji}, \qquad E^{ijrs} = E^{jirs} = E^{ijsr} = E^{rsij}. \tag{5.3.3}$$

The functions E^{ij}, E^{ijrs} do not depend on γ_{ij}, and since W is invariant they must be tensors. If W is taken to be zero when the body is in the initial state B_0 in which $\gamma_{ij} = 0$ then the constant is zero.

From (5.3.1), (5.3.2), and (5.3.3) we have

$$\tau^{ij} = E^{ij} + E^{ijrs}\gamma_{rs}. \tag{5.3.4}$$

We confine our attention to a body which at time $t = 0$ is unstrained and unstressed.† Hence E^{ij} is zero and

$$\tau^{ij} = E^{ijrs}\gamma_{rs}. \tag{5.3.5}$$

The expression (5.3.5) gives us a general relation between stress and (infinitesimal) strain, and we call the coefficients E^{ijrs} the elastic coefficients of the body. They depend on the metric tensor g_{ij} of the unstrained body B_0 and on the physical properties of B_0. The functions E^{ijrs} are coefficients in an expansion of the elastic potential in terms of components of strain, and W exists (at least) for an elastic body which undergoes either isothermal or adiabatic changes of state. In general the corresponding physical constants (and therefore elastic coefficients) are different for these two cases. Numerical differences do occur in the elastic coefficients which have been found by experiments involving either isothermal or adiabatic changes of state, but for many purposes these differences are not great and can be ignored.‡

Using (5.3.5) we see that the elastic potential W can be written in the forms

$$W = \tfrac{1}{2}E^{ijrs}\gamma_{ij}\gamma_{rs} = \tfrac{1}{2}\tau^{ij}\gamma_{ij}. \tag{5.3.6}$$

Since associated tensors for infinitesimal displacement and displacement gradients are related with the help of the metric tensors g_{ij}, g^{ij} of the unstrained body, we can express the elastic potential (5.3.6) and the stress–strain relations (5.3.5) in the alternative forms

† This means, for example, that no stresses are induced in B_0 by its manufacture or by the action of body forces at $t = 0$.

‡ See A. E. H. Love, loc. cit., p. 148.

$$W = \tfrac{1}{2}E^{ij}_{rs}\gamma_{ij}\gamma^{rs} = \tfrac{1}{2}E_{ijrs}\gamma^{ij}\gamma^{rs} = \tfrac{1}{2}\tau_{ij}\gamma^{ij} = \tfrac{1}{2}\tau^i_j\gamma^j_i, \tag{5.3.7}$$

$$\tau^{ij} = E^{ij}_{rs}\gamma^{rs}, \qquad \tau_{ij} = E_{ijrs}\gamma^{rs} = E^{rs}_{ij}\gamma_{rs}. \tag{5.3.8}$$

The associated tensors E^{ij}_{rs}, E_{ijrs} possess symmetric properties similar to those given in (5.3.3). For example

$$E^{ij}_{rs} = g_{rm}g_{sn}E^{ijmn} = E^{ji}_{rs} = E^{ij}_{sr} = E^{ji}_{sr}, \tag{5.3.9}$$

the order of the pair of suffixes i, j, in relation to the pair of suffixes r, s in E^{ij}_{rs} being immaterial.

In order to examine the properties of the elastic coefficients in more detail, we refer all functions to the rectangular coordinates x_i in the unstrained body B_0. In this system of coordinates there is no distinction between contravariant, covariant, and mixed tensors and we denote the strain tensor† by e_{ij}, the stress tensor by t_{ij}, and the elastic coefficients by c^{ij}_{rs}. Because of the symmetric properties (5.3.9),

$$c^{ij}_{rs} = c^{ji}_{rs} = c^{ij}_{sr} = c^{ji}_{sr} = c^{rs}_{ij}, \tag{5.3.10}$$

the additional relation $\qquad c^{ij}_{rs} = c^{rs}_{ij}$

holding because the coordinates are rectangular. The elastic potential function (5.3.7) becomes

$$W = \tfrac{1}{2}c^{ij}_{rs}e_{ij}e_{rs}, \tag{5.3.11}$$

and

$$t_{ij} = c^{ij}_{rs}e_{rs}. \tag{5.3.12}$$

If the density of the original body B_0 is constant and if the elastic coefficients c^{ij}_{rs} are constants, we say that the body is *elastically homogeneous*. In this case the covariant derivatives of the general elastic coefficients E^{ij}_{rs} are zero, so that

$$E^{ij}_{rs}|_t = 0, \qquad E^{ijrs}|_t = 0, \qquad E_{ijrs}|_t = 0. \tag{5.3.13}$$

When the body is homogeneous the coefficients c^{ij}_{rs} are constants and they define the physical properties of the body. In general these constants depend on the orientation of the rectangular axes x_i at each point of the body. If, however, the expression for the elastic potential W always has the same form whatever the orientation of the rectangular axes x_i, then the elastic coefficients are constants which are independent of the axes and the body is said to be *elastically isotropic* or, more simply, *isotropic*. Otherwise the body is said to be *aeolotropic* or *anisotropic*. The above definitions of homogeneous and isotropic bodies for infinitesimal theory are equivalent to those given in Chapter 2.

In the remainder of this chapter we restrict our attention to bodies

† The strain tensor e_{ij} need not be confused with e_{ijk} defined in Chapter 1.

which are homogeneous in the above sense so that equations (5.3.13) hold. Similar results, however, hold for non-homogeneous bodies. The assumption of homogeneity and isotropy when applied to a body leads to reasonable agreement between experiments and theoretical results for a wide variety of materials, despite the fact that most structural materials are formed of crystalline substances and very small portions of such substances are not isotropic. The agreement is explained by the fact that the dimensions of most crystals are small compared with the dimensions of the entire body and that they are distributed chaotically, so the behaviour of the entire body is isotropic.

Materials such as natural crystals and wood are aeolotropic. Symmetry of structure of the general aeolotropic body may introduce relations among the elastic coefficients and we now discuss some particular cases of elastic symmetry.†

5.4. Elastic coefficients

In general, remembering the symmetric properties (5.3.10) we have 21 elastic coefficients c^{ij}_{rs} for an aeolotropic body. The 21 coefficients may conveniently be represented in the following table:

$$\begin{array}{cccccc} c^{11}_{11} & c^{11}_{22} & c^{11}_{33} & c^{11}_{23} & c^{11}_{13} & c^{11}_{12} \\ & c^{22}_{22} & c^{22}_{33} & c^{22}_{23} & c^{22}_{13} & c^{22}_{12} \\ & & c^{33}_{33} & c^{33}_{23} & c^{33}_{13} & c^{33}_{12} \\ & & & c^{23}_{23} & c^{23}_{13} & c^{23}_{12} \\ & & & & c^{13}_{13} & c^{13}_{12} \\ & & & & & c^{12}_{12} \end{array} \tag{5.4.1}$$

If the coordinate system x_i is changed to another rectangular cartesian system $\bar{x}_i$ then the coefficients c^{ij}_{rs} become $\bar{c}^{ij}_{rs}$, and, in view of their tensor character,

$$\bar{c}^{ij}_{mn} = \frac{\partial \bar{x}_i}{\partial x_r}\frac{\partial \bar{x}_j}{\partial x_s}\frac{\partial \bar{x}_m}{\partial x_p}\frac{\partial \bar{x}_n}{\partial x_t}\, c^{rs}_{pt}, \tag{5.4.2}$$

if we remember (1.8.2). Also, the relation between the components of strain e_{ij} referred to the x_i-axes, and the components $\bar{e}_{rs}$ referred to the $\bar{x}_i$-axes, is

$$\bar{e}_{rs} = \frac{\partial \bar{x}_r}{\partial x_i}\frac{\partial \bar{x}_s}{\partial x_j}\, e_{ij}. \tag{5.4.3}$$

† Full references may be found in A. E. H. Love, loc. cit., p. 148; p. 149, for other types of symmetry. A complete discussion of the form of the strain energy for non-linear elasticity for all the crystal classes is given by A. E. Green and J. E. Adkins, loc. cit., p. 64.

For a general aeolotropic body the elastic potential W is a function of all the components of strain so that

$$W = W(e_{11}, e_{22}, e_{33}, e_{12}, e_{23}, e_{31}). \tag{5.4.4}$$

(*a*) ***Symmetry with respect to a plane*** (13 ***coefficients***)

Consider a change of axes

$$\bar{x}_1 = x_1, \qquad \bar{x}_2 = x_2, \qquad \bar{x}_3 = -x_3. \tag{5.4.5}$$

With this change of axes

$$\frac{\partial \bar{x}_r}{\partial x_\alpha} = \delta^r_\alpha \quad (\alpha = 1, 2), \qquad \frac{\partial \bar{x}_r}{\partial x_3} = -\delta^r_3,$$

and $\quad \bar{e}_{\alpha\beta} = e_{\alpha\beta}, \quad \bar{e}_{33} = e_{33}, \quad \bar{e}_{\alpha 3} = -e_{\alpha 3} \quad (\alpha, \beta = 1, 2).$

Also, the elastic potential W, which is an invariant, can now be expressed in terms of the strain components $\bar{e}_{rs}$. If the expression for W in terms of $\bar{e}_{rs}$ can be obtained from (5.4.4) simply by replacing e_{rs} by $\bar{e}_{rs}$ then the body is said to be elastically symmetrical with respect to the plane $x_3 = 0$. In this case W must be restricted to the form

$$W = W(e_{11}, e_{22}, e_{33}, e_{13}^2, e_{23}^2, e_{13}\,e_{23}, e_{12}). \tag{5.4.6}$$

Hence, from (5.3.11),

$$c^{\alpha\beta}_{3\delta} = c^{3\delta}_{\alpha\beta} = c^{\alpha 3}_{33} = c^{33}_{\alpha 3} = 0 \quad (\alpha, \beta, \delta = 1, 2). \tag{5.4.7}$$

In other words, all coefficients which contain the index 3 either once or three times, vanish. The scheme of coefficients now becomes

$$\begin{array}{cccccc}
c^{11}_{11} & c^{11}_{22} & c^{11}_{33} & 0 & 0 & c^{11}_{12} \\
 & c^{22}_{22} & c^{22}_{33} & 0 & 0 & c^{22}_{12} \\
 & & c^{33}_{33} & 0 & 0 & c^{33}_{12} \\
 & & & c^{23}_{23} & c^{23}_{13} & 0 \\
 & & & & c^{13}_{13} & 0 \\
 & & & & & c^{12}_{12}
\end{array} \tag{5.4.8}$$

(*b*) ***Symmetry with respect to two orthogonal planes: Orthotropy*** (9 ***coefficients***)

A body is said to be elastically symmetrical with respect to two perpendicular planes $x_3 = 0$ and $x_1 = 0$ if, when the axes are changed to

$$\bar{x}_1 = -x_1, \qquad \bar{x}_2 = x_2, \qquad \bar{x}_3 = x_3, \tag{5.4.9}$$

in addition to the transformation (5.4.5), the elastic potential W in terms of $\bar{e}_{rs}$ can be obtained from (5.4.6) simply by replacing e_{rs} by $\bar{e}_{rs}$.

This implies that W as given by (5.4.6) must now be restricted to the functional form

$$W = W(e_{11}, e_{22}, e_{33}, e_{12}^2, e_{23}^2, e_{31}^2, e_{12}\, e_{23}\, e_{31}). \tag{5.4.10}$$

In particular, when W is given by (5.3.11), all coefficients c_{rs}^{ij} in the scheme (5.4.8) which contain the index 1 either once or three times must vanish. The coefficients therefore reduce to 9 given by

$$\begin{matrix} c_{11}^{11} & c_{22}^{11} & c_{33}^{11} & 0 & 0 & 0 \\ & c_{22}^{22} & c_{33}^{22} & 0 & 0 & 0 \\ & & c_{33}^{33} & 0 & 0 & 0 \\ & & & c_{23}^{23} & 0 & 0 \\ & & & & c_{13}^{13} & 0 \\ & & & & & c_{12}^{12} \end{matrix} \tag{5.4.11}$$

An examination of (5.4.10) shows that elastic symmetry with respect to two orthogonal planes implies elastic symmetry with respect to a third plane which is orthogonal to both the original planes. Also, since the body is homogeneous, it follows that it then has three orthogonal planes of symmetry at each point, and such a body is said to be *orthotropic*. Wood materials may often be thought of as orthotropic. This is not an exact physical property of wood, but the assumption that wood is an orthotropic body is a good approximation for many purposes.

(c) *Transverse isotropy* (5 *coefficients*)

This is obtained from an orthotropic body if, after a change of axes of the form

$$\left.\begin{aligned} \bar{x}_1 &= x_1 \cos\alpha + x_2 \sin\alpha \\ \bar{x}_2 &= -x_1 \sin\alpha + x_2 \cos\alpha \\ \bar{x}_3 &= x_3 \end{aligned}\right\}, \tag{5.4.12}$$

the elastic potential W in terms of $\bar{e}_{rs}$ can be obtained from (5.4.10) simply by replacing e_{rs} by $\bar{e}_{rs}$. The values of $\partial\bar{x}_i/\partial x_r$ which correspond to (5.4.12) are

$$\frac{\partial\bar{x}_1}{\partial x_1} = \frac{\partial\bar{x}_2}{\partial x_2} = \cos\alpha, \qquad \frac{\partial\bar{x}_1}{\partial x_2} = -\frac{\partial\bar{x}_2}{\partial x_1} = \sin\alpha,$$

$$\frac{\partial\bar{x}_1}{\partial x_3} = \frac{\partial\bar{x}_2}{\partial x_3} = \frac{\partial\bar{x}_3}{\partial x_1} = \frac{\partial\bar{x}_3}{\partial x_2} = 0, \qquad \frac{\partial\bar{x}_3}{\partial x_3} = 1,$$

and, from (5.4.3),

$$\left.\begin{aligned}\bar{e}_{11} &= e_{11}\cos^2\alpha+2e_{12}\cos\alpha\sin\alpha+e_{22}\sin^2\alpha\\ \bar{e}_{22} &= e_{11}\sin^2\alpha-2e_{12}\cos\alpha\sin\alpha+e_{22}\cos^2\alpha\\ \bar{e}_{12} &= (e_{22}-e_{11})\cos\alpha\sin\alpha+e_{12}(\cos^2\alpha-\sin^2\alpha)\\ \bar{e}_{13} &= e_{13}\cos\alpha+e_{23}\sin\alpha\\ \bar{e}_{23} &= -e_{13}\sin\alpha+e_{23}\cos\alpha\\ \bar{e}_{33} &= e_{33}\end{aligned}\right\}. \tag{5.4.13}$$

It follows that, for all values of the angle α,

$$\left.\begin{aligned}\bar{e}_{11}+\bar{e}_{22} = e_{11}+e_{22}, \qquad \bar{e}_{11}\bar{e}_{22}-\bar{e}_{12}^2 = e_{11}e_{22}-e_{12}^2\\ \bar{e}_{13}^2+\bar{e}_{23}^2 = e_{13}^2+e_{23}^2, \qquad |\bar{e}_{ij}| = |e_{ij}|\end{aligned}\right\}, \tag{5.4.14}$$

and therefore W, as given by (5.4.10), must reduce to the form

$$W = W(e_{11}+e_{22}, e_{11}e_{22}-e_{12}^2, e_{33}, e_{13}^2+e_{23}^2, |e_{ij}|). \tag{5.4.15}$$

When W is given by (5.3.11) and the elastic coefficients are those given in (5.4.11), then the form (5.4.15) for W implies that

$$\left.\begin{aligned}c_{33}^{11} = c_{33}^{22}, \qquad c_{13}^{13} = c_{23}^{23}\\ c_{11}^{11} = c_{22}^{22}, \qquad c_{12}^{12} = \tfrac{1}{2}(c_{11}^{11}-c_{22}^{11})\end{aligned}\right\}. \tag{5.4.16}$$

For transverse isotropy the scheme of coefficients (5.4.11) becomes

$$\begin{array}{cccccc} c_{11}^{11} & c_{22}^{11} & c_{33}^{11} & 0 & 0 & 0\\ & c_{11}^{11} & c_{33}^{11} & 0 & 0 & 0\\ & & c_{33}^{33} & 0 & 0 & 0\\ & & & c_{13}^{13} & 0 & 0\\ & & & & c_{13}^{13} & 0\\ & & & & & \tfrac{1}{2}(c_{11}^{11}-c_{22}^{11}) \end{array} \tag{5.4.17}$$

(*d*) *Isotropic bodies* (2 *coefficients*)

If the elastic potential W is unaltered in form under all possible changes to other rectangular cartesian systems of axes, we have already said that the body is *isotropic*, W in this case being a function of the strain invariants. Alternatively, if we specialize from the system in (*c*), W must be unaltered in form under the transformations

$$\left.\begin{aligned}\bar{x}_3 &= x_3\cos\alpha+x_1\sin\alpha\\ \bar{x}_1 &= -x_3\sin\alpha+x_1\cos\alpha\\ \bar{x}_2 &= x_2\end{aligned}\right\}, \tag{5.4.18}$$

and

$$\left.\begin{aligned}\bar{x}_2 &= x_2\cos\alpha+x_3\sin\alpha\\ \bar{x}_3 &= -x_2\sin\alpha+x_3\cos\alpha\\ \bar{x}_1 &= x_1\end{aligned}\right\}. \tag{5.4.19}$$

In other words, W when expressed in terms of $\bar{e}_{rs}$ must be obtained from (5.4.15) simply by replacing e_{rs} by $\bar{e}_{rs}$. By analogy with section (c) it is seen that for this to be true under the transformation (5.4.18)

$$\left.\begin{aligned} c^{11}_{22} = c^{22}_{33}, \qquad c^{11}_{11} = c^{33}_{33} \\ c^{12}_{12} = c^{13}_{13} = \tfrac{1}{2}(c^{11}_{11}-c^{11}_{22}) \end{aligned}\right\}. \tag{5.4.20}$$

It follows automatically that W is unaltered in form under the transformation (5.4.19). The coefficients now reduce to two, namely, c^{11}_{11} and c^{11}_{22} in the scheme

$$\begin{matrix} c^{11}_{11} & c^{11}_{22} & c^{11}_{22} & 0 & 0 & 0 \\ & c^{11}_{11} & c^{11}_{22} & 0 & 0 & 0 \\ & & c^{11}_{11} & 0 & 0 & 0 \\ & & & \tfrac{1}{2}(c^{11}_{11}-c^{11}_{22}) & 0 & 0 \\ & & & & \tfrac{1}{2}(c^{11}_{11}-c^{11}_{22}) & 0 \\ & & & & & \tfrac{1}{2}(c^{11}_{11}-c^{11}_{22}) \end{matrix} \tag{5.4.21}$$

It can now be verified that W is unaltered in form under *all* possible changes to other rectangular coordinate systems, i.e. it is the same function of $\bar{e}_{rs}$ as it is of e_{rs} when x_i is changed to $\bar{x}_j$.

The elastic coefficients c^{ij}_{rs} for an isotropic body referred to rectangular cartesian coordinates in the unstrained body, have been obtained in the form (5.4.21) which may be expressed in the alternative notation

$$\begin{matrix} \lambda+2\mu & \lambda & \lambda & 0 & 0 & 0 \\ & \lambda+2\mu & \lambda & 0 & 0 & 0 \\ & & \lambda+2\mu & 0 & 0 & 0 \\ & & & \mu & 0 & 0 \\ & & & & \mu & 0 \\ & & & & & \mu \end{matrix} \tag{5.4.22}$$

where
$$\lambda = c^{11}_{22}, \qquad \mu = \tfrac{1}{2}(c^{11}_{11}-c^{11}_{22}). \tag{5.4.23}$$

The elastic coefficients E^{ij}_{rs}, in general coordinates, are now given by the tensor transformation

$$E^{ij}_{rs} = \frac{\partial\theta^i}{\partial x^m}\frac{\partial\theta^j}{\partial x^n}\frac{\partial x^p}{\partial\theta^r}\frac{\partial x^t}{\partial\theta^s}c^{mn}_{pt}. \tag{5.4.24}$$

From the table (5.4.22) and the symmetric properties (5.3.10) we see that we may express the coefficients c^{ij}_{rs} in the compact form

$$c^{ij}_{rs} = \lambda\delta^{ij}\delta_{rs}+\mu(\delta^i_r\delta^j_s+\delta^i_s\delta^j_r). \tag{5.4.25}$$

Hence, in general coordinates, we have†

$$E^{ij}_{rs} = \lambda g^{ij}g_{rs}+\mu(\delta^i_r\delta^j_s+\delta^i_s\delta^j_r), \tag{5.4.26}$$

† This result may also be obtained from the assumption that, for homogeneous isotropic bodies, the elastic potential W is a function of the strain invariants.

since this tensor equation reduces to (5.4.25) when the coordinates are rectangular cartesian coordinates. This result also follows immediately from (5.4.24) and (5.4.25), and is the general form for the elastic coefficients for a homogeneous isotropic body. The constants λ, μ are the elastic constants of Lamé and these are sometimes put in the alternative forms

$$\lambda = \frac{E\eta}{(1+\eta)(1-2\eta)}, \qquad \mu = \frac{E}{2(1+\eta)}, \tag{5.4.27}$$

where the constants E and η are called Young's modulus and Poisson's ratio respectively. The Lamé constant μ is known as the shear modulus and the relation (5.4.26) is often expressed in terms of μ and η. Thus

$$E^{ij}_{rs} = \mu\left(\delta^i_r\delta^j_s+\delta^i_s\delta^j_r+\frac{2\eta}{1-2\eta}g^{ij}g_{rs}\right), \tag{5.4.28}$$

since

$$\lambda = \frac{2\eta\mu}{1-2\eta}. \tag{5.4.29}$$

From (5.3.5) and (5.4.28) we find that the stress–strain relation for an isotropic body takes the form

$$\tau^{ij} = \mu\left(g^{ir}g^{js}+g^{is}g^{jr}+\frac{2\eta}{1-2\eta}g^{ij}g^{rs}\right)\gamma_{rs}, \tag{5.4.30}$$

and this may also be written in the alternative forms

$$\left.\begin{aligned} \tau^{ij} &= 2\mu\left(\gamma^{ij}+\frac{\eta}{1-2\eta}g^{ij}\gamma^r_r\right) \\ \tau^i_j &= 2\mu\left(\gamma^i_j+\frac{\eta}{1-2\eta}\delta^i_j\gamma^r_r\right) \\ \tau_{ij} &= 2\mu\left(\gamma_{ij}+\frac{\eta}{1-2\eta}g_{ij}\gamma^r_r\right) \end{aligned}\right\}. \tag{5.4.31}$$

We may also express the components of strain in terms of the components of stress. Thus

$$\left.\begin{aligned} E\gamma_{rs} &= \{(1+\eta)g_{ir}g_{js}-\eta g_{ij}g_{rs}\}\tau^{ij} \\ &= (1+\eta)\tau_{rs}-\eta g_{rs}\tau^i_i \\ E\gamma^r_s &= (1+\eta)\tau^r_s-\eta\delta^r_s\tau^i_i \\ E\gamma^{rs} &= (1+\eta)\tau^{rs}-\eta g^{rs}\tau^i_i \end{aligned}\right\}. \tag{5.4.32}$$

From (5.4.31) and (5.1.6) we have immediately

$$\left.\begin{aligned} \tau^r_r &= 2\mu\frac{1+\eta}{1-2\eta}\gamma^r_r = 2\mu\frac{1+\eta}{1-2\eta}v^r|_r \\ \text{or} \qquad v^r|_r &= \gamma^r_r = \frac{1-2\eta}{2\mu(1+\eta)}\tau^r_r \end{aligned}\right\}, \tag{5.4.33}$$

which gives an important relation between the strain invariant $\gamma_r^r = v^r|_r$ and the stress invariant τ_r^r.

When the material has no symmetrical properties it is still, in general, possible to express components of strain in terms of components of stress by solving equations (5.3.5) or (5.3.8) for γ_{rs}. Thus

$$\gamma_{rs} = F_{ijrs}\tau^{ij} = F_{rs}^{ij}\tau_{ij}, \tag{5.4.34}$$

where
$$E^{ijrs}F_{ijmn} = E_{ij}^{rs}F_{mn}^{ij} = \tfrac{1}{2}(\delta_m^r\delta_n^s+\delta_n^r\delta_m^s). \tag{5.4.35}$$

The symmetrical properties of the tensors F_{ijrs}, F_{rs}^{ij} are similar to those of E_{ijrs}, E_{rs}^{ij} so we shall not consider them in detail. We observe, however, that if the elastic coefficients F_{rs}^{ij} are denoted by s_{rs}^{ij} when referred to rectangular cartesian axes then

$$s_{rs}^{ij} = s_{rs}^{ji} = s_{sr}^{ij} = s_{sr}^{ji} = s_{ij}^{rs}, \tag{5.4.36}$$

and the strain–stress relation (5.4.34) becomes

$$e_{rs} = s_{ij}^{rs}t_{ij}. \tag{5.4.37}$$

Also the elastic potential W is

$$W = \tfrac{1}{2}s_{ij}^{rs}t_{rs}t_{ij}. \tag{5.4.38}$$

5.5. Differential equations for an isotropic body

In the previous sections we have found that the classical theory of elasticity is governed by the following equations: the equations of motion, which we repeat here in the form

$$\tau^{ij}|_i+\rho F^j = \rho f^j, \tag{5.5.1}$$

and the stress–strain relations

$$\tau^{ij} = E^{ijrs}\gamma_{rs}, \tag{5.5.2}$$

where, for an isotropic material,

$$E^{ijrs} = \mu\left(g^{ir}g^{js}+g^{is}g^{jr}+\frac{2\eta}{1-2\eta}g^{ij}g^{rs}\right). \tag{5.5.3}$$

Also, the components of the strain tensor γ_{rs} may be expressed in terms of the displacements by the formula

$$\gamma_{rs} = \tfrac{1}{2}(v_r|_s+v_s|_r). \tag{5.5.4}$$

The system of equations (5.5.1)–(5.5.4) must be satisfied at every interior point of the body. At the bounding surface of the body certain boundary conditions must be fulfilled. When surface tractions are prescribed at the surface,

$$\tau^{ij}n_i = P^j, \tag{5.5.5}$$

and this is usually described as the first boundary-value problem.

Alternatively, we may prescribe displacements at the surface of the body and this gives us the second boundary-value problem. We have a mixed boundary-value problem when surface tractions are prescribed over one part of the surface and displacements are prescribed over the remaining surface of the body.

We note that on account of the linearity of the equations (5.5.1)–(5.5.4) the principle of superposition is applicable.† Thus, suppose that under the influence of external forces $F^{(1)j}$ and accelerations $f^{(1)j}$ we find stresses and displacements $\tau^{(1)ij}$, $v_j^{(1)}$; and corresponding to external forces $F^{(2)j}$ and accelerations $f^{(2)j}$ we obtain stresses and displacements $\tau^{(2)ij}$, $v_j^{(2)}$. Then, because the fundamental equations are linear, it follows that stresses $\tau^{(1)ij}+\tau^{(2)ij}$ and displacements $v_j^{(1)}+v_j^{(2)}$ correspond to body forces $F^{(1)j}+F^{(2)j}$ and accelerations $f^{(1)j}+f^{(2)j}$.

Consider a bounded elastic body in equilibrium governed by equations (5.5.1) with $f^i = 0$, (5.5.2), and (5.5.4), and subject to one of the sets of boundary conditions mentioned above. If, under suitable smoothness assumptions, a solution of these equations exists, then it can be proved‡ that the solution is unique if the strain energy is positive definite. (For given surface tractions on the boundary the displacements are unique apart from rigid-body displacements.) If part or the whole of the region extends to infinity the uniqueness problem requires further examination but this will not be considered here. Also extra conditions must be imposed if singularities are allowed.§

It is possible to express the equations (5.5.1)–(5.5.4) in alternative

† It must be emphasized that the principle of superposition is not valid for finite displacements considered in Chapter 2, because the equations are not linear.

‡ These proofs can be found in most of the books already mentioned, e.g. see I. S. Sokolnikoff, p. 86; loc. cit., p. 148 for references. The proofs usually assume an isotropic material but this is not necessary.

§ E. Sternberg and R. A. Eubanks, *J. rat. Mech. Analysis* **4** (1955) 135. For further references on uniqueness see, e.g.:

G. Kirchhoff, *J. reine angew. Math.* **56** (1859);

E. and F. Cosserat, *C. r. hebd. Séanc. Acad. Sci., Paris* **126** (1898) 1089; **127** (1898) 315;

E. Almansi, *Atti Accad. naz. Lincei Rc. Sed. solen.* **16** (1907) 863;

G. Fichera, *Annali Scu. norm. sup., Pisa* **4** (1950) 35;

N. E. Muskhelishvili, *Some basic problems of elasticity* (Noordhoff, 1953);

J. L. Ericksen and R. A. Toupin, *Can. J. Math.* **8** (1956) 432;

R. Hill, *J. Mech. Phys. Solids* **9** (1961) 114;

J. H. Bramble and L. E. Payne, *Proc. 4th U.S. Congr. appl. Mech.* **1** (1962) 469;

M. A. Hayes, *Proc. R. Soc.* **A274** (1963) 500; *Arch. ration. Mech. Analysis* **16** (1964) 238;

R. J. Knops, *Arch. ration. Mech. Analysis* **18** (1965) 107;

M. E. Gurtin and R. A. Toupin, *Q. appl. Math.* **23** (1965) 79.

In some of these papers the strain energy function is not necessarily positive definite and the region of space may be unbounded.

forms, some of which we now consider. We first substitute (5.5.2) into (5.5.1) and use (5.5.4), noting also from (5.5.3) that

$$E^{ijrs}|_t = 0.$$

Thus
$$E^{ijrs}v_s|_{ri}+\rho F^j = \rho f^j, \tag{5.5.6}$$

which is a set of three differential equations for the three components v_s of the displacement vector. If we now use (5.5.3), equations (5.5.6) become

$$v^j|^r_r+\frac{v^r|^j_r}{1-2\eta}+\frac{\rho F^j}{\mu} = \frac{\rho f^j}{\mu} \tag{5.5.7}$$

if we use the notation described in (1.12.18). Taking the covariant derivative of (5.5.7) with respect to θ_j we get

$$v^j|^r_{rj}+\frac{v^r|^j_{rj}}{1-2\eta}+\frac{\rho F^j|_j}{\mu} = \frac{\rho f^j|_j}{\mu},$$

assuming that ρ is constant. Since

$$v^j|^r_{rj} = v^j|^r_{jr} = v^r|^j_{rj},$$

this last equation reduces to

$$\frac{2(1-\eta)v^j|^r_{rj}}{1-2\eta}+\frac{\rho F^j|_j}{\mu} = \frac{\rho f^j|_j}{\mu}. \tag{5.5.8}$$

If we further assume that the body forces and accelerations obey the laws

$$F^j|_j = 0, \qquad f^j|_j = 0, \tag{5.5.9}$$

then
$$v^j|^r_{rj} = 0, \tag{5.5.10}$$

or, using (5.1.6),
$$\gamma^i_i|^r_r = 0. \tag{5.5.11}$$

From (5.4.33) and (5.5.11) we also have

$$\tau^i_i|^r_r = 0. \tag{5.5.12}$$

On comparing equations (5.5.11) and (5.5.12) with (1.12.29) we see that the dilatation γ^i_i and the stress invariant τ^i_i are harmonic functions provided equations (5.5.9) are satisfied.

Again, we find from (5.5.7), that

$$v^j|^{ri}_{ri}+\frac{v^r|^{ji}_{ri}}{1-2\eta}+\frac{\rho F^j|^i_i}{\mu} = \frac{\rho f^j|^i_i}{\mu}. \tag{5.5.13}$$

If all components of the body forces and accelerations in rectangular cartesian coordinates are constants then

$$F^j|_i = 0, \quad f^j|_i = 0 \tag{5.5.14}$$

and hence
$$v^j|_{ri}^{ri} + \frac{v^r|_{ri}^{ji}}{1-2\eta} = 0.$$

But, from (5.5.10), we find that

$$v^r|_{ri}^{ji} = 0,$$

therefore
$$v^j|_{ri}^{ri} = 0. \tag{5.5.15}$$

This shows that when equations (5.5.14) are satisfied each component of the displacement in rectangular cartesian coordinates is a biharmonic function (see (1.12.32)).

So far we have obtained equations which involve only the components of the displacement vector. We now proceed to obtain equations which contain only components of the stress tensor. For this purpose we combine (5.4.32) and (5.5.4) to give

$$\mu(v_s|_r + v_r|_s) = \tau_{rs} - \frac{\eta}{1+\eta} g_{rs} \tau_i^i,$$

and raising the suffixes r, s in both sides of this equation we get

$$\mu(v^r|^s + v^s|^r) = \tau^{rs} - \frac{\eta}{1+\eta} g^{rs} \tau_i^i. \tag{5.5.16}$$

This leads at once to

$$\mu(v^r|_i^{si} + v^s|_i^{ri}) = \tau^{rs}|_i^i - \frac{\eta}{1+\eta} g^{rs} \tau_m^m|_i^i. \tag{5.5.17}$$

From (5.4.33) and (5.5.7) we find, respectively,

$$2\mu v^i|_i^{rs} = \frac{1-2\eta}{1+\eta} \tau_i^i|^{rs}, \tag{5.5.18}$$

$$v^s|_i^{ir} = -\frac{v^i|_i^{sr}}{1-2\eta} - \frac{\rho}{\mu}(F^s|^r - f^s|^r), \tag{5.5.19}$$

and, interchanging r and s,

$$v^r|_i^{is} = -\frac{v^i|_i^{rs}}{1-2\eta} - \frac{\rho}{\mu}(F^r|^s - f^r|^s). \tag{5.5.20}$$

The displacement v^r can be eliminated from equations (5.5.17)–(5.5.20) to give

$$\tau^{rs}|^i_i+\frac{\tau^i_i|^{rs}}{1+\eta}-\frac{\eta g^{rs}\tau^m_m|^i_i}{1+\eta} = -\rho(F^s|^r+F^r|^s-f^s|^r-f^r|^s), \quad (5.5.21)$$

if we observe that $$v^r|^{si}_i = v^r|^{is}_i.$$

Equations (5.5.21) are a set of differential equations† for the components of stress τ^{rs}. Equations (5.5.21) may be expressed in a slightly different form if we eliminate $\tau^m_m|^i_i$. Thus, if we lower the suffix r in (5.5.21) and contract by putting $r = s$, we obtain

$$\frac{1-\eta}{1+\eta}\tau^r_r|^i_i = -\rho(F^r|_r-f^r|_r), \quad (5.5.22)$$

and, substituting for $\tau^r_r|^i_i$ in (5.5.21), we have

$$\tau^{rs}|^i_i+\frac{\tau^i_i|^{rs}}{1+\eta} = -\frac{\eta\rho}{1-\eta}g^{rs}(F^i|_i-f^i|_i)-\rho(F^r|^s+F^s|^r-f^r|^s-f^s|^r). \quad (5.5.23)$$

If the relations (5.5.14) hold, then, using (5.5.12), we obtain from (5.5.23) the equations
$$\tau^{rs}|^{ij}_{ij} = 0,$$
and hence, using (5.4.32), $$\gamma^{rs}|^{ij}_{ij} = 0.$$

When, therefore, equations (5.5.14) are satisfied, all components of the stress and strain tensors, referred to rectangular cartesian coordinates, are biharmonic functions.

5.6. Solution for isotropic bodies in terms of potential functions‡

When body forces are zero and the strained body is in equilibrium, the equations (5.5.7) for the displacements in an isotropic body become

$$(1-2\eta)v^i|^j_j+v^j|^i_j = 0, \quad (5.6.1)$$

† These may also be obtained from the equations of compatibility (5.1.5) and the equations (5.4.32) and (5.5.1).

‡ See J. Boussinesq, *Applications des Potentials* (Paris, 1885); J. Dougall, *Trans. R. Soc. Edinb.* **41** (1904) 129; **49** (1914) 895; H. Neuber, *Kerbspannungslehre* (Berlin, 1937; Michigan, 1944). A number of writers have used these potential functions for special problems. Among recent contributions may be mentioned papers on the stress concentration around an ellipsoidal cavity by M. A. Sadowsky and E. Sternberg, *J. appl. Mech.* **14** (1947) 191; **16** (1949) 149. Here we only list possible potential function solutions. For discussions of the important question of completeness of stress function solutions of the elastic equations for bounded regions see

R. D. Mindlin, *Bull. Am. math. Soc.* **42** (1936) 373;
K. Marguerre, *Z. angew. Math. Mech.* **35** (1955) 242;
R. D. Duffin, *J. rat. Mech. Analysis* **5** (1956) 939;
R. A. Eubanks and E. Sternberg, ibid. **5** (1956) 735;
C. Truesdell, *Arch. ration. Mech. Analysis* **4** (1959) 1;
E. Sternberg and M. E. Gurtin, *Proc. 4th U.S. Nat. Congr. appl. Mech.*;
E. Sternberg, *Structural Mechanics* (Pergamon, 1960).

and, with the help of (5.1.1) and (5.4.30), stresses may be expressed directly in terms of displacements by the formulae

$$\left.\begin{aligned} \tau^{ij} &= \mu\left(g^{ir}g^{js}+\frac{\eta}{1-2\eta}g^{ij}g^{rs}\right)(v_r|_s+v_s|_r) \\ \frac{\tau^{ij}}{\mu} &= g^{js}v^i|_s+g^{ir}v^j|_r+\frac{2\eta}{1-2\eta}g^{ij}\Delta, \qquad \Delta = v^r|_r \end{aligned}\right\}. \tag{5.6.2}$$

In the special case of rectangular cartesian coordinates $x_i = (x,y,z)$ in which the components of displacement are denoted by (u_x, u_y, u_z), we have, from (5.6.1),

$$\left.\begin{aligned} (1-2\eta)\nabla^2(u_x,u_y,u_z)+\left(\frac{\partial}{\partial x},\frac{\partial}{\partial y},\frac{\partial}{\partial z}\right)\Delta = 0 \\ \Delta = \frac{\partial u_x}{\partial x}+\frac{\partial u_y}{\partial y}+\frac{\partial u_z}{\partial z} \end{aligned}\right\}. \tag{5.6.3}$$

Also, the components of the stress tensor τ^{ij} become physical components of stress σ_{ij}, and we adopt the notation $\sigma_{xx}, \sigma_{yy}, \sigma_{zz}, \sigma_{yz}, \sigma_{zx}, \sigma_{xy}$. Hence, from (5.6.2), remembering that $g^{ij} = \delta^{ij}$ for rectangular cartesian axes,

$$\left.\begin{aligned} \frac{\sigma_{xx}}{2\mu} &= \frac{\eta\Delta}{1-2\eta}+\frac{\partial u_x}{\partial x}, & \frac{\sigma_{yz}}{\mu} &= \frac{\partial u_z}{\partial y}+\frac{\partial u_y}{\partial z} \\ \frac{\sigma_{yy}}{2\mu} &= \frac{\eta\Delta}{1-2\eta}+\frac{\partial u_y}{\partial y}, & \frac{\sigma_{zx}}{\mu} &= \frac{\partial u_x}{\partial z}+\frac{\partial u_z}{\partial x} \\ \frac{\sigma_{zz}}{2\mu} &= \frac{\eta\Delta}{1-2\eta}+\frac{\partial u_z}{\partial z}, & \frac{\sigma_{xy}}{\mu} &= \frac{\partial u_y}{\partial x}+\frac{\partial u_x}{\partial y} \\ & \Delta = \frac{\partial u_x}{\partial x}+\frac{\partial u_y}{\partial y}+\frac{\partial u_z}{\partial z} \end{aligned}\right\}. \tag{5.6.4}$$

Similarly, we may obtain expressions for stresses in terms of displacements referred to any other suitable system of axes. We record results for one other important special case, namely, cylindrical polar coordinates (r, θ, z) for which

$$\left.\begin{aligned} g_{ij} = \begin{bmatrix} 1 & 0 & 0 \\ 0 & r^2 & 0 \\ 0 & 0 & 1 \end{bmatrix}, \quad g^{ij} = \begin{bmatrix} 1 & 0 & 0 \\ 0 & \frac{1}{r^2} & 0 \\ 0 & 0 & 1 \end{bmatrix}, \quad \sqrt{g} = r \\ \Gamma^1_{22} = -r, \qquad \Gamma^2_{12} = \Gamma^2_{21} = \frac{1}{r} \end{aligned}\right\}, \tag{5.6.5}$$

the remaining Christoffel symbols being zero. We denote the physical components of displacement by (u_r, u_θ, u_z) so that

$$u_r = v^1, \qquad u_\theta = rv^2, \qquad u_z = v^3. \tag{5.6.6}$$

The physical components of stress σ_{ij} are given by (2.2.17), and we replace σ_{ij} by the notation σ_{rr}, $\sigma_{\theta\theta}$, σ_{zz}, $\sigma_{\theta z}$, σ_{zr}, $\sigma_{r\theta}$, so that

$$\left.\begin{aligned} \sigma_{rr} &= \tau^{11}, & \sigma_{\theta\theta} &= r^2\tau^{22}, & \sigma_{zz} &= \tau^{33} \\ \sigma_{\theta z} &= r\tau^{23}, & \sigma_{zr} &= \tau^{31}, & \sigma_{r\theta} &= r\tau^{12} \end{aligned}\right\}. \tag{5.6.7}$$

Hence, from (5.6.2), (5.6.5), (5.6.6), and (5.6.7),

$$\left.\begin{aligned} \frac{\sigma_{rr}}{2\mu} &= \frac{\eta\Delta}{1-2\eta}+\frac{\partial u_r}{\partial r}, & \frac{\sigma_{\theta z}}{\mu} &= \frac{1}{r}\frac{\partial u_z}{\partial \theta}+\frac{\partial u_\theta}{\partial z} \\ \frac{\sigma_{\theta\theta}}{2\mu} &= \frac{\eta\Delta}{1-2\eta}+\frac{1}{r}\left(\frac{\partial u_\theta}{\partial \theta}+u_r\right), & \frac{\sigma_{rz}}{\mu} &= \frac{\partial u_r}{\partial z}+\frac{\partial u_z}{\partial r} \\ \frac{\sigma_{zz}}{2\mu} &= \frac{\eta\Delta}{1-2\eta}+\frac{\partial u_z}{\partial z}, & \frac{\sigma_{r\theta}}{\mu} &= \frac{1}{r}\frac{\partial u_r}{\partial \theta}+\frac{\partial u_\theta}{\partial r}-\frac{u_\theta}{r} \\ & \Delta = \frac{\partial u_r}{\partial r}+\frac{u_r}{r}+\frac{1}{r}\frac{\partial u_\theta}{\partial \theta}+\frac{\partial u_z}{\partial z} \end{aligned}\right\}, \tag{5.6.8}$$

Solutions of equations (5.6.3), and therefore of (5.6.1), can be found in a variety of forms in terms of harmonic functions, and we shall refer to these as basic solutions. A simple calculation enables us to verify at once that these solutions satisfy (5.6.3).

Solution A

$$\left.\begin{aligned} 2\mu(u_x, u_y, u_z) &= \left(\frac{\partial}{\partial x}, \frac{\partial}{\partial y}, \frac{\partial}{\partial z}\right)F \\ \nabla^2 F = 0, &\qquad \Delta = 0 \end{aligned}\right\}. \tag{5.6.9}$$

In general vector and tensor forms this solution becomes

$$\left.\begin{aligned} 2\mu\mathbf{v} &= \operatorname{grad} F \\ 2\mu v_i &= F_{,i} \end{aligned}\right\}. \tag{5.6.10}$$

We record the corresponding stresses for rectangular and cylindrical polar coordinates:

$$\left.\begin{aligned} \sigma_{xx} = \frac{\partial^2 F}{\partial x^2}, \qquad \sigma_{xx}+\sigma_{yy} = -\frac{\partial^2 F}{\partial z^2}, \qquad \sigma_{zz} = \frac{\partial^2 F}{\partial z^2} \\ \sigma_{yz} = \frac{\partial^2 F}{\partial y\partial z}, \qquad \sigma_{zx} = \frac{\partial^2 F}{\partial z\partial x}, \qquad \sigma_{xy} = \frac{\partial^2 F}{\partial x\partial y} \\ 2\mu(u_r, u_\theta, u_z) = \left(\frac{\partial}{\partial r}, \frac{1}{r}\frac{\partial}{\partial \theta}, \frac{\partial}{\partial z}\right)F \\ \sigma_{rr} = \frac{\partial^2 F}{\partial r^2}, \qquad \sigma_{rr}+\sigma_{\theta\theta} = -\frac{\partial^2 F}{\partial z^2} \\ \sigma_{\theta z} = \frac{1}{r}\frac{\partial^2 F}{\partial \theta\partial z}, \qquad \sigma_{rz} = \frac{\partial^2 F}{\partial r\partial z}, \qquad \sigma_{r\theta} = \frac{1}{r}\frac{\partial^2 F}{\partial r\partial \theta}-\frac{1}{r^2}\frac{\partial F}{\partial \theta} \end{aligned}\right\}. \tag{5.6.11}$$

Solution B

$$\left.\begin{aligned} 2\mu(u_x, u_y, u_z) &= z\left(\frac{\partial}{\partial x}, \frac{\partial}{\partial y}, \frac{\partial}{\partial z}\right)Z-(0,0,(3-4\eta)Z) \\ \nabla^2 Z = 0, &\qquad \mu\Delta = -(1-2\eta)\frac{\partial Z}{\partial z} \end{aligned}\right\}. \tag{5.6.12}$$

In general coordinates

$$2\mu v_i = zZ_{,i}-(3-4\eta)Z\frac{\partial z}{\partial\theta^i}. \tag{5.6.13}$$

Hence

$$\left.\begin{aligned} &2\mu u_r = z\frac{\partial Z}{\partial r}, \qquad 2\mu u_\theta = \frac{z}{r}\frac{\partial Z}{\partial\theta} \\ &\sigma_{xx} = z\frac{\partial^2 Z}{\partial x^2}-2\eta\frac{\partial Z}{\partial z}, \qquad \sigma_{xx}+\sigma_{yy} = -z\frac{\partial^2 Z}{\partial z^2}-4\eta\frac{\partial Z}{\partial z} \\ &\sigma_{zz} = z\frac{\partial^2 Z}{\partial z^2}-2(1-\eta)\frac{\partial Z}{\partial z}, \qquad \sigma_{xy} = z\frac{\partial^2 Z}{\partial x\partial y} \\ &\sigma_{yz} = z\frac{\partial^2 Z}{\partial y\partial z}-(1-2\eta)\frac{\partial Z}{\partial y}, \qquad \sigma_{zx} = z\frac{\partial^2 Z}{\partial x\partial z}-(1-2\eta)\frac{\partial Z}{\partial x} \\ &\sigma_{rr} = z\frac{\partial^2 Z}{\partial r^2}-2\eta\frac{\partial Z}{\partial z}, \qquad \sigma_{rr}+\sigma_{\theta\theta} = -z\frac{\partial^2 Z}{\partial z^2}-4\eta\frac{\partial Z}{\partial z} \\ &\sigma_{\theta z} = \frac{z}{r}\frac{\partial^2 Z}{\partial\theta\partial z}-\frac{(1-2\eta)}{r}\frac{\partial Z}{\partial\theta}, \qquad \sigma_{rz} = z\frac{\partial^2 Z}{\partial r\partial z}-(1-2\eta)\frac{\partial Z}{\partial r} \\ &\sigma_{r\theta} = \frac{z}{r}\left(\frac{\partial^2 Z}{\partial r\partial\theta}-\frac{1}{r}\frac{\partial Z}{\partial\theta}\right) \end{aligned}\right\}. \tag{5.6.14}$$

Solution C

$$\left.\begin{aligned} 2\mu(u_x, u_y, u_z) &= x\left(\frac{\partial}{\partial x}, \frac{\partial}{\partial y}, \frac{\partial}{\partial z}\right)X-((3-4\eta)X, 0, 0) \\ \nabla^2 X = 0, &\qquad \mu\Delta = -(1-2\eta)\frac{\partial X}{\partial x} \\ 2\mu v_i &= xX_{,i}-(3-4\eta)X\frac{\partial x}{\partial\theta^i} \end{aligned}\right\}. \tag{5.6.15}$$

Solution D

$$\left.\begin{aligned} 2\mu(u_x, u_y, u_z) &= y\left(\frac{\partial}{\partial x}, \frac{\partial}{\partial y}, \frac{\partial}{\partial z}\right)Y-(0, (3-4\eta)Y, 0) \\ \nabla^2 Y = 0, &\qquad \mu\Delta = -(1-2\eta)\frac{\partial Y}{\partial y} \\ 2\mu v_i &= yY_{,i}-(3-4\eta)Y\frac{\partial y}{\partial\theta^i} \end{aligned}\right\}. \tag{5.6.16}$$

The stress components corresponding to solutions C and D are omitted but may be obtained if required from (5.6.4) and (5.6.8).

Solution E

$$\left.\begin{aligned} \mu(u_x, u_y, u_z) &= \left(\frac{\partial\psi}{\partial y}, -\frac{\partial\psi}{\partial x}, 0\right) \\ \nabla^2\psi = 0, \qquad & \Delta = 0 \\ \mu\mathbf{v} &= \operatorname{curl}(\psi z_{,k}\,\mathbf{g}^k) \\ \mu v^i &= \epsilon^{ijk}\psi_{,j}z_{,k} \end{aligned}\right\}. \tag{5.6.17}$$

$$\left.\begin{aligned} &\mu u_r = \frac{1}{r}\frac{\partial\psi}{\partial\theta}, \qquad \mu u_\theta = -\frac{\partial\psi}{\partial r} \\ &\sigma_{xx} = 2\frac{\partial^2\psi}{\partial x\partial y}, \qquad \sigma_{yy} = -2\frac{\partial^2\psi}{\partial x\partial y}, \qquad \sigma_{zz} = 0 \\ &\sigma_{yz} = -\frac{\partial^2\psi}{\partial z\partial x}, \qquad \sigma_{zx} = \frac{\partial^2\psi}{\partial y\partial z}, \qquad \sigma_{xy} = \frac{\partial^2\psi}{\partial y^2}-\frac{\partial^2\psi}{\partial x^2} \\ &\sigma_{rr} = \frac{2}{r}\frac{\partial^2\psi}{\partial r\partial\theta}-\frac{2}{r^2}\frac{\partial\psi}{\partial\theta}, \qquad \sigma_{rr}+\sigma_{\theta\theta} = 0 \\ &\sigma_{\theta z} = -\frac{\partial^2\psi}{\partial r\partial z}, \qquad \sigma_{rz} = \frac{1}{r}\frac{\partial^2\psi}{\partial\theta\partial z} \\ &\sigma_{r\theta} = \frac{1}{r}\frac{\partial\psi}{\partial r}-\frac{\partial^2\psi}{\partial r^2}+\frac{1}{r^2}\frac{\partial^2\psi}{\partial\theta^2} \end{aligned}\right\}. \tag{5.6.18}$$

Other solutions of type E exist in which the displacement vector is the curl of a vector with components $(\psi, 0, 0)$ or $(0, \psi, 0)$ referred to axes (x, y, z).

5.7. Problems depending on one harmonic function†

An interesting class of problems in which the shearing stress vanishes at all points in a plane, say the plane $z = 0$, can be reduced to classical problems in potential theory. Thus, if we combine solutions A and B and put

$$F = (1-2\eta)\phi, \qquad Z = \frac{\partial\phi}{\partial z}, \qquad \nabla^2\phi = 0, \tag{5.7.1}$$

† The combination of solutions A, B, and E can be obtained by direct integration of the equations of equilibrium (5.6.1) using a method similar to that in §§ 4.4, 5.12. The exact conditions under which such a complete solution of (5.6.1) exists in this form, including the region of space in which it applies, need further discussion.

which is possible since both F and Z are harmonic functions, we have

$$\left.\begin{aligned}
2\mu u_x &= z\frac{\partial^2\phi}{\partial x\partial z}+(1-2\eta)\frac{\partial\phi}{\partial x}\\
2\mu u_y &= z\frac{\partial^2\phi}{\partial y\partial z}+(1-2\eta)\frac{\partial\phi}{\partial y}\\
2\mu u_z &= z\frac{\partial^2\phi}{\partial z^2}-2(1-\eta)\frac{\partial\phi}{\partial z}\\
2\mu u_r &= z\frac{\partial^2\phi}{\partial r\partial z}+(1-2\eta)\frac{\partial\phi}{\partial r}\\
2\mu u_\theta &= \frac{z}{r}\frac{\partial^2\phi}{\partial\theta\partial z}+(1-2\eta)\frac{\partial\phi}{r\partial\theta}\\
\sigma_{xx} &= z\frac{\partial^3\phi}{\partial x^2\partial z}+\frac{\partial^2\phi}{\partial x^2}+2\eta\frac{\partial^2\phi}{\partial y^2}\\
\sigma_{xx}+\sigma_{yy} = \sigma_{rr}+\sigma_{\theta\theta} &= -z\frac{\partial^3\phi}{\partial z^3}-(1+2\eta)\frac{\partial^2\phi}{\partial z^2}\\
\sigma_{xy} &= z\frac{\partial^3\phi}{\partial x\partial y\partial z}+(1-2\eta)\frac{\partial^2\phi}{\partial x\partial y}\\
\sigma_{zz} &= z\frac{\partial^3\phi}{\partial z^3}-\frac{\partial^2\phi}{\partial z^2}\\
\sigma_{yz} &= z\frac{\partial^3\phi}{\partial y\partial z^2},\qquad \sigma_{zx} = z\frac{\partial^3\phi}{\partial x\partial z^2}\\
\sigma_{rr} &= z\frac{\partial^3\phi}{\partial r^2\partial z}+\frac{\partial^2\phi}{\partial r^2}-2\eta\left(\frac{\partial^2\phi}{\partial r^2}+\frac{\partial^2\phi}{\partial z^2}\right)
\end{aligned}\right\}, \tag{5.7.2}$$

From (5.7.2) we see that, provided the various limits exist as $z \to 0$, the values of the stresses and displacements on the plane $z = 0$ are

$$\left.\begin{aligned}
\sigma_{yz} &= \sigma_{xz} = 0\\
\sigma_{zz} &= -\frac{\partial^2\phi}{\partial z^2}\\
2\mu u_z &= -2(1-\eta)\frac{\partial\phi}{\partial z}
\end{aligned}\right\}. \tag{5.7.3}$$

If, therefore, the shear stresses vanish on the plane $z = 0$, and either the normal stress σ_{zz}, or the normal displacement u_z, has prescribed values on $z = 0$, we see that the problem reduces to the determination of a potential function $\partial\phi/\partial z$ with prescribed values for $\partial\phi/\partial z$, or $\partial^2\phi/\partial z^2$, on $z = 0$. The solution of some problems may be completed with the help of Hankel transforms.† Perhaps the most interesting cases arise when the boundary values at $z = 0$ are of a mixed type, and we consider briefly, in the next section, the problem of the indentation of a plane surface of a semi-infinite medium by a rigid punch. The solution is, however, obtained without the aid of Hankel transforms.

5.8. Indentation problems

We shall suppose that a perfectly rigid solid of revolution of prescribed shape, whose axis of revolution coincides with the z-axis, is pressed normally against the plane $z = 0$ of a semi-infinite elastic medium $z \geqslant 0$. The strained surface of the elastic medium will fit the rigid body over the part between the lowest point and a circular section of radius a. We assume that the shearing stress vanishes at all points of the boundary $z = 0$, the z-component of surface displacement is prescribed over the region $r \leqslant a$, $z = 0$, and the normal stress is zero on the remainder of the boundary. Thus, for $z = 0$,

$$\left.\begin{aligned} u_z &= f(r) \qquad (0 \leqslant r \leqslant a) \\ \sigma_{zz} &= 0 \qquad\quad (r > a) \\ \sigma_{zx} &= \sigma_{zy} = 0 \quad (0 \leqslant r < \infty) \end{aligned}\right\}. \tag{5.8.1}$$

Also, all components of stress and displacement are $O(R^{-2})$ and $O(R^{-1})$ respectively as $R = (r^2+z^2)^{\frac{1}{2}} \to \infty$.

The third boundary condition in (5.8.1) is satisfied by the stress system of § 5.7, and the first two boundary conditions give

$$\left.\begin{aligned} \frac{\partial \omega}{\partial z} &= 0 \qquad (r > a) \\ \omega &= f(r) \quad (0 \leqslant r \leqslant a) \end{aligned}\right\}, \tag{5.8.2}$$

where
$$\omega(r,z) = -\frac{1-\eta}{\mu}\frac{\partial \phi}{\partial z}, \qquad \nabla^2 \omega = 0. \tag{5.8.3}$$

We see, therefore, that the problem is reduced to an equivalent problem in potential theory in which we are required to find the potential due to a perfectly conducting disk $0 \leqslant r \leqslant a$, $z = 0$ which is maintained

† An account of problems solved by this method has been given by I. N. Sneddon, *Fourier Transforms* (New York, 1951).

at a given potential symmetrically distributed about its centre. To solve this problem† we consider the function

$$\omega(r,z) = \frac{1}{2}\int\limits_0^a \frac{g(t)\,dt}{\sqrt{\{r^2+(z+it)^2\}}}+\frac{1}{2}\int\limits_0^a \frac{g(t)\,dt}{\sqrt{\{r^2+(z-it)^2\}}}$$

$$= \frac{1}{2}\int\limits_{-a}^a \frac{g(t)\,dt}{\sqrt{\{r^2+(z+it)^2\}}}, \tag{5.8.4}$$

where $g(t)$ is a real continuous even function of t and where, for $0 \leqslant t \leqslant a$,

$$\left.\begin{aligned} \sqrt{\{r^2+(z+it)^2\}} = \xi e^{\frac{1}{2}i\eta}, \qquad \sqrt{\{r^2+(z-it)^2\}} = \xi e^{-\frac{1}{2}i\eta} \quad (\xi \geqslant 0) \\ \xi^2\cos\eta = r^2+z^2-t^2, \qquad \xi^2 \sin\eta = 2zt \quad (0 \leqslant \eta \leqslant \pi) \end{aligned}\right\}. \tag{5.8.5}$$

We also add the notation

$$\left.\begin{aligned} \sqrt{\{r^2+(z+ia)^2\}} = R_2, \qquad \sqrt{\{r^2+(z-ia)^2\}} = R_1 \\ R_2 = \bar{R}_1 = \rho e^{\frac{1}{2}iv} \quad (\rho \geqslant 0,\ 0 \leqslant v \leqslant \pi) \\ \rho^2 \cos v = r^2+z^2-a^2, \qquad \rho^2 \sin v = 2az \end{aligned}\right\}. \tag{5.8.6}$$

When $z = 0$ we observe that

$$\left.\begin{aligned} \sqrt{\{r^2+(z+it)^2\}} = \sqrt{\{r^2+(z-it)^2\}} = \sqrt{(r^2-t^2)} \quad (r \geqslant t) \\ R_2 = R_1 = \sqrt{(r^2-a^2)} \quad (r \geqslant a) \\ \sqrt{\{r^2+(z+it)^2\}} = -\sqrt{\{r^2+(z-it)^2\}} = i\sqrt{(t^2-r^2)} \quad (r \leqslant t) \\ R_2 = -R_1 = i\sqrt{(a^2-r^2)} \quad (r \leqslant a) \end{aligned}\right\}, \tag{5.8.7}$$

and, when $r = 0$,

$$\left.\begin{aligned} \sqrt{\{r^2+(z+it)^2\}} = z+it, \qquad \sqrt{\{r^2+(z-it)^2\}} = z-it \\ R_2 = \sigma e^{iu} = z+ia, \qquad R_1 = \sigma e^{-iu} = z-ia \end{aligned}\right\}. \tag{5.8.8}$$

The function $\omega(r,z)$ given by (5.8.4) is real and the Laplacian of this function can be expressed by means of the appropriate differentiations within the integral sign, provided the integrands so obtained are continuous functions of (r,z,t). This is so provided no denominators vanish, that is

$$t \neq \pm iz \pm r \quad (0 \leqslant t \leqslant a),$$

and these conditions are satisfied everywhere except on the disk $z = 0$, $r \leqslant a$. Hence $\omega(r,z)$ is harmonic everywhere, except on the disk $z = 0$, $r \leqslant a$, since the Laplacian of the integrand is zero for each value of t.

† The method of solution in this section is derived by combining ideas from the following papers: E. T. Copson, *Proc. Edinb. math. Soc.* **8** (1947) 14; A. E. H. Love, *Q. J. Math.* **10** (1939) 161; A. E. Green, *Proc. Camb. phil. Soc. math. phys. Sci.* **45** (1949) 251; E. R. Love, *Q. J. Mech. appl. Math.* **2** (1949) 428.

We now show that $\omega(r,z)$ is continuous for normal approach to the disk as $z \to +0$. Suppose first $0 < r \leqslant a$. Then

$$|r^2+(z\pm it)^2| \geqslant z^2+r^2-t^2 \quad \text{if} \quad t^2 < r^2$$
$$\geqslant r^2-t^2,$$
$$|r^2+(z\pm it)^2| \geqslant t^2-r^2 \quad \text{if} \quad t^2 > r^2,$$

whatever real value z may have. Thus, for $0 \leqslant t \leqslant a$,

$$\left|\frac{g(t)}{\sqrt{\{r^2+(z\pm it)^2\}}}\right| \leqslant \frac{|g(t)|}{\sqrt{|r^2-t^2|}},$$

and so by the theorem of dominated convergence, as $z \to +0$,

$$\int_0^a \frac{g(t)\,dt}{\sqrt{\{r^2+(z\pm it)^2\}}} \to \int_0^a \frac{g(t)\,dt}{\sqrt{(r^2-t^2)}}. \tag{5.8.9}$$

This establishes the continuity of $\omega(r,z)$ on the disk, except at $r = 0$. The square root on the right of (5.8.9) is to be understood as the limit of that on the left; it is thus the positive root if $t < r$ but $\pm i\sqrt{(t^2-r^2)}$ if $t > r$. It follows that (5.8.4) reduces to

$$\int_0^r \frac{g(t)\,dt}{\sqrt{(r^2-t^2)}} = \int_0^{\frac{1}{2}\pi} g(r\cos\phi)\,d\phi \quad (0 < r \leqslant a), \tag{5.8.10}$$

when $z \to 0$. In the case $r = 0$, (5.8.4) becomes

$$\frac{1}{2}\int_0^a \frac{g(t)\,dt}{z+it} + \frac{1}{2}\int_0^a \frac{g(t)\,dt}{z-it} = \int_0^a \frac{zg(t)\,dt}{z^2+t^2} \to \tfrac{1}{2}\pi g(0),$$

as $z \to 0$, by the usual argument.† This shows that at $r = 0$ (5.8.10) still gives the correct interpretation of (5.8.4) to make it continuous at all points on the disk.

We further require that $\omega(r,z)$ be continuously differentiable in the region $z \geqslant 0$, except at the edge of the disk. The zeros of the function

$$r^2+(z+it)^2 \quad (t = t_1+it_2),$$

regarded as a function of t, lie in the upper half t-plane when $z > 0$. We can therefore displace the path of integration in (5.8.4) from $-a$ to a away from the real axis into the lower half t-plane for $z > 0$, provided we assume $g(t)$ to be an analytic function in a simply-connected domain which contains both the old and new paths of integration. In the new regions of definition of $\omega(r,z)$ the integrands are continuously

† T. J. I'A. Bromwich, *An Introduction to the Theory of Infinite Series* (London, 1908), p. 443.

differentiable functions of r and z, other than on the edge of the disk $z = 0, r = a$. Therefore derivatives of $\omega(r,z)$ with respect to r and z exist and are continuous in the neighbourhood of the disk, and can be calculated by differentiation under the integral sign. Thus $\omega(r,z)$ is harmonic in this neighbourhood and tends to finite limits as the point (r,z) approaches a point on the disk, including the point $(0,0)$, in any manner through values of z greater than 0. Since $\omega(r,z)$ admits this harmonic extension it must in its original region of definition be continuously differentiable in $z \geqslant 0$, except at the edge of the disk.

Finally, since $|r^2+(z\pm it)^2| \geqslant r^2+z^2-a^2$ when $0 \leqslant t \leqslant a$, we see that

$$|\omega(r,z)| \leqslant \frac{1}{\sqrt{(r^2+z^2-a^2)}} \int_0^a |g(t)|\, dt,$$

and hence
$$\omega(r,z) = O\{(r^2+z^2)^{-\frac{1}{2}}\}$$
at infinity.

Remembering (5.8.7) we see at once that when $z = 0$ the first boundary condition in (5.8.2) is satisfied. In view of the continuity of $\omega(r,z)$ on the disk the second boundary condition in (5.8.2) gives

$$f(r) = \int_0^r \frac{g(t)\, dt}{\sqrt{(r^2-t^2)}} \quad (0 \leqslant r \leqslant a). \tag{5.8.11}$$

If we assume that $f(r)$ is continuously differentiable in $0 \leqslant r \leqslant a$ this integral equation can be solved by elementary analysis. Thus

$$\begin{aligned}\int_0^u \frac{rf(r)\, dr}{\sqrt{(u^2-r^2)}} &= \int_0^u \frac{r\, dr}{\sqrt{(u^2-r^2)}} \int_0^r \frac{g(t)\, dt}{\sqrt{(r^2-t^2)}} \\ &= \int_0^u g(t)\, dt \int_t^u \frac{r\, dr}{\sqrt{\{(u^2-r^2)(r^2-t^2)\}}} \\ &= \tfrac{1}{2}\pi \int_0^u g(t)\, dt,\end{aligned}$$

the change of order of the integrals being valid under the conditions stated. Hence

$$g(t) = \frac{2}{\pi}\frac{d}{dt} \int_0^t \frac{rf(r)\, dr}{\sqrt{(t^2-r^2)}} = \frac{2}{\pi} f(0) + \frac{2t}{\pi} \int_0^t \frac{f'(r)\, dr}{\sqrt{(t^2-r^2)}}. \tag{5.8.12}$$

When $g(t)$ has been found from this equation the value of $\omega(r,z)$ follows from (5.8.4) and the displacements and stresses from (5.7.2) and

(5.8.3). In particular, the distribution of normal stress under the punch is

$$(\sigma_{zz})_{z=0} = \frac{\mu}{1-\eta}\left\{\frac{\partial\omega(r,z)}{\partial z}\right\}_{z=0}. \tag{5.8.13}$$

This formula may be evaluated by observing from (5.8.4), for points not on the disk, that

$$\frac{\partial\omega}{\partial z} = \frac{1}{2r}\frac{\partial}{\partial r}\left[\int\limits_0^a \frac{(z+it)g(t)\,dt}{\sqrt{\{r^2+(z+it)^2\}}}+\int\limits_0^a \frac{(z-it)g(t)\,dt}{\sqrt{\{r^2+(z-it)^2\}}}\right]$$

and by arguments similar to those used earlier in the section we may show that, as $z \to 0$ on the disk,

$$\left(\frac{\partial\omega}{\partial z}\right)_{z\to 0} = \frac{1}{r}\frac{\partial}{\partial r}\int\limits_r^a \frac{tg(t)\,dt}{\sqrt{(t^2-r^2)}}.$$

Hence
$$(\sigma_{zz})_{z=0} = \frac{\mu}{(1-\eta)r}\frac{\partial}{\partial r}\int\limits_r^a \frac{tg(t)\,dt}{\sqrt{(t^2-r^2)}}. \tag{5.8.14}$$

5.9. Indentation by flat-ended cylinder

The simplest example of the theory of the previous section is that of a flat-ended circular cylinder of radius a which penetrates a small depth ϵ below the level of the undisturbed boundary $z = 0$. Thus

$$f(r) = \epsilon \quad (0 \leqslant r \leqslant a), \tag{5.9.1}$$

so that, from (5.8.12),
$$g(t) = \frac{2\epsilon}{\pi}, \tag{5.9.2}$$

and, from (5.8.14),

$$(\sigma_{zz})_{z=0} = -\frac{2\mu\epsilon}{\pi(1-\eta)\sqrt{(a^2-r^2)}} = -\frac{E\epsilon}{\pi(1-\eta^2)\sqrt{(a^2-r^2)}}, \tag{5.9.3}$$

where E is Young's modulus. The total force exerted by the cylinder on the elastic medium is

$$P = \pi a^2 p = -2\pi\int\limits_0^a (\sigma_{zz})_{z=0}\, r\,dr = \frac{2E\epsilon a}{1-\eta^2}, \tag{5.9.4}$$

and we may therefore express (5.9.3) in the form

$$(\sigma_{zz})_{z=0} = -\frac{pa}{2\sqrt{(a^2-r^2)}} \quad (0 \leqslant r < a). \tag{5.9.5}$$

In order to evaluate stresses at any point in the medium we have, from (5.8.4) and (5.9.2),

$$\omega(r,z) = \frac{\epsilon}{\pi}\int\limits_0^a \frac{dt}{\sqrt{\{r^2+(z+it)^2\}}}+\frac{\epsilon}{\pi}\int\limits_0^a \frac{dt}{\sqrt{\{r^2+(z-it)^2\}}}$$

$$= -\frac{\epsilon i}{\pi}\ln\frac{R_2+z+ia}{R_1+z-ia} \tag{5.9.6}$$

if we use the notation of (5.8.6), and hence, using (5.8.3),

$$\phi(r,z) = \frac{\mu\epsilon i}{\pi(1-\eta)}\{(z+ia)\ln(R_2+z+ia)-R_2-(z-ia)\ln(R_1+z-ia)+R_1\} \tag{5.9.7}$$

if we remember the conditions to be satisfied at infinity. It is now a simple matter to express all the required derivatives of ϕ in terms of the real quantities ρ, v, σ, u defined in (5.8.6) and (5.8.8). The reader is, however, referred to Sneddon† for a more detailed discussion of this and other similar problems.

5.10. Distribution of stress in neighbourhood of a circular crack

We consider an infinite elastic medium containing a circular crack $0 \leqslant r \leqslant a$, in the plane $z = 0$, which is opened by a symmetrical normal pressure. The distribution of stress is the same as that in a semi-infinite body $z \geqslant 0$ when its free surface is subject to the boundary conditions

$$\left.\begin{aligned} &\sigma_{xz} = \sigma_{yz} = 0 \\ &\sigma_{zz} = -p(r) \quad (0 \leqslant r \leqslant a) \\ &u_z = 0 \qquad\quad (r > a) \end{aligned}\right\} \tag{5.10.1}$$

on $z = 0$. The first condition is satisfied by the stress system of § 5.7, whilst the remaining boundary conditions give, on $z = 0$,

$$\left.\begin{aligned} &\frac{\partial\phi}{\partial z} = 0 \qquad (r > a) \\ &\frac{\partial^2\phi}{\partial z^2} = p(r) \quad (0 \leqslant r \leqslant a) \end{aligned}\right\}. \tag{5.10.2}$$

† loc. cit., p. 169.

We also assume that all displacements and stresses vanish at infinity, at least like $O(R^{-2})$ and $O(R^{-3})$ respectively, where $R^2 = r^2+z^2$.

Consider

$$\omega(r,z) = \frac{\partial\phi}{\partial z} = \frac{1}{2i}\int\limits_{-a}^{a} \frac{g(t)\,dt}{\sqrt{\{r^2+(z+it)^2\}}}, \tag{5.10.3}$$

where $g(t)$ is a real continuous odd function of t. By methods similar to that used in § 5.8 we can show that ω is a real harmonic function in $z > 0$ and $O(R^{-2})$ at infinity. Also it is continuously differentiable in $z \geqslant 0$ except at the edge of the crack $z = 0$, $r = a$. Moreover, the first condition in (5.10.2) is automatically satisfied.

At any point (r, z) not on the crack

$$\frac{\partial^2\phi}{\partial z^2} = -\frac{1}{2i}\int\limits_{-a}^{a} \frac{(z+it)g(t)\,dt}{\{r^2+(z+it)^2\}^{\frac{3}{2}}}.$$

We cannot take the limit as $z \to 0$ $(0 \leqslant r \leqslant a)$ of this expression as it stands but we can form the integral

$$\int\limits_{0}^{r} \rho\frac{\partial^2\phi(\rho,z)}{\partial z^2}\,d\rho = \frac{1}{2i}\int\limits_{-a}^{a} (z+it)g(t)\left[\frac{1}{\sqrt{\{\rho^2+(z+it)^2\}}}\right]_0^r dt$$

$$= \frac{1}{2i}\int\limits_{-a}^{a} \frac{(z+it)g(t)\,dt}{\sqrt{\{r^2+(z+it)^2\}}},$$

the lower limit not contributing since $g(t)$ is an odd function. We can now take the limit of this integral as $z \to 0+$ $(0 \leqslant r \leqslant a)$ to obtain

$$\int\limits_{0}^{r} \frac{tg(t)\,dt}{\sqrt{(r^2-t^2)}} = \int\limits_{0}^{r} \rho p(\rho)\,d\rho \quad (0 \leqslant r \leqslant a). \tag{5.10.4}$$

The solution of this equation is

$$g(t) = \frac{2}{\pi}\int\limits_{0}^{t} \frac{rp(r)\,dr}{\sqrt{(t^2-r^2)}} \quad (0 \leqslant t \leqslant a). \tag{5.10.5}$$

We observe that the normal displacement at the crack is

$$(u_z)_{z=0} = -\frac{(1-\eta)}{\mu}\left(\frac{\partial\phi}{\partial z}\right)_{z=0}$$

$$= \frac{(1-\eta)}{\mu}\int\limits_{r}^{a} \frac{g(t)\,dt}{\sqrt{(t^2-r^2)}} \quad (0 \leqslant r \leqslant a). \tag{5.10.6}$$

We consider briefly one example of a circular crack opened by a uniform normal pressure p_0 so that (5.10.5) gives

$$g(t) = \frac{2tp_0}{\pi}, \tag{5.10.7}$$

and

$$(u_z)_{z=0} = \frac{2(1-\eta)p_0}{\pi\mu}(a^2-r^2)^{\frac{1}{2}}. \tag{5.10.8}$$

The solution of the problem may now be completed by straightforward calculation but details are left to the reader who may also refer to Sneddon.† This problem has been used by Sneddon in connexion with the Griffith theory of rupture in three dimensions.‡

The analysis of § 5.7 may be used to solve problems in which there is no rotational symmetry about the z-axis.§

5.11. Notation

The values of the strain components e_{rs} used in §§ 5.3, 5.4, when referred to rectangular cartesian coordinates x_i differ from those defined by many writers,|| when $r \neq s$. If we denote the axes by Ox, Oy, Oz and the components of displacement by (u_x, u_y, u_z), then strain components are often defined as

$$\left.\begin{aligned} e_{xx} &= \frac{\partial u_x}{\partial x}, & e_{yy} &= \frac{\partial u_y}{\partial y}, & e_{zz} &= \frac{\partial u_z}{\partial z} \\ e_{yz} &= \frac{\partial u_y}{\partial z}+\frac{\partial u_z}{\partial y}, & e_{zx} &= \frac{\partial u_z}{\partial x}+\frac{\partial u_x}{\partial z}, & e_{xy} &= \frac{\partial u_x}{\partial y}+\frac{\partial u_y}{\partial x} \end{aligned}\right\}. \tag{5.11.1}$$

Also, if the stress components are denoted by $\sigma_{xx}, \sigma_{yy}, \sigma_{zz}, \sigma_{yz}, \sigma_{zx}, \sigma_{xy}$, then the general stress–strain relations for an aeolotropic body (see (5.3.12) or (5.4.37)) are expressed in the matrix forms

$$(e_{xx}, e_{yy}, e_{zz}, e_{yz}, e_{zx}, e_{xy}) = \begin{bmatrix} s_{11} & s_{12} & s_{13} & s_{14} & s_{15} & s_{16} \\ s_{21} & s_{22} & s_{23} & s_{24} & s_{25} & s_{26} \\ s_{31} & s_{32} & s_{33} & s_{34} & s_{35} & s_{36} \\ s_{41} & s_{42} & s_{43} & s_{44} & s_{45} & s_{46} \\ s_{51} & s_{52} & s_{53} & s_{54} & s_{55} & s_{56} \\ s_{61} & s_{62} & s_{63} & s_{64} & s_{65} & s_{66} \end{bmatrix} \begin{bmatrix} \sigma_{xx} \\ \sigma_{yy} \\ \sigma_{zz} \\ \sigma_{yz} \\ \sigma_{zx} \\ \sigma_{xy} \end{bmatrix}, \tag{5.11.2}$$

† loc. cit., p. 169. ‡ See also R. A. Sack, *Proc. phys. Soc.* **58** (1946) 729.

§ See A. E. Green and I. N. Sneddon, *Proc. Camb. phil. Soc. math. phys. Sci.* **46** (1950) 159; A. E. Green, loc. cit., p. 170.

|| See A. E. H. Love, loc. cit., p. 149.

or

$$(\sigma_{xx}, \sigma_{yy}, \sigma_{zz}, \sigma_{yz}, \sigma_{zx}, \sigma_{xy}) = \begin{bmatrix} c_{11} & c_{12} & c_{13} & c_{14} & c_{15} & c_{16} \\ c_{21} & c_{22} & c_{23} & c_{24} & c_{25} & c_{26} \\ c_{31} & c_{32} & c_{33} & c_{34} & c_{35} & c_{36} \\ c_{41} & c_{42} & c_{43} & c_{44} & c_{45} & c_{46} \\ c_{51} & c_{52} & c_{53} & c_{54} & c_{55} & c_{56} \\ c_{61} & c_{62} & c_{63} & c_{64} & c_{65} & c_{66} \end{bmatrix} \begin{bmatrix} e_{xx} \\ e_{yy} \\ e_{zz} \\ e_{yz} \\ e_{zx} \\ e_{xy} \end{bmatrix}, \tag{5.11.3}$$

where $s_{ij} = s_{ji}$ and $c_{ij} = c_{ji}$.

If we compare these equations with (5.3.12) and (5.4.37), and use (5.11.1), we see that

$$\left.\begin{array}{llll} s_{11} = s_{11}^{11}, & s_{12} = s_{22}^{11}, & s_{13} = s_{33}^{11} & \\ s_{14} = 2s_{23}^{11}, & s_{15} = 2s_{13}^{11}, & s_{16} = 2s_{12}^{11} & \\ s_{22} = s_{22}^{22}, & s_{23} = s_{33}^{22}, & s_{24} = 2s_{23}^{22} & \\ s_{25} = 2s_{13}^{22}, & s_{26} = 2s_{12}^{22} & & \\ s_{33} = s_{33}^{33}, & s_{34} = 2s_{23}^{33}, & s_{35} = 2s_{13}^{33}, & s_{36} = 2s_{12}^{33} \\ s_{44} = 4s_{23}^{23}, & s_{45} = 4s_{13}^{23}, & s_{46} = 4s_{12}^{23} & \\ s_{55} = 4s_{13}^{13}, & s_{56} = 4s_{12}^{13}, & s_{66} = 4s_{12}^{12} & \end{array}\right\}, \tag{5.11.4}$$

and

$$\left.\begin{array}{llll} c_{11} = c_{11}^{11}, & c_{12} = c_{22}^{11}, & c_{13} = c_{33}^{11} & \\ c_{14} = c_{23}^{11}, & c_{15} = c_{13}^{11}, & c_{16} = c_{12}^{11} & \\ c_{22} = c_{22}^{22}, & c_{23} = c_{33}^{22}, & c_{24} = c_{23}^{22} & \\ c_{25} = c_{13}^{22}, & c_{26} = c_{12}^{22} & & \\ c_{33} = c_{33}^{33}, & c_{34} = c_{23}^{33}, & c_{35} = c_{13}^{33}, & c_{36} = c_{12}^{33} \\ c_{44} = c_{23}^{23}, & c_{45} = c_{13}^{23}, & c_{46} = c_{12}^{23} & \\ c_{55} = c_{13}^{13}, & c_{56} = c_{12}^{13}, & c_{66} = c_{12}^{12} & \end{array}\right\}. \tag{5.11.5}$$

5.12. Potential solutions for transversely isotropic bodies

We adopt rectangular cartesian coordinates (x, y, z) with components of displacement (u_x, u_y, u_z) and components of stress $\sigma_{xx}, \ldots, \sigma_{xy}$. Then, using the scheme (5.4.17) of elastic coefficients for a transversely isotropic body, and the notation of § 5.11, we have

$$\left.\begin{aligned}
\sigma_{xx} &= c_{11}\frac{\partial u_x}{\partial x}+c_{12}\frac{\partial u_y}{\partial y}+c_{13}\frac{\partial u_z}{\partial z}, & \sigma_{yz} &= c_{44}\left(\frac{\partial u_y}{\partial z}+\frac{\partial u_z}{\partial y}\right)\\
\sigma_{yy} &= c_{12}\frac{\partial u_x}{\partial x}+c_{11}\frac{\partial u_y}{\partial y}+c_{13}\frac{\partial u_z}{\partial z}, & \sigma_{zx} &= c_{44}\left(\frac{\partial u_z}{\partial x}+\frac{\partial u_x}{\partial z}\right)\\
\sigma_{zz} &= c_{13}\left(\frac{\partial u_x}{\partial x}+\frac{\partial u_y}{\partial y}\right)+c_{33}\frac{\partial u_z}{\partial z}, & \sigma_{xy} &= \tfrac{1}{2}(c_{11}-c_{12})\left(\frac{\partial u_x}{\partial y}+\frac{\partial u_y}{\partial x}\right)
\end{aligned}\right\}. \tag{5.12.1}$$

Substituting these values in the equations of equilibrium

$$\left.\begin{aligned}
\frac{\partial \sigma_{xx}}{\partial x}+\frac{\partial \sigma_{xy}}{\partial y}+\frac{\partial \sigma_{xz}}{\partial z} &= 0\\
\frac{\partial \sigma_{xy}}{\partial x}+\frac{\partial \sigma_{yy}}{\partial y}+\frac{\partial \sigma_{yz}}{\partial z} &= 0\\
\frac{\partial \sigma_{xz}}{\partial x}+\frac{\partial \sigma_{yz}}{\partial y}+\frac{\partial \sigma_{zz}}{\partial z} &= 0
\end{aligned}\right\}, \tag{5.12.2}$$

we have

$$c_{11}\frac{\partial^2 u_x}{\partial x^2}+\tfrac{1}{2}(c_{11}-c_{12})\frac{\partial^2 u_x}{\partial y^2}+c_{44}\frac{\partial^2 u_x}{\partial z^2}+ \\ +\frac{\partial}{\partial x}\left\{\tfrac{1}{2}(c_{11}+c_{12})\frac{\partial u_y}{\partial y}+(c_{13}+c_{44})\frac{\partial u_z}{\partial z}\right\} = 0, \tag{5.12.3}$$

$$\tfrac{1}{2}(c_{11}-c_{12})\frac{\partial^2 u_y}{\partial x^2}+c_{11}\frac{\partial^2 u_y}{\partial y^2}+c_{44}\frac{\partial^2 u_y}{\partial z^2}+ \\ +\frac{\partial}{\partial y}\left\{\tfrac{1}{2}(c_{11}+c_{12})\frac{\partial u_x}{\partial x}+(c_{13}+c_{44})\frac{\partial u_z}{\partial z}\right\} = 0, \tag{5.12.4}$$

$$c_{44}\left(\frac{\partial^2 u_z}{\partial x^2}+\frac{\partial^2 u_z}{\partial y^2}\right)+c_{33}\frac{\partial^2 u_z}{\partial z^2}+(c_{13}+c_{44})\frac{\partial}{\partial z}\left(\frac{\partial u_x}{\partial x}+\frac{\partial u_y}{\partial y}\right) = 0. \tag{5.12.5}$$

We assume that a solution of these equations may be found in the form (4.4.11). Following the method used in § 4.4, and omitting terms which make no contribution to the displacements, we see that (5.12.3), (5.12.4), and (5.12.5) are satisfied if

$$\tfrac{1}{2}(c_{11}-c_{12})\nabla_1^2\phi_3+c_{44}\frac{\partial^2\phi_3}{\partial z^2} = 0, \tag{5.12.6}$$

$$\left.\begin{aligned}
c_{11}\nabla_1^2\phi_1+c_{44}\frac{\partial^2\phi_1}{\partial z^2}+(c_{13}+c_{44})\frac{\partial^2\phi_2}{\partial z^2} &= 0\\
(c_{13}+c_{44})\nabla_1^2\phi_1+c_{44}\nabla_1^2\phi_2+c_{33}\frac{\partial^2\phi_2}{\partial z^2} &= 0
\end{aligned}\right\}. \tag{5.12.7}$$

To complete the solution we put

$$\frac{c_{11}\nu - c_{44}}{c_{13}+c_{44}} = \frac{(c_{13}+c_{44})\nu}{c_{33}-c_{44}\nu} = k, \tag{5.12.8}$$

so that $$c_{11}c_{44}\nu^2+\{c_{13}(2c_{44}+c_{13})-c_{11}c_{33}\}\nu+c_{33}c_{44} = 0. \tag{5.12.9}$$

We assume that there are two distinct roots ν_1, ν_2 of these equations with corresponding values k_1, k_2 of k. Again following the work of § 4.4 we see that

$$\phi_1 = \chi_1+\chi_2, \qquad \phi_2 = k_1\chi_1+k_2\chi_2, \tag{5.12.10}$$

where $$\left(\nabla_1^2+\nu_\alpha\frac{\partial^2}{\partial z^2}\right)\chi_\alpha = 0 \quad (\alpha = 1,2;\ \alpha \text{ not summed}). \tag{5.12.11}$$

To complete the notation we put

$$\phi_3 = \chi_3, \tag{5.12.12}$$

so that $$\left(\nabla_1^2+\nu_3\frac{\partial^2}{\partial z^2}\right)\chi_3 = 0, \qquad \nu_3 = \frac{2c_{44}}{c_{11}-c_{12}}. \tag{5.12.13}$$

The complete forms for the displacements are

$$u_x = \frac{\partial}{\partial x}(\chi_1+\chi_2)+\frac{\partial\chi_3}{\partial y}, \qquad u_y = \frac{\partial}{\partial y}(\chi_1+\chi_2)-\frac{\partial\chi_3}{\partial x},$$

$$u_z = k_1\frac{\partial\chi_1}{\partial z}+k_2\frac{\partial\chi_2}{\partial z}, \tag{5.12.14}$$

and the corresponding stresses, found from (5.12.1), are

$$\left.\begin{aligned}
\sigma_{xx} &= \left(c_{11}\frac{\partial^2}{\partial x^2}+c_{12}\frac{\partial^2}{\partial y^2}\right)(\chi_1+\chi_2)+c_{13}\left(k_1\frac{\partial^2\chi_1}{\partial z^2}+k_2\frac{\partial^2\chi_2}{\partial z^2}\right)+\\
&\qquad\qquad +(c_{11}-c_{12})\frac{\partial^2\chi_3}{\partial x\partial y}\\
\sigma_{yy} &= \left(c_{12}\frac{\partial^2}{\partial x^2}+c_{11}\frac{\partial^2}{\partial y^2}\right)(\chi_1+\chi_2)+c_{13}\left(k_1\frac{\partial^2\chi_1}{\partial z^2}+k_2\frac{\partial^2\chi_2}{\partial z^2}\right)-\\
&\qquad\qquad -(c_{11}-c_{12})\frac{\partial^2\chi_3}{\partial x\partial y}\\
\sigma_{zz} &= (k_1c_{33}-\nu_1c_{13})\frac{\partial^2\chi_1}{\partial z^2}+(k_2c_{33}-\nu_2c_{13})\frac{\partial^2\chi_2}{\partial z^2}
\end{aligned}\right\}, \tag{5.12.15}$$

$$\left.\begin{aligned}
\sigma_{xy} &= (c_{11}-c_{12})\left\{\frac{\partial^2}{\partial x\partial y}(\chi_1+\chi_2)+\frac{1}{2}\frac{\partial^2\chi_3}{\partial y^2}-\frac{1}{2}\frac{\partial^2\chi_3}{\partial x^2}\right\}\\
\sigma_{yz} &= c_{44}\left\{(1+k_1)\frac{\partial^2\chi_1}{\partial y\partial z}+(1+k_2)\frac{\partial^2\chi_2}{\partial y\partial z}-\frac{\partial^2\chi_3}{\partial x\partial z}\right\}\\
\sigma_{zx} &= c_{44}\left\{(1+k_1)\frac{\partial^2\chi_1}{\partial x\partial z}+(1+k_2)\frac{\partial^2\chi_2}{\partial x\partial z}+\frac{\partial^2\chi_3}{\partial y\partial z}\right\}
\end{aligned}\right\}. \tag{5.12.16}$$

The roots of equation (5.12.8) may be either real (with the same sign) or complex conjugates; they are real, for example, in the case of magnesium but are complex conjugates for zinc.† The same statement holds for the corresponding values of k_1, k_2. Using the equations (5.12.15) and (5.12.16) with $\chi_3 = 0$, a number of problems have been discussed by H. A. Elliott† and R. T. Shield.‡ Here we restrict our attention to indentation and crack problems and show that they can be reduced to the solution of a classical potential problem, as was found to be possible for isotropic bodies.

If we put§

$$\chi_1 = \frac{\sqrt{\nu_1}}{1+k_1}\Phi(x,y,z_1), \qquad \chi_2 = -\frac{\sqrt{\nu_2}}{1+k_2}\Phi(x,y,z_2), \qquad \chi_3 = 0, \quad (5.12.17)$$

where

$$\nabla^2\Phi(x,y,z) = 0, \qquad z_\alpha = z/\sqrt{\nu_\alpha}, \quad (5.12.18)$$

then, from (5.12.16),

$$\sigma_{yz} = c_{44}\frac{\partial}{\partial y}\left[\frac{\partial\Phi(x,y,z_1)}{\partial z_1} - \frac{\partial\Phi(x,y,z_2)}{\partial z_2}\right],$$

$$\sigma_{xz} = c_{44}\frac{\partial}{\partial x}\left[\frac{\partial\Phi(x,y,z_1)}{\partial z_1} - \frac{\partial\Phi(x,y,z_2)}{\partial z_2}\right].$$

Hence, when $z = 0$ (and therefore $z_1 = z_2 = 0$), we see that the shearing stresses σ_{yz}, σ_{xz} are zero. The remaining stresses and displacements may be expressed in terms of Φ, and we observe, in particular, that on $z = 0$,

$$\left.\begin{aligned} \sigma_{zz} &= \beta\frac{\partial^2\Phi(x,y,z)}{\partial z^2} \\ u_z &= \alpha\frac{\partial\Phi(x,y,z)}{\partial z} \end{aligned}\right\}, \quad (5.12.19)$$

where

$$\beta = \frac{k_1 c_{33} - \nu_1 c_{13}}{(1+k_1)\sqrt{\nu_1}} - \frac{k_2 c_{33} - \nu_2 c_{13}}{(1+k_2)\sqrt{\nu_2}}, \qquad \alpha = \frac{k_1}{1+k_1} - \frac{k_2}{1+k_2}. \quad (5.12.20)$$

Since $\Phi(x,y,z)$ is a harmonic function the expressions (5.12.19) for the values of the normal stress and displacement on the plane are now

† See H. A. Elliott, *Proc. Camb. phil. Soc. math. phys. Sci.* **44** (1948) 522; **45** (1949) 621.

‡ R. T. Shield, ibid. **47** (1951) 401.

§ When ν_1, ν_2 are negative or complex conjugates we choose $\sqrt{\nu_1}$, $\sqrt{\nu_2}$ to be complex conjugates with positive (or zero) real parts.

similar in form to those given in (5.7.3) for an isotropic material. The solutions of indentation and crack problems in §§ 5.8, 5.9 may therefore be made available at once for the corresponding problems in hexagonal aeolotropic bodies, in order to determine the function $\Phi(x, y, z)$.†

When $\sqrt{\nu_1}$, $\sqrt{\nu_2}$ are complex conjugates (and also k_1, k_2) care must be taken to check that the final expressions for displacements and stresses are real, but this usually presents no difficulty.

† Punch and crack problems for transversely isotropic bodies have been solved by H. A. Elliott, *Proc. Camb. phil. Soc. math. phys. Sci.* **45** (1949), 621 using Hankel transforms. The method of this section was used by R. T. Shield, loc. cit., p. 180 for elliptical crack and punch problems.

6 Theory of Plane Strain

In elasticity, as in other branches of mathematical physics, it is often difficult to find an explicit solution of a given three-dimensional problem, but its analogue in two dimensions can frequently be solved. The theories of plane strain and generalized plane stress, and the classical theory of the transverse flexure of thin plates are well-known examples of two-dimensional theories. In these theories, as far as isotropic bodies are concerned, the biharmonic equation has a central place, and many solutions of this equation have been obtained with the help of real variable analysis.† The presentation of two-dimensional elasticity, and the solution of special problems is, however, greatly simplified by the use of complex variable techniques. Moreover, the range of problems which can be solved is greatly extended. The first use and development of the methods of complex function theory in two-dimensional elastic problems was made by M. Kolossoff and N. I. Muskhelishvili,‡ and their ideas are expounded by the latter writer in two books.§ Much of the earlier work of these writers was overlooked in this country so that independent development of the use of complex variable methods for elasticity by subsequent writers contained some duplications.||

† For example, see A. E. H. Love, loc. cit., p. 149; E. G. Coker and L. N. G. Filon, *A Treatise on Photo-elasticity* (Cambridge, 1931); S. Timoshenko and J. N. Goodier, loc. cit., p. 148.

‡ M. Kolossoff, *C. r. hebd. Séanc. Acad. Sci., Paris* **146** (1908) 522; **148** (1909) 1242; **181** (1925) 24; **184** (1927) 512; **193** (1931) 389; idem, *Z. Math. Phys.* **62** (1914) 384. M. Kolossoff and N. I. Muskhelishvili, *Annls Inst. électrotech. Pétrogr.* **12** (1915) 39. N. I. Muskhelishvili, *Bull. Acad. Sci. Russ.* **13** (1919) 663; *Math. Annln* **107** (1932) 282; *Z. angew. Math. Mech.* **13** (1933) 264; *Dokl. Akad. Nauk SSSR* **3** (1934) 7, 73, 141.

§ N. I. Muskhelishvili, *Some Fundamental Problems of the Mathematical Theory of Elasticity*, 2nd ed. (Moscow, 1949); *Singular Integral Equations*, 2nd ed. (Moscow, 1946), translated by J. R. M. Radok and W. G. Woolnough (Melbourne, 1949). Further references are given in these books. Unfortunately some of the publications and the book on elasticity are not readily accessible in this country. The present authors have, however, received great help from a copy of the translation of *Singular Integral Equations*.

|| See, e.g., A. C. Stevenson, *Proc. R. Soc.* **A184** (1945) 129, 218 (publication delayed by the war); *Phil. Mag.* **33** (1942) 639; **34** (1943) 766; **36** (1945) 178. L. M. Milne-Thomson, *J. Lond. math. Soc.* **17** (1942) 115. A. E. Green, *Proc. R. Soc.* **A180** (1942) 173; **A184** (1945) 231 (publication delayed by the war); *Proc. Camb. phil. Soc. math. phys. Sci.* **41** (1945) 224. S. Holgate, *Proc. R. Soc.* **A185** (1946) 35; *Proc. Camb. phil. Soc. math. phys. Sci.* **40** (1944) 172.

The development of two-dimensional theories and the solution of problems for aeolotropic bodies was somewhat later than for isotropic bodies and pioneer work in the use of complex variable techniques was due to Lechnitzky, Savin, and other Russian authors.† This work was also not readily available in this country so that independent developments by other writers did not contain any reference to it.‡

For convenience the two-dimensional work of this book has been put into four chapters, 6, 7, 8, and 9. In the present chapter attention is restricted to the fundamental equations for plane strain for both isotropic and aeolotropic materials, the solution of special problems being postponed until Chapters 8 and 9. Chapter 7 is concerned with plate theory. Tensor calculus notations are retained as long as possible so that the complex variable aspects of the theory then appear naturally by specializing the coordinate system to complex coordinates.

6.1. Coordinate system

We use the notations of § 1.13 and express the position vector $\mathbf{R}$ of any point of the unstrained body§ in the special form

$$\mathbf{R} = \mathbf{r}(\theta_1, \theta_2)+\theta_3\,\mathbf{a}_3, \tag{6.1.1}$$

where $\mathbf{a}_3$ is a *constant* unit vector perpendicular to the *plane* surface $\theta_3 = 0$. We also choose the rectangular axes x_i so that

$$\theta_3 = x_3, \tag{6.1.2}$$

and the origins of the vectors $\mathbf{R}$, $\mathbf{r}$ and the axes x_i coincide. The θ_α-curves then lie in the (x_1, x_2) plane. All the formulae developed in § 1.13 are

† S. G. Lechnitzky, *Prikl. Mat. Mekh.* **3** (1936) no. 1; **1** (1937, N.S.) no. 1; **2** (1938) 181; **5** (1941) 71; *Dokl. Akad. Nauk SSSR* **3** (1936) 3; **4** (1936) 111; *Anisotropic Plates* (Moscow, 1947); *Theory of Elasticity of an Anisotropic Body* (Moscow, 1950). These publications have not been available to the present writers. Other references to work by Russian authors (including G. N. Savin and I. N. Vekua) for aeolotropic and isotropic materials are contained in a paper and book by I. S. Sokolnikoff, *Bull. Am. math. Soc.* **48** (1942) 539; loc. cit., p. 148.

‡ A. E. Green, *Proc. R. Soc.* **A180** (1942) 173; *Phil. Mag.* **34** (1943) 416; *Proc. R. Soc.* **A184** (1945) 231, 289, 301 (publication delayed by the war); *Proc. Camb. phil. Soc. math. phys. Sci.* **41** (1945) 224. C. B. Smith, *U.S. Dept. Agric., For. Prod. Lab.* Mimeo. 1510 (1944). D. Sherman, *Prikl. Mat. Mekh.* **9** (1945) 347. V. Morkovin, *Q. appl. Math.* **1** (1943) 116; **2** (1945) 350. S. Holgate, *Proc. R. Soc.* **A185** (1946) 50; *Proc. Camb. phil. Soc. math. phys. Sci.* **40** (1944) 172. See also G. H. Livens and R. M. Morris, *Phil. Mag.* **38** (1947) 153. For earlier work using real variable analysis see A. E. Green and G. I. Taylor, *Proc. R. Soc.* **A173** (1939) 162; **A181** (1945) 181 (publication delayed by the war). A. E. Green, *Proc. R. Soc.* **A173** (1939) 173. The first of these papers contains references to other writers who mostly solved problems for a restricted type of aeolotropy.

§ Since we are using infinitesimal theory here we refer all functions to the unstrained body, so that $\mathbf{R}$ need not be confused with the notation of Chapter 2.

available here, with the extra conditions (1.13.57) and (1.13.58) that

$$\mathbf{a}_{3,\alpha} = 0, \qquad b_{\alpha\beta} = 0, \qquad \bar{R}_{1212} = 0. \tag{6.1.3}$$

Since the Riemann–Christoffel tensor for the plane surface $\theta_3 = 0$ is zero, it follows that the order of covariant differentiation with respect to the surface coordinates θ_α is immaterial. For reference we repeat the results (1.13.59):

$$\left.\begin{gathered} \mathbf{g}_\alpha = \mathbf{r}_{,\alpha} = \mathbf{a}_\alpha, \qquad \mathbf{g}_3 = \mathbf{a}_3 = \mathbf{a}^3 \\ g_{\alpha\beta} = \mathbf{a}_\alpha . \mathbf{a}_\beta = a_{\alpha\beta}, \qquad g_{\alpha 3} = \mathbf{a}_\alpha . \mathbf{a}_3 = 0, \qquad g_{33} = \mathbf{a}_3 . \mathbf{a}_3 = 1 \\ g^{\alpha\beta} = \mathbf{a}^\alpha . \mathbf{a}^\beta = a^{\alpha\beta}, \qquad g^{\alpha 3} = \mathbf{a}^\alpha . \mathbf{a}^3 = 0, \qquad g^{33} = \mathbf{a}^3 . \mathbf{a}^3 = 1 \\ g = |g_{ik}| = |a_{\alpha\beta}| = a, \qquad \bar{\Gamma}^3_{ik} = 0, \qquad \bar{\Gamma}^\alpha_{\beta 3} = 0 \\ \mathbf{a}_{\alpha,\beta} = \Gamma^\lambda_{\alpha\beta}\mathbf{a}_\lambda, \qquad \mathbf{a}^\alpha{}_{,\beta} = -\Gamma^\alpha_{\beta\lambda}\mathbf{a}^\lambda \end{gathered}\right\}, \tag{6.1.4}$$

and we observe that all geometrical quantities (except **R**) are independent of x_3. Christoffel symbols with respect to the $\theta_3 = 0$ surface are denoted without ambiguity by $\Gamma^\alpha_{\beta\gamma}$, with the bar omitted, where

$$\left.\begin{gathered} \Gamma^\alpha_{\beta\gamma} = a^{\alpha\lambda}\Gamma_{\beta\gamma\lambda}, \qquad \Gamma_{\beta\gamma\lambda} = \tfrac{1}{2}(a_{\lambda\beta,\gamma} + a_{\lambda\gamma,\beta} - a_{\beta\gamma,\lambda}) \\ \Gamma^\lambda_{\lambda\alpha} = \frac{1}{\sqrt{a}}\frac{\partial\sqrt{a}}{\partial\theta^\alpha} \end{gathered}\right\}. \tag{6.1.5}$$

6.2. Airy's stress function

Consider the state of stress in a body which is such that

$$\tau^{\alpha 3} = 0, \tag{6.2.1}$$

and all other components $\tau^{\alpha\beta}$, τ^{33} of the stress tensor are independent of θ_3. We shall see in § 6.3 that such a state of stress arises under conditions of plane strain. Then, from (5.2.5),

$$\mathbf{T}_\alpha = \sqrt{(a)}\tau^{\alpha\beta}\mathbf{a}_\beta, \qquad \mathbf{T}_3 = \sqrt{(a)}\tau^{33}\mathbf{a}_3, \tag{6.2.2}$$

and since the functions $\mathbf{T}_i$ are independent of θ_3, the equations of equilibrium (5.2.6) become

$$\mathbf{T}_{\alpha,\alpha} + \rho\sqrt{(a)}\mathbf{F} = \mathbf{0}, \tag{6.2.3}$$

where **F** must now be a two-dimensional body force vector in the variables θ_α. In many practical problems the body force vector may be derived from a potential function, so we assume here that

$$\mathbf{F} = -U_{,\alpha}\mathbf{a}^\alpha, \tag{6.2.4}$$

where U is a scalar invariant function of the coordinates θ_1 and θ_2. The equations of equilibrium (6.2.3) may now be written in the form

$$\mathbf{T}_{\alpha,\alpha} - \rho\sqrt{(a)}U_{,\alpha}\mathbf{a}^\alpha = \mathbf{0},$$

or

$$\{\mathbf{T}_\alpha - \rho\sqrt{(a)}U\mathbf{a}^\alpha\}_{,\alpha} = \mathbf{0}, \tag{6.2.5}$$

since, from (6.1.4) and (6.1.5),

$$\{\sqrt{(a)}U\mathbf{a}^\alpha\}_{,\alpha} = \{U_{,\alpha}\mathbf{a}^\alpha + U\Gamma^\lambda_{\lambda\alpha}\mathbf{a}^\alpha - U\Gamma^\alpha_{\alpha\lambda}\mathbf{a}^\lambda\}\sqrt{a}$$
$$= \sqrt{(a)}U_{,\alpha}\mathbf{a}^\alpha.$$

From (6.2.5) we see that we may express $\mathbf{T}_\alpha$ in the form

$$\mathbf{T}_\alpha = \sqrt{(a)}\epsilon^{\gamma\alpha}\boldsymbol{\chi}_{,\gamma} + \rho\sqrt{(a)}U\mathbf{a}^\alpha, \tag{6.2.6}$$

where $\boldsymbol{\chi}$ is a vector in the plane $\theta_3 = 0$. If

$$\boldsymbol{\chi} = \chi^\beta \mathbf{a}_\beta, \tag{6.2.7}$$

then

$$\boldsymbol{\chi}_{,\gamma} = \chi^\beta|_\gamma \mathbf{a}_\beta, \tag{6.2.8}$$

and hence

$$\mathbf{T}_\alpha = \sqrt{(a)}\epsilon^{\gamma\alpha}\chi^\beta|_\gamma \mathbf{a}_\beta + \rho\sqrt{(a)}U\mathbf{a}^\alpha. \tag{6.2.9}$$

Therefore, with (6.2.2) we have

$$\tau^{\alpha\beta}\mathbf{a}_\beta = \epsilon^{\gamma\alpha}\chi^\beta|_\gamma \mathbf{a}_\beta + \rho U\mathbf{a}^\alpha.$$

We take the scalar product of both sides of this equation with the base vector $\mathbf{a}^\lambda$ and, replacing λ by β, we obtain

$$\tau^{\alpha\beta} = \epsilon^{\gamma\alpha}\chi^\beta|_\gamma + \rho U a^{\alpha\beta}. \tag{6.2.10}$$

The tensor $\tau^{\alpha\beta}$ is, however, symmetric, so that $\epsilon^{\gamma\alpha}\chi^\beta|_\gamma = \epsilon^{\gamma\beta}\chi^\alpha|_\gamma$, and we may therefore put

$$\chi^\beta = \epsilon^{\rho\beta}\phi_{,\rho}. \tag{6.2.11}$$

Hence

$$\tau^{\alpha\beta} = \epsilon^{\gamma\alpha}\epsilon^{\rho\beta}\phi|_{\rho\gamma} + \rho U a^{\alpha\beta} = \epsilon^{\alpha\gamma}\epsilon^{\beta\rho}\phi|_{\gamma\rho} + \rho U a^{\alpha\beta}, \tag{6.2.12}$$

where ϕ is a scalar invariant function of θ_1, θ_2 and we recall that the order of covariant differentiation is immaterial. The function ϕ is known as Airy's stress function. Equations (6.2.12) may also be written in the alternative forms

$$\left.\begin{aligned} \tau^\alpha_\beta &= \delta^{\alpha\gamma}_{\beta\rho}\phi|^\rho_\gamma + \rho U\delta^\alpha_\beta \\ \phi|_{\alpha\beta} &= \epsilon_{\alpha\gamma}\epsilon_{\beta\rho}\tau^{\gamma\rho} - \rho U a_{\alpha\beta} \\ \phi|^\alpha_\beta &= \delta^{\alpha\gamma}_{\beta\rho}\tau^\rho_\gamma - \rho U\delta^\alpha_\beta \end{aligned}\right\}. \tag{6.2.13}$$

The vector $\boldsymbol{\chi}$ may be interpreted physically in the following way. Consider any curve AB which lies in the plane $x_3 = 0$ and does not intersect itself, and suppose that positive direction along the curve is from A to B (Fig. 6.1). The curve AB separates the plane into two regions 1 and 2 which lie immediately adjacent to AB, and the force exerted by the region 1 on the region 2 across an element ds of AB is $-\mathbf{t}\,ds$ measured per unit length of the x_3-axis. Since the stress vector $\mathbf{t}$ lies in the plane $x_3 = 0$ we see, from (5.2.4), that

$$\mathbf{t} = \frac{n_\alpha \mathbf{T}_\alpha}{\sqrt{a}},$$

where n_α are the covariant components of the unit normal vector $\mathbf{n}$ at any point of AB, directed from the region 1 to the region 2. The

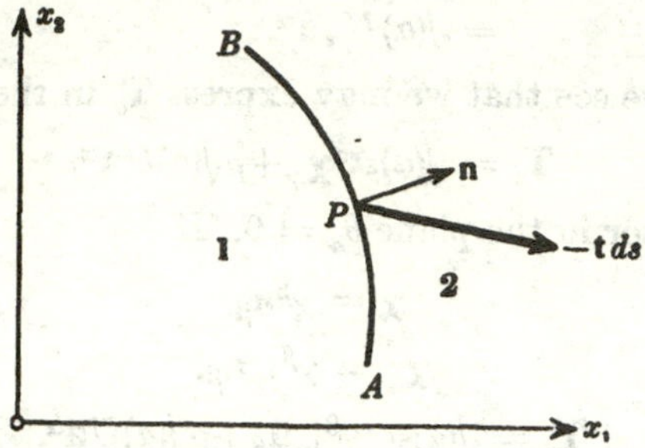

FIG. 6.1. Force vector on curve separating regions 1 and 2.

total force $\mathbf{P}$ exerted across a part AP of the curve AB, measured per unit length of the x_3-axis, is therefore

$$\mathbf{P} = -\int_A^P \frac{n_\alpha \mathbf{T}_\alpha}{\sqrt{a}}\, ds$$

$$= -\int_A^P n_\alpha(\epsilon^{\gamma\alpha}\boldsymbol{\chi}_{,\gamma} + \rho U \mathbf{a}^\alpha)\, ds, \tag{6.2.14}$$

if we use (6.2.6). Referred to cartesian axes x_α the components of $\mathbf{n}$ are $(dx_2/ds, -dx_1/ds)$, and therefore, by a tensor transformation,

$$n_\alpha = \epsilon_{\alpha\beta}\frac{d\theta^\beta}{ds}. \tag{6.2.15}$$

The expression (6.2.14) for $\mathbf{P}$ now becomes

$$\mathbf{P} = -\int_A^P \epsilon_{\alpha\beta}(\epsilon^{\gamma\alpha}\boldsymbol{\chi}_{,\gamma} + \rho U \mathbf{a}^\alpha)\frac{d\theta^\beta}{ds}\, ds$$

$$= \int_A^P \boldsymbol{\chi}_{,\beta}\frac{d\theta^\beta}{ds}\, ds - \rho\int_A^P U\mathbf{n}\, ds$$

$$= \boldsymbol{\chi} - \rho\int_A^P U\mathbf{n}\, ds, \tag{6.2.16}$$

apart from an arbitrary constant vector which may be absorbed into $\boldsymbol{\chi}$ since this does not affect the stresses. In the absence of body forces we have the simple formula

$$\mathbf{P} = \boldsymbol{\chi}. \tag{6.2.17}$$

The moment about the x_3-axis of the forces exerted by the region 1 on the region 2 across AP, measured per unit length of the x_3-axis, is

$$\mathbf{M} = \int_A^P [\mathbf{r}\times\boldsymbol{\chi}_{,\beta}]\frac{d\theta^\beta}{ds}\,ds - \rho\int_A^P U[\mathbf{r}\times\mathbf{n}]\,ds$$

$$= [\mathbf{r}\times\boldsymbol{\chi}]_A^P - \int_A^P [\mathbf{a}_\alpha\times\boldsymbol{\chi}]\frac{d\theta^\alpha}{ds}\,ds - \rho\int_A^P U\epsilon_{\alpha\beta}[\mathbf{r}\times\mathbf{a}^\alpha]\frac{d\theta^\beta}{ds}\,ds.$$

Now, from (6.2.7), (6.2.11), and (1.13.30),

$$\mathbf{a}_\alpha\times\boldsymbol{\chi} = \chi^\beta[\mathbf{a}_\alpha\times\mathbf{a}_\beta] = \epsilon_{\alpha\beta}\chi^\beta\mathbf{a}^3$$
$$= \epsilon_{\alpha\beta}\epsilon^{\rho\beta}\phi_{,\rho}\,\mathbf{a}_3 = \phi_{,\alpha}\,\mathbf{a}_3,$$

and

$$\mathbf{r}\times\boldsymbol{\chi} = \epsilon_{\alpha\beta3}r^\alpha\chi^\beta\mathbf{a}^3 = \epsilon_{\alpha\beta}\epsilon^{\rho\beta}r^\alpha\phi_{,\rho}\,\mathbf{a}_3$$
$$= r^\alpha\phi_{,\alpha}\,\mathbf{a}_3,$$

where

$$\mathbf{r} = r^\alpha\mathbf{a}_\alpha = r_\alpha\mathbf{a}^\alpha. \qquad (6.2.18)$$

Also

$$\epsilon_{\alpha\beta}\mathbf{r}\times\mathbf{a}^\alpha = \epsilon_{\alpha\beta}r_\lambda[\mathbf{a}^\lambda\times\mathbf{a}^\alpha]$$
$$= \epsilon_{\alpha\beta}r_\lambda\epsilon^{\lambda\alpha}\mathbf{a}_3$$
$$= -r_\beta\,\mathbf{a}_3.$$

Hence, apart from an arbitrary constant which may be absorbed into ϕ without affecting the stress distribution,

$$\mathbf{M} = \left\{r^\alpha\phi_{,\alpha} - \phi + \rho\int_A^P Ur_\beta\frac{d\theta^\beta}{ds}\,ds\right\}\mathbf{a}_3, \qquad (6.2.19)$$

and is therefore a couple of magnitude

$$M = r^\alpha\phi_{,\alpha} - \phi + \rho\int_A^P Ur_\beta\frac{d\theta^\beta}{ds}\,ds \qquad (6.2.20)$$

about the x_3-axis. When the body forces are zero this becomes

$$M = r^\alpha\phi_{,\alpha} - \phi. \qquad (6.2.21)$$

If the curve AB is a single bounding curve of a body and this curve is entirely free from applied forces then, in the absence of body forces,

$$\boldsymbol{\chi} = 0, \qquad r^\alpha\phi_{,\alpha} - \phi = 0, \qquad (6.2.22)$$

at all points of this curve. From (6.2.7) and (6.2.11) we see that $\boldsymbol{\chi} = 0$ gives

$$\phi_{,1} = 0, \qquad \phi_{,2} = 0, \qquad (6.2.23)$$

at all points of AB and this implies that ϕ is constant on AB. By taking this constant to be zero we see that the conditions (6.2.23) may replace (6.2.22) at all points of a single bounding curve of a body which is free from applied stress, provided body forces are zero.

6.3. Isotropic bodies

From (5.1.1) and (5.4.31) we see that the stress tensor τ^i_k is related to the displacement vector v^i by the formulae

$$\tau^i_k = 2\mu\left(\gamma^i_k + \frac{\eta}{1-2\eta}\delta^i_k \gamma^r_r\right), \tag{6.3.1}$$

$$\gamma^i_k = \tfrac{1}{2}g^{ir}(v_r|_k + v_k|_r), \tag{6.3.2}$$

when the body is homogeneous and isotropic, and the deformations are infinitesimal.

We now consider a state of strain called *plane strain* in which

$$v^\alpha = v^\alpha(\theta_1, \theta_2), \qquad v^3 = 0. \tag{6.3.3}$$

It follows from (6.3.1), (6.3.2), and (6.1.4) that

$$\left.\begin{aligned} \gamma_{r3} = \gamma^3_r = \gamma^r_3 = \gamma^{r3} = 0, \qquad \gamma^r_r = \gamma^\alpha_\alpha \\ \tau^{\alpha 3} = \tau^\alpha_3 = \tau^3_\alpha = 0 \end{aligned}\right\}, \tag{6.3.4}$$

all other components of the stress and strain tensors being independent of x_3, and that

$$\left.\begin{aligned} \tau^\alpha_\beta &= 2\mu\left(\gamma^\alpha_\beta + \frac{\eta}{1-2\eta}\delta^\alpha_\beta \gamma^\lambda_\lambda\right) \\ \gamma^\alpha_\beta &= \tfrac{1}{2}a^{\alpha\lambda}(v_\lambda|_\beta + v_\beta|_\lambda) \\ &= \tfrac{1}{2}(v^\alpha|_\beta + v_\beta|^\alpha) \\ \tau^3_3 &= \frac{2\mu\eta}{1-2\eta}\gamma^\lambda_\lambda \end{aligned}\right\}, \tag{6.3.5}$$

covariant differentiation being with respect to the two variables θ_α with Christoffel symbols given by (6.1.5). By contracting with respect to the indices α, β in the first two equations of (6.3.5) we have

$$\tau^\alpha_\alpha = \frac{2\mu}{1-2\eta}\gamma^\alpha_\alpha = \frac{2\mu}{1-2\eta}v^\alpha|_\alpha, \tag{6.3.6}$$

and hence, solving for γ^α_β,

$$2\mu\gamma^\alpha_\beta = \tau^\alpha_\beta - \eta\delta^\alpha_\beta \tau^\lambda_\lambda = \mu(v^\alpha|_\beta + v_\beta|^\alpha). \tag{6.3.7}$$

Also, from (6.3.5) and (6.3.6),

$$\tau^3_3 = \eta\tau^\alpha_\alpha. \tag{6.3.8}$$

Covariant differentiation of (6.3.7) with respect to a contravariant index β and covariant index α gives

$$\tau^\alpha_\beta|^\beta_\alpha - \eta\delta^\alpha_\beta \tau^\lambda_\lambda|^\beta_\alpha = \mu(v^\alpha|^\beta_{\beta\alpha} + v_\beta|^{\alpha\beta}_\alpha),$$

or, since $v_\beta|_\alpha^{\alpha\beta} = v^\alpha|_{\beta\alpha}^\beta = v^\alpha|_{\alpha\beta}^\beta$,

$$\tau_\beta^\alpha|_\alpha^\beta - \eta\tau_\alpha^\alpha|_\beta^\beta = 2\mu v^\alpha|_{\beta\alpha}^\beta = (1-2\eta)\tau_\alpha^\alpha|_\beta^\beta,$$

if we use (6.3.6). Hence

$$\tau_\beta^\alpha|_\alpha^\beta = (1-\eta)\tau_\alpha^\alpha|_\beta^\beta. \tag{6.3.9}$$

Since $\tau^{\alpha 3}$ is zero and all other components of stress are independent of θ_3, we may use the results of § 6.2 and, from (6.2.13), we have

$$\tau_\beta^\alpha = \delta_{\beta\rho}^{\alpha\gamma}\phi|_\gamma^\rho + \rho U\delta_\beta^\alpha. \tag{6.3.10}$$

From this equation it follows that

$$\left.\begin{aligned} \tau_\alpha^\alpha &= \phi|_\alpha^\alpha + 2\rho U \\ \tau_\alpha^\alpha|_\beta^\beta &= \phi|_{\alpha\beta}^{\alpha\beta} + 2\rho U|_\alpha^\alpha \\ \tau_\beta^\alpha|_\alpha^\beta &= \delta_{\beta\rho}^{\alpha\gamma}\phi|_{\alpha\gamma}^{\beta\rho} + \rho U|_\alpha^\alpha = \rho U|_\alpha^\alpha \end{aligned}\right\}, \tag{6.3.11}$$

the term $\delta_{\beta\rho}^{\alpha\gamma}\phi|_{\alpha\gamma}^{\beta\rho}$ being zero since $\phi|_{\alpha\gamma}^{\beta\rho}$ is symmetrical in γ and α (or β and ρ). If we now substitute the results (6.3.11) in (6.3.9) we obtain the differential equation

$$\phi|_{\alpha\beta}^{\alpha\beta} + \frac{\rho(1-2\eta)}{1-\eta}U|_\alpha^\alpha = 0 \tag{6.3.12}$$

for ϕ. Also, from (6.3.8) and (6.3.11)

$$\tau_3^3 = \eta\tau_\alpha^\alpha = \eta(\phi|_\alpha^\alpha + 2\rho U). \tag{6.3.13}$$

In the absence of body forces (6.3.12) becomes

$$\phi|_{\alpha\beta}^{\alpha\beta} = 0, \tag{6.3.14}$$

provided $\eta \neq 1$, so that in this case ϕ is a plane biharmonic function. In the general case we may put

$$\phi = -\frac{\rho(1-2\eta)}{1-\eta}U' + \psi, \tag{6.3.15}$$

where ψ is a scalar biharmonic function so that

$$\psi|_{\alpha\beta}^{\alpha\beta} = 0, \tag{6.3.16}$$

and where U' is a scalar which is a particular integral of the equation

$$U'|_\alpha^\alpha = U. \tag{6.3.17}$$

In order to solve problems of plane strain we must find a function which satisfies the biharmonic equation (6.3.16) and also suitable boundary conditions. In many cases the solution of problems is greatly simplified with the help of complex variable notations which we introduce in the next section.

6.4. Complex variable

We define complex coordinates $(z, \bar{z})$ by the formulae

$$z = x_1 + ix_2, \qquad \bar{z} = x_1 - ix_2, \tag{6.4.1}$$

and denote covariant and contravariant base vectors in this system of coordinates by $\mathbf{A}_\alpha$ and $\mathbf{A}^\alpha$ respectively. The position vector $\mathbf{r}$ may then be written

$$\mathbf{r} = z^\alpha \mathbf{A}_\alpha = z_\alpha \mathbf{A}^\alpha. \tag{6.4.2}$$

For completeness we also use the notation $z^3 = x_3$. By tensor transformations

$$z^1 = \frac{\partial z}{\partial x_1} x_1 + \frac{\partial z}{\partial x_2} x_2 = x_1 + ix_2 = z,$$

$$z^2 = \frac{\partial \bar{z}}{\partial x_1} x_1 + \frac{\partial \bar{z}}{\partial x_2} x_2 = x_1 - ix_2 = \bar{z},$$

so that the complex coordinates $(z, \bar{z})$ may also be denoted by z^α. If $A_{\alpha\beta}$ denotes the covariant metric tensor in complex coordinates, then the square of the line element has the form

$$ds^2 = dx_i\, dx_i = dz d\bar{z} + dx_3\, dx_3$$
$$= A_{\alpha\beta}\, dz^\alpha dz^\beta + dz^3 dz^3.$$

Hence

$$\left.\begin{array}{lll} A_{12} = \frac{1}{2}, & A^{12} = 2, & \sqrt{A} = \frac{1}{2}i \\ A_{11} = A_{22} = A^{11} = A^{22} = 0, & & A = |A_{\alpha\beta}| \end{array}\right\}, \tag{6.4.3}$$

where $A^{\alpha\beta}$ is the contravariant metric tensor in complex coordinates.

In the rectangular cartesian coordinates x_i we denote the components of the displacement vector by u_i $(= u^i)$ and the stress tensor by t_{rs} $(= t^r_s = t^{rs})$ and we observe that these functions are also the physical components of displacement and stress. In the system of complex coordinates z^i we denote the contravariant components of the displacement vector by D^i, the covariant components of the displacement vector by D_i, and the contravariant, mixed, and covariant stress tensors by T^{rs}, T^r_s, T_{rs} respectively. We shall be concerned mainly with T^{rs} and D^i but the associated forms of these tensors may be obtained immediately with the help of the metric tensors (6.4.3).

The tensors D^i, T^{rs} can be expressed in terms of u_i and t_{rs} respectively by the appropriate tensor transformations. Thus

$$\left.\begin{array}{l} D^i = \dfrac{\partial z^i}{\partial x^k} u^k \\[2ex] T^{rs} = \dfrac{\partial z^r}{\partial x^m} \dfrac{\partial z^s}{\partial x^n} t^{mn} \end{array}\right\}, \tag{6.4.4}$$

and written out in full these become

$$D = D^1 = u_1+iu_2, \quad D^2 = \bar{D} = u_1-iu_2, \quad w = D^3 = u_3, \tag{6.4.5}$$

$$\left.\begin{aligned} T^{11} &= t_{11}-t_{22}+2it_{12} \\ T^{22} &= t_{11}-t_{22}-2it_{12} \\ T^{12} &= t_{11}+t_{22} \\ T^{13} &= t_{13}+it_{23} \\ T^{23} &= t_{13}-it_{23} \\ T^{33} &= t_{33} \end{aligned}\right\}, \tag{6.4.6}$$

if we remember that $u^i = u_i$, $t^{rs} = t_{rs}$. We observe from (6.4.5) and (6.4.6) that when the indices 1 and 2 are interchanged in D^i and T^{rs} the functions are changed into their complex conjugates, and, for convenience, we have introduced the notation D, $\bar{D}$ for D^1, D^2 respectively, where a bar denotes the complex conjugate.

In applications to special problems we shall usually modify the above notation and replace the coordinates x_i by

$$x_1 = x, \qquad x_2 = y, \qquad x_3 = Z, \tag{6.4.7}$$

where Z is used to avoid confusion with the complex variable

$$z = x+iy, \qquad \bar{z} = x-iy. \tag{6.4.8}$$

We shall denote t_{rs} by $\sigma_{xx},\ldots, \sigma_{xy}$ but in these stress components we write z instead of Z since there is no risk of confusion and since, in two-dimensional elasticity, we mainly use only the components σ_{xx}, σ_{yy}, σ_{xy}. We also replace T^{12}, T^{11}, T^{22}, u_1, u_2 by Θ, Φ, $\bar{\Phi}$, u_x, u_y respectively so that

$$\left.\begin{aligned} \Theta &= T^{12} = \sigma_{xx}+\sigma_{yy} \\ \Phi &= T^{11} = \sigma_{xx}-\sigma_{yy}+2i\sigma_{xy} \\ \Psi &= T^{13} = \sigma_{xz}+i\sigma_{yz} \\ D &= u_x+iu_y \end{aligned}\right\}. \tag{6.4.9}$$

The metric tensor which corresponds to the complex coordinates (6.4.1) has constant components so that covariant differentiation in this coordinate system reduces to partial differentiation. In particular, equations (6.3.17) and (6.3.16) become

$$A^{\alpha\beta}U'_{,\alpha\beta} = U, \qquad A^{\alpha\lambda}A^{\beta\mu}\psi_{,\alpha\lambda\beta\mu} = 0,$$

and using (6.4.3) these may be written

$$4\frac{\partial^2 U'}{\partial z \partial \bar{z}} = U, \tag{6.4.10}$$

$$\frac{\partial^4 \psi}{\partial z \partial z \partial \bar{z} \partial \bar{z}} = 0. \tag{6.4.11}$$

Equation (6.4.11) may be integrated immediately to give

$$\psi = z\bar{\Omega}(\bar{z}) + \bar{z}\Omega(z) + \omega(z) + \bar{\omega}(\bar{z}), \tag{6.4.12}$$

where $\Omega(z)$, $\omega(z)$ are arbitrary functions of the complex variable z, and $\bar{\Omega}(\bar{z})$, $\bar{\omega}(\bar{z})$ are respectively their complex conjugates, making ψ a real function. We shall frequently refer to $\Omega(z)$, $\omega(z)$ as complex potentials. The complete expression for ϕ is

$$\phi = z\bar{\Omega}(\bar{z}) + \bar{z}\Omega(z) + \omega(z) + \bar{\omega}(\bar{z}) - \frac{\rho(1-2\eta)}{1-\eta} U'. \tag{6.4.13}$$

The contravariant form of (6.3.10) is

$$\tau^{\alpha\beta} = \epsilon^{\alpha\gamma}\epsilon^{\beta\rho}\phi|_{\gamma\rho} + \rho U a^{\alpha\beta}, \tag{6.4.14}$$

and, from (6.3.5),

$$\tau^{\alpha\beta} = \frac{2\mu\eta}{1-2\eta} a^{\alpha\beta} v^\gamma|_\gamma + \mu(a^{\beta\gamma}v^\alpha|_\gamma + a^{\alpha\gamma}v^\beta|_\gamma). \tag{6.4.15}$$

If we now interpret equations (6.4.14) and (6.4.15) in complex coordinates we obtain

$$\left.\begin{aligned} \Phi = T^{11} &= -4\frac{\partial^2 \phi}{\partial \bar{z} \partial \bar{z}} = 4\mu \frac{\partial D}{\partial \bar{z}} \\ \Theta = T^{12} &= 4\frac{\partial^2 \phi}{\partial z \partial \bar{z}} + 2\rho U = \frac{2\mu}{1-2\eta}\left(\frac{\partial D}{\partial z} + \frac{\partial \bar{D}}{\partial \bar{z}}\right) \end{aligned}\right\}, \tag{6.4.16}$$

and with (6.4.13) and (6.4.10) these give

$$\mu \frac{\partial D}{\partial \bar{z}} = \frac{\rho(1-2\eta)}{1-\eta} \frac{\partial^2 U'}{\partial \bar{z} \partial \bar{z}} - z\bar{\Omega}''(\bar{z}) - \bar{\omega}''(\bar{z}),$$

$$\mu\left(\frac{\partial D}{\partial z} + \frac{\partial \bar{D}}{\partial \bar{z}}\right) = \frac{2\rho(1-2\eta)}{1-\eta} \frac{\partial^2 U'}{\partial z \partial \bar{z}} + 2(1-2\eta)\{\Omega'(z) + \bar{\Omega}'(\bar{z})\},$$

where $\quad \Omega'(z) = \dfrac{d\Omega(z)}{dz}, \quad \bar{\Omega}'(\bar{z}) = \dfrac{d\bar{\Omega}(\bar{z})}{d\bar{z}},$ etc.

These equations may be integrated† to give

$$\mu D = \frac{\rho(1-2\eta)}{1-\eta}\frac{\partial U'}{\partial \bar{z}} + (3-4\eta)\Omega(z) - z\bar{\Omega}'(\bar{z}) - \bar{\omega}'(\bar{z}). \tag{6.4.17}$$

From (6.4.13) and (6.4.16) we find that the complex stress tensor is given by

$$\left.\begin{aligned}\Phi &= \frac{4\rho(1-2\eta)}{1-\eta}\frac{\partial^2 U'}{\partial \bar{z}\partial \bar{z}} - 4\{z\bar{\Omega}''(\bar{z}) + \bar{\omega}''(\bar{z})\}\\ \Theta &= \frac{\rho U}{1-\eta} + 4\{\Omega'(z) + \bar{\Omega}'(\bar{z})\}\end{aligned}\right\}. \tag{6.4.18}$$

Also, from (6.3.13) and (6.4.6),

$$T^{33} = t_{33} = \eta T^{12} = \eta\Theta. \tag{6.4.19}$$

We recall from (6.2.23) that, when body forces are zero, a single bounding curve of the body is entirely free from applied stress if

$$\phi_{,1} = \phi_{,2} = 0.$$

In complex variable notation this can be replaced by the single condition

$$\frac{\partial \phi}{\partial z} = 0,$$

or

$$\bar{z}\Omega'(z) + \bar{\Omega}(\bar{z}) + \omega'(z) = 0. \tag{6.4.20}$$

Returning to the case when body forces are present we may express the resultant force (6.2.16) and resultant couple (about the origin) (6.2.20) in complex variable notation. If the resultant force $\mathbf{P}$ has components (X, Y) along the x_1- and x_2-axes respectively then a simple tensor transformation gives

$$\mathbf{P} = (X+iY)\mathbf{A}_1 + (X-iY)\mathbf{A}_2 = P\mathbf{A}_1 + \bar{P}\,\mathbf{A}_2, \tag{6.4.21}$$

and the base vectors $\mathbf{A}_\alpha$ are constant in magnitude and direction. Also, from (6.2.7) and (6.2.11),

$$\chi = 2i\left(\frac{\partial \phi}{\partial \bar{z}}\mathbf{A}_1 - \frac{\partial \phi}{\partial z}\mathbf{A}_2\right). \tag{6.4.22}$$

† Alternatively, equations (6.4.16) may be integrated to give the results (6.4.13) and (6.4.17) for ϕ and D directly. Thus, from the first equation of (6.4.16),

$$\mu D + \frac{\partial \phi}{\partial \bar{z}} = 4(1-\eta)\Omega(z),$$

and, taking its conjugate,

$$\mu \bar{D} + \frac{\partial \phi}{\partial z} = 4(1-\eta)\bar{\Omega}(\bar{z}).$$

We may now eliminate D, $\bar{D}$ from these equations and the second equation of (6.4.16) to give

$$\frac{\partial^2 \phi}{\partial z \partial \bar{z}} = -\frac{\rho(1-2\eta)}{1-\eta}\frac{\partial^2 U'}{\partial z \partial \bar{z}} + \Omega'(z) + \bar{\Omega}'(\bar{z}).$$

Hence we obtain the result (6.4.13) for ϕ and then (6.4.17) for D.

If we now interpret (6.2.16) in complex coordinates we find that

$$\mathbf{P} = 2i\mathbf{A}_1\frac{\partial\phi}{\partial\bar{z}} - 2i\mathbf{A}_2\frac{\partial\phi}{\partial z} + \rho i\mathbf{A}_1\int_A^P U\,dz - \rho i\mathbf{A}_2\int_A^P U\,d\bar{z}, \tag{6.4.23}$$

and hence

$$P = X+iY = 2i\frac{\partial\phi}{\partial\bar{z}} + \rho i\int_A^P U\,dz$$

$$= 2i\{z\bar{\Omega}'(\bar{z})+\Omega(z)+\bar{\omega}'(\bar{z})\} - \frac{2i\rho(1-2\eta)}{1-\eta}\frac{\partial U'}{\partial\bar{z}} + \rho i\int_A^P U\,dz. \tag{6.4.24}$$

Again, interpreting (6.2.20) in complex coordinates gives

$$M = z\frac{\partial\phi}{\partial z} + \bar{z}\frac{\partial\phi}{\partial\bar{z}} - \phi + \tfrac{1}{2}\rho\int_A^P U(z\,d\bar{z}+\bar{z}\,dz), \tag{6.4.25}$$

or

$$M = z\bar{z}\{\Omega'(z)+\bar{\Omega}'(\bar{z})\} + z\omega'(z) + \bar{z}\bar{\omega}'(\bar{z}) - \omega(z) - \bar{\omega}(\bar{z}) +$$

$$+\frac{\rho(1-2\eta)}{1-\eta}\left(U' - z\frac{\partial U'}{\partial z} - \bar{z}\frac{\partial U'}{\partial\bar{z}}\right) + \tfrac{1}{2}\rho\int_A^P U(z\,d\bar{z}+\bar{z}\,dz). \tag{6.4.26}$$

In many problems we require stresses and displacements to be single-valued, and, from (6.4.17) and (6.4.18), we see that the conditions for this are

$$\left.\begin{aligned} [\Omega'(z)+\bar{\Omega}'(\bar{z})]_A^A &= 0 \\ [z\bar{\Omega}''(\bar{z})+\bar{\omega}''(\bar{z})]_A^A &= 0 \\ [(3-4\eta)\Omega(z)-z\bar{\Omega}'(\bar{z})-\bar{\omega}'(\bar{z})]_A^A &= 0 \end{aligned}\right\}, \tag{6.4.27}$$

provided the particular integral U' together with its derivatives up to the second order are single-valued. In (6.4.27) the square brackets $[\;]_A^A$ denote the change in value of the function inside on passing once round a contour in the conventional positive sense which keeps the area enclosed on the left, the contour lying entirely in the body. By differentiating the last equation† in (6.4.27) with respect to z and $\bar{z}$ we see that the three conditions may be replaced by

$$\left.\begin{aligned} [\Omega'(z)]_A^A = 0, \qquad [\omega''(z)]_A^A = 0 \\ (3-4\eta)[\Omega(z)]_A^A = [\bar{\omega}'(\bar{z})]_A^A \end{aligned}\right\}. \tag{6.4.28}$$

When these conditions for single-valued stresses and displacements are satisfied, equations (6.4.24) for P and (6.4.26) for M, when applied

† We have already assumed that these derivatives exist and it may be proved that differentiation may be carried out inside the square brackets.

to a contour, reduce to

$$\left.\begin{aligned} P &= 8i(1-\eta)[\Omega(z)]_A^A - 2\rho \int_S \frac{\partial U}{\partial \bar{z}}\, dS \\ M &= [z\omega'(z)+\bar{z}\bar{\omega}'(\bar{z})-\omega(z)-\bar{\omega}(\bar{z})]_A^A + \rho i \int_S \left(\bar{z}\frac{\partial U}{\partial \bar{z}} - z\frac{\partial U}{\partial z}\right) dS \end{aligned}\right\}, \quad (6.4.29)$$

where we have assumed that U satisfies sufficient conditions for the application of Stokes's theorem, the integrals in (6.4.29) extending over the area S enclosed by the contour.

In the above work we have assumed that $\partial\phi/\partial z$, $\partial\phi/\partial\bar{z}$ are continuous everywhere in the elastic body.† If these functions have finite discontinuities at a point C of a boundary but remain one-valued and continuous in the neighbourhood of this point in the body, then C is a point of application of a concentrated force. If we cut an indefinitely small curve $C'DC''$ from the body (Fig. 6.2) then the resulting force acting on $C'DC''$ becomes in the limit

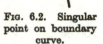

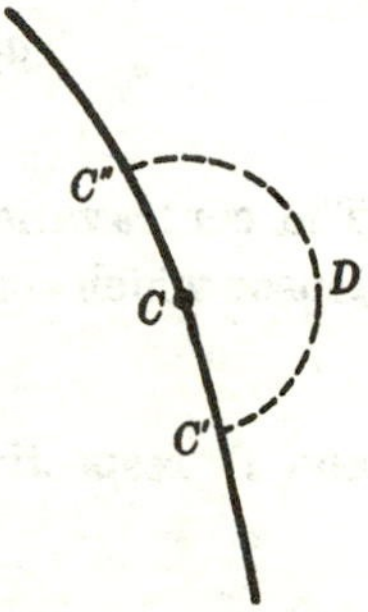

FIG. 6.2. Singular point on boundary curve.

$$P_C = X_C + iY_C = 2i\left[\frac{\partial\phi}{\partial\bar{z}}\right]_{C'}^{C''} = 2i[z\bar{\Omega}'(\bar{z})+\Omega(z)+\bar{\omega}'(\bar{z})]_{C'}^{C''}, \quad (6.4.30)$$

if we assume that the body forces give no discontinuity at C. Also, if ϕ is continuous at C there is no concentrated couple there since, from (6.4.25),

$$M = \left[z\frac{\partial\phi}{\partial z}+\bar{z}\frac{\partial\phi}{\partial\bar{z}}-\phi\right]_{C'}^{C''} = \tfrac{1}{2}i\{z\bar{P}_C-\bar{z}P_C\} = x_C Y_C - y_C X_C,$$

so that M is the moment of the force (X_C, Y_C) about the origin.

6.5. Conformal transformation

The usefulness of the complex variable solution of the plane strain problem may be extended by the help of conformal transformation. For this purpose we suppose that the θ_α-curves form an orthogonal system which can be derived by conformal transformation from some known orthogonal system (e.g. the x_α-curves), and we put

$$\left.\begin{aligned} &\zeta = \zeta^1 = \theta_1 + i\theta_2, \quad \bar{\zeta} = \zeta^2 = \theta_1 - i\theta_2, \quad \xi = \theta_1, \quad \eta = \theta_2 \\ &\zeta^3 = \theta_3 = x_3 = z^3, \quad z = z(\zeta), \quad \frac{dz}{d\zeta} = Je^{i\epsilon} \end{aligned}\right\}, \quad (6.5.1)$$

† Except possibly for a finite discontinuity after a complete circuit of a contour.

where J and ϵ are real. The coordinates (ξ, η) are used instead of (θ_1, θ_2) for convenience in later work. The square of the line element becomes

$$\begin{aligned} ds^2 &= J^2\, d\zeta^1 d\zeta^2 + d\zeta^3 d\zeta^3 \\ &= J^2(d\theta^1 d\theta^1 + d\theta^2 d\theta^2) + dz^3 dz^3. \end{aligned} \tag{6.5.2}$$

The metric tensors $a_{\alpha\beta}$, $a^{\alpha\beta}$ which correspond to the θ_α-curves therefore have the values

$$\left.\begin{aligned} a_{11} = a_{22} = J^2, \quad a_{12} = 0, \quad a = J^4 \\ a^{11} = a^{22} = 1/J^2, \quad a^{12} = 0 \end{aligned}\right\}. \tag{6.5.3}$$

The contravariant displacement vector and the contravariant stress tensor which correspond to the complex coordinates ζ^i are denoted† by

$$V^1 = V, \qquad V^2 = \bar{V}, \qquad V^3 = w$$

and Υ^{rs} respectively, so that

$$V^i = \frac{\partial \zeta^i}{\partial \theta^k} v^k, \qquad \Upsilon^{rs} = \frac{\partial \zeta^r}{\partial \theta^m}\frac{\partial \zeta^s}{\partial \theta^n}\tau^{mn}. \tag{6.5.4}$$

Hence $$V^1 = v^1 + iv^2, \qquad V^2 = v^1 - iv^2, \qquad V^3 = v^3, \tag{6.5.5}$$

$$\left.\begin{aligned} \Upsilon^{11} &= \tau^{11} - \tau^{22} + 2i\tau^{12} \\ \Upsilon^{22} &= \tau^{11} - \tau^{22} - 2i\tau^{12} \\ \Upsilon^{12} &= \tau^{11} + \tau^{22} \\ \Upsilon^{13} &= \tau^{13} + i\tau^{23} \\ \Upsilon^{23} &= \tau^{13} - i\tau^{23} \\ \Upsilon^{33} &= \tau^{33} \end{aligned}\right\}. \tag{6.5.6}$$

Also, tensor transformations give

$$D^i = \frac{\partial z^i}{\partial \zeta^k} V^k, \qquad T^{rs} = \frac{\partial z^r}{\partial \zeta^m}\frac{\partial z^s}{\partial \zeta^n}\Upsilon^{mn}, \tag{6.5.7}$$

so that

$$\left.\begin{aligned} D = Je^{i\epsilon}V^1 = Je^{i\epsilon}V, \qquad \bar{D} = Je^{-i\epsilon}V^2 = Je^{-i\epsilon}\bar{V} \\ \Phi = T^{11} = J^2 e^{2i\epsilon}\Upsilon^{11}, \qquad \bar{\Phi} = T^{22} = J^2 e^{-2i\epsilon}\Upsilon^{22}, \qquad T^{33} = \tau^{33} \\ \Theta = T^{12} = J^2\Upsilon^{12}, \qquad \Psi = T^{13} = Je^{i\epsilon}\Upsilon^{13}, \qquad \bar{\Psi} = T^{23} = Je^{-i\epsilon}\Upsilon^{23} \end{aligned}\right\}. \tag{6.5.8}$$

Physical components of stress referred to the θ_α-curves, which are denoted by σ_{ik}, are given by (2.2.17) in the form

$$\sigma_{ik} = \sqrt{\left(\frac{g_{kk}}{g^{ii}}\right)}\tau^{ik}, \tag{6.5.9}$$

† V^1, V^2 have different meanings from the same symbols used in Chapter 2.

and therefore, using (6.5.3) and (6.1.4),

$$\left.\begin{array}{lll}\sigma_{\xi\xi} = \sigma_{11} = J^2\tau^{11}, & \sigma_{\eta\eta} = \sigma_{22} = J^2\tau^{22}, & \sigma_{\xi\eta} = \sigma_{12} = J^2\tau^{12} \\ \sigma_{\xi z} = \sigma_{13} = J\tau^{13}, & \sigma_{\eta z} = \sigma_{23} = J\tau^{23}, & \sigma_{zz} = \sigma_{33} = \tau^{33}\end{array}\right\}, \tag{6.5.10}$$

where suffixes ξ, η, z will usually be used instead of 1, 2, 3 in applications in which the coordinate system is (ξ, η, Z). We now see from (6.4.9), (6.5.6), (6.5.8), and (6.5.10) that

$$\Theta' = \Theta, \qquad \Phi' = e^{-2i\epsilon}\Phi, \qquad \Psi' = e^{-i\epsilon}\Psi, \tag{6.5.11}$$

where

$$\left.\begin{array}{l}\Theta' = \sigma_{11}+\sigma_{22} = \sigma_{\xi\xi}+\sigma_{\eta\eta} \\ \Phi' = \sigma_{11}-\sigma_{22}+2i\sigma_{12} = \sigma_{\xi\xi}-\sigma_{\eta\eta}+2i\sigma_{\xi\eta} \\ \Psi' = \sigma_{13}+i\sigma_{23} = \sigma_{\xi z}+i\sigma_{\eta z}\end{array}\right\}. \tag{6.5.12}$$

Finally, we denote the physical components of displacement along the θ_1, θ_2 (or ξ, η) curves by u_ξ, u_η so that

$$u_\xi = v^1\surd g_{11} = Jv^1, \qquad u_\eta = v^2\surd g_{22} = Jv^2. \tag{6.5.13}$$

Equations (6.4.5), (6.5.5), (6.5.8), and (6.5.13) then give

$$u_\xi+iu_\eta = JV^1 = e^{-i\epsilon}D = e^{-i\epsilon}(u_x+iu_y). \tag{6.5.14}$$

We are now able to express the physical stress and displacement components referred to θ_α-curves in terms of complex functions. If

$$\Omega(z) = f(\zeta), \qquad \omega(z) = g(\zeta),$$

then, from (6.4.17), (6.4.18), (6.5.12), (6.5.13), and (6.5.14),

$$\begin{aligned}\sigma_{\xi\xi}+\sigma_{\eta\eta} &= \frac{\rho U}{1-\eta}+4\{\Omega'(z)+\bar{\Omega}'(\bar{z})\} \\ &= \frac{\rho U}{1-\eta}+4\left\{f'(\zeta)\frac{d\zeta}{dz}+\bar{f}'(\bar{\zeta})\frac{d\bar{\zeta}}{d\bar{z}}\right\},\end{aligned} \tag{6.5.15}$$

$$\begin{aligned}\sigma_{\xi\xi}-\sigma_{\eta\eta}+2i\sigma_{\xi\eta} &= \frac{d\zeta}{dz}\frac{d\bar{z}}{d\bar{\zeta}}\left\{\frac{4\rho(1-2\eta)}{1-\eta}\frac{\partial^2 U'}{\partial\bar{z}\partial\bar{z}}-4z\bar{\Omega}''(\bar{z})-4\bar{\omega}''(\bar{z})\right\} \\ &= \frac{4\rho(1-2\eta)}{1-\eta}\frac{d\zeta}{dz}\frac{\partial}{\partial\bar{\zeta}}\left\{\frac{\partial U'}{\partial\bar{z}}\right\}-4\frac{d\zeta}{dz}\frac{\partial}{\partial\bar{\zeta}}\left\{z\frac{d\bar{\zeta}}{d\bar{z}}\bar{f}'(\bar{\zeta})+\frac{d\bar{\zeta}}{d\bar{z}}\bar{g}'(\bar{\zeta})\right\},\end{aligned} \tag{6.5.16}$$

$$\mu(u_\xi+iu_\eta) = \left(\frac{d\zeta}{dz}\frac{d\bar{z}}{d\bar{\zeta}}\right)^{\frac{1}{2}}\left\{\frac{\rho(1-2\eta)}{1-\eta}\frac{\partial U'}{\partial\bar{z}}+(3-4\eta)f(\zeta)-z\frac{d\bar{\zeta}}{d\bar{z}}\bar{f}'(\bar{\zeta})-\frac{d\bar{\zeta}}{d\bar{z}}\bar{g}'(\bar{\zeta})\right\}. \tag{6.5.17}$$

From (6.2.2), (6.2.14), and (6.2.15) the resultant force acting across the arc AP of any curve is given by

$$\mathbf{P} = -\int_A^P \epsilon_{\alpha\beta}\tau^{\alpha\lambda}\mathbf{a}_\lambda\, d\theta^\beta. \tag{6.5.18}$$

If $\mathbf{A}_\mu$ are the covariant base vectors referred to complex coordinates z, $\bar{z}$ then

$$\mathbf{a}_\lambda = \mathbf{A}_\mu \frac{\partial z^\mu}{\partial\theta^\lambda}, \tag{6.5.19}$$

so that
$$\mathbf{P} = -\int_A^P \epsilon_{\alpha\beta}\tau^{\alpha\lambda}\frac{\partial z^\mu}{\partial\theta^\lambda}\mathbf{A}_\mu\, d\theta^\beta.$$

Hence, using (6.4.21),

$$P = X+iY = -\int_A^P \epsilon_{\alpha\beta}\tau^{\alpha\lambda}\frac{\partial z^1}{\partial\theta^\lambda}\, d\theta^\beta. \tag{6.5.20}$$

If we now specialize the θ_α-coordinates to the orthogonal system (ξ, η) defined in (6.5.1), and use (6.5.10) we have

$$P = -\int_A^P \left\{\left(\sigma_{\xi\xi}\frac{\partial z}{\partial\xi}+\sigma_{\xi\eta}\frac{\partial z}{\partial\eta}\right)d\eta-\left(\sigma_{\xi\eta}\frac{\partial z}{\partial\xi}+\sigma_{\eta\eta}\frac{\partial z}{\partial\eta}\right)d\xi\right\}. \tag{6.5.21}$$

In particular the force exerted across an arc of the coordinate curve ξ = constant becomes

$$P(\eta) = i\int_A^P (\sigma_{\xi\xi}+i\sigma_{\xi\eta})\frac{dz}{d\eta}\, d\eta, \tag{6.5.22}$$

since on this curve
$$\frac{\partial z}{\partial\xi} = -i\frac{\partial z}{\partial\eta}.$$

6.6. Properties of complex potentials for isotropic bodies

From § 6.4 we see that if the stress field is given, the real part of $\Omega'(z)$ is determined and hence $\Omega'(z)$ is determined except for a purely imaginary constant iC; and $\Omega(z)$ except for additional terms

$$iCz+\alpha+i\beta.$$

It follows that $\omega''(z)$ is completely determined and hence $\omega(z)$ except for additional terms
$$(\alpha'+i\beta')z+\alpha''+i\beta''.$$

If we replace $\Omega(z)$ by $\quad \Omega(z)+iCz+\alpha+i\beta$

and $\omega(z)$ by $\quad \omega(z)+(\alpha'+i\beta')z+\alpha''+i\beta''$, the stress system is unaltered, but, from (6.4.17),

$$\mu D = \text{original } \mu D + 4(1-\eta)iCz - \alpha' + i\beta' + (3-4\eta)(\alpha+i\beta), \quad (6.6.1)$$

the additional displacement representing a rigid-body displacement of the whole body. If the displacements are given everywhere, not all five constants C, α, β, α', β' are disposable and only one of the pairs α, β and α', β' can be chosen arbitrarily.

When the region S occupied by the (two-dimensional) body is finite and simply connected, then we suppose that the state of stress is such that $\Omega(z)$, $\omega(z)$ are holomorphic in the *open* region S; the stresses and displacements are then single-valued. When S is multiply-connected, suppose S is bounded by one or more smooth non-intersecting contours $L_0, L_1,..., L_n$ of which L_0 contains all the others. The contour L_0 may be absent (i.e. it may be reducible to the point at infinity) in which case the region S will be infinite with certain contours $L_1, L_2,..., L_n$ as internal boundaries. If we require stresses and displacements to be single-valued then in view of (6.4.28) we assume that

$$\Omega'(z) = \sum_{k=1}^{n} \frac{a_k+ib_k}{2\pi(z-z_k)} + \Omega_0'(z),$$

where the point z_k is inside the contour L_k (and therefore outside S) and where $\Omega_0(z)$ is holomorphic in S except possibly at infinity. Also a_k and b_k are real constants. Hence

$$\Omega(z) = \frac{1}{2\pi} \sum_{k=1}^{n} (a_k+ib_k)\ln(z-z_k) + \Omega_0(z).$$

Around any contour L_k' enclosing L_k, and not cutting the other contours, $\Omega(z)$ increases by $i(a_k+ib_k)$. Recalling (6.4.28), we assume that

$$\omega'(z) = -\frac{3-4\eta}{2\pi} \sum_{k=1}^{n} (a_k-ib_k)\ln(z-z_k) + \omega_0'(z),$$

where $\omega_0'(z)$ is holomorphic in S except possibly at infinity.

Now let the contours L_k' tend to the contours L_k except that points of discontinuity of $\Omega(z)$, $\Omega'(z)$, $\omega'(z)$ on L_k (at which isolated forces may act) are excluded by small curves which tend to zero (see Fig. 6.2). Then the resultant forces exerted on the contours L_k, given by (6.4.29), are

$$P_k = X_k + iY_k = -8(1-\eta)(a_k+ib_k),$$

when body forces are absent, and therefore

$$\left.\begin{aligned}\Omega(z) &= -\frac{1}{16\pi(1-\eta)}\sum_{k=1}^{n} P_k \ln(z-z_k)+\Omega_0(z)\\ \omega'(z) &= \frac{3-4\eta}{16\pi(1-\eta)}\sum_{k=1}^{n}\bar{P}_k \ln(z-z_k)+\omega_0'(z)\end{aligned}\right\}. \tag{6.6.2}$$

When the region S is infinite then, for sufficiently large $|z|$,

$$\Omega(z) = -\frac{P\ln z}{16\pi(1-\eta)}+\phi_0(z),$$

$$\omega'(z) = \frac{(3-4\eta)\bar{P}\ln z}{16\pi(1-\eta)}+\psi_0'(z),$$

where
$$P = X+iY = \sum_{k=1}^{n} P_k, \tag{6.6.3}$$

and (X,Y) is the resultant force acting on all the boundaries $L_1, L_2,\ldots, L_n$. The functions $\phi_0(z)$ and $\psi_0'(z)$ denote holomorphic functions which for sufficiently large $|z|$ may be developed in Laurent series. If the stresses are to remain bounded at infinity,

$$\phi_0(z) = (B+iC)z+\alpha+i\beta+\frac{\alpha_1+i\beta_1}{z}+\ldots,$$

$$\psi_0'(z) = (B'+iC')z+\alpha'+i\beta'+\frac{\alpha_1'+i\beta_1'}{z}+\ldots.$$

Thus, for large $|z|$,

$$\left.\begin{aligned}\Omega(z) &= (B+iC)z-\frac{P\ln z}{16\pi(1-\eta)}+\alpha+i\beta+O\left(\frac{1}{z}\right)\\ \omega'(z) &= (B'+iC')z+\frac{(3-4\eta)\bar{P}\ln z}{16\pi(1-\eta)}+\alpha'+i\beta'+O\left(\frac{1}{z}\right)\end{aligned}\right\}, \tag{6.6.4}$$

and from (6.4.18), in the absence of body forces,

$$\Theta = 8B = N_1+N_2,$$

$$\Phi = -4(B'-iC') = (N_1-N_2)e^{2i\alpha},$$

where N_1, N_2 are the principal stresses at infinity and α is the angle between N_1 and the axis Ox.

If ω_0 is the rotation at infinity,

$$\omega_0 = \frac{1}{2}\left(\frac{\partial v}{\partial x}-\frac{\partial u}{\partial y}\right) = \frac{1}{2i}\left(\frac{\partial D}{\partial z}-\frac{\partial \bar{D}}{\partial \bar{z}}\right)_{|z|\to\infty},$$

so that
$$\omega_0 = \frac{4(1-\eta)C}{\mu}. \tag{6.6.5}$$

We observe that the displacements are only bounded at infinity if

$$B = C = B' = C' = X = Y = 0,$$

and then $\Omega(z)$, $\omega'(z)$ are holomorphic at infinity. The displacements are not bounded at infinity if the resultant of the external forces on the boundaries is different from zero.

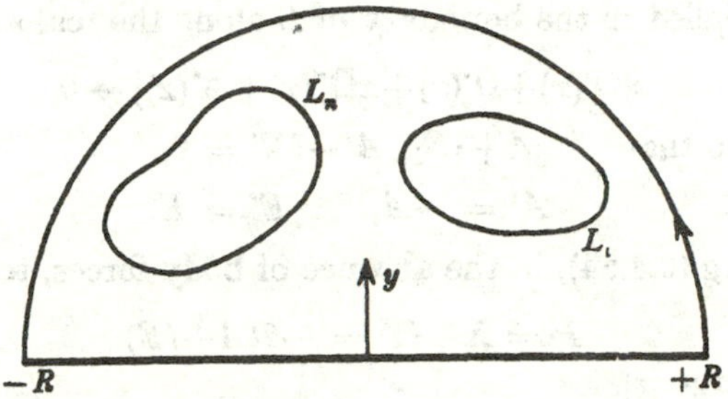

FIG. 6.3. Half-plane bounded internally by smooth non-intersecting contours.

If the stresses at infinity are not bounded but have known values we can modify the above by adding suitable terms

$$\sum_{r=2}^{m} (B_r+iC_r)z^r, \qquad \sum_{r=2}^{m} (B'_r+iC'_r)z^r,$$

to $\Omega(z)$, $\omega'(z)$ respectively.

When the region S is bounded by a single closed curve L and extends to infinity, then, if the stresses and rotation are zero at infinity,

$$\left.\begin{aligned} \Omega'(z) &= -\frac{P}{16\pi(1-\eta)z}+O\left(\frac{1}{z^2}\right) \\ \omega''(z) &= \frac{(3-4\eta)\bar{P}}{16\pi(1-\eta)z}+O\left(\frac{1}{z^2}\right) \end{aligned}\right\}, \tag{6.6.6}$$

for large $|z|$.

The results (6.6.4) and (6.6.6) do not in general hold if part of the boundary of the region S is in the finite part of the plane and the rest of the boundary extends to infinity. For example, suppose S is bounded internally by the contours $L_1, L_2,..., L_n$ and externally by the real axis $y = 0$, from $x = -R$ to $x = R$, and the semicircle $|z| = R$ in the upper half-plane, where $R \to \infty$ (Fig. 6.3). Suppose the resultant stress over $L_1, L_2,..., L_n$ and the real axis from $x = -R$ to R is $P = X+iY$, then the resultant stress exerted over the semicircle $|z| = R$ is $-P$ as $R \to \infty$. Alternatively, the force exerted by the material inside the semicircle across its boundary is P. For large $|z|$, we assume that

$$\Omega'(z) = \frac{A+iE}{2\pi z}+O\left(\frac{1}{z^2}\right), \quad \Omega''(z) = -\frac{A+iE}{2\pi z^2}+O\left(\frac{1}{z^3}\right),$$

$$\omega''(z) = \frac{A'+iE'}{2\pi z}+O\left(\frac{1}{z^2}\right),$$

if stresses and rotation vanish at infinity. If, in addition, any stresses which are applied to the boundary of S along the real axis are $O(1/x^2)$, then

$$x\{\Omega'(x)+\bar{\Omega}'(x)+x\bar{\Omega}''(x)+\bar{\omega}''(x)\} \to 0,$$

as $|x| \to \infty$, so that $\quad A+iE+A'-iE' = 0,$

or $\quad A' = -A, \quad E' = E.$

Also, applying (6.4.24), in the absence of body forces, to the semicircle $|z| = R$ gives

$$P = X+iY = -2(A+iE).$$

Hence, for large $|z|$,

$$\left.\begin{aligned} \Omega'(z) &= -\frac{P}{4\pi z}+O\left(\frac{1}{z^2}\right) \\ \omega''(z) &= \frac{\bar{P}}{4\pi z}+O\left(\frac{1}{z^2}\right) \end{aligned}\right\}. \tag{6.6.7}$$

We now derive the corresponding theory for aeolotropic bodies.

6.7. Aeolotropic bodies

We recall from (5.3.12) and (5.3.10) that the stress–strain relations for a homogeneous aeolotropic body referred to rectangular cartesian axes x_i are

$$t_{ik} = c_{ik}^{rs} e_{rs}, \tag{6.7.1}$$

where

$$c_{rs}^{ik} = c_{rs}^{ki} = c_{sr}^{ki} = c_{sr}^{ik} = c_{ik}^{rs}. \tag{6.7.2}$$

We now restrict our attention to bodies which are elastically symmetrical with respect to the plane $x_3 = 0$ so that, from (5.4.7),

$$c_{\lambda\mu}^{3\beta} = c_{3\beta}^{\lambda\mu} = c_{3\mu}^{33} = c_{33}^{3\mu} = 0, \tag{6.7.3}$$

and the elastic coefficients c_{rs}^{ik} reduce to the 13 independent coefficients given in table (5.4.8).

When the body is in a state of plane strain perpendicular to the x_3-axis, the displacements, referred to general coordinates, are given by (6.3.3) so that, referred to x_i-axes,

$$u_1 = u_1(x_1, x_2), \qquad u_2 = u_2(x_1, x_2), \qquad u_3 = 0, \tag{6.7.4}$$

and therefore

$$e_{3i} = 0, \tag{6.7.5}$$

all other components of strain $e_{\alpha\beta}$ being independent of x_3. It follows that all the components of stress are independent of x_3 and, remembering (6.7.3),

$$t_{\alpha\beta} = c^{\lambda\mu}_{\alpha\beta} e_{\lambda\mu}, \qquad t_{3\alpha} = 0, \qquad t_{33} = c^{\lambda\mu}_{33} e_{\lambda\mu}, \tag{6.7.6}$$

the independent coefficients $c^{\alpha\beta}_{\lambda\mu}$ being

$$c^{11}_{11},\quad c^{11}_{22},\quad c^{11}_{12},\quad c^{12}_{12},\quad c^{12}_{22},\quad c^{22}_{22}. \tag{6.7.7}$$

If we define six independent coefficients

$$s^{11}_{11},\quad s^{11}_{22},\quad s^{11}_{12},\quad s^{12}_{12},\quad s^{12}_{22},\quad s^{22}_{22}, \tag{6.7.8}$$

by the equations

$$c^{\alpha\beta}_{\lambda\mu} s^{\rho\gamma}_{\alpha\beta} = \tfrac{1}{2}(\delta^{\rho}_{\lambda}\,\delta^{\gamma}_{\mu}+\delta^{\rho}_{\mu}\,\delta^{\gamma}_{\lambda}), \tag{6.7.9}$$

then equations (6.7.6) may be solved for $e_{\alpha\beta}$ in the form

$$e_{\alpha\beta} = s^{\lambda\mu}_{\alpha\beta} t_{\lambda\mu}. \tag{6.7.10}$$

This result may now be written in general form by referring the stress and strain to the two-dimensional general coordinates θ_α with corresponding metric tensors $a_{\alpha\beta}$, $a^{\alpha\beta}$. Thus

$$\gamma^{\alpha\beta} = F^{\alpha\beta}_{\lambda\mu}\tau^{\lambda\mu}, \qquad \gamma^{\alpha\beta} = \tfrac{1}{2}(a^{\alpha\lambda}v^{\beta}|_{\lambda}+a^{\beta\lambda}v^{\alpha}|_{\lambda}), \tag{6.7.11}$$

and we also remember that the stress tensor can be expressed in terms of Airy's stress function ϕ by (6.4.14), so that

$$\tau^{\alpha\beta} = \epsilon^{\alpha\gamma}\epsilon^{\beta\rho}\phi|_{\gamma\rho}+\rho U a^{\alpha\beta}. \tag{6.7.12}$$

For aeolotropic bodies it is now convenient to develop the theory by working with complex variable coordinates which were defined in § 6.4. If we use the notation of § 6.4 and also denote the value of the tensor $F^{\alpha\beta}_{\lambda\mu}$ by $S^{\alpha\beta}_{\lambda\mu}$ referred to complex coordinates then

$$S^{\alpha\beta}_{\lambda\mu} = s^{\rho\gamma}_{\xi\eta}\frac{\partial z^{\alpha}}{\partial x^{\rho}}\frac{\partial z^{\beta}}{\partial x^{\gamma}}\frac{\partial x^{\xi}}{\partial z^{\lambda}}\frac{\partial x^{\eta}}{\partial z^{\mu}}. \tag{6.7.13}$$

With the help of (6.4.3), equations (6.7.12) and (6.7.11) may now be interpreted in complex coordinates. Thus

$$\Phi = -4\frac{\partial^2\phi}{\partial\bar{z}\partial\bar{z}}, \qquad \bar{\Phi} = -4\frac{\partial^2\phi}{\partial z\partial z}, \qquad \Theta = 4\frac{\partial^2\phi}{\partial z\partial\bar{z}}+2\rho U, \tag{6.7.14}$$

$$\left.\begin{aligned} 2\frac{\partial D}{\partial\bar{z}} &= S^{11}_{11}\Phi+S^{11}_{22}\bar{\Phi}+2S^{11}_{12}\Theta\\ 2\frac{\partial\bar{D}}{\partial z} &= S^{22}_{11}\Phi+S^{22}_{22}\bar{\Phi}+2S^{22}_{12}\Theta\\ \frac{\partial D}{\partial z}+\frac{\partial\bar{D}}{\partial\bar{z}} &= S^{12}_{11}\Phi+S^{12}_{22}\bar{\Phi}+2S^{12}_{12}\Theta \end{aligned}\right\}. \tag{6.7.15}$$

The coefficients $S^{\alpha\beta}_{\lambda\mu}$ satisfy the symmetry conditions

$$S^{\alpha\beta}_{\lambda\mu} = S^{\beta\alpha}_{\lambda\mu} = S^{\beta\alpha}_{\mu\lambda} = S^{\alpha\beta}_{\mu\lambda}, \tag{6.7.16}$$

and, from (6.7.13) and (6.4.1),

$$\left.\begin{aligned} S^{11}_{11} &= S^{22}_{22} = \tfrac{1}{4}(s^{11}_{11}+s^{22}_{22}+4s^{12}_{12}-2s^{11}_{22}) \\ S^{11}_{22} &= \tfrac{1}{4}(s^{11}_{11}+s^{22}_{22}-4s^{12}_{12}-2s^{11}_{22}+4is^{11}_{12}-4is^{12}_{22}) \\ S^{12}_{22} &= S^{11}_{12} = \tfrac{1}{4}(s^{11}_{11}-s^{22}_{22}+2is^{11}_{12}+2is^{12}_{22}) \\ S^{12}_{12} &= \tfrac{1}{4}(s^{11}_{11}+s^{22}_{22}+2s^{11}_{22}) \\ S^{22}_{11} &= \bar{S}^{11}_{22}, \qquad S^{22}_{12} = S^{12}_{11} = \bar{S}^{11}_{12} \end{aligned}\right\}, \tag{6.7.17}$$

where a bar denotes the complex conjugate.

Eliminating D from equations (6.7.15) gives

$$\frac{\partial^2}{\partial z^2}(S^{11}_{11}\Phi+S^{11}_{22}\bar{\Phi}+2S^{11}_{12}\Theta)+\frac{\partial^2}{\partial \bar{z}^2}(S^{22}_{11}\Phi+S^{22}_{22}\bar{\Phi}+2S^{22}_{12}\Theta)- \\ -2\frac{\partial^2}{\partial z\partial \bar{z}}(S^{12}_{11}\Phi+S^{12}_{22}\bar{\Phi}+2S^{12}_{12}\Theta) = 0,$$

and, substituting from (6.7.14),

$$S^{11}_{22}\frac{\partial^4\phi}{\partial z^4}-4S^{11}_{12}\frac{\partial^4\phi}{\partial z^3\partial \bar{z}}+2(S^{11}_{11}+2S^{12}_{12})\frac{\partial^4\phi}{\partial z^2\partial \bar{z}^2}-4S^{12}_{11}\frac{\partial^4\phi}{\partial z\partial \bar{z}^3}+S^{22}_{11}\frac{\partial^4\phi}{\partial \bar{z}^4} \\ = \rho\left(S^{11}_{12}\frac{\partial^2 U}{\partial z^2}+S^{22}_{12}\frac{\partial^2 U}{\partial \bar{z}^2}-2S^{12}_{12}\frac{\partial^2 U}{\partial z\partial \bar{z}}\right). \tag{6.7.18}$$

When body forces are absent the equation for ϕ reduces to

$$S^{11}_{22}\frac{\partial^4\phi}{\partial z^4}-4S^{11}_{12}\frac{\partial^4\phi}{\partial z^3\partial \bar{z}}+2(S^{11}_{11}+2S^{12}_{12})\frac{\partial^4\phi}{\partial z^2\partial \bar{z}^2}-4S^{12}_{11}\frac{\partial^4\phi}{\partial z\partial \bar{z}^3}+S^{22}_{11}\frac{\partial^4\phi}{\partial \bar{z}^4} = 0, \tag{6.7.19}$$

and in the rest of this chapter we shall only consider this case. A solution of (6.7.19) of the form $\phi(z+\gamma\bar{z})$ exists provided the complex constant γ is a root of the equation

$$\bar{S}^{11}_{22}\gamma^4-4\bar{S}^{11}_{12}\gamma^3+2(S^{11}_{11}+2S^{12}_{12})\gamma^2-4S^{11}_{12}\gamma+S^{11}_{22} = 0. \tag{6.7.20}$$

We observe that if γ is a root of this equation so is $1/\bar{\gamma}$ and we may, in general, denote the roots by

$$\gamma_1, \quad \gamma_2, \quad 1/\bar{\gamma}_1, \quad 1/\bar{\gamma}_2, \tag{6.7.21}$$

provided

$$\gamma_1\bar{\gamma}_1 \neq 1, \qquad \gamma_2\bar{\gamma}_2 \neq 1. \tag{6.7.22}$$

Also, since ϕ is a real function the general solution of (6.7.19) is

$$\phi = \Omega(z_1)+\bar{\Omega}(\bar{z}_1)+\omega(z_2)+\bar{\omega}(\bar{z}_2), \tag{6.7.23}$$

where†

$$\left.\begin{aligned} z_1 &= z+\gamma_1\bar{z}, & \bar{z}_1 &= \bar{z}+\bar{\gamma}_1 z \\ z_2 &= z+\gamma_2\bar{z}, & \bar{z}_2 &= \bar{z}+\bar{\gamma}_2 z \end{aligned}\right\}, \tag{6.7.24}$$

and where the roots γ_1, γ_2 of (6.7.20) can be selected to be those roots with modulus less than unity so that

$$|\gamma_1| < 1, \qquad |\gamma_2| < 1. \tag{6.7.25}$$

When condition (6.7.22) holds the equation (6.7.19) is of the elliptic type and (6.7.23) represents a satisfactory general solution. If, however, either γ_1 or γ_2 is complex and has modulus equal to unity, then the solution (6.7.23) changes its form since the equation (6.7.19) becomes hyperbolic. We can, however, show that condition (6.7.22) must always be satisfied. The elastic potential W can be expressed in the form

$$W = \tfrac{1}{2}\tau^{ik}\gamma_{ik},$$

and since γ_{i3} is zero here

$$W = \tfrac{1}{2}\tau^{\alpha\beta}\gamma_{\alpha\beta} = \tfrac{1}{2}a_{\alpha\gamma}\,a_{\beta\rho}\,F^{\alpha\beta}_{\lambda\mu}\,\tau^{\lambda\mu}\tau^{\gamma\rho}.$$

In complex variable notation this becomes

$$8W = \bar{S}^{11}_{22}\Phi^2+S^{11}_{22}\bar{\Phi}^2+4S^{12}_{12}\Theta^2+2S^{11}_{11}\Phi\bar{\Phi}+4\bar{S}^{11}_{12}\Theta\Phi+4S^{11}_{12}\Theta\bar{\Phi}. \tag{6.7.26}$$

Now suppose that one root of (6.7.20) has modulus equal to unity and put

$$\gamma = -\left(\frac{\Phi}{\bar{\Phi}}\right)^{\frac{1}{2}} = -\frac{\Theta}{\bar{\Phi}}, \qquad \Theta^2 = \Phi\bar{\Phi},$$

where Φ (and therefore Θ) is non-zero. Hence, from (6.7.20),

$$\bar{S}^{11}_{22}\Phi^2+S^{11}_{22}\bar{\Phi}^2+4S^{12}_{12}\Theta^2+2S^{11}_{11}\Phi\bar{\Phi}+4\bar{S}^{11}_{12}\Theta\Phi+4S^{11}_{12}\Theta\bar{\Phi} = 0,$$

for some non-zero value of Φ (and therefore of $\bar{\Phi}$ and Θ). But, from (6.7.26), this means that W vanishes for a non-zero value of Φ, $\bar{\Phi}$, and Θ, which is impossible if W is assumed to be essentially positive. We see that (6.7.20) cannot have any roots with moduli equal to unity, and therefore condition (6.7.22) is always satisfied.

It may happen that $\gamma_1 = \gamma_2$, in which case the form of solution (6.7.23) must be modified. We shall not, however, deal with this case separately since, in special problems, we may deduce by a limiting process any special results which are required.

One special case is, however, worthy of notice. If the body is orthotropic with respect to the three coordinate planes $x_i = 0$, then c^{11}_{12} and c^{12}_{22} are zero and the coefficients (6.7.7) reduce to four, namely,

$$c^{11}_{11}, \quad c^{11}_{22}, \quad c^{12}_{12}, \quad c^{22}_{22}. \tag{6.7.27}$$

† z_1, z_2 must not be confused with those defined in (6.4.2).

It follows that we then have, in general, four non-zero coefficients

$$s^{11}_{11},\quad s^{11}_{22},\quad s^{12}_{12},\quad s^{22}_{22}, \tag{6.7.28}$$

and the coefficients (6.7.17) reduce to

$$\left.\begin{aligned} S^{11}_{11} &= S^{22}_{22} = \tfrac{1}{4}(s^{11}_{11}+s^{22}_{22}+4s^{12}_{12}-2s^{11}_{22}) \\ S^{22}_{11} &= S^{11}_{22} = \tfrac{1}{4}(s^{11}_{11}+s^{22}_{22}-4s^{12}_{12}-2s^{11}_{22}) \\ S^{11}_{12} &= S^{22}_{12} = S^{12}_{22} = S^{12}_{11} = \tfrac{1}{4}(s^{11}_{11}-s^{22}_{22}) \\ S^{12}_{12} &= \tfrac{1}{4}(s^{11}_{11}+s^{22}_{22}+2s^{11}_{22}) \end{aligned}\right\}. \tag{6.7.29}$$

The equation (6.7.20) for γ then has real coefficients and becomes a reciprocal equation so that either

$$(a)\quad \bar{\gamma}_2 = \gamma_1,\qquad \bar{\gamma}_1 = \gamma_2,$$

or

$$(b)\quad \bar{\gamma}_1 = \gamma_1,\qquad \bar{\gamma}_2 = \gamma_2,$$

that is γ_1 and γ_2 are either complex conjugates or real.

We return to the general solution (6.7.23) and substitute this value of ϕ in (6.7.15) with the help of (6.7.14), remembering that $U = 0$. Thus

$$\begin{aligned} \frac{\partial D}{\partial \bar{z}} = &-2(S^{11}_{11}\gamma_1^2-2S^{11}_{12}\gamma_1+S^{11}_{22})\Omega''(z+\gamma_1\bar{z})- \\ &\qquad -2(S^{11}_{11}-2S^{11}_{12}\bar{\gamma}_1+S^{11}_{22}\bar{\gamma}_1^2)\bar{\Omega}''(\bar{z}+\bar{\gamma}_1 z)- \\ &-2(S^{11}_{11}\gamma_2^2-2S^{11}_{12}\gamma_2+S^{11}_{22})\omega''(z+\gamma_2\bar{z})- \\ &\qquad -2(S^{11}_{11}-2S^{11}_{12}\bar{\gamma}_2+S^{11}_{22}\bar{\gamma}_2^2)\bar{\omega}''(\bar{z}+\bar{\gamma}_2 z), \end{aligned}$$

$$\begin{aligned} \frac{\partial D}{\partial z}+\frac{\partial \bar{D}}{\partial \bar{z}} = &-4(S^{12}_{11}\gamma_1^2-2S^{12}_{12}\gamma_1+S^{12}_{22})\Omega''(z+\gamma_1\bar{z})- \\ &\qquad -4(S^{12}_{11}-2S^{12}_{12}\bar{\gamma}_1+S^{12}_{22}\bar{\gamma}_1^2)\bar{\Omega}''(\bar{z}+\bar{\gamma}_1 z)- \\ &-4(S^{12}_{11}\gamma_2^2-2S^{12}_{12}\gamma_2+S^{12}_{22})\omega''(z+\gamma_2\bar{z})- \\ &\qquad -4(S^{12}_{11}-2S^{12}_{12}\bar{\gamma}_2+S^{12}_{22}\bar{\gamma}_2^2)\bar{\omega}''(\bar{z}+\bar{\gamma}_2 z). \end{aligned}$$

If we remember that γ_1 and γ_2 satisfy (6.7.20) we may integrate these equations to give

$$D = \delta_1\Omega'(z_1)+\rho_1\bar{\Omega}'(\bar{z}_1)+\delta_2\,\omega'(z_2)+\rho_2\,\bar{\omega}'(\bar{z}_2), \tag{6.7.30}$$

where

$$\left.\begin{aligned} \delta_1 &= -2(S^{11}_{11}\gamma_1^2-2S^{11}_{12}\gamma_1+S^{11}_{22})/\gamma_1, &\quad \rho_1 &= -2(S^{11}_{11}-2S^{11}_{12}\bar{\gamma}_1+S^{11}_{22}\bar{\gamma}_1^2) \\ \delta_2 &= -2(S^{11}_{11}\gamma_2^2-2S^{11}_{12}\gamma_2+S^{11}_{22})/\gamma_2, &\quad \rho_2 &= -2(S^{11}_{11}-2S^{11}_{12}\bar{\gamma}_2+S^{11}_{22}\bar{\gamma}_2^2) \end{aligned}\right\}. \tag{6.7.31}$$

Also, from (6.7.14) and (6.7.23),

$$\left.\begin{aligned} \Phi &= -4\gamma_1^2\Omega''(z_1)-4\bar{\Omega}''(\bar{z}_1)-4\gamma_2^2\,\omega''(z_2)-4\bar{\omega}''(\bar{z}_2) \\ \Theta &= 4\gamma_1\Omega''(z_1)+4\bar{\gamma}_1\bar{\Omega}''(\bar{z}_1)+4\gamma_2\,\omega''(z_2)+4\bar{\gamma}_2\,\bar{\omega}''(\bar{z}_2) \end{aligned}\right\}. \tag{6.7.32}$$

The resultant force and couple exerted by the material inside a closed curve on the material outside, across an arc AP, are found from (6.4.21), (6.4.23), and (6.4.25). Thus

$$P = X+iY = 2i\frac{\partial\phi}{\partial\bar{z}} = 2i\{\gamma_1\Omega'(z_1)+\bar{\Omega}'(\bar{z}_1)+\gamma_2\,\omega'(z_2)+\bar{\omega}'(\bar{z}_2)\}, \tag{6.7.33}$$

$$M = z\frac{\partial\phi}{\partial z}+\bar{z}\frac{\partial\phi}{\partial\bar{z}}-\phi = z_1\Omega'(z_1)+\bar{z}_1\bar{\Omega}'(\bar{z}_1)-\Omega(z_1)-\bar{\Omega}(\bar{z}_1)+ +z_2\,\omega'(z_2)+\bar{z}_2\,\bar{\omega}'(\bar{z}_2)-\omega(z_2)-\bar{\omega}(\bar{z}_2). \tag{6.7.34}$$

If the displacements and stresses are single-valued then

$$[D]_A^A = 0, \qquad [\Phi]_A^A = 0, \qquad [\Theta]_A^A = 0. \tag{6.7.35}$$

It follows also that†

$$\left[\frac{\partial D}{\partial z}\right]_A^A = 0, \qquad \left[\frac{\partial D}{\partial\bar{z}}\right]_A^A = 0,$$

and if these conditions are combined with the last two in (6.7.35) we see that, in general,

$$[\Omega''(z_1)]_A^A = 0, \qquad [\omega''(z_2)]_A^A = 0. \tag{6.7.36}$$

As in § 6.4 a concentrated force acts at a point C of a boundary if $\partial\phi/\partial z$, $\partial\phi/\partial\bar{z}$ have finite discontinuities at C but ϕ is continuous there.

For later use we see from (6.7.17), (6.7.20), and (6.7.21) that

$$\left.\begin{aligned} \frac{\gamma_1\gamma_2}{\bar{\gamma}_1\bar{\gamma}_2} &= \frac{S_{22}^{11}}{S_{11}^{22}} = \frac{S_{22}^{11}}{\bar{S}_{22}^{11}} \\ 4S_{12}^{11} &= S_{22}^{11}(\bar{\gamma}_1+\bar{\gamma}_2+1/\gamma_1+1/\gamma_2) \end{aligned}\right\}. \tag{6.7.37}$$

Also, from (6.7.31),

$$\left.\begin{aligned} \frac{\delta_1-\delta_2}{\gamma_1-\gamma_2} &= \frac{\bar{\delta}_1-\bar{\delta}_2}{\bar{\gamma}_1-\bar{\gamma}_2} = -2\left(S_{11}^{11}-\frac{S_{22}^{11}}{\gamma_1\gamma_2}\right) \\ \frac{\gamma_2\bar{\rho}_1-\gamma_1\bar{\rho}_2}{\gamma_1-\gamma_2} &= \frac{\bar{\gamma}_2\rho_1-\bar{\gamma}_1\rho_2}{\bar{\gamma}_1-\bar{\gamma}_2} = 2(S_{11}^{11}-S_{22}^{11}\bar{\gamma}_1\bar{\gamma}_2) \\ \frac{\gamma_2\delta_1-\gamma_1\delta_2}{\gamma_1-\gamma_2} &= \frac{\rho_1-\rho_2}{\bar{\gamma}_1-\bar{\gamma}_2} = \frac{S_{22}^{11}}{\gamma_1\gamma_2}\{\gamma_1+\gamma_2-\gamma_1\gamma_2(\bar{\gamma}_1+\bar{\gamma}_2)\} \end{aligned}\right\}, \tag{6.7.38}$$

and

$$\left.\begin{aligned} \delta_1-\gamma_1\bar{\rho}_1 &= (\gamma_1-\gamma_2)(1-\gamma_1\bar{\gamma}_1)(1-\gamma_1\bar{\gamma}_2)S_{22}^{11}/(\gamma_1\gamma_2) \\ \delta_2-\gamma_2\bar{\rho}_2 &= -(\gamma_1-\gamma_2)(1-\bar{\gamma}_1\gamma_2)(1-\gamma_2\bar{\gamma}_2)S_{22}^{11}/(\gamma_1\gamma_2) \\ \bar{\delta}_1-\bar{\gamma}_1\rho_1 &= (\bar{\gamma}_1-\bar{\gamma}_2)(1-\gamma_1\bar{\gamma}_1)(1-\bar{\gamma}_1\gamma_2)S_{22}^{11}/(\gamma_1\gamma_2) \\ \bar{\delta}_2-\bar{\gamma}_2\rho_2 &= -(\bar{\gamma}_1-\bar{\gamma}_2)(1-\gamma_1\bar{\gamma}_2)(1-\gamma_2\bar{\gamma}_2)S_{22}^{11}/(\gamma_1\gamma_2) \end{aligned}\right\}. \tag{6.7.39}$$

† We have already assumed that these derivatives exist.

6.8. Properties of complex potentials for aeolotropic bodies

If the stress field is given we see from (6.7.32) that

$$\gamma_1\Omega''(z_1)+\gamma_2\omega''(z_2)$$

is determined except for a purely imaginary constant iC. Hence, again using (6.7.32), $\Omega'(z_1)$ is determined except for a term

$$\frac{iC(1-\gamma_2\bar{\gamma}_2)(1-\bar{\gamma}_1\gamma_2)(z+\gamma_1\bar{z})}{(\gamma_1-\gamma_2)(1-\gamma_1\bar{\gamma}_1\gamma_2\bar{\gamma}_2)}+\alpha+i\beta, \tag{6.8.1}$$

and $\omega'(z_2)$ is determined except for a term

$$-\frac{iC(1-\gamma_1\bar{\gamma}_1)(1-\gamma_1\bar{\gamma}_2)(z+\gamma_2\bar{z})}{(\gamma_1-\gamma_2)(1-\gamma_1\bar{\gamma}_1\gamma_2\bar{\gamma}_2)}+\alpha'+i\beta', \tag{6.8.2}$$

where α, α', β, β' are real constants. If we replace $\Omega'(z_1)$ by $\Omega'(z_1)$ plus (6.8.1), and $\omega'(z_2)$ by $\omega'(z_2)$ plus (6.8.2), it can be verified from (6.7.32) that the stress system is unaltered. From (6.7.30), however, it is found that the displacement D contains an additional term

$$\frac{iS_{22}^{11}KCz}{\gamma_1\gamma_2}+a'\alpha+b'\alpha'+c'\beta+d'\beta', \tag{6.8.3}$$

where a', b', c', d' are constants which can be expressed in terms of the elastic constants of the material, and K is a real constant which can be expressed in terms of γ_1 and γ_2. From (6.7.37) we see that $S_{22}^{11}/(\gamma_1\gamma_2)$ is real. The additional displacement terms (6.8.3) therefore represent a rigid-body displacement. If the displacements are given everywhere, not all five constants C, α, α', β, β' are disposable and only one of the pairs α, β and α', β' can be chosen arbitrarily.

When the region S occupied by the body is finite and simply-connected, then we suppose the state of stress is such that $\Omega(z_1)$, $\omega(z_2)$ are holomorphic functions of z_1, z_2 respectively in the open regions S_1, S_2 of the z_1- and z_2-planes which correspond to S; the stresses and displacements are then single-valued. When S is multiply-connected, suppose S is bounded by one or more smooth non-intersecting contours $L_0, L_1,..., L_n$ as described in § 6.6. If we require stresses and displacements to be single-valued then in view of (6.7.36) we assume that

$$\Omega''(z_1) = \sum_{r=1}^{n}\frac{H_r}{2\pi(z_1-z_{1r})}+\Omega_0''(z_1),$$

$$\omega''(z_2) = \sum_{r=1}^{n}\frac{K_r}{2\pi(z_2-z_{2r})}+\omega_0''(z_2),$$

where $z_{1r} = z_r+\gamma_1\bar{z}_r$, $z_{2r} = z_r+\gamma_2\bar{z}_r$, and z_r is inside the contour L_r.

Also H_r, K_r are complex constants and $\Omega_0'(z_1)$, $\omega_0'(z_2)$ are holomorphic in the regions S_1, S_2 of the z_1- and z_2-planes respectively which correspond to S, except possibly at infinity. Hence

$$\left.\begin{aligned}\Omega'(z_1) &= \frac{1}{2\pi}\sum_{r=1}^{n} H_r \ln(z_1 - z_{1r}) + \Omega_0'(z_1)\\ \omega'(z_2) &= \frac{1}{2\pi}\sum_{r=1}^{n} K_r \ln(z_2 - z_{2r}) + \omega_0'(z_2)\end{aligned}\right\}. \tag{6.8.4}$$

Around any contour L_r' enclosing L_r, and not cutting the other contours, $\Omega'(z_1)$ increases† by iH_r and $\omega'(z_2)$ increases by iK_r. Since D is single-valued on each contour L_r' it follows from (6.7.30) that

$$\delta_1 H_r - \rho_1 \bar{H}_r + \delta_2 K_r - \rho_2 \bar{K}_r = 0 \quad (r = 1, 2, \ldots, n). \tag{6.8.5}$$

Now let the contours L_r' tend to the contours L_r except that points of discontinuity of $\Omega'(z_1)$ and $\omega'(z_2)$ on L_r (at which isolated forces may act) are excluded by small semicircles whose radii tend to zero (see Fig. 6.2). Then the forces on L_r, given by (6.7.33), are

$$P_r = X_r + iY_r = -2(\gamma_1 H_r - \bar{H}_r + \gamma_2 K_r - \bar{K}_r). \tag{6.8.6}$$

Equations (6.8.5) and (6.8.6), together with their complex conjugates, are in general sufficient to determine the constants H_r, K_r, in terms of the forces P_r.

When the region S is infinite then, for sufficiently large $|z|$,

$$\Omega'(z_1) = \frac{H}{2\pi}\ln z_1 + \phi_0'(z_1),$$

$$\omega'(z_2) = \frac{K}{2\pi}\ln z_2 + \psi_0'(z_2),$$

where $$H = \sum_{r=1}^{n} H_r, \qquad K = \sum_{r=1}^{n} K_r, \tag{6.8.7}$$

and $\phi_0'(z_1)$, $\psi_0'(z_2)$ denote holomorphic functions in the regions S_1, S_2 respectively which for sufficiently large $|z|$ may be developed in Laurent series. If the stresses are to remain bounded at infinity,

$$\phi_0'(z_1) = (B+iC)z_1 + \alpha + i\beta + \frac{\alpha_1 + i\beta_1}{z_1} + \ldots,$$

$$\psi_0'(z_2) = (B'+iC')z_2 + \alpha' + i\beta' + \frac{\alpha_1' + i\beta_1'}{z_2} + \ldots,$$

† We remember that $|\gamma_1| < 1$, $|\gamma_2| < 1$.

where B, C, B', C' are real constants. Thus, for large $|z|$,

$$\left.\begin{aligned}\Omega'(z_1) &= (B+iC)z_1+\frac{H}{2\pi}\ln z_1+\alpha+i\beta+O\left(\frac{1}{z_1}\right)\\ \omega'(z_2) &= (B'+iC')z_2+\frac{K}{2\pi}\ln z_2+\alpha'+i\beta'+O\left(\frac{1}{z_2}\right)\end{aligned}\right\}, \tag{6.8.8}$$

and from (6.7.32)

$$\left.\begin{aligned}N_1+N_2 &= 4\gamma_1(B+iC)+4\bar{\gamma}_1(B-iC)+4\gamma_2(B'+iC')+4\bar{\gamma}_2(B'-iC')\\ (N_1-N_2)e^{2i\alpha} &= -4\gamma_1^2(B+iC)-4(B-iC)-4\gamma_2^2(B'+iC')-4(B'-iC')\end{aligned}\right\}, \tag{6.8.9}$$

where N_1, N_2 are the principal stresses at infinity and α is the angle between N_1 and the x-axis. Also, if ω_0 is the rotation at infinity,

$$\omega_0 = \frac{1}{2i}\left(\frac{\partial D}{\partial z}-\frac{\partial \bar{D}}{\partial \bar{z}}\right)_{|z|\to\infty}$$

so that

$$\begin{aligned}2i\omega_0 = (\delta_1-\gamma_1\bar{\rho}_1)(B+iC)+(\bar{\gamma}_1\rho_1-\bar{\delta}_1)(B-iC)+\\ +(\delta_2-\gamma_2\bar{\rho}_2)(B'+iC')+(\bar{\gamma}_2\rho_2-\bar{\delta}_2)(B'-iC').\end{aligned} \tag{6.8.10}$$

Equations (6.8.9) and (6.8.10) are, in general, sufficient to express B, C, B', C' in terms of N_1, N_2, α, and ω_0. We observe that the displacements are only bounded at infinity if

$$B = C = B' = C' = 0, \qquad H = K = 0,$$

or

$$N_1 = N_2 = \omega_0 = 0, \qquad H = K = 0,$$

and then $\Omega'(z_1)$, $\omega'(z_2)$ are holomorphic at infinity. The displacements are not bounded at infinity if the external forces on the boundaries have a resultant different from zero.

If the stresses at infinity are not bounded but have known values, we can modify the above by adding suitable terms

$$\sum_{r=2}^{m}(B_r+iC_r)z_1^r, \qquad \sum_{r=2}^{m}(B'_r+iC'_r)z_2^r,$$

to $\Omega'(z_1)$ and $\omega'(z_2)$ respectively.

When the region is bounded by a single closed curve L and extends to infinity, then if the stresses and rotation are zero at infinity

$$\left.\begin{aligned}\Omega''(z_1) &= \frac{H}{2\pi z_1}+O\left(\frac{1}{z_1^2}\right)\\ \omega''(z_2) &= \frac{K}{2\pi z_2}+O\left(\frac{1}{z_2^2}\right)\end{aligned}\right\}, \tag{6.8.11}$$

where

$$\left.\begin{aligned}\delta_1 H-\rho_1\bar{H}+\delta_2 K-\rho_2\bar{K} = 0\\ P = X+iY = 2(\bar{H}-\gamma_1 H+\bar{K}-\gamma_2 K)\end{aligned}\right\}, \tag{6.8.12}$$

and $P = X+iY$ is the resultant force on L.

The results (6.8.8) and (6.8.11) must be modified when part of the boundary of the region S is in the finite part of the plane and the rest of the boundary extends to infinity. For example, as in § 6.6, we consider the region represented by Fig. 6.3, and we use the notation described for this figure. Then, for large $|z|$, we assume that

$$\left.\begin{aligned}\Omega''(z_1) &= \frac{A}{2\pi z_1}+O\left(\frac{1}{z_1^2}\right)\\ \omega''(z_2) &= \frac{E}{2\pi z_2}+O\left(\frac{1}{z_2^2}\right)\end{aligned}\right\}, \tag{6.8.13}$$

if stresses and rotation vanish at infinity. If, in addition, any stresses which are applied to the boundary of S along the real axis are $O(1/x^2)$, then

$$\lim_{|x|\to\infty} x(\Theta-\Phi) = 0,$$

so that, from (6.7.32),

$$\gamma_1 A+\bar{A}+\gamma_2 E+\bar{E} = 0. \tag{6.8.14}$$

Also applying (6.7.33) to the infinite semicircle $|z| = R$ in the upper half-plane, using (6.8.13), gives

$$P = -\gamma_1 A+\bar{A}-\gamma_2 E+\bar{E}.$$

With (6.8.14) this becomes

$$P = -2(\gamma_1 A+\gamma_2 E) = 2(\bar{A}+\bar{E}). \tag{6.8.15}$$

Equation (6.8.15) may be solved for A, E, to give

$$A = -\frac{P+\gamma_2\bar{P}}{2(\gamma_1-\gamma_2)}, \qquad E = \frac{P+\gamma_1\bar{P}}{2(\gamma_1-\gamma_2)}. \tag{6.8.16}$$

Hence, for large $|z|$,

$$\left.\begin{aligned}\Omega''(z_1) &= -\frac{P+\gamma_2\bar{P}}{4\pi(\gamma_1-\gamma_2)z_1}+O\left(\frac{1}{z_1^2}\right)\\ \omega''(z_2) &= \frac{P+\gamma_1\bar{P}}{4\pi(\gamma_1-\gamma_2)z_2}+O\left(\frac{1}{z_2^2}\right)\end{aligned}\right\}. \tag{6.8.17}$$

6.9. Orthotropic bodies

In § 6.7 we indicated that when a body is orthotropic with respect to the coordinate planes $x_i = 0$ then the values obtained for γ are either real or else complex conjugates. For such bodies certain simplifications can be made. If we put

$$\alpha = \frac{1+\gamma}{1-\gamma}, \qquad \gamma = \frac{\alpha-1}{\alpha+1}, \qquad \alpha_\lambda = \frac{1+\gamma_\lambda}{1-\gamma_\lambda} \quad (\lambda = 1, 2), \tag{6.9.1}$$

then from (6.7.29) we see that the equation (6.7.20) for γ reduces to

$$s_{22}^{22}\alpha^4-2(s_{22}^{11}+2s_{12}^{12})\alpha^2+s_{11}^{11} = 0, \tag{6.9.2}$$

and we see that

$$\left.\begin{aligned}\bar{\alpha}_1^2\bar{\alpha}_2^2 = \alpha_1^2\alpha_2^2 = s_{11}^{11}/s_{22}^{22}\\ \bar{\alpha}_1^2+\bar{\alpha}_2^2 = \alpha_1^2+\alpha_2^2 = 2(s_{22}^{11}+2s_{12}^{12})/s_{22}^{22}\end{aligned}\right\}. \tag{6.9.3}$$

Also, from (6.7.29), (6.9.1), and (6.9.3), the constants δ_1, δ_2, ρ_1, ρ_2 in (6.7.31) reduce to

$$\left.\begin{aligned}\delta_1 = (1+\gamma_1)\beta_2-(1-\gamma_1)\beta_1, \qquad \delta_2 = (1+\gamma_2)\beta_1-(1-\gamma_2)\beta_2\\ \bar{\rho}_1 = (1+\gamma_1)\beta_2+(1-\gamma_1)\beta_1, \qquad \bar{\rho}_2 = (1+\gamma_2)\beta_1+(1-\gamma_2)\beta_2\end{aligned}\right\}, \tag{6.9.4}$$

where

$$\beta_1 = s_{22}^{11}-s_{22}^{22}\alpha_1^2, \qquad \beta_2 = s_{22}^{11}-s_{22}^{22}\alpha_2^2. \tag{6.9.5}$$

In general, as shown in § 6.7, two cases arise:

(*a*) $\bar{\gamma}_2 = \gamma_1, \qquad \bar{\gamma}_1 = \gamma_2$

so that γ_1, γ_2 are complex conjugates, and we also recall that $|\gamma_1| < 1$, $|\gamma_2| < 1$. It follows that the corresponding roots α_1, α_2 of (6.9.2) are complex conjugates and have positive real parts. Also

$$\bar{\beta}_1 = \beta_2, \qquad \bar{\delta}_1 = \delta_2, \qquad \bar{\rho}_1 = \rho_2. \tag{6.9.6}$$

(*b*) $\bar{\gamma}_1 = \gamma_1, \qquad \bar{\gamma}_2 = \gamma_2$

so that γ_1, γ_2 are real and must also be numerically less than unity. It follows that the corresponding values of α_1, α_2 are real and positive and that β_1, β_2, δ_1, δ_2, ρ_1, ρ_2 are all real.

For both cases we may solve equations (6.8.12) for H, K and express the results in the form

$$\left.\begin{aligned}4(\alpha_1^2-\alpha_2^2)s_{22}^{22}H &= (1+\alpha_1)(\beta_2 X+i\beta_1 Y/\alpha_1)\\ 4(\alpha_1^2-\alpha_2^2)s_{22}^{22}K &= -(1+\alpha_2)(\beta_1 X+i\beta_2 Y/\alpha_2)\end{aligned}\right\}. \tag{6.9.7}$$

Also, if the rotation ω_0 at infinity is zero, then equations (6.8.9) and (6.8.10) may be solved for B, B', C, C', and in both cases (*a*) and (*b*) we find that†

$$\left.\begin{aligned}\frac{32(\alpha_1^2-\alpha_2^2)(B+iC)}{(\alpha_1+1)^2} &= i(\alpha_1^2-\alpha_2^2)(N_1-N_2)\alpha_1^{-1}\sin 2\alpha'+\\ &\quad +(\alpha_2+1)^2\{(N_1+N_2)(1+\gamma_2^2)+2\gamma_2(N_1-N_2)\cos 2\alpha'\}\\ \frac{32(\alpha_1^2-\alpha_2^2)(B'+iC')}{(\alpha_2+1)^2} &= i(\alpha_1^2-\alpha_2^2)(N_1-N_2)\alpha_2^{-1}\sin 2\alpha'-\\ &\quad -(\alpha_1+1)^2\{(N_1+N_2)(1+\gamma_1^2)+2\gamma_1(N_1-N_2)\cos 2\alpha'\}\end{aligned}\right\}. \tag{6.9.8}$$

† In these equations the angle between N_1 and the x-axis is now denoted by α' instead of α.

7 Plate Theory

THE general three-dimensional theory of stress systems in plane plates is one of considerable analytical difficulty, but some useful practical information about the distribution of stress can be obtained by introducing approximations which reduce the problem to one of two dimensions. We consider two-dimensional theories of stretching and bending of plane plates in this chapter.† We again retain tensor notations as long as possible and specialize later to complex variables. The reader is referred to the introduction of Chapter 6 for a brief account of original work in the use of complex function theory for two-dimensional elasticity.

We adopt the notations of § 6.1 and consider a plane plate which is bounded by the plane faces

$$\theta_3 = x_3 = \pm h = \pm \tfrac{1}{2}t,$$

where we restrict our attention to plates of constant thickness $t = 2h$.

7.1. Stress resultants and stress couples

We now suppose that the plate is in a state of stress and that the displacements are such that we may use the infinitesimal theory of Chapter 5. Instead of examining the state of stress in detail at every point of the plate we introduce stress resultants and stress couples. For this purpose consider the coordinate surface $\theta_1 =$ constant. From § 5.2 we see that the force acting on an element of area in this surface is

$$\mathbf{T}_1\, d\theta^2 d\theta^3,$$

where

$$\mathbf{T}_1 = \sqrt{(gg^{11})}\mathbf{t}_1 = \sqrt{(g)}\tau^{1k}\mathbf{g}_k, \tag{7.1.1}$$

† Some complete three-dimensional solutions of special problems indicate that generalized plane stress theory for the stretching of an isotropic plate is likely to give reasonably accurate values for the average stresses. See A. E. Green, *Phil. Trans. R. Soc.* **A240** (1948) 561; A. E. Green and T. J. Willmore, *Proc. R. Soc.* **A193** (1948) 229; E. Sternberg and M. A. Sadowsky, *J. appl. Mech.* **16** (1949) 27. See also A. E. Green, *Proc. R. Soc.* **A195** (1948) 533. J. B. Alblass, *Theorie van de Driedimensional Spanningstoestand in Een Doorboorde Plaat* (Amsterdam, 1957).

For a study of Plate Theory via asymptotic expansions of the three-dimensional equations see K. O. Friedrich and F. R. Dressler, *Communs pure appl. Math.* **14** (1961) 1; E. L. Reiss and S. Locke, *Q. appl. Math.* **19** (1961) 195; A. L. Gol'denveizer, *PMM* (Translation of *Prikl. Mat. Mekh.*) **26** (1962) 1000; N. Laws, *Proc. Camb. phil. Soc. math. phys. Sci.* **62** (1966) 313. Compare with asymptotic expansions for shell theory given in Chapter 16.

and the length of the corresponding line element of the middle plane $x_3 = 0$ is

$$\sqrt{(a_{22})}\,d\theta^2 = \sqrt{(aa^{11})}\,d\theta^2.$$

The stress across the surface $\theta_1 =$ constant may therefore be replaced by a physical stress resultant $\mathbf{n}_1$ and a physical stress couple $\mathbf{m}_1$, measured per unit length of the middle line of the plane $x_3 = 0$, which is in the surface $\theta_1 =$ constant, where

$$\left.\begin{aligned} \mathbf{n}_1 &= \frac{1}{\sqrt{(aa^{11})}} \int\limits_{-h}^{h} \mathbf{T}_1\,dx_3 \\ \mathbf{m}_1 &= \frac{1}{\sqrt{(aa^{11})}} \int\limits_{-h}^{h} (\mathbf{a}_3 \times \mathbf{T}_1) x_3\,dx_3 \end{aligned}\right\}. \tag{7.1.2}$$

Similarly, stress resultants and stress couples may be defined for the surface $\theta_2 =$ constant. Stress resultants and stress couples over the surface $\theta_\alpha =$ constant, measured per unit length of the line $\theta_\alpha =$ constant, $x_3 = 0$, may be defined by the general formulae

$$\mathbf{n}_\alpha = \frac{\mathbf{N}_\alpha}{\sqrt{(aa^{\alpha\alpha})}}, \qquad \mathbf{m}_\alpha = \frac{\mathbf{M}_\alpha}{\sqrt{(aa^{\alpha\alpha})}}, \tag{7.1.3}$$

where $$\mathbf{N}_\alpha = \int\limits_{-h}^{h} \mathbf{T}_\alpha\,dx_3, \qquad \mathbf{M}_\alpha = \int\limits_{-h}^{h} (\mathbf{a}_3 \times \mathbf{T}_\alpha) x_3\,dx_3. \tag{7.1.4}$$

The stress resultant $\mathbf{n}$ and stress couple $\mathbf{m}$ per unit length of a line of the middle plane, whose unit normal in the plane is

$$\mathbf{u} = u_\alpha \mathbf{a}^\alpha, \tag{7.1.5}$$

may be obtained by a process similar to that used in Chapter 2 for the three-dimensional stress vector $\mathbf{t}$, the corresponding results being

$$\mathbf{n} = \sum_{\alpha=1}^{2} u_\alpha \mathbf{n}_\alpha \sqrt{a^{\alpha\alpha}}, \qquad \mathbf{m} = \sum_{\alpha=1}^{2} u_\alpha \mathbf{m}_\alpha \sqrt{a^{\alpha\alpha}}. \tag{7.1.6}$$

These formulae show that $\mathbf{n}_\alpha \sqrt{a^{\alpha\alpha}}$ and $\mathbf{m}_\alpha \sqrt{a^{\alpha\alpha}}$ transform according to the contravariant type of transformation for changes of surface coordinates θ_α.

From (5.2.5) and (6.1.4),

$$\mathbf{T}_i = \sqrt{(a)}\,\tau^{ik}\mathbf{a}_k,$$

so that substituting this in (7.1.4) gives

$$\mathbf{n}_\alpha \sqrt{a^{\alpha\alpha}} = n^{\alpha\rho}\mathbf{a}_\rho + q^\alpha \mathbf{a}_3, \qquad \mathbf{N}_\alpha = N^{\alpha\rho}\mathbf{a}_\rho + Q^\alpha \mathbf{a}_3, \tag{7.1.7}$$

$$\mathbf{m}_\alpha \sqrt{a^{\alpha\alpha}} = m^{\alpha\rho}\mathbf{a}_3 \times \mathbf{a}_\rho, \qquad \mathbf{M}_\alpha = M^{\alpha\rho}\mathbf{a}_3 \times \mathbf{a}_\rho, \tag{7.1.8}$$

where $$N^{\alpha\rho} = n^{\alpha\rho}\sqrt{a}, \qquad M^{\alpha\rho} = m^{\alpha\rho}\sqrt{a}, \qquad Q^\alpha = q^\alpha\sqrt{a}, \tag{7.1.9}$$

and

$$n^{\alpha\rho} = \int\limits_{-h}^{h} \tau^{\alpha\rho}\,dx_3, \tag{7.1.10}$$

$$m^{\alpha\rho} = \int\limits_{-h}^{h} \tau^{\alpha\rho}x_3\,dx_3, \tag{7.1.11}$$

$$q^{\alpha} = \int\limits_{-h}^{h} \tau^{\alpha 3}\,dx_3. \tag{7.1.12}$$

We observe that $n^{\alpha\rho}$, $m^{\alpha\rho}$ are both symmetric in α and ρ. Also, from (7.1.6), (7.1.7), and (7.1.8),

$$\mathbf{n} = u_\alpha(n^{\alpha\rho}\mathbf{a}_\rho + q^\alpha \mathbf{a}_3), \qquad \mathbf{m} = u_\alpha m^{\alpha\rho}\mathbf{a}_3 \times \mathbf{a}_\rho. \tag{7.1.13}$$

With the help of (1.13.30) the expressions (7.1.8) may be transformed into

$$\mathbf{m}_\alpha \sqrt{a^{\alpha\alpha}} = \epsilon_{\rho\lambda} m^{\alpha\rho}\mathbf{a}^\lambda, \qquad \mathbf{M}_\alpha = \epsilon_{\rho\lambda} M^{\alpha\rho}\mathbf{a}^\lambda, \tag{7.1.14}$$

which, written in full, become

$$\left.\begin{aligned} \mathbf{m}_1 &= (m^{11}\mathbf{a}^2 - m^{12}\mathbf{a}^1)\sqrt{(a/a^{11})}, \qquad & \mathbf{M}_1 &= (M^{11}\mathbf{a}^2 - M^{12}\mathbf{a}^1)\sqrt{a} \\ \mathbf{m}_2 &= (m^{12}\mathbf{a}^2 - m^{22}\mathbf{a}^1)\sqrt{(a/a^{22})}, \qquad & \mathbf{M}_2 &= (M^{12}\mathbf{a}^2 - M^{22}\mathbf{a}^1)\sqrt{a} \end{aligned}\right\}. \tag{7.1.15}$$

The functions defined in (7.1.10) to (7.1.12) are all surface tensors, the surface here being a plane. The components of the symmetric contravariant tensor $n^{\alpha\rho}$ and the components of the contravariant tensor q^α are called stress resultants and shearing forces respectively, and are related to the (vector) stress resultants $\mathbf{n}_\alpha$ by the formulae (7.1.7). The components of the symmetric contravariant tensor $m^{\alpha\rho}$ are called stress couples and are related to the (vector) stress couples $\mathbf{m}_\alpha$ by (7.1.8). Mixed and covariant tensors n^α_β, $n_{\alpha\beta}$, m^α_β, $m_{\alpha\beta}$, q_α may be formed with the help of the metric tensors $a_{\alpha\beta}$, $a^{\alpha\beta}$. Since $n^{\alpha\rho}$, $m^{\alpha\rho}$ are symmetric we may write the mixed forms without ambiguity as n^α_β, m^α_β.

In order to find the physical components of $\mathbf{n}_\alpha$ and $\mathbf{m}_\alpha$ we express these vectors in terms of unit base vectors along the coordinate curves $\theta_\alpha =$ constant. The physical stress resultants, physical shearing forces, and physical stress couples are denoted respectively by $n_{(\alpha\beta)}$, $q_{(\alpha)}$, $m_{(\alpha\beta)}$, the brackets indicating that these physical components are not tensors. Thus

$$\left.\begin{aligned} \mathbf{n}_\alpha &= n_{(\alpha 1)}\frac{\mathbf{a}_1}{\sqrt{a_{11}}} + n_{(\alpha 2)}\frac{\mathbf{a}_2}{\sqrt{a_{22}}} + q_{(\alpha)}\mathbf{a}_3 \\ \mathbf{m}_\alpha &= m_{(\alpha 1)}\frac{\mathbf{a}^2}{\sqrt{a^{22}}} + m_{(\alpha 2)}\frac{\mathbf{a}^1}{\sqrt{a^{11}}} \end{aligned}\right\}. \tag{7.1.16}$$

Comparison between these equations and equations (7.1.7) and (7.1.15) gives immediately

$$n_{(\alpha\beta)} = n^{\alpha\beta}\sqrt{(a_{\beta\beta}/a^{\alpha\alpha})}, \qquad q_{(\alpha)} = q^{\alpha}/\sqrt{a^{\alpha\alpha}}, \tag{7.1.17}$$

$$\left.\begin{aligned} m_{(11)} &= m^{11}\sqrt{(aa^{22}/a^{11})} = m^{11}\sqrt{(a_{11}/a^{11})} = m^{11}\sqrt{(aa_{11}/a_{22})} \\ m_{(12)} &= -m^{12}\sqrt{a}, \qquad m_{(21)} = m^{12}\sqrt{a} \\ m_{(22)} &= -m^{22}\sqrt{(aa^{11}/a^{22})} = -m^{22}\sqrt{(a_{22}/a^{22})} = -m^{22}\sqrt{(aa_{22}/a_{11})} \end{aligned}\right\}. \tag{7.1.18}$$

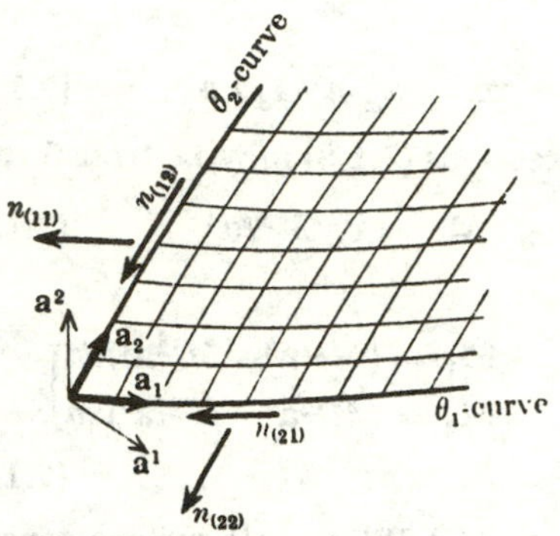

Fig. 7.1. Physical stress resultants.

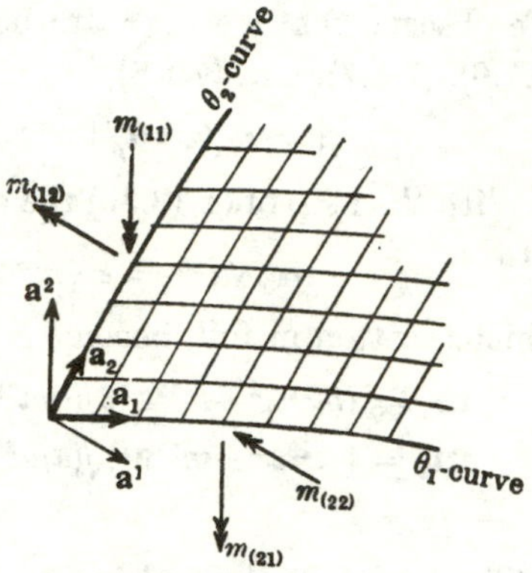

Fig. 7.2. Physical stress couples.

The meaning of these physical stress resultants and couples is seen by examining (7.1.16), and they are illustrated geometrically in Figs. 7.1 and 7.2. The physical stress resultants $n_{(\alpha\beta)}$ are directed along the covariant base vectors of the middle plane of the plate, and $q_{(\alpha)}$ are at right angles to the plate. The physical stress couples $m_{(\alpha\beta)}$ are couples about the contravariant base vectors of the middle plane. $n_{(\alpha\beta)}$ is symmetric in α and β, and $m_{(12)} = -m_{(21)}$.

7.2. Loads

We consider the external forces acting on the faces of the plate. The boundary conditions at these faces require that the forces per unit area acting on the faces should be

$$[\mathbf{t}_3]_{x_3=h}, \qquad [-\mathbf{t}_3]_{x_3=-h}.$$

We replace these surface forces by a resultant force

$$\mathbf{p} = [\mathbf{t}_3]_{x_3=h} - [\mathbf{t}_3]_{x_3=-h} = [\mathbf{t}_3]^{h}_{-h} \tag{7.2.1}$$

and a couple

$$\mathbf{s} = [(\mathbf{a}_3 \times \mathbf{t}_3)x_3]^{h}_{-h} \tag{7.2.2}$$

about the point $(\theta_1, \theta_2, 0)$ of the middle plane, where $\mathbf{p}$ and $\mathbf{s}$ are measured per unit area of the middle plane of the plate. From (5.2.5) and (6.1.4) we see that (7.2.1) and (7.2.2) may be replaced by

$$\mathbf{p} = \frac{\mathbf{P}}{\sqrt{a}}, \qquad \mathbf{s} = \frac{\mathbf{S}}{\sqrt{a}}, \tag{7.2.3}$$

where
$$\mathbf{P} = [\mathbf{T}_3]^h_{-h}, \qquad \mathbf{S} = [(\mathbf{a}_3 \times \mathbf{T}_3)x_3]^h_{-h}. \tag{7.2.4}$$

Also, as
$$\mathbf{T}_3 = \sqrt{(a)}\tau^{3k}\mathbf{a}_k,$$

it follows that

$$\mathbf{p} = p^\lambda \mathbf{a}_\lambda + p\mathbf{a}_3, \qquad \mathbf{P} = P^\lambda \mathbf{a}_\lambda + P\mathbf{a}_3, \tag{7.2.5}$$

$$\left.\begin{aligned} \mathbf{s} &= (\mathbf{a}_3 \times \mathbf{a}_\beta)s^\beta = \epsilon_{\beta\lambda} s^\beta \mathbf{a}^\lambda = S^1\mathbf{a}^2 - S^2\mathbf{a}^1 \\ \mathbf{S} &= (\mathbf{a}_3 \times \mathbf{a}_\beta)S^\beta = \epsilon_{\beta\lambda} S^\beta \mathbf{a}^\lambda \end{aligned}\right\}, \tag{7.2.6}$$

where

$$P^\alpha = p^\alpha \sqrt{a}, \qquad P = p\sqrt{a}, \qquad S^\alpha = s^\alpha \sqrt{a}, \tag{7.2.7}$$

$$p^\alpha = [\tau^{3\alpha}]^h_{-h}, \qquad p = [\tau^{33}]^h_{-h}, \qquad s^\alpha = [\tau^{3\alpha}x_3]^h_{-h}. \tag{7.2.8}$$

The function p is a surface invariant and p^α, s^α are surface contravariant tensors.

The components of $\mathbf{p}$, $\mathbf{s}$ referred to unit base vectors are denoted by $p_{(i)}$, $s_{(\alpha)}$ respectively and are called the physical components of the loads. Thus

$$\left.\begin{aligned} \mathbf{p} &= p_{(1)}\frac{\mathbf{a}_1}{\sqrt{a_{11}}} + p_{(2)}\frac{\mathbf{a}_2}{\sqrt{a_{22}}} + p_{(3)}\mathbf{a}_3 \\ \mathbf{s} &= s_{(2)}\frac{\mathbf{a}^1}{\sqrt{a^{11}}} + s_{(1)}\frac{\mathbf{a}^2}{\sqrt{a^{22}}} \end{aligned}\right\}. \tag{7.2.9}$$

Comparing these with (7.2.5) and (7.2.6) we obtain

$$p_{(\alpha)} = p^\alpha \sqrt{a_{\alpha\alpha}}, \qquad p_{(3)} = p, \tag{7.2.10}$$

$$s_{(1)} = s^1\sqrt{(aa^{22})} = s^1\sqrt{a_{11}}, \qquad s_{(2)} = -s^2\sqrt{(aa^{11})} = -s^2\sqrt{a_{22}}. \tag{7.2.11}$$

The physical components $p_{(\alpha)}$, $s_{(\alpha)}$ are not components of tensors and this is again indicated by enclosing the suffixes in a bracket.

7.3. Equations of equilibrium

In the absence of body forces the equations of equilibrium (5.2.6) are

$$\mathbf{T}_{i,i} = 0. \tag{7.3.1}$$

Equations of equilibrium in terms of stress resultants and stress couples may be obtained directly from (7.3.1). If we integrate (7.3.1) with respect to x_3 between the limits $\pm h$ we obtain

$$\int\limits_{-h}^{h} \mathbf{T}_{\alpha,\alpha}\, dx_3 + [\mathbf{T}_3]^h_{-h} = 0,$$

and using (7.1.4) and (7.2.4) this becomes

$$\mathbf{N}_{\alpha,\alpha}+\mathbf{P} = \mathbf{0}. \tag{7.3.2}$$

With the help of (7.1.7) and (7.2.5) equation (7.3.2) takes the form

$$N^{\alpha\beta}{}_{,\alpha}\mathbf{a}_\beta+N^{\alpha\beta}\mathbf{a}_{\beta,\alpha}+Q^\alpha{}_{,\alpha}\mathbf{a}_3+P^\beta\mathbf{a}_\beta+P\mathbf{a}_3 = \mathbf{0}. \tag{7.3.3}$$

From (6.1.4) $\mathbf{a}_{\beta,\alpha}$ can be expressed in terms of $\mathbf{a}_\lambda$, so, equating to zero the coefficients of $\mathbf{a}_i$ in (7.3.3), we have

$$\left.\begin{aligned} N^{\alpha\beta}{}_{,\alpha}+\Gamma^\beta_{\rho\alpha}N^{\alpha\rho}+P^\beta = 0 \\ Q^\alpha{}_{,\alpha}+P = 0 \end{aligned}\right\}. \tag{7.3.4}$$

Using (6.1.5) and (7.1.9) we see that the equations (7.3.4) may be reduced to

$$n^{\alpha\beta}|_\alpha+p^\beta = 0, \tag{7.3.5}$$

$$q^\alpha|_\alpha+p = 0, \tag{7.3.6}$$

where we have the usual notation

$$\left.\begin{aligned} n^{\alpha\beta}|_\alpha &= n^{\alpha\beta}{}_{,\alpha}+\Gamma^\beta_{\alpha\rho}n^{\alpha\rho}+\Gamma^\alpha_{\alpha\rho}n^{\rho\beta} \\ q^\alpha|_\alpha &= q^\alpha{}_{,\alpha}+\Gamma^\alpha_{\alpha\beta}q^\beta \end{aligned}\right\}, \tag{7.3.7}$$

for covariant differentiation with respect to coordinate curves $\theta_\alpha =$ constant in the middle plane.

We now take the vector product of (7.3.1) with $x_3\mathbf{a}_3$ and integrate with respect to x_3. After an integration by parts this gives

$$\int_{-h}^{h}[\mathbf{a}_3\times\mathbf{T}_{\alpha,\alpha}]x_3\,dx_3-\int_{-h}^{h}[\mathbf{a}_3\times\mathbf{T}_3]\,dx_3+[(\mathbf{a}_3\times\mathbf{T}_3)x_3]^h_{-h} = \mathbf{0}.$$

Hence, using the definitions (7.1.4), (7.1.9), (7.1.12), and (7.2.4) we have

$$\mathbf{M}_{\alpha,\alpha}-[\mathbf{a}_3\times\mathbf{a}_\alpha]Q^\alpha+\mathbf{S} = \mathbf{0}. \tag{7.3.8}$$

From (7.1.8) and (7.1.9)

$$\begin{aligned} \mathbf{M}_{\alpha,\alpha} &= M^{\alpha\rho}{}_{,\alpha}[\mathbf{a}_3\times\mathbf{a}_\rho]+M^{\alpha\rho}[\mathbf{a}_3\times\mathbf{a}_{\rho,\alpha}] \\ &= [\mathbf{a}_3\times\mathbf{a}_\rho](M^{\alpha\rho}{}_{,\alpha}+M^{\alpha\lambda}\Gamma^\rho_{\alpha\lambda}) \\ &= [\mathbf{a}_3\times\mathbf{a}_\rho]m^{\alpha\rho}|_\alpha\sqrt{a}, \end{aligned} \tag{7.3.9}$$

where, as before, covariant differentiation is with respect to coordinate curves in the middle plane. Hence, since

$$\mathbf{S} = [\mathbf{a}_3\times\mathbf{a}_\rho]s^\rho\sqrt{a},$$

we find from equations (7.3.8) and (7.3.9) that

$$m^{\alpha\beta}|_\alpha-q^\beta+s^\beta = 0. \tag{7.3.10}$$

Equations (7.3.5), (7.3.6), and (7.3.10) can also be obtained by first finding the results in rectangular cartesian coordinates and then generalizing to curvilinear coordinates by tensor rules. The above method of derivation, however, provides other formulae (e.g. (7.3.2) and (7.3.8)) which are sometimes useful.

For convenience we collect together the equations of equilibrium (7.3.5), (7.3.6), and (7.3.10) for stress resultants and couples:

$$\left.\begin{aligned} n^{\alpha\beta}|_\alpha + p^\beta &= 0 \\ q^\alpha|_\alpha + p &= 0 \\ m^{\alpha\beta}|_\alpha - q^\beta + s^\beta &= 0 \end{aligned}\right\}. \tag{7.3.11}$$

7.4. Stress–strain relations for isotropic plates

Stress–strain relations have been given in a variety of forms in Chapter 5, the most convenient for our purpose here being

$$2\mu\gamma_{rs} = \left\{g_{ir}\,g_{ks} - \frac{\eta}{1+\eta}\,g_{ik}\,g_{rs}\right\}\tau^{ik}, \tag{7.4.1}$$

or

$$2\mu\gamma^r_s = \tau^r_s - \frac{\eta}{1+\eta}\,\delta^r_s\,\tau^i_i, \tag{7.4.2}$$

where the strain tensor is given by

$$\gamma_{rs} = \tfrac{1}{2}(V_r|_s + V_s|_r). \tag{7.4.3}$$

The displacement vector is here denoted by

$$\mathbf{v} = V_r\,\mathbf{g}^r,$$

the components being denoted by V_r instead of v_r, since v_r is used for another purpose in this chapter.

When the metric tensor has the special form (6.1.4) we see that

$$\left.\begin{aligned} 2\mu\gamma_{\alpha\beta} &= \left(a_{\alpha\lambda}\,a_{\beta\mu} - \frac{\eta}{1+\eta}\,a_{\alpha\beta}\,a_{\lambda\mu}\right)\tau^{\lambda\mu} - \frac{\eta}{1+\eta}\,a_{\alpha\beta}\,\tau^{33} \\ 2\mu\gamma^\alpha_\beta &= \tau^\alpha_\beta - \frac{\eta}{1+\eta}\,\delta^\alpha_\beta(\tau^\lambda_\lambda + \tau^3_3) \end{aligned}\right\}, \tag{7.4.4}$$

$$\left.\begin{aligned} 2\mu\gamma_{\alpha 3} = a_{\alpha\lambda}\,\tau^{\lambda 3}, \qquad 2\mu\gamma^\alpha_3 &= \tau^\alpha_3 \\ 2\mu\gamma_{33} = \tau^{33} - \frac{\eta}{1+\eta}\,(a_{\alpha\beta}\,\tau^{\alpha\beta} &+ \tau^{33}) \end{aligned}\right\}, \tag{7.4.5}$$

and

$$\gamma_{\alpha\beta} = \tfrac{1}{2}(V_\alpha|_\beta + V_\beta|_\alpha), \tag{7.4.6}$$

$$\gamma_{\alpha 3} = \tfrac{1}{2}(V_{\alpha,3} + V_{3,\alpha}). \tag{7.4.7}$$

V_α is a covariant surface tensor, V_3 is a surface invariant, and the vertical line denotes covariant differentiation with respect to the (plane) surface coordinates θ_α using Christoffel symbols (6.1.5). In particular we observe that $V_3|_\alpha = V_{3,\alpha}$.

7.5. Stretching of isotropic plates: generalized plane stress

We consider a plane isotropic plate which is acted on by forces in its plane so that the plate is displaced symmetrically about its middle plane $x_3 = 0$. We also suppose that the faces of the plate are free from applied stress. Then, from (7.3.2) and (7.3.11),

$$\mathbf{N}_{\alpha,\alpha} = \mathbf{0} \quad \text{or} \quad n^{\alpha\beta}|_{\alpha} = 0. \tag{7.5.1}$$

The equation satisfied by $\mathbf{N}_\alpha$ is the same as the equation (6.2.3) satisfied by $\mathbf{T}_\alpha$ in plane strain when body forces are zero. Also, since the displacement is symmetrical about $x_3 = 0$,

$$q^\alpha = Q^\alpha = 0, \tag{7.5.2}$$

so that

$$\mathbf{N}_\alpha = N^{\alpha\beta}\mathbf{a}_\beta = \sqrt{(a)}n^{\alpha\rho}\mathbf{a}_\rho. \tag{7.5.3}$$

The relation (7.5.3) is of the same form as (6.2.2) if $\mathbf{T}_\alpha$ is replaced by $\mathbf{N}_\alpha$ and $\tau^{\alpha\beta}$ by $n^{\alpha\beta}$. We see therefore that all the work of § 6.2 on Airy's stress function when body forces are zero has its analogue here. Hence, quoting the corresponding formulae, we have

$$\mathbf{N}_\alpha = \sqrt{(a)}\epsilon^{\gamma\alpha}\boldsymbol{\chi}_{,\gamma} = \sqrt{(a)}\epsilon^{\gamma\alpha}\chi^\beta|_\gamma \mathbf{a}_\beta, \tag{7.5.4}$$

$$\left.\begin{aligned} n^{\alpha\beta} &= \epsilon^{\alpha\gamma}\epsilon^{\beta\rho}\phi|_{\gamma\rho} \\ n^\alpha_\beta &= \delta^{\alpha\gamma}_{\beta\rho}\phi|^\rho_\gamma \\ \phi|_{\alpha\beta} &= \epsilon_{\alpha\gamma}\epsilon_{\beta\rho}n^{\gamma\rho} \\ \phi|^\alpha_\beta &= \delta^{\alpha\gamma}_{\beta\rho}n^\rho_\gamma \end{aligned}\right\}, \tag{7.5.5}$$

where

$$\boldsymbol{\chi} = \chi^\beta\mathbf{a}_\beta = \epsilon^{\rho\beta}\phi_{,\rho}\mathbf{a}_\beta. \tag{7.5.6}$$

Also the resultant force and couple on any section of the plate normal to its plane are

$$\left.\begin{aligned} \mathbf{P} &= \boldsymbol{\chi} \\ M &= r^\alpha\phi_{,\alpha} - \phi \end{aligned}\right\}. \tag{7.5.7}$$

If we now integrate equations (7.4.4) with respect to x_3 through the thickness of the plate, and use (7.1.10) we have

$$\left.\begin{aligned} 2\mu\bar{\gamma}_{\alpha\beta} &= (a_{\alpha\lambda}a_{\beta\mu} - \sigma a_{\alpha\beta}a_{\lambda\mu})n^{\lambda\mu} - \sigma a_{\alpha\beta}q^3 \\ \text{or} \quad 2\mu\bar{\gamma}^\alpha_\beta &= n^\alpha_\beta - \sigma\delta^\alpha_\beta(n^\lambda_\lambda + q^3) \end{aligned}\right\}, \tag{7.5.8}$$

where

$$q^3 = \int_{-h}^{h} \tau^{33}\,dx_3, \qquad \bar{\gamma}_{\alpha\beta} = \tfrac{1}{2}\int_{-h}^{h} (V_\alpha|_\beta + V_\beta|_\alpha)\,dx_3, \tag{7.5.9}$$

and

$$\sigma = \frac{\eta}{1+\eta}. \tag{7.5.10}$$

Also, from (7.4.6),

$$\left.\begin{aligned}\bar{\gamma}_{\alpha\beta} &= \tfrac{1}{2}(v_\alpha|_\beta + v_\beta|_\alpha)\\ \bar{\gamma}^\alpha_\beta &= a^{\alpha\lambda}\bar{\gamma}_{\lambda\beta}\end{aligned}\right\}, \tag{7.5.11}$$

where

$$v_\alpha = \int\limits_{-h}^{h} V_\alpha\, dx_3. \tag{7.5.12}$$

In order to make further progress we must introduce some additional assumption. We suppose that the state of stress is such that q^3 is small compared with $n^{\alpha\beta}$, so that, from (7.5.8),

$$2\mu\bar{\gamma}^\alpha_\beta = n^\alpha_\beta - \sigma\delta^\alpha_\beta n^\lambda_\lambda, \tag{7.5.13}$$

which is seen to have the same form as equation (6.3.7) provided τ^α_β, γ^α_β, and the constant η are replaced by n^α_β, $\bar{\gamma}^\alpha_\beta$, and σ respectively. The plate is then said to be in a state of generalized plane stress.

We may now conclude that all the analysis for plane strain (in the absence of body forces) can be applied at once to generalized plane stress by a simple change of notation and by replacing η by $\sigma = \eta/(1+\eta)$, keeping μ unaltered.

7.6. Stretching of aeolotropic plates: generalized plane stress

We consider a plate acted on by the same system of forces as described in § 7.5 so that all the results (7.5.1) to (7.5.7) apply at once to the present case when the plate is no longer isotropic. We also restrict our attention to those plates which are elastically symmetrical about the middle plane $x_3 = 0$. It follows from (5.4.37) that, referred to cartesian axes x_i, the stress–strain relations are

$$e_{rs} = s^{ik}_{rs} t_{ik}, \tag{7.6.1}$$

where

$$s^{ik}_{rs} = s^{ki}_{rs} = s^{ki}_{sr} = s^{ik}_{sr} = s^{rs}_{ik}, \tag{7.6.2}$$

and, by analogy with (5.4.7),

$$s^{3\beta}_{\lambda\mu} = s^{\lambda\mu}_{3\beta} = s^{33}_{3\mu} = s^{3\mu}_{33} = 0. \tag{7.6.3}$$

The conditions (7.6.3) are satisfied because of symmetry about the plane $x_3 = 0$ and reduce the number of independent coefficients to 13. If we transform to general coordinates θ_α in the (x_1, x_2) plane, still retaining the third axis θ_3, then (7.6.1) becomes

$$\gamma^{ik} = F^{ik}_{rs}\tau^{rs}, \tag{7.6.4}$$

where F^{ik}_{rs} satisfies the conditions

$$\left.\begin{aligned}F^{ik}_{rs} = F^{ki}_{rs} = F^{ki}_{sr} = F^{ik}_{sr}\\ F^{3\beta}_{\lambda\mu} = F^{\lambda\mu}_{3\beta} = F^{33}_{3\mu} = F^{3\mu}_{33} = 0\end{aligned}\right\}. \tag{7.6.5}$$

By integrating equations (7.6.4) with respect to x_3 from $-h$ to h we find that

$$\bar{\gamma}^{\alpha\beta} = F^{\alpha\beta}_{\lambda\mu} n^{\lambda\mu} + F^{\alpha\beta}_{33} q^3, \tag{7.6.6}$$

where

$$\bar{\gamma}^{\alpha\beta} = a^{\alpha\lambda} a^{\beta\mu} \bar{\gamma}_{\lambda\mu},$$

and $\bar{\gamma}_{\lambda\mu}$ is given by (7.5.11). If we now neglect q^3 compared with the stress resultants $n^{\alpha\beta}$, equation (7.6.6) reduces to

$$\bar{\gamma}^{\alpha\beta} = F^{\alpha\beta}_{\lambda\mu} n^{\lambda\mu}. \tag{7.6.7}$$

This equation is identical in form with (6.7.11) which was obtained for plane strain provided we replace $\gamma^{\alpha\beta}$, $\tau^{\alpha\beta}$ by $\bar{\gamma}^{\alpha\beta}$, $n^{\alpha\beta}$, but it must be remembered that the coefficients $F^{\alpha\beta}_{\lambda\mu}$ (and $s^{\alpha\beta}_{\lambda\mu}$) have different values from those defined in § 6.7. From this point the theory may be developed in exactly the same way as the theory of § 6.7 for plane strain and results for plane strain are immediately applicable to generalized plane stress for plates by a simple change in the values of the elastic coefficients.

7.7. Transverse flexure of isotropic plates: Reissner's theory†

Since the equations of infinitesimal elasticity are linear, the stress distributions which are symmetric about the plane $x_3 = 0$ can be considered separately from those which are antisymmetric about $x_3 = 0$. Here we imagine that the plate is loaded in a general antisymmetric manner on its faces and we evaluate the stress-couples $m^{\alpha\beta}$ and the shear stresses q^α which are associated with the transverse bending of the plate. From (7.3.11) the relevant equations of equilibrium are

$$\left.\begin{aligned} m^{\alpha\beta}|_\alpha - q^\beta + s^\beta &= 0 \\ q^\alpha|_\alpha + p &= 0 \end{aligned}\right\}. \tag{7.7.1}$$

We also make use of one of the equations of equilibrium for stresses, namely,

$$\tau^{3\alpha}|_\alpha + \tau^{33}{}_{,3} = 0. \tag{7.7.2}$$

We assume that the shear stresses $\tau^{3\alpha}$ through the plate may be represented approximately by a quadratic expression in x_3, so that, remembering the boundary conditions (7.2.8) and the formulae (7.1.12), we put

$$\tau^{3\alpha} = \frac{3q^\alpha}{4h}\left(1 - \frac{x_3^2}{h^2}\right) - \frac{s^\alpha}{4h}\left(1 - \frac{3x_3^2}{h^2}\right). \tag{7.7.3}$$

† This theory was originally obtained by E. Reissner by a variational method: *J. Math. Phys.* **23** (1944) 184; *J. appl. Mech.* **12** (1945) 68; *Q. appl. Math.* **5** (1947) 55. An alternative derivation was given by A. E. Green, *Q. appl. Math.* **7** (1949) 223.

It follows from (7.7.2), the second of equations (7.7.1), and the boundary conditions (7.2.8) for τ^{33} that

$$\tau^{33} = \frac{3p}{4}\left(\frac{x_3}{h}-\frac{x_3^3}{3h^3}\right)+\frac{s^\alpha|_\alpha}{4}\left(\frac{x_3}{h}-\frac{x_3^3}{h^3}\right). \tag{7.7.4}$$

The stress–strain relations (7.4.4) and (7.4.5) may be solved for $\tau^{\alpha\beta}$, $\tau^{\alpha 3}$ in the form

$$\left.\begin{aligned}\tau^{\alpha\beta} &= 2\mu\left(\gamma^{\alpha\beta}+\frac{\eta}{1-\eta}a^{\alpha\beta}\gamma^\lambda_\lambda\right)+\frac{\eta}{1-\eta}a^{\alpha\beta}\tau^{33}\\ \tau^{\alpha 3} &= 2\mu\gamma^{\alpha 3}\end{aligned}\right\}, \tag{7.7.5}$$

where $$\gamma^{\alpha\beta} = \tfrac{1}{2}a^{\alpha\lambda}a^{\beta\mu}(V_\lambda|_\mu+V_\mu|_\lambda). \tag{7.7.6}$$

From the first of equations (7.7.5) we may evaluate $m^{\alpha\beta}$, which is defined by (7.1.11), in terms of certain 'weighted displacements'

$$w_\alpha = \frac{3}{2h^3}\int\limits_{-h}^{h} V_\alpha x_3\,dx_3, \qquad w^\alpha = a^{\alpha\lambda}w_\lambda, \tag{7.7.7}$$

where the definition (7.7.7) has been chosen so that if V_α is approximately proportional to x_3, then the factor of proportionality is equal to w_α. Thus, using also (7.7.4),

$$m^{\alpha\beta} = \frac{2\mu h^3}{3}\left\{a^{\alpha\lambda}a^{\beta\mu}(w_\lambda|_\mu+w_\mu|_\lambda)+\frac{2\eta}{1-\eta}a^{\alpha\beta}w^\lambda|_\lambda\right\}+ \\ +\frac{\eta a^{\alpha\beta}}{1-\eta}\left(\frac{2h^2p}{5}+\frac{h^2s^\lambda|_\lambda}{15}\right). \tag{7.7.8}$$

The second of equations (7.7.5) may be written

$$\tau^{\alpha 3} = \mu a^{\alpha\lambda}(V_{3,\lambda}+V_{\lambda,3}). \tag{7.7.9}$$

We have already introduced 'weighted displacements' (7.7.7) and this suggests that equation (7.7.9) should be multiplied by a term x_3^2 and then integrated through the plate, since, after an integration by parts, we reproduce the weighted displacements w_α. This process, however, also introduces unknown values of V_λ at the faces of the plate, so to avoid this we multiply (7.7.9) by $1-x_3^2/h^2$ before integration. Then, using (7.7.3) we obtain, after an integration by parts,

$$q^\alpha = \tfrac{1}{5}s^\alpha+\frac{5\mu h a^{\alpha\lambda}}{3}(w_{,\lambda}+w_\lambda), \tag{7.7.10}$$

where $$w = \frac{3}{4h}\int\limits_{-h}^{h}\left(1-\frac{x_3^2}{h^2}\right)V_3\,dx_3. \tag{7.7.11}$$

The numerical factor outside the integral in (7.7.11) has been chosen so that if V_3 is constant then this constant is equal to w.

From (7.7.10)
$$w^\lambda = -w|^\lambda + \frac{3}{5\mu h}\left(q^\lambda - \frac{s^\lambda}{6}\right), \tag{7.7.12}$$

and, if we substitute this in (7.7.8) and use (7.7.1), we have

$$m^{\alpha\beta} = -\frac{4\mu h^3}{3}\left(w|^{\alpha\beta} + \frac{\eta}{1-\eta}a^{\alpha\beta}w|^\lambda_\lambda\right) + \\ + \frac{h^2}{15}(6q^\alpha|^\beta + 6q^\beta|^\alpha - s^\alpha|^\beta - s^\beta|^\alpha) - \frac{\eta h^2 a^{\alpha\beta}(6p + s^\lambda|_\lambda)}{15(1-\eta)}. \tag{7.7.13}$$

From this and from equations (7.7.1) we obtain

$$q^\beta - \frac{2h^2}{5}q^\beta|^\alpha_\alpha = s^\beta - \frac{h^2}{15(1-\eta)}\{6p|^\beta + s^\alpha|^\beta_\alpha + (1-\eta)s^\beta|^\alpha_\alpha\} - \frac{4\mu h^3}{3(1-\eta)}w|^{\alpha\beta}_\alpha, \tag{7.7.14}$$

and differentiating this covariantly with respect to β and again using (7.7.1) gives

$$\frac{4\mu h^3}{3(1-\eta)}w|^{\alpha\beta}_{\alpha\beta} - p - s^\alpha|_\alpha + \frac{(2-\eta)h^2(6p|^\alpha_\alpha + s^\alpha|^\beta_{\alpha\beta})}{15(1-\eta)} = 0. \tag{7.7.15}$$

In many problems the derivatives of any order of the loads p and s^α with respect to θ_β are of the same order of magnitude in the thickness of the plate as the loads themselves, so that, if h is small, we may write (7.7.15) and (7.7.14) approximately as

$$\left.\begin{aligned} Dw|^{\alpha\beta}_{\alpha\beta} - p - s^\alpha|_\alpha &= 0 \\ q^\beta - \frac{2h^2}{5}q^\beta|^\alpha_\alpha &= s^\beta - Dw|^{\alpha\beta}_\alpha \end{aligned}\right\}, \tag{7.7.16}$$

where
$$D = \frac{4\mu h^3}{3(1-\eta)}. \tag{7.7.17}$$

Also (7.7.13) and (7.7.12) become

$$m^{\alpha\beta} = -D\{(1-\eta)w|^{\alpha\beta} + \eta a^{\alpha\beta}w|^\lambda_\lambda\} + \frac{2h^2}{5}(q^\alpha|^\beta + q^\beta|^\alpha), \tag{7.7.18}$$

$$w^\lambda = -w|^\lambda + \frac{3q^\lambda}{5\mu h}. \tag{7.7.19}$$

This approximation is equivalent to neglecting τ^{33} in the stress–strain relations.

The second equation in (7.7.1) shows that we may express q^β in the form
$$q^\beta = R|^{\beta\lambda}_\lambda + \epsilon^{\beta\lambda}\chi_{,\lambda}, \tag{7.7.20}$$

where χ is a scalar and R is a particular integral of the equation

$$p = -R|^{\beta\lambda}_{\beta\lambda}. \tag{7.7.21}$$

If we substitute (7.7.20) in (7.7.16) we find that

$$\epsilon^{\beta\lambda}\left(\chi-\frac{2h^2}{5}\chi|^{\alpha}_{\alpha}\right)\Big|_{\lambda}+R|^{\beta\lambda}_{\lambda} = s^{\beta}-Dw|^{\alpha\beta}_{\alpha}, \tag{7.7.22}$$

where we have neglected $\frac{2}{5}h^2p|^{\beta}$ compared with $R|^{\beta\lambda}_{\lambda}$. Alternatively

$$\left(\chi-\frac{2h^2}{5}\chi|^{\alpha}_{\alpha}\right)\Big|_{\lambda} = \epsilon_{\beta\lambda}(s^{\beta}-R|^{\beta\alpha}_{\alpha}-Dw|^{\alpha\beta}_{\alpha}), \tag{7.7.23}$$

and hence

$$\left(\chi-\frac{2h^2}{5}\chi|^{\alpha}_{\alpha}\right)\Big|^{\lambda}_{\lambda} = \epsilon_{\beta\lambda}\, s^{\beta}|^{\lambda}, \tag{7.7.24}$$

since

$$\epsilon_{\beta\lambda}(R|^{\beta\alpha\lambda}_{\alpha}+Dw|^{\alpha\beta\lambda}_{\alpha}) = 0.$$

In the rest of the chapter we restrict attention to problems for which s^{β} is zero. Equation (7.7.24) then shows that χ satisfies either of the equations

$$\chi|^{\lambda}_{\lambda} = 0, \qquad \chi-\frac{2h^2}{5}\chi|^{\lambda}_{\lambda} = 0.$$

We now replace equation (7.7.20) by

$$q^{\beta} = R|^{\beta\lambda}_{\lambda}+\epsilon^{\beta\lambda}\psi_{,\lambda}+\phi|^{\beta}, \tag{7.7.25}$$

where ψ and ϕ are scalars,† and

$$\left.\begin{aligned} \chi_{,\lambda} &= \psi_{,\lambda}+\epsilon_{\beta\lambda}\phi|^{\beta} \\ \phi|^{\lambda}_{\lambda} &= 0 \\ \psi-\frac{2h^2}{5}\psi|^{\lambda}_{\lambda} &= 0 \end{aligned}\right\}. \tag{7.7.26}$$

Remembering that $s^{\beta} = 0$, equation (7.7.22) becomes

$$\phi|^{\beta}+R|^{\beta\lambda}_{\lambda} = -Dw|^{\alpha\beta}_{\alpha},$$

and therefore

$$Dw|^{\alpha}_{\alpha} = -\phi-R|^{\alpha}_{\alpha}, \tag{7.7.27}$$

if we absorb an arbitrary constant in the function ϕ. Also, from (7.7.18) and (7.7.25), neglecting the term of order $h^2R|^{\alpha\beta\lambda}_{\lambda}$,

$$m^{\alpha\beta} = -D\{(1-\eta)w|^{\alpha\beta}+\eta a^{\alpha\beta}w|^{\lambda}_{\lambda}\}+\frac{2h^2}{5}(2\phi|^{\alpha\beta}+\epsilon^{\alpha\lambda}\psi|^{\beta}_{\lambda}+\epsilon^{\beta\lambda}\psi|^{\alpha}_{\lambda}),$$

but, since ϕ and h^3w are the same order of magnitude in the thickness h, we may write this approximately as

$$m^{\alpha\beta} = -D\{(1-\eta)w|^{\alpha\beta}+\eta a^{\alpha\beta}w|^{\lambda}_{\lambda}\}+\frac{2h^2}{5}(\epsilon^{\alpha\lambda}\psi|^{\beta}_{\lambda}+\epsilon^{\beta\lambda}\psi|^{\alpha}_{\lambda}). \tag{7.7.28}$$

† ϕ need not be confused with Airy's stress function used earlier in this chapter for the stretching of plates.

With the help of (7.7.27) the equation (7.7.25) for q^β may be written in the alternative form

$$q^\beta = \epsilon^{\beta\lambda}\psi_{,\lambda} - Dw|^{\beta\lambda}_{\lambda}. \tag{7.7.29}$$

Finally, putting $s^\lambda = 0$, we may write (7.7.19) approximately in the form

$$w^\beta = -w|^\beta + \frac{3\epsilon^{\beta\lambda}\psi_{,\lambda}}{5\mu h}, \tag{7.7.30}$$

where we have neglected terms of the type $h^2 w|^{\beta\lambda}_{\lambda}$ compared with $w|^\beta$.

For transverse flexure of the plate we see, from (7.1.6) and (7.1.7), that the force per unit length exerted across any curve in the plate whose unit normal is given by (7.1.5) is†

$$\mathbf{n} = q^\alpha u_\alpha \mathbf{a}_3, \qquad u_\alpha = \epsilon_{\alpha\beta}\frac{d\theta^\beta}{ds}, \tag{7.7.31}$$

and this is a force of magnitude

$$N = q^\alpha u_\alpha$$

normal to the plate. The resultant force normal to the plate of the forces on an arc AP of a contour c in the plate, acting on the material outside c (see Fig. 6.1), is therefore of magnitude‡

$$F = \int_A^P q^\alpha u_\alpha \, ds, \tag{7.7.32}$$

which, with (7.7.20), becomes

$$F = \int_A^P R|^{\lambda\alpha}_{\lambda} u_\alpha \, ds + \int_A^P \epsilon^{\alpha\lambda}\epsilon_{\alpha\beta}\,\chi_{,\lambda}\frac{d\theta^\beta}{ds}\, ds$$

$$= \int_A^P R|^{\lambda\alpha}_{\lambda} u_\alpha \, ds + \int_A^P \chi_{,\beta}\, d\theta^\beta,$$

or

$$F = \int_A^P R|^{\lambda\alpha}_{\lambda} u_\alpha \, ds + \chi, \tag{7.7.33}$$

apart from an arbitrary constant which may be absorbed by χ since this does not affect the stress distribution.

If the distribution of load p on and inside a contour c is continuous, then we can choose an integral R of (7.7.21) which is continuous and which has continuous derivatives of the fourth order. The resultant force on the complete contour c then becomes, by Green's theorem,

$$F = [\chi]_A^A - \int_S p \, dS. \tag{7.7.34}$$

† $n^{\alpha\rho} = 0$ here.

‡ The force is directed along $-\mathbf{a}_3$.

The second term in (7.7.34) represents the total load on the portion of the plate enclosed by c so that, in this case,

$$[\chi]_A^A = 0.$$

If, however, we can find a neighbourhood of the contour c throughout which there is zero load, we can choose the particular integral R of (7.7.21) to be identically zero throughout this neighbourhood, so that the resultant force on the contour c becomes

$$F = [\chi]_A^A. \tag{7.7.35}$$

If the total load inside the contour c is non-zero, then $[\chi]_A^A$ no longer vanishes. For example, the load distribution may be zero everywhere inside c except at an isolated point at which an isolated load acts. The corresponding value of R is zero everywhere, except at the isolated load, and the function χ must contain a singularity at the load which is such that $-[\chi]_A^A$, taken around a contour c enclosing the load, is equal to the load.

The moment about axes at the origin of the forces and couples, exerted by the region 2 on the region 1 across AP (Fig. 6.1), is

$$\mathbf{K} = \int\limits_A^P (\mathbf{m}+\mathbf{r}\times\mathbf{n})\, ds,$$

where, from (7.1.6), (7.1.8), and (7.7.31),

$$\begin{aligned}\mathbf{m}+\mathbf{r}\times\mathbf{n} &= u_\alpha m^{\alpha\lambda}\epsilon_{\lambda\rho}\mathbf{a}^\rho + r^\lambda q^\alpha u_\alpha \mathbf{a}_\lambda\times\mathbf{a}_3\\ &= u_\alpha \epsilon_{\lambda\rho}(m^{\alpha\lambda}-q^\alpha r^\lambda)\mathbf{a}^\rho\\ &= u_\alpha(m^{\alpha\lambda}-q^\alpha r^\lambda)a_{\lambda\rho}\epsilon^{\rho\mu}\mathbf{a}_\mu,\end{aligned}$$

so that

$$\mathbf{K} = \int\limits_A^P \delta^{\rho\mu}_{\alpha\beta}(m^{\alpha\lambda}-q^\alpha r^\lambda)a_{\lambda\rho}\mathbf{a}_\mu\, d\theta^\beta. \tag{7.7.36}$$

7.8. Notation

As mentioned in Chapter 6 we frequently abandon tensor notations in applications to special problems and we use rectangular coordinaets x, y, Z, and curvilinear coordinates (ξ, η) in the (x, y)-plane. Also complex variable notations are again used so that

$$z = x+iy, \quad \bar{z} = x-iy, \quad \zeta = \xi+i\eta, \quad \bar{\zeta} = \xi-i\eta, \tag{7.8.1}$$

and we suppose that z, ζ are connected by the transformation

$$z = z(\zeta), \qquad \frac{dz}{d\zeta} = Je^{i\epsilon}. \tag{7.8.2}$$

For convenience we repeat the formulae (6.4.9), (6.5.11), (6.5.12), and (6.5.14):

$$\left.\begin{aligned} \Theta &= \sigma_{xx}+\sigma_{yy}, \quad \Phi = \sigma_{xx}-\sigma_{yy}+2i\sigma_{xy}, \quad \Psi = \sigma_{xz}+i\sigma_{yz} \\ \Theta' &= \sigma_{\xi\xi}+\sigma_{\eta\eta}, \quad \Phi' = \sigma_{\xi\xi}-\sigma_{\eta\eta}+2i\sigma_{\xi\eta}, \quad \Psi' = \sigma_{\xi z}+i\sigma_{\eta z} \\ &\Theta' = \Theta, \qquad \Phi' = \Phi e^{-2i\epsilon} \\ \Psi' &= \Psi e^{-i\epsilon}, \qquad u_\xi+iu_\eta = (u_x+iu_y)e^{-i\epsilon} \end{aligned}\right\}. \tag{7.8.3}$$

In rectangular coordinates x, y, Z the stress couples $m^{\alpha\beta}$ and shear stresses q^α are denoted by

$$\left.\begin{aligned} G_x &= \int_{-h}^{h} \sigma_{xx} Z\,dZ, \quad G_y = \int_{-h}^{h} \sigma_{yy} Z\,dZ, \quad H_{xy} = \int_{-h}^{h} \sigma_{xy} Z\,dZ \\ N_x &= \int_{-h}^{h} \sigma_{xz}\,dZ, \qquad N_y = \int_{-h}^{h} \sigma_{yz}\,dZ \end{aligned}\right\}. \tag{7.8.4}$$

The corresponding (physical) stress couples and shear stresses in curvilinear coordinates are

$$\left.\begin{aligned} G_\xi &= \int_{-h}^{h} \sigma_{\xi\xi} Z\,dZ, \quad G_\eta = \int_{-h}^{h} \sigma_{\eta\eta} Z\,dZ, \quad H_{\xi\eta} = \int_{-h}^{h} \sigma_{\xi\eta} Z\,dZ \\ N_\xi &= \int_{-h}^{h} \sigma_{\xi z}\,dZ, \qquad N_\eta = \int_{-h}^{h} \sigma_{\eta z}\,dZ \end{aligned}\right\}. \tag{7.8.5}$$

We also use the abbreviations

$$\left.\begin{aligned} \Lambda &= \int_{-h}^{h} \Theta Z\,dZ = G_x+G_y \\ \Gamma &= \int_{-h}^{h} \Phi Z\,dZ = G_x-G_y+2iH_{xy} \\ \Psi_0 &= \int_{-h}^{h} \Psi\,dZ = N_x+iN_y \end{aligned}\right\}, \tag{7.8.6}$$

$$\left.\begin{aligned} \Lambda' &= \int_{-h}^{h} \Theta' Z\,dZ = G_\xi+G_\eta \\ \Gamma' &= \int_{-h}^{h} \Phi' Z\,dZ = G_\xi-G_\eta+2iH_{\xi\eta} \\ \Psi_0' &= \int_{-h}^{h} \Psi'\,dZ = N_\xi+iN_\eta \end{aligned}\right\}, \tag{7.8.7}$$

so that $$\Lambda' = \Lambda, \qquad \Gamma' = \Gamma e^{-2i\epsilon}, \qquad \Psi_0' = \Psi_0 e^{-i\epsilon}. \tag{7.8.8}$$

7.9. Integration in terms of complex potentials†

The equations for the small transverse deflexion of an isotropic plate may be partly integrated in terms of two complex potentials. Thus, using the notation of § 6.4, the second of equations (7.7.26) may be written

$$\frac{\partial^2 \phi}{\partial z \partial \bar{z}} = 0,$$

and therefore
$$\phi = 4D\{\Omega'(z)+\bar{\Omega}'(\bar{z})\}, \tag{7.9.1}$$

where $\Omega(z)$ is a general complex function of z and the constant factor has been chosen for convenience in later stages of the integration.
From (7.7.27) and (7.9.1), using § 6.4, we obtain

$$\frac{\partial^2 w}{\partial z \partial \bar{z}} = -\frac{1}{D}\frac{\partial^2 R}{\partial z \partial \bar{z}} - \Omega'(z) - \bar{\Omega}'(\bar{z}),$$

and therefore
$$w = -R/D - z\bar{\Omega}(\bar{z}) - \bar{z}\Omega(z) - \omega(z) - \bar{\omega}(\bar{z}), \tag{7.9.2}$$

where $\omega(z)$ is an arbitrary function of z.

We are now able to write down the complex components of the stress couples $m^{\alpha\beta}$, shear stresses q^{α}, and displacements by interpreting (7.7.28), (7.7.29), and (7.7.30) in complex coordinates z, $\bar{z}$, and we may then transform to general complex coordinates ζ, $\bar{\zeta}$ by an exactly similar process to that used in § 6.5 for plane strain. Using the metric components (6.4.3) which correspond to complex coordinates z, $\bar{z}$ and observing that the components of $m^{\alpha\beta}$ in these coordinates are Γ, Λ, $\bar{\Gamma}$ we have, from (7.7.28), (7.9.1), and (7.9.2),

$$\Lambda = -4D(1+\eta)\frac{\partial^2 w}{\partial z \partial \bar{z}} = (1+\eta)\nabla_1^2 R + 4D(1+\eta)\{\Omega'(z)+\bar{\Omega}'(\bar{z})\}, \tag{7.9.3}$$

$$\Gamma = -4D(1-\eta)\frac{\partial^2 w}{\partial \bar{z}^2} - \frac{16h^2 i}{5}\frac{\partial^2 \psi}{\partial \bar{z}^2}$$

$$= 4(1-\eta)\frac{\partial^2 R}{\partial \bar{z}^2} + 4D(1-\eta)\{z\bar{\Omega}''(\bar{z})+\bar{\omega}''(\bar{z})\} - \frac{16h^2 i}{5}\frac{\partial^2 \psi}{\partial \bar{z}^2}, \tag{7.9.4}$$

where, in (7.9.3), ∇_1^2 denotes the two-dimensional Laplace operator $4(\partial^2/\partial z \partial \bar{z})$. Also, from (7.7.29) and (7.9.2),

$$\Psi_0 = -2i\frac{\partial \psi}{\partial \bar{z}} - 2D\frac{\partial \nabla_1^2 w}{\partial \bar{z}} = -2i\frac{\partial \psi}{\partial \bar{z}} + 2\frac{\partial \nabla_1^2 R}{\partial \bar{z}} + 8D\bar{\Omega}''(\bar{z}), \tag{7.9.5}$$

† See A. E. Green, loc. cit., p. 222.

and, denoting the complex components of w^{α} by $\bar{u}_x \pm i\bar{u}_y$, we have, from (7.7.30),

$$\bar{u}_x + i\bar{u}_y = -2\frac{\partial w}{\partial \bar{z}} - \frac{6i}{5\mu h}\frac{\partial \psi}{\partial \bar{z}}. \tag{7.9.6}$$

If we interpret (7.7.26) in complex coordinates and use (7.9.1), we have

$$\frac{\partial \chi}{\partial z} = \frac{\partial \psi}{\partial z} - 4iD\Omega''(z),$$

$$\frac{\partial \chi}{\partial \bar{z}} = \frac{\partial \psi}{\partial \bar{z}} + 4iD\bar{\Omega}''(\bar{z}),$$

so that

$$\chi = \psi + 4iD\{\bar{\Omega}'(\bar{z}) - \Omega'(z)\}, \tag{7.9.7}$$

apart from an arbitrary additional constant.

The resultant force (7.7.33) on an arc AP of a contour becomes, in terms of complex coordinates,

$$F = 4i\int_A^P \left(\frac{\partial^3 R}{\partial z \partial \bar{z}^2}\,d\bar{z} - \frac{\partial^3 R}{\partial z^2 \partial \bar{z}}\,dz\right) + \psi + 4iD\{\bar{\Omega}'(\bar{z}) - \Omega'(z)\}. \tag{7.9.8}$$

If the arc AP lies in a neighbourhood of the plate throughout which the load is zero and the corresponding value of R vanishes, then (7.9.8) reduces to

$$F = \psi + 4iD\{\bar{\Omega}'(\bar{z}) - \Omega'(z)\}. \tag{7.9.9}$$

Again, if the components of the couple $\mathbf{K}$ about the x- and y-axes are (L, L'), and if $\mathbf{A}_\alpha$ are the complex covariant base vectors, then

$$\mathbf{K} = (L + iL')\mathbf{A}_1 + (L - iL')\mathbf{A}_2 = K\mathbf{A}_1 + \bar{K}\mathbf{A}_2, \tag{7.9.10}$$

and, from (7.7.36) and (6.4.2),

$$K = L + iL' = \tfrac{1}{2}\int_A^P \{(\Lambda - z\bar{\Psi}_0)\,dz - (\Gamma - z\Psi_0)\,d\bar{z}\}. \tag{7.9.11}$$

Substituting for Λ, Γ, Ψ_0 from (7.9.3)–(7.9.5) gives

$$K = \int_A^P \left\{2(1+\eta)\frac{\partial^2 R}{\partial z \partial \bar{z}} - 4z\frac{\partial^3 R}{\partial z^2 \partial \bar{z}}\right\} dz - \int_A^P \left\{2(1-\eta)\frac{\partial^2 R}{\partial \bar{z}^2} - 4z\frac{\partial^3 R}{\partial z \partial \bar{z}^2}\right\} d\bar{z} +$$

$$+ D\int_A^P [2(1+\eta)\{\Omega'(z) + \bar{\Omega}'(\bar{z})\} - 4z\Omega''(z)]\,dz -$$

$$- D\int_A^P [2(1-\eta)\{z\bar{\Omega}''(\bar{z}) + \bar{\omega}''(\bar{z})\} - 4z\bar{\Omega}''(\bar{z})]\,d\bar{z} -$$

$$- i\int_A^P \left\{z\frac{\partial \psi}{\partial z}\,dz + z\frac{\partial \psi}{\partial \bar{z}}\,d\bar{z} - \frac{8h^2}{5}\frac{\partial^2 \psi}{\partial \bar{z}^2}\,d\bar{z}\right\},$$

and apart from an arbitrary additional constant† this reduces to

$$K = -2(1-\eta)\frac{\partial R}{\partial \bar{z}} + 4z\frac{\partial^2 R}{\partial z \partial \bar{z}} - 8\int\limits_A^P z\frac{\partial^3 R}{\partial z^2 \partial \bar{z}}\,dz +$$

$$+2D\{(3+\eta)\Omega(z) - 2z\Omega'(z) + (1+\eta)z\bar{\Omega}'(\bar{z}) - (1-\eta)\bar{\omega}'(\bar{z})\} +$$

$$+i\left(\frac{8h^2}{5}\frac{\partial\psi}{\partial\bar{z}} - z\psi\right), \tag{7.9.12}$$

if we use the last equation in (7.7.26) in the form

$$\psi - \frac{8h^2}{5}\frac{\partial^2\psi}{\partial z \partial \bar{z}} = 0. \tag{7.9.13}$$

If the distribution of load p over the interior of a contour c is continuous then the complex couple K exerted across c by the material outside reduces to

$$K = i\int\limits_S pz\,dS + i\left[\frac{8h^2}{5}\frac{\partial\psi}{\partial\bar{z}} - z\psi\right]_A^A +$$

$$+2D[(3+\eta)\Omega(z) - 2z\Omega'(z) + (1+\eta)z\bar{\Omega}'(\bar{z}) - (1-\eta)\bar{\omega}'(\bar{z})]_A^A. \tag{7.9.14}$$

The first term is an integral over the area enclosed by the contour c and represents the (complex) moment of the load p about the axes.

If the arc AP lies entirely in a neighbourhood of the plate which has no load on its face, then $R = 0$ and the complex couple (7.9.12) becomes

$$K = i\left(\frac{8h^2}{5}\frac{\partial\psi}{\partial\bar{z}} - z\psi\right) +$$

$$+2D\{(3+\eta)\Omega(z) - 2z\Omega'(z) + (1+\eta)z\bar{\Omega}'(\bar{z}) - (1-\eta)\bar{\omega}'(\bar{z})\}. \tag{7.9.15}$$

The above theory expresses the stress couples, shear stresses, and displacements in terms of two harmonic functions $\Omega(z)$, $\omega(z)$ and a function ψ which satisfies equation (7.9.13), so that on each boundary in the plane of the plate three boundary conditions are required. For example, if the displacement is given on any boundary, then $\bar{u}_x + i\bar{u}_y$ and w take prescribed values. On the other hand, if the applied stresses are prescribed on the curve ξ = constant, then

$$\Psi_0' + \bar{\Psi}_0' = 2N_\xi, \qquad \Lambda' + \Gamma' = 2(G_\xi + iH_{\xi\eta}) \tag{7.9.16}$$

† This constant can be absorbed into the complex potentials. See § 7.10.

are known on this curve. Alternatively, we may express these boundary conditions in terms of the resultant force and couple exerted across an arc AP of a contour, using (7.9.8) and (7.9.12). It is, however, more convenient to replace (7.9.12) by

$$K+izF = 2D[(3+\eta)\Omega(z)-(1-\eta)\{z\overline{\Omega}'(\bar{z})+\bar{\omega}'(\bar{z})\}]+$$

$$+\frac{8ih^2}{5}\frac{\partial\psi}{\partial\bar{z}}-2(1-\eta)\frac{\partial R}{\partial\bar{z}}-8\int_A^P z\frac{\partial^3 R}{\partial z^2\partial\bar{z}}\,dz+8z\int_A^P \frac{\partial^3 R}{\partial z^2\partial\bar{z}}\,dz. \quad (7.9.17)$$

If the plate is unloaded on its faces and a single contour c in the plate is free from applied stresses, then from (7.9.8) and (7.9.17) we have

$$\left.\begin{aligned} \psi+4iD\{\overline{\Omega}'(\bar{z})-\Omega'(z)\} &= 0 \\ \frac{4h^2i}{5}\frac{\partial\psi}{\partial\bar{z}}+D[(3+\eta)\Omega(z)-(1-\eta)\{z\overline{\Omega}'(\bar{z})+\bar{\omega}'(\bar{z})\}] &= 0 \end{aligned}\right\}. \quad (7.9.18)$$

7.10. Properties of the complex potentials

If the complex potential $\Omega(z)$ is replaced by $\Omega(z)+iCz+\alpha+i\beta$ and $\omega(z)$ is replaced by $\omega(z)+(\alpha'+i\beta')z+\alpha''+i\beta''$, where α, α', β, β', α'', β'', C are real constants, the stress system given by (7.9.3) to (7.9.5) is unaltered, but, from (7.9.2) and (7.9.6), w and $\bar{u}_x+i\bar{u}_y$ become

$$w = \text{original } w-(\alpha+\alpha'-i\beta+i\beta')z-(\alpha+\alpha'+i\beta-i\beta')\bar{z}-2\alpha'',$$

$$\bar{u}_x+i\bar{u}_y = \text{original } (\bar{u}_x+i\bar{u}_y)+2(\alpha+\alpha'+i\beta-i\beta'),$$

the additional displacement representing a rigid-body displacement of the whole plate. If the displacement is given everywhere not all the constants α, α', α'', β, β' are disposable and only one of the pairs α, β and α', β' can be chosen arbitrarily. The constants C, β'' can always be chosen arbitrarily since they do not affect the stresses or displacements, and we shall usually replace these constants by zeros.

We assume that the load on the plate is distributed over a finite number of non-overlapping simply-connected regions, so that p is continuous inside each region and zero outside. Inside each loaded region the function R, which is a particular integral of equation (7.7.21), together with its derivatives up to the fourth order, may be chosen to be continuous and single-valued. Outside these regions R is identically zero. If follows from (7.9.2) to (7.9.6) that, if the stresses and displacements are to be single-valued, then

$$\left.\begin{aligned} \left[\frac{\partial\psi}{\partial z}\right]_A^A = 0, \quad \left[\frac{\partial\psi}{\partial \bar{z}}\right]_A^A = 0, \quad \left[\frac{\partial^2\psi}{\partial \bar{z}^2}\right]_A^A = 0 \\ [\Omega''(z)]_A^A = 0, \quad [\Omega'(z)+\bar{\Omega}'(\bar{z})]_A^A = 0, \quad [\omega''(z)]_A^A = 0 \\ [\omega(z)+\bar{\omega}(\bar{z})+z\bar{\Omega}(\bar{z})+\bar{z}\Omega(z)]_A^A = 0 \\ [\omega'(z)+\bar{\Omega}(\bar{z})+\bar{z}\Omega'(z)]_A^A = 0 \end{aligned}\right\}, \tag{7.10.1}$$

where $[\,]_A^A$ denotes the change in value of the function around any contour lying entirely inside a loaded region of the plate, or entirely inside an unloaded region of the plate. Also, partial derivatives of the above functions, if they exist, satisfy the same conditions. Hence

$$\left[\frac{\partial^2\psi}{\partial z\partial\bar{z}}\right]_A^A = 0,$$

and since, from (7.7.26), at points inside the plate,

$$\psi - \frac{8h^2}{5}\frac{\partial^2\psi}{\partial z\partial\bar{z}} = 0,$$

it follows that $[\psi]_A^A = 0$. Therefore ψ and its derivatives up to the second order (at least) must be single-valued in each loaded or unloaded region of the plate.

When the above conditions for single-valued stresses and displacements are satisfied, the resultant force and couple on a contour c lying entirely in a loaded region of the plate become, from § 7.9,

$$\left.\begin{aligned} F &= -\int\limits_S p\,dS - 8iD[\Omega'(z)]_A^A \\ K &= i\int\limits_S pz\,dS - 8D[\bar{\omega}'(\bar{z})]_A^A \end{aligned}\right\}, \tag{7.10.2}$$

if the load is continuously distributed over all the interior of c. If the contour c lies entirely in an unloaded part of the plate, the corresponding results are

$$\left.\begin{aligned} F &= -8iD[\Omega'(z)]_A^A \\ K &= -8D[\bar{\omega}'(\bar{z})]_A^A \end{aligned}\right\}. \tag{7.10.3}$$

When the plate S is finite and simply-connected and has a continuous distribution of load over its entire surface, then we suppose that $\Omega(z)$, $\omega(z)$ are holomorphic in the *open* region S, and that ψ and its partial derivatives up to the second order are continuous and single-valued in the open region S; the stresses and displacements are then single-valued. The case when the load is distributed discontinuously over a finite simply-connected plate may be included in the following discussion on multiply-connected regions.

When S is multiply-connected, suppose S is bounded by one or more smooth non-intersecting contours $L_0, L_1, \ldots, L_n$ of which L_0 contains all the others. The contour L_0 may be absent (i.e. it may be reducible to the point at infinity) in which case the region S will be infinite with contours $L_1, L_2, \ldots, L_n$ as internal boundaries. For the present we assume that the region S is unloaded. If we require the stresses and displacements to be single-valued in the open region S, then, in view of (7.10.1), we assume that ψ and its derivatives up to the second order are continuous and single-valued and

$$\Omega''(z) = \sum_{r=1}^{n} \frac{a_r+ib_r}{2\pi(z-z_r)} + \psi_0''(z),$$

where the point z_r is inside the contour L_r (and therefore outside S) and where $\psi_0'(z)$ is holomorphic in S except possibly at infinity. Also a_r, b_r are real constants. Hence

$$\Omega'(z) = \sum_{r=1}^{n} \frac{a_r+ib_r}{2\pi}\ln(z-z_r)+\psi_0'(z),$$

and since $[\Omega'(z)+\bar{\Omega}'(\bar{z})]_A^A = 0$ it follows that $b_r = 0$ and

$$\Omega(z) = \sum_{r=1}^{n} \frac{a_r z}{2\pi}\ln(z-z_r)+\sum_{r=1}^{n} \frac{c_r+id_r}{2\pi}\ln(z-z_r)+\Omega_0(z), \quad (7.10.4)$$

where c_r, d_r are real constants and $\Omega_0(z)$ is holomorphic in S except possibly at infinity. Using the last but one result in (7.10.1) we find that

$$\omega(z) = \sum_{r=1}^{n} \frac{c_r-id_r}{2\pi} z\ln(z-z_r)+\sum_{r=1}^{n} \frac{a_r'}{2\pi}\ln(z-z_r)+\omega_0(z), \quad (7.10.5)$$

where a_r' are real constants and $\omega_0(z)$ is holomorphic in S except possibly at infinity.

Around any contour L_r' enclosing L_r and not cutting the other contours, $\Omega'(z)$ increases by ia_r and $\bar{\omega}'(\bar{z})$ increases by $-i(c_r+id_r)$. Now let the contours L_r' tend to L_r except that points of discontinuity of ψ, $\Omega(z)$, $\omega(z)$ on L_r are excluded by small semicircles. Then, since the plate is unloaded, the total forces F_r and couples K_r on each contour are, by (7.10.3),

$$\left.\begin{aligned} F_r &= 8Da_r \\ K_r &= 8iD(c_r+id_r) \end{aligned}\right\}. \tag{7.10.6}$$

Hence, (7.10.4) and (7.10.5) may be written in the forms

$$\left.\begin{aligned} \Omega(z) &= \sum_{r=1}^{n} \frac{F_r}{16\pi D} z\ln(z-z_r) - \sum_{r=1}^{n} \frac{iK_r}{16\pi D}\ln(z-z_r)+\Omega_0(z) \\ \omega(z) &= \sum_{r=1}^{n} \frac{i\bar{K}_r}{16\pi D} z\ln(z-z_r) + \sum_{r=1}^{n} \frac{a'_r}{2\pi}\ln(z-z_r)+\omega_0(z) \end{aligned}\right\}. \tag{7.10.7}$$

When the region S is infinite, then, for sufficiently large $|z|$,

$$\left.\begin{aligned} \Omega(z) &= \frac{1}{16\pi D}(Fz\ln z - iK\ln z)+\Omega_1(z) \\ \omega(z) &= \frac{i\bar{K}}{16\pi D} z\ln z + K'\ln z+\omega_1(z) \end{aligned}\right\}, \tag{7.10.8}$$

where

$$F = \sum_{r=1}^{n} F_r, \qquad K = \sum_{r=1}^{n} K_r, \tag{7.10.9}$$

so that F is the resultant force normal to the plate acting on all the boundaries $L_1, L_2,..., L_n$. Also K represents the total (complex) moment about the x- and y-axes of the couple acting on these contours. The constant K' is real and the functions $\Omega_1(z)$, $\omega_1(z)$ denote holomorphic functions which, for sufficiently large $|z|$, may be developed in Laurent series. From (7.9.3), (7.9.4), (7.9.5), and (7.10.8) we see that, in general, the stress system will become infinite at infinity unless the resultant transverse force vanishes. If the resultant force F vanishes the displacement w given by (7.9.2) becomes infinite at infinity when the resultant couple K is non-zero; but even when both F and K vanish the displacement w may still become infinite at infinity due to the presence of a term $K'\ln z$ in $\omega(z)$ which contributes nothing to the stress system at infinity.

Suppose now that the resultant transverse force F_r on each boundary L_r vanishes. Then $a_r = F = 0$ and if the stresses are to remain bounded at infinity we have, for sufficiently large $|z|$,

$$\left.\begin{aligned} \Omega(z) &= -\frac{iK\ln z}{16\pi D}+Bz+\alpha+i\beta+\frac{\alpha_1+i\beta_1}{z}+O\left(\frac{1}{z^2}\right) \\ \omega(z) &= \frac{i\bar{K}z\ln z}{16\pi D}+K'\ln z+ \\ &\qquad +\tfrac{1}{2}(B'+iC')z^2+(\alpha'+i\beta')z+\alpha''+O\left(\frac{1}{z}\right) \end{aligned}\right\}, \tag{7.10.10}$$

the constants $B, B', \ldots$ in (7.10.10) being real. If M_1, M_2 are the principal stress couples at infinity and α is the angle between M_1 and the axis Ox, then, provided ψ and its derivatives vanish at infinity,

$$\left.\begin{aligned} M_1+M_2 &= 8(1+\eta)DB \\ (M_1-M_2)e^{2i\alpha} &= 4(1-\eta)D(B'-iC') \end{aligned}\right\}. \tag{7.10.11}$$

If the stress couples at infinity are not bounded but have known values we can modify (7.10.10) accordingly.

When the region S is bounded by a single closed curve L and extends to infinity, then, if the stresses at infinity are zero,

$$\left.\begin{aligned} \Omega'(z) &= -\frac{iK}{16\pi Dz}+O\left(\frac{1}{z^2}\right) \\ \omega''(z) &= \frac{i\bar{K}}{16\pi Dz}+O\left(\frac{1}{z^2}\right) \end{aligned}\right\}, \tag{7.10.12}$$

for large $|z|$.

The results (7.10.4) to (7.10.12) may also be used for a plate bounded externally by a contour L_0 and internally by contours $L_1, L_2, \ldots, L_m$ and loaded over a finite number of non-intersecting simply-connected regions which are bounded by contours $L_{m+1}, \ldots, L_n$. The previous formulae then apply to the region S bounded externally by the contour L_0 and internally by $L_1, L_2, \ldots, L_n$.

If the whole surface of a multiply-connected plate is loaded, some modifications are required in the formulae (7.10.6), but the rest of the discussion of the complex potentials is similar to that given above.

The above results are also modified if part of the boundary of the region S is in the finite part of the plane and the rest of the boundary extends to infinity. Suppose S consists of the (unloaded) region of Fig. 6.3, which was described in § 6.6, and let $K = L+iL'$ be the resultant couple exerted over $L_1, L_2, \ldots, L_n$, and the real axis, the resultant transverse force over the same boundaries being assumed to be zero. The resultant couple exerted on the material inside the semicircle $|z| = R$ across its boundary is then $K = L+iL'$. If stress resultants and stress couples vanish at infinity, then, for large $|z|$, we assume that

$$\Omega'(z) = \frac{A+iE}{2\pi z}+O\left(\frac{1}{z^2}\right), \qquad \Omega''(z) = -\frac{A+iE}{2\pi z^2}+O\left(\frac{1}{z^3}\right),$$

$$\omega''(z) = \frac{A'+iE'}{2\pi z}+O\left(\frac{1}{z^2}\right),$$

where A, E, A', E' are real constants. We also assume that $z\psi$ and its derivatives up to the second order all vanish for large $|z|$. If, in addition, we anticipate (7.13.7) and (7.13.10) of classical theory in § 7.13, and assume that G_y and $\chi - H_{xy}$ are $O(1/x^2)$ on $y = 0$, then

$$\lim_{|x|\to\infty} xG_y = 0, \qquad \lim_{|x|\to\infty} x(\chi - H_{xy}) = 0,$$

so that, from (7.9.3), (7.9.4), and (7.9.7),

$$(3+\eta)A = (1-\eta)A', \qquad (3+\eta)E = -(1-\eta)E'.$$

By applying (7.9.12) to the semicircle $|z| = R$ in the upper half-plane, and remembering the conditions imposed on ψ for large $|z|$, we obtain

$$K = Di\{(3+\eta)(A+iE)+(1-\eta)(A'-iE')\}.$$

Hence
$$A+iE = \frac{K}{2iD(3+\eta)},$$

$$A'+iE' = -\frac{\bar{K}}{2iD(1-\eta)},$$

and, for large $|z|$,

$$\left.\begin{aligned} \Omega'(z) &= \frac{K}{4i\pi D(3+\eta)z}+O\left(\frac{1}{z^2}\right) \\ \omega''(z) &= -\frac{\bar{K}}{4i\pi D(1-\eta)z}+O\left(\frac{1}{z^2}\right) \end{aligned}\right\}. \tag{7.10.13}$$

For many practical problems a further simplification of the theory can be made which we consider in the next section.

7.11. Classical bending theory: isotropic plates

If the shear resultants q^α are such that derivatives of q^α with respect to θ_β are of the same order of magnitude in h as q^α itself, then we can write the second equation of (7.7.16) approximately in the form

$$q^\alpha = s^\alpha - Dw|_{\beta}^{\alpha\beta}, \tag{7.11.1}$$

and then (7.7.18) becomes

$$m^{\alpha\beta} = -D\{(1-\eta)w|^{\alpha\beta}+\eta a^{\alpha\beta}w|_{\lambda}^{\lambda}\}, \tag{7.11.2}$$

whilst w satisfies the equation

$$Dw|_{\alpha\beta}^{\alpha\beta}-p-s^\alpha|_\alpha = 0. \tag{7.11.3}$$

These approximations mean that we have neglected terms in $m^{\alpha\beta}$ containing the shear resultants q^α which appear after we have substituted for w^λ in terms of $w|^\lambda$ and q^λ. We may therefore write (7.7.19) in the approximate form

$$w^\lambda = -w|^\lambda, \tag{7.11.4}$$

when shear loads s^λ are zero.

An equivalent statement of the above theory is obtained if we neglect the function ψ so that, when shear loads s^α are zero, the formulae of § 7.9 become

$$\left.\begin{aligned} w &= -R/D - z\bar{\Omega}(\bar{z}) - \bar{z}\Omega(z) - \omega(z) - \bar{\omega}(\bar{z}) \\ \Lambda &= (1+\eta)\nabla_1^2 R + 4D(1+\eta)\{\Omega'(z) + \bar{\Omega}'(\bar{z})\} \\ \Gamma &= 4(1-\eta)\frac{\partial^2 R}{\partial \bar{z}^2} + 4D(1-\eta)\{z\bar{\Omega}''(\bar{z}) + \bar{\omega}''(\bar{z})\} \\ \Psi_0 &= 2\frac{\partial \nabla_1^2 R}{\partial \bar{z}} + 8D\bar{\Omega}''(\bar{z}) \end{aligned}\right\}. \tag{7.11.5}$$

Since we now have only two potential functions at our disposal we can no longer impose three boundary conditions at each boundary of the plate. If, for example, the plate is clamped at its edges, then w and $\bar{u}_x$, $\bar{u}_y$ vanish there. From (7.9.6), when $\psi \equiv 0$, the vanishing of $\bar{u}_x$ and $\bar{u}_y$ implies that

$$\frac{\partial w}{\partial z} = 0, \qquad \frac{\partial w}{\partial \bar{z}} = 0, \tag{7.11.6}$$

and, for a single boundary, these conditions include the condition that w is zero. We may therefore replace the boundary conditions by the single (complex) condition

$$\frac{\partial w}{\partial z} = 0, \tag{7.11.7}$$

or

$$\frac{1}{D}\frac{\partial R}{\partial z} + \bar{\Omega}(\bar{z}) + \bar{z}\Omega'(z) + \omega'(z) = 0. \tag{7.11.8}$$

Alternatively, at a clamped edge, we may write the boundary conditions as

$$w = 0, \qquad \frac{\partial w}{\partial n} = 0, \tag{7.11.9}$$

where $\partial w/\partial n$ is the outward normal derivative of w at the boundary.

When the plate is unloaded on its face but is bent by forces over its edges, then the condition (7.11.8) at a clamped edge becomes

$$\bar{\Omega}(\bar{z}) + \bar{z}\Omega'(z) + \omega'(z) = 0. \tag{7.11.10}$$

Comparing this with the boundary condition (6.4.20) at a stress-free edge in plane strain (or generalized plane stress) we see that there is an analogy between problems of the bending of thin plates which are clamped at their edges and problems of stretching of thin plates which have stress-free boundaries. If, however, the plate extends to infinity, some differences arise in the forms of the potential functions for large $|z|$.

We encounter difficulties in the boundary conditions when given stresses are applied at the edges since the three conditions (7.9.16) do not obviously reduce to two. In order to obtain the correct boundary conditions we consider the rate of work equation for the flexure of a plate.

7.12. Rate of work and elastic potential for flexure

If the displacements V_α are approximately proportional to the x_3-coordinate then, in orthogonal cartesian coordinates x_α, the sections $x_2 = \text{constant}$, $x_1 = \text{constant}$ in the plate rotate by amounts $-w_2$, w_1 respectively during pure flexure of the plate (see (7.7.7)). Hence, in general θ_α-coordinates a section of the plate receives a vector rotation

$$\mathbf{\Theta} = \mathbf{a}_3 \times \mathbf{w}, \tag{7.12.1}$$

where

$$\mathbf{w} = w_\alpha \mathbf{a}^\alpha. \tag{7.12.2}$$

Also, if the transverse displacement is approximately constant then the middle surface of the plate receives a displacement $w\mathbf{a}_3$ during flexure.

Now consider the velocity $\dot{w}\mathbf{a}_3$ and angular velocity

$$\dot{\mathbf{\Theta}} = \mathbf{a}_3 \times \dot{\mathbf{w}} \tag{7.12.3}$$

of the middle plane of the plate, where

$$\dot{\mathbf{w}} = \dot{w}_\alpha \mathbf{a}^\alpha. \tag{7.12.4}$$

We take the scalar product of equation (7.3.2) with $\dot{w}\mathbf{a}_3$ and the scalar product of (7.3.8) with $\dot{\mathbf{\Theta}}$ and add and integrate over the middle plane of the plate. Thus

$$\iint \{(\mathbf{N}_{\alpha,\alpha}+\mathbf{P})\,.\,\mathbf{a}_3\dot{w}+\mathbf{M}_{\alpha,\alpha}\,.\,\dot{\mathbf{\Theta}}+\mathbf{S}\,.\,\dot{\mathbf{\Theta}}-Q^\alpha[\mathbf{a}_3\times\mathbf{a}_\alpha]\,.\,\dot{\mathbf{\Theta}}\}\frac{dS}{\sqrt{a}} = 0, \tag{7.12.5}$$

where the surface element of the middle plane is

$$dS = \sqrt{(a)}\,d\theta^1 d\theta^2.$$

Equations (7.12.5) may be written in the form

$$\iint \{(\mathbf{N}_\alpha\,.\,\mathbf{a}_3\dot{w})_{,\alpha}+(\mathbf{M}_\alpha\,.\,\dot{\mathbf{\Theta}})_{,\alpha}+\mathbf{P}\,.\,\mathbf{a}_3\dot{w}+\mathbf{S}\,.\,\dot{\mathbf{\Theta}}-$$
$$-\mathbf{N}_\alpha\,.\,\mathbf{a}_3\dot{w}_{,\alpha}-\mathbf{M}_\alpha\,.\,\dot{\mathbf{\Theta}}_{,\alpha}-Q^\alpha[\mathbf{a}_3\times\mathbf{a}_\alpha]\,.\,\dot{\mathbf{\Theta}}\}\frac{dS}{\sqrt{a}} = 0.$$

We now suppose that the middle plane of the plate is bounded by one or more closed curves which we denote by c, the unit normal at any point of c being $\mathbf{u}$, where

$$\mathbf{u} = u_\alpha \mathbf{a}^\alpha.$$

From (7.1.3) and (7.1.6),

$$\mathbf{n} = \frac{u_\alpha \mathbf{N}_\alpha}{\sqrt{a}}, \qquad \mathbf{m} = \frac{u_\alpha \mathbf{M}_\alpha}{\sqrt{a}},$$

and using these results and Green's theorem, equation (7.12.5) becomes

$$\int (\mathbf{n}'.\mathbf{a}_3\dot{w}+\mathbf{m}'.\dot{\boldsymbol{\theta}})\,ds+\iint \mathbf{P}.\mathbf{a}_3\dot{w}\frac{dS}{\sqrt{a}}+\iint \mathbf{S}.\dot{\boldsymbol{\theta}}\frac{dS}{\sqrt{a}}$$
$$= \iint \{\mathbf{N}_\alpha.\mathbf{a}_3\dot{w}_{,\alpha}+\mathbf{M}_\alpha.\dot{\boldsymbol{\theta}}_{,\alpha}+Q^\alpha[\mathbf{a}_3\times\mathbf{a}_\alpha].\dot{\boldsymbol{\theta}}\}\frac{dS}{\sqrt{a}}, \quad (7.12.6)$$

where ds is the line element along c and $\mathbf{n}'$, $\mathbf{m}'$ are the values of the stress resultant and stress couple at the curve c. Alternatively, equation (7.12.6) may be written

$$R = R_c+R_p = \dot{U}, \quad (7.12.7)$$

where
$$R_c = \int (\mathbf{n}'.\mathbf{a}_3\dot{w}+\mathbf{m}'.\dot{\boldsymbol{\theta}})\,ds \quad (7.12.8)$$

is the rate of work of the forces and couples along c,

$$R_p = \iint p\dot{w}\,dS+\iint \mathbf{s}.\dot{\boldsymbol{\theta}}\,dS \quad (7.12.9)$$

is the rate of work of the loads acting on the middle surface of the plate, and

$$\dot{U} = \iint \{\mathbf{N}_\alpha.\mathbf{a}_3\dot{w}_{,\alpha}+\mathbf{M}_\alpha.\dot{\boldsymbol{\theta}}_{,\alpha}+Q^\alpha[\mathbf{a}_3\times\mathbf{a}_\alpha].\dot{\boldsymbol{\theta}}\}\frac{dS}{\sqrt{a}}. \quad (7.12.10)$$

The total rate of work of the external forces is therefore R and U is the total potential energy of the body due to straining.†

From (7.1.7), (7.1.8), (7.1.9), and (7.12.3),

$$\mathbf{N}_\alpha.\mathbf{a}_3\dot{w}_{,\alpha} = Q^\alpha\dot{w}_{,\alpha},$$
$$\mathbf{M}_\alpha.\dot{\boldsymbol{\theta}}_{,\alpha} = \sqrt{(a)}m^{\alpha\rho}[\mathbf{a}_3\times\mathbf{a}_\rho].[\mathbf{a}_3\times\dot{\mathbf{w}}_{,\alpha}]$$
$$= \sqrt{(a)}m^{\alpha\rho}\mathbf{a}_\rho.\dot{\mathbf{w}}_{,\alpha}$$
$$= \tfrac{1}{2}\sqrt{(a)}m^{\alpha\rho}[\mathbf{a}_\rho.\dot{\mathbf{w}}_{,\alpha}+\mathbf{a}_\alpha.\dot{\mathbf{w}}_{,\rho}],$$

the last result following because $m^{\alpha\rho}$ is symmetric.

Hence
$$\mathbf{M}_\alpha.\dot{\boldsymbol{\theta}}_{,\alpha} = \sqrt{(a)}m^{\alpha\rho}\dot{\eta}_{\alpha\rho},$$

where
$$\eta_{\alpha\rho} = \tfrac{1}{2}[\mathbf{a}_\alpha.\mathbf{w}_{,\rho}+\mathbf{a}_\rho.\mathbf{w}_{,\alpha}]$$
$$= \tfrac{1}{2}(w_\alpha|_\rho+w_\rho|_\alpha). \quad (7.12.11)$$

Also
$$Q^\alpha[\mathbf{a}_3\times\mathbf{a}_\alpha].\dot{\boldsymbol{\theta}} = Q^\alpha\mathbf{a}_\alpha.\dot{\mathbf{w}} = Q^\alpha\dot{w}_\alpha.$$

Equation (7.12.10) therefore becomes

$$\dot{U} = \iint \{m^{\alpha\rho}\dot{\eta}_{\alpha\rho}+q^\alpha(\dot{w}_\alpha+\dot{w}_{,\alpha})\}\,dS. \quad (7.12.12)$$

† Under either isentropic or isothermal conditions. See Chapter 2.

We now restrict further discussion to the classical theory of bending for which

$$w_\alpha + w_{,\alpha} = 0. \tag{7.12.13}$$

Also, from (7.12.11) and (7.12.13),

$$\eta_{\alpha\beta} = -w|_{\alpha\beta}. \tag{7.12.14}$$

Hence, equation (7.12.12) becomes

$$\dot{U} = \iint \dot{W}\sqrt{(a)}\, d\theta^1 d\theta^2, \tag{7.12.15}$$

where

$$\dot{W} = m^{\alpha\beta}\dot{\eta}_{\alpha\beta}, \tag{7.12.16}$$

W being the elastic potential per unit area of the middle surface of the plate. Since the plate is elastic W is a function of $\eta_{\alpha\beta}$ and

$$\dot{W} = \frac{\partial W}{\partial \eta_{\alpha\beta}}\dot{\eta}_{\alpha\beta}, \qquad \dot{\eta}_{\alpha\beta} = \dot{\eta}_{\beta\alpha}.$$

Also

$$U = \iint W\sqrt{(a)}\, d\theta^1 d\theta^2, \tag{7.12.17}$$

and

$$m^{\alpha\beta} = \frac{1}{2}\left(\frac{\partial W}{\partial \eta_{\alpha\beta}} + \frac{\partial W}{\partial \eta_{\beta\alpha}}\right). \tag{7.12.18}$$

From (7.11.2), (7.12.14), and (7.12.18) we find that the elastic potential for the flexure of an isotropic plate is

$$W = \tfrac{1}{2}D\{(1-\eta)a^{\alpha\lambda}a^{\beta\mu} + \eta a^{\alpha\beta}a^{\lambda\mu}\}w|_{\alpha\beta}\,w|_{\lambda\mu}, \tag{7.12.19}$$

where D is given by (7.7.17). Also we may express W in the form

$$W = \tfrac{1}{2}m^{\alpha\beta}\eta_{\alpha\beta}. \tag{7.12.20}$$

7.13. Boundary conditions

In § 7.11 it was pointed out that difficulties occur in finding the correct boundary conditions at a free edge of the plate in classical bending theory. We now determine these conditions by examining the rate of work R_c of the forces and couples acting along an edge c of the plate (see 7.12.8).

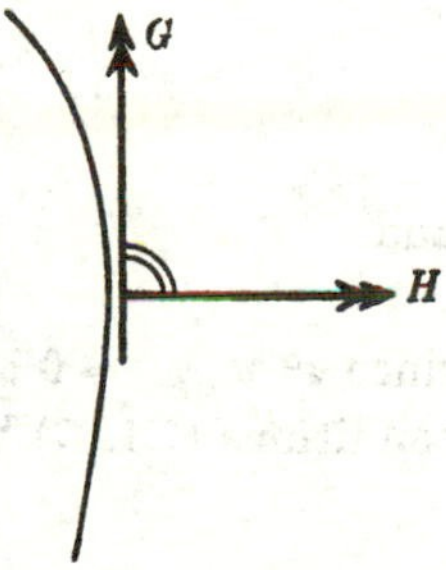

FIG. 7.3. Stress couples on boundary

For this purpose it is convenient to express **m** in terms of a tangential and normal component at each point of the curve c (see Fig. 7.3). Thus

$$\mathbf{m} = G[\mathbf{a}_3 \times \mathbf{u}] + H\mathbf{u}, \tag{7.13.1}$$

and we denote the value of **n** by

$$\mathbf{n} = N\mathbf{a}_3. \tag{7.13.2}$$

Also if $\partial/\partial s$, $\partial/\partial n$ denote derivatives along the tangent and normal to c, then

$$w_{,\alpha} = u_\alpha \frac{\partial w}{\partial n} - \epsilon_{\alpha\beta}u^\beta \frac{\partial w}{\partial s}, \tag{7.13.3}$$

and, using the approximation $w_\alpha = -w_{,\alpha}$, equation (7.12.1) becomes

$$\boldsymbol{\Theta} = -\frac{\partial w}{\partial n}[\mathbf{a}_3 \times \mathbf{u}] + \frac{\partial w}{\partial s}\mathbf{u}. \tag{7.13.4}$$

Hence

$$\mathbf{m}.\dot{\boldsymbol{\Theta}} + \mathbf{n}.\mathbf{a}_3 \dot{w} = -G\frac{\partial \dot{w}}{\partial n} + H\frac{\partial \dot{w}}{\partial s} + N\dot{w}$$

$$= -G\frac{\partial \dot{w}}{\partial n} + \frac{\partial}{\partial s}(H\dot{w}) + \left(N - \frac{\partial H}{\partial s}\right)\dot{w}, \tag{7.13.5}$$

and

$$R_c = \int\limits_c \left\{\left(N' - \frac{\partial H'}{\partial s}\right)\dot{w} - G'\frac{\partial \dot{w}}{\partial n}\right\} ds, \tag{7.13.6}$$

since $H'\dot{w}$ is single-valued. Formula (7.13.6) shows that the rate of work of the external stresses is equivalent to the rate of work of a flexural couple G' and the rate of work of a force $N' - (\partial H'/\partial s)$. This indicates that at a boundary where stresses are prescribed

$$G = G', \qquad N - \frac{\partial H}{\partial s} = N' - \frac{\partial H'}{\partial s}, \tag{7.13.7}$$

where G', H', N' are the externally applied stress couples and shearing forces.

When the faces of the plate are entirely free from stress then, from (7.7.20),

$$q^\alpha = \epsilon^{\alpha\beta}\chi_{,\beta} = \epsilon^{\alpha\beta}\left(u_\beta \frac{\partial \chi}{\partial n} - \epsilon_{\beta\lambda} u^\lambda \frac{\partial \chi}{\partial s}\right) = \epsilon^{\alpha\beta} u_\beta \frac{\partial \chi}{\partial n} + u^\alpha \frac{\partial \chi}{\partial s}, \tag{7.13.8}$$

and

$$N = u_\alpha q^\alpha = \frac{\partial \chi}{\partial s}, \tag{7.13.9}$$

since $\epsilon^{\alpha\beta} u_\alpha u_\beta = 0$ and $u_\alpha u^\alpha = 1$. It follows that we may replace the conditions (7.13.7) by

$$G = G', \qquad \chi - H = \chi' - H' \tag{7.13.10}$$

at a boundary where stresses are prescribed.

The values of the resultant force F and couple $\mathbf{K}$ acting on an arc AP and given by (7.7.32) and (7.7.36) respectively are altered for the classical bending theory. Thus the resultant force F normal to the plate acting on an arc AP of a contour c is

$$F = \int\limits_A^P N\, ds - \int\limits_A^P \frac{\partial H}{\partial s}\, ds, \tag{7.13.11}$$

and using (7.7.33) this becomes

$$F = \int_A^P R|_\lambda^{\lambda\alpha} u_\alpha \, ds + \chi - H. \tag{7.13.12}$$

Since ψ is now zero, χ in (7.13.12) is, from (7.9.7), given by

$$\chi = 4iD\{\overline{\Omega}'(\bar{z}) - \Omega'(z)\}. \tag{7.13.13}$$

The couple **K** about axes at the origin acting on the arc AP consists of contributions from tangential couples of magnitude G together with couples about the axes due to the stresses $N-(\partial H/\partial s)$ normal to the plate. Thus

$$\mathbf{K} = \int \left\{\mathbf{m} - H\mathbf{u} + \mathbf{r} \times \left(\mathbf{n} - \frac{\partial H}{\partial s}\mathbf{a}_3\right)\right\} ds. \tag{7.13.14}$$

Now

$$\int \mathbf{r} \times \mathbf{a}_3 \frac{\partial H}{\partial s} \, ds + \int H\mathbf{u} \, ds$$

$$= [H\mathbf{r} \times \mathbf{a}_3]_A^P - \int_A^P H\mathbf{a}_\beta \times \mathbf{a}_3 \frac{d\theta^\beta}{ds} \, ds + \int_A^P H\mathbf{u} \, ds = [H\mathbf{r} \times \mathbf{a}_3]_A^P,$$

and the other terms in (7.13.14) have been reduced previously to the form (7.7.36). Therefore

$$\mathbf{K} = \int_A^P \delta_{\alpha\beta}^{\rho\mu}(m^{\alpha\lambda} - q^\alpha r^\lambda) a_{\lambda\rho} \mathbf{a}_\mu \, d\theta^\beta - [H\mathbf{r} \times \mathbf{a}_3]_A^P. \tag{7.13.15}$$

When stresses are prescribed at the boundary of a contour, then from (7.13.11) and (7.13.14) we must have

$$F = \int_A^P N \, ds - H = \int_A^P N' \, ds - H', \tag{7.13.16}$$

$$\mathbf{K} = \int_A^P \{\mathbf{m} + N[\mathbf{r} \times \mathbf{a}_3]\} \, ds - H[\mathbf{r} \times \mathbf{a}_3]$$

$$= \int_A^P \{\mathbf{m}' + N'[\mathbf{r} \times \mathbf{a}_3]\} \, ds - H'[\mathbf{r} \times \mathbf{a}_3], \tag{7.13.17}$$

where a dash denotes the values of the externally applied forces and, except in the integrals, **r** is the position vector of the point P. We have taken $H = H'$ at the point A since from (7.13.7) we see that the addition of an arbitrary constant to H does not affect the boundary conditions. At first sight equations (7.13.16) and (7.13.17) appear to give us three boundary conditions instead of two, but only two of these are independent. We replace (7.13.17) by

$$\mathbf{K}-F[\mathbf{r}\times\mathbf{a}_3] = \int_A^P \{\mathbf{m}+N[\mathbf{r}\times\mathbf{a}_3]\}\,ds-[\mathbf{r}\times\mathbf{a}_3]\int_A^P N\,ds$$

$$= \int_A^P \{\mathbf{m}'+N'[\mathbf{r}\times\mathbf{a}_3]\}\,ds-[\mathbf{r}\times\mathbf{a}_3]\int_A^P N'\,ds. \qquad (7.13.18)$$

Since this equation is true for all points P on the contour, it follows on differentiation with respect to arc length s that

$$\frac{\partial}{\partial s}\{\mathbf{K}-F[\mathbf{r}\times\mathbf{a}_3]\} = \mathbf{m}-[\mathbf{a}_\alpha\times\mathbf{a}_3]\frac{d\theta^\alpha}{ds}\int_A^P N\,ds$$

$$= \mathbf{m}'-[\mathbf{a}_\alpha\times\mathbf{a}_3]\frac{d\theta^\alpha}{ds}\int_A^P N'\,ds,$$

or

$$\frac{\partial}{\partial s}\{\mathbf{K}-F[\mathbf{r}\times\mathbf{a}_3]\} = \mathbf{m}-\mathbf{u}\int_A^P N\,ds$$

$$= \mathbf{m}'-\mathbf{u}\int_A^P N'\,ds, \qquad (7.13.19)$$

where $\mathbf{u}$ is the unit normal to the contour at P and

$$\mathbf{u} = \epsilon_{\alpha\beta}\frac{d\theta^\beta}{ds}\mathbf{a}^\alpha = \mathbf{a}_\beta\times\mathbf{a}_3\frac{d\theta^\beta}{ds}.$$

If we now take the scalar product of equation (7.13.19) with $\mathbf{u}$, we obtain

$$H-\int_A^P N\,ds = H'-\int_A^P N'\,ds,$$

since $H = \mathbf{m}.\mathbf{u}$, and this is the third boundary condition (7.13.16). The two equations (7.13.18) are therefore sufficient when stresses are prescribed at a boundary.

The function $\mathbf{K}-F[\mathbf{r}\times\mathbf{a}_3]$ has components $K+izF$ and $\bar{K}-i\bar{z}F$ when referred to complex base vectors $\mathbf{A}_1$, $\mathbf{A}_2$ and these have already been evaluated in (7.9.17) for the general bending theory of an isotropic plate.† For the classical bending theory

$$K+izF = 2D[(3+\eta)\Omega(z)-(1-\eta)\{z\overline{\Omega}'(\bar{z})+\bar{\omega}'(\bar{z})\}]-2(1-\eta)\frac{\partial R}{\partial \bar{z}}-$$

$$-8\int_A^P z\frac{\partial^3 R}{\partial z^2\partial\bar{z}}\,dz+8z\int_A^P \frac{\partial^3 R}{\partial z^2\partial\bar{z}}\,dz, \quad (7.13.20)$$

on putting $\psi \equiv 0$ in (7.9.17).

When the stresses and displacements in the plate are single-valued, then the formulae (7.10.2) and (7.10.3) for the force and couple acting on a contour c in the plate are still valid for the classical bending theory. The discussion of the complex potentials in § 7.10 also remains true for the classical theory since it is independent of the function ψ.

When the transverse load on a plate is zero and the plate is bent by given tractions at its edges we see, from (7.13.20), that the value of

$$\frac{3+\eta}{1-\eta}\Omega(z)-z\overline{\Omega}'(\bar{z})-\bar{\omega}'(\bar{z})$$

is prescribed at the edges of the plate. On the other hand, if body forces in plane strain are zero and the *displacements* at the edges of the plate are prescribed, then

$$(3-4\eta)\Omega(z)-z\overline{\Omega}'(\bar{z})-\bar{\omega}'(\bar{z})$$

takes given values at the edges of the plate. The two problems are therefore analytically similar. If the plates extend to infinity, some differences arise in the forms of the potential functions for large $|z|$, but this does not introduce any essential difference in the methods of solution of the two problems.

7.14. Bending of aeolotropic plates

We again consider plates which are elastically symmetrical about the middle plane $x_3 = 0$ so that we may use equations (7.6.1) to (7.6.4). We restrict our attention here to plates which are bent about the plane $x_3 = 0$ by antisymmetrical normal pressures only so that $s^\beta = 0$ and equations (7.7.1), which are valid for aeolotropic plates, become

$$\left.\begin{aligned} m^{\alpha\beta}|_\alpha - q^\beta &= 0 \\ q^\alpha|_\alpha + p &= 0 \end{aligned}\right\}. \quad (7.14.1)$$

† We see from (7.13.18) that $\mathbf{K}-F[\mathbf{r}\times\mathbf{a}_3]$, when expressed in terms of stress resultants and stress couples, has the same forms for general and classical bending theory.

As in (7.7.20), the second equation in (7.14.1) may be satisfied by

$$q^{\alpha} = R|_{\lambda}^{\alpha\lambda}+\epsilon^{\alpha\lambda}\chi_{,\lambda}, \tag{7.14.2}$$

where

$$p = -R|_{\alpha\lambda}^{\alpha\lambda}. \tag{7.14.3}$$

Eliminating q^{α} from (7.14.1) gives

$$m^{\alpha\beta}|_{\alpha\beta}+p = 0. \tag{7.14.4}$$

We shall also only develop here a theory which corresponds to the classical bending theory of isotropic plates. From (7.6.4) and (7.6.5),

$$\left.\begin{aligned}\gamma^{\alpha\beta} &= F^{\alpha\beta}_{\lambda\mu}\,\tau^{\lambda\mu}+F^{\alpha\beta}_{33}\,\tau^{33}\\ \gamma^{\alpha3} &= 2F^{\alpha3}_{\lambda3}\,\tau^{\lambda3}\end{aligned}\right\}, \tag{7.14.5}$$

and

$$\left.\begin{aligned}\gamma^{\alpha\beta} &= \tfrac{1}{2}a^{\alpha\lambda}a^{\beta\mu}(V_{\lambda}|_{\mu}+V_{\mu}|_{\lambda})\\ \gamma^{\alpha3} &= \tfrac{1}{2}a^{\alpha\lambda}(V_{3,\lambda}+V_{\lambda,3})\end{aligned}\right\}. \tag{7.14.6}$$

We multiply the first equation in (7.14.5) by x_3 and integrate with respect to x_3 from $-h$ to h. Thus, using (7.1.11) and (7.7.7),

$$\frac{2h^3\eta^{\alpha\beta}}{3} = F^{\alpha\beta}_{\lambda\mu}\,m^{\lambda\mu}, \tag{7.14.7}$$

if we neglect $\int_{-h}^{h}\tau^{33}x_3\,dx_3$ compared with $m^{\lambda\mu}$, where

$$\left.\begin{aligned}\eta^{\alpha\beta} &= a^{\alpha\lambda}a^{\beta\mu}\eta_{\lambda\mu}\\ \eta_{\alpha\beta} &= \tfrac{1}{2}(w_{\alpha}|_{\beta}+w_{\beta}|_{\alpha})\end{aligned}\right\}. \tag{7.14.8}$$

We observe that this definition of $\eta_{\alpha\beta}$ agrees with (7.12.11). We also multiply the second equation in (7.14.5) by $(3/4h)\{1-(x_3^2/h^2)\}$ and integrate through the plate so that, after an integration by parts, we have, using (7.7.7) and (7.7.11),

$$\tfrac{1}{2}a^{\alpha\lambda}(w_{,\lambda}+w_{\lambda}) = \frac{3}{2h}\int_{-h}^{h}\left(1-\frac{x_3^2}{h^2}\right)F^{\alpha3}_{\lambda3}\,\tau^{\lambda3}\,dx_3$$

or

$$w_{,\lambda}+w_{\lambda} = \frac{3a_{\alpha\lambda}}{h}\int_{-h}^{h}\left(1-\frac{x_3^2}{h^2}\right)F^{\alpha3}_{\lambda3}\,\tau^{\lambda3}\,dx_3.$$

If we now substitute w_{λ} from this equation into (7.14.7) and neglect terms of the order

$$h^2\int_{-h}^{h}\left(1-\frac{x_3^2}{h^2}\right)F^{\alpha3}_{\lambda3}\,\tau^{\lambda3}\,dx_3$$

compared with $m^{\lambda\mu}$ we see that this is equivalent to the assumption

$$w_\lambda = -w_{,\lambda} \tag{7.14.9}$$

which we found to hold in the classical bending theory for isotropic plates. Using the result (7.14.9) we find, from (7.14.8), that

$$\left.\begin{aligned}\eta_{\alpha\beta} &= -w|_{\alpha\beta}\\ \eta^{\alpha\beta} &= -a^{\alpha\lambda}a^{\beta\mu}w|_{\lambda\mu} = -w|^{\alpha\beta}\end{aligned}\right\}. \tag{7.14.10}$$

For most purposes it is convenient to express $m^{\alpha\beta}$ in terms of $\eta^{\lambda\mu}$. To do this we define elastic constants $E^{\alpha\beta}_{\lambda\mu}$ by the equations

$$E^{\alpha\beta}_{\rho\delta}F^{\rho\delta}_{\lambda\mu} = \tfrac{1}{2}(\delta^\alpha_\lambda\delta^\beta_\mu+\delta^\alpha_\mu\delta^\beta_\lambda), \tag{7.14.11}$$

so that, from (7.14.7),

$$m^{\alpha\beta} = \frac{2h^3}{3}E^{\alpha\beta}_{\lambda\mu}\eta^{\lambda\mu} = -\frac{2h^3}{3}E^{\alpha\beta}_{\lambda\mu}w|^{\lambda\mu}. \tag{7.14.12}$$

For applications to practical problems it is useful to notice that the formulae corresponding to (7.14.11) in rectangular coordinates are

$$s^{\alpha\beta}_{\rho\delta}c^{\rho\delta}_{\lambda\mu} = \tfrac{1}{2}(\delta^\alpha_\lambda\delta^\beta_\mu+\delta^\alpha_\mu\delta^\beta_\lambda), \tag{7.14.13}$$

so that we have six coefficients

$$c^{11}_{11},\quad c^{11}_{22},\quad c^{11}_{12},\quad c^{12}_{12},\quad c^{12}_{22},\quad c^{22}_{22}, \tag{7.14.14}$$

defined in terms of six independent coefficients

$$s^{11}_{11},\quad s^{11}_{22},\quad s^{11}_{12},\quad s^{12}_{12},\quad s^{12}_{22},\quad s^{22}_{22}. \tag{7.14.15}$$

We also observe that

$$c^{\alpha\beta}_{\lambda\mu} = c^{\beta\alpha}_{\lambda\mu} = c^{\beta\alpha}_{\mu\lambda} = c^{\alpha\beta}_{\mu\lambda} = c^{\lambda\mu}_{\alpha\beta}. \tag{7.14.16}$$

The coefficients $E^{\alpha\beta}_{\lambda\mu}$, $c^{\alpha\beta}_{\lambda\mu}$ are not the same as those defined in Chapter 5.

From (7.14.4) and (7.14.12), we have, at once,

$$\frac{2h^3}{3}E^{\alpha\beta}_{\lambda\mu}w|^{\lambda\mu}_{\alpha\beta} = p, \tag{7.14.17}$$

since the coefficients $E^{\alpha\beta}_{\lambda\mu}$ are such that $E^{\alpha\beta}_{\lambda\mu}|_\rho = 0$, and this is a fourth-order differential equation for w. This equation may be integrated with the help of the complex variable notations of § 6.4. If $C^{\alpha\beta}_{\lambda\mu}$ denotes the values of the elastic coefficients $E^{\alpha\beta}_{\lambda\mu}$ in complex coordinates, then

$$C^{\alpha\beta}_{\lambda\mu} = \frac{\partial z^\alpha}{\partial x^\rho}\frac{\partial z^\beta}{\partial x^\gamma}\frac{\partial x^\xi}{\partial z^\lambda}\frac{\partial x^\eta}{\partial z^\mu}c^{\rho\gamma}_{\xi\eta}, \tag{7.14.18}$$

so that
$$C^{\alpha\beta}_{\lambda\mu} = C^{\beta\alpha}_{\lambda\mu} = C^{\beta\alpha}_{\mu\lambda} = C^{\alpha\beta}_{\mu\lambda}, \tag{7.14.19}$$
and
$$\left.\begin{aligned} C^{11}_{11} &= C^{22}_{22} = \tfrac{1}{4}(c^{11}_{11}+c^{22}_{22}+4c^{12}_{12}-2c^{11}_{22}) \\ C^{11}_{22} &= \tfrac{1}{4}(c^{11}_{11}+c^{22}_{22}-4c^{12}_{12}-2c^{11}_{22}+4ic^{11}_{12}-4ic^{12}_{22}) \\ C^{12}_{22} &= C^{11}_{12} = \tfrac{1}{4}(c^{11}_{11}-c^{22}_{22}+2ic^{11}_{12}+2ic^{12}_{22}) \\ C^{12}_{12} &= \tfrac{1}{4}(c^{11}_{11}+c^{22}_{22}+2c^{11}_{22}) \\ C^{22}_{11} &= \bar{C}^{11}_{22}, \qquad C^{22}_{12} = C^{12}_{11} = \bar{C}^{11}_{12} \end{aligned}\right\}. \tag{7.14.20}$$

Equation (7.14.17) may be written in the form
$$\frac{2h^3}{3} a^{\lambda\xi} a^{\mu\eta} E^{\alpha\beta}_{\lambda\mu} w|_{\alpha\beta\xi\eta} = p, \tag{7.14.21}$$
and interpreting this in complex coordinates with the help of (6.4.3) gives
$$C^{11}_{22}\frac{\partial^4 w}{\partial z^4} + 4C^{11}_{12}\frac{\partial^4 w}{\partial z^3 \partial \bar{z}} + 2(C^{11}_{11}+2C^{12}_{12})\frac{\partial^4 w}{\partial z^2 \partial \bar{z}^2} +$$
$$+4C^{12}_{11}\frac{\partial^4 w}{\partial z \partial \bar{z}^3} + C^{22}_{11}\frac{\partial^4 w}{\partial \bar{z}^4} = \frac{3p}{8h^3}. \tag{7.14.22}$$

When $p = 0$ this equation is similar to equation (6.7.19) (apart from a change of sign in two terms) and therefore the complementary function of (7.14.22) may be written
$$w = \Omega(z+\gamma_1 \bar{z}) + \bar{\Omega}(\bar{z}+\bar{\gamma}_1 z) + \omega(z+\gamma_2 \bar{z}) + \bar{\omega}(\bar{z}+\bar{\gamma}_2 z), \tag{7.14.23}$$
where γ_1, γ_2, $1/\bar{\gamma}_1$, $1/\bar{\gamma}_2$ are roots of the equation
$$C^{22}_{11}\gamma^4 + 4C^{12}_{11}\gamma^3 + 2(C^{11}_{11}+2C^{12}_{12})\gamma^2 + 4C^{11}_{12}\gamma + C^{11}_{22} = 0, \tag{7.14.24}$$
and γ_1, γ_2 are chosen so that $|\gamma_1| < 1$, $|\gamma_2| < 1$. By an argument similar to that used in § 6.7 we can show that $\gamma_1 \bar{\gamma}_1 \neq 1$, $\gamma_2 \bar{\gamma}_2 \neq 1$, a condition which is necessary for the equation (7.14.22) to be elliptic. As in § 6.7 special cases of the general result (7.14.23) may occur in practice but these can be considered in an exactly similar fashion.

Using the notation of § 7.8 and interpreting (7.14.12) in complex coordinates we have
$$\left.\begin{aligned} \Lambda &= -\frac{8h^3}{3}\left(C^{12}_{22}\frac{\partial^2 w}{\partial z^2} + 2C^{12}_{12}\frac{\partial^2 w}{\partial z \partial \bar{z}} + C^{12}_{11}\frac{\partial^2 w}{\partial \bar{z}^2}\right) \\ \Gamma &= -\frac{8h^3}{3}\left(C^{11}_{22}\frac{\partial^2 w}{\partial z^2} + 2C^{11}_{12}\frac{\partial^2 w}{\partial z \partial \bar{z}} + C^{11}_{11}\frac{\partial^2 w}{\partial \bar{z}^2}\right) \end{aligned}\right\}, \tag{7.14.25}$$
whilst, from (7.14.1),
$$\Psi_0 = \frac{\partial \Gamma}{\partial z} + \frac{\partial \Lambda}{\partial \bar{z}}. \tag{7.14.26}$$

Hence, when the load on the plate is zero,

$$\Lambda = -\frac{8h^3}{3}\{(C^{12}_{22}+2C^{12}_{12}\gamma_1+C^{12}_{11}\gamma_1^2)\Omega''(z_1)+(C^{12}_{22}\bar{\gamma}_1^2+2C^{12}_{12}\bar{\gamma}_1+C^{12}_{11})\bar{\Omega}''(\bar{z}_1)+$$
$$+(C^{12}_{22}+2C^{12}_{12}\gamma_2+C^{12}_{11}\gamma_2^2)\omega''(z_2)+(C^{12}_{22}\bar{\gamma}_2^2+2C^{12}_{12}\bar{\gamma}_2+C^{12}_{11})\bar{\omega}''(\bar{z}_2)\}, \quad (7.14.27)$$

$$\Gamma = -\frac{8h^3}{3}\{(C^{11}_{22}+2C^{11}_{12}\gamma_1+C^{11}_{11}\gamma_1^2)\Omega''(z_1)+(C^{11}_{22}\bar{\gamma}_1^2+2C^{11}_{12}\bar{\gamma}_1+C^{11}_{11})\bar{\Omega}''(\bar{z}_1)+$$
$$+(C^{11}_{22}+2C^{11}_{12}\gamma_2+C^{11}_{11}\gamma_2^2)\omega''(z_2)+(C^{11}_{22}\bar{\gamma}_2^2+2C^{11}_{12}\bar{\gamma}_2+C^{11}_{11})\bar{\omega}''(\bar{z}_2)\}, \quad (7.14.28)$$

$$\Psi_0 = -\frac{8h^3}{3}\{\gamma_1 k_1\Omega'''(z_1)-\bar{k}_1\bar{\Omega}'''(\bar{z}_1)+\gamma_2 k_2\omega'''(z_2)-\bar{k}_2\bar{\omega}'''(\bar{z}_2)\}, \quad (7.14.29)$$

where

$$\left.\begin{aligned} k_1 &= \frac{C^{11}_{22}}{4}\left(\frac{1}{\gamma_1}-\frac{1}{\gamma_2}-\bar{\gamma}_1-\bar{\gamma}_2\right)-\frac{C^{22}_{11}\gamma_1^2}{4}\left(\gamma_1-\gamma_2-\frac{1}{\bar{\gamma}_1}-\frac{1}{\bar{\gamma}_2}\right)\\ k_2 &= \frac{C^{11}_{22}}{4}\left(\frac{1}{\gamma_2}-\frac{1}{\gamma_1}-\bar{\gamma}_1-\bar{\gamma}_2\right)-\frac{C^{22}_{11}\gamma_2^2}{4}\left(\gamma_2-\gamma_1-\frac{1}{\bar{\gamma}_1}-\frac{1}{\bar{\gamma}_2}\right)\\ z_1 &= z+\gamma_1\bar{z}, \qquad z_2 = z+\gamma_2\bar{z} \end{aligned}\right\}. \quad (7.14.30)$$

In deriving (7.14.29) we have used equation (7.14.24) and the relations

$$\left.\begin{aligned} -\frac{4C^{12}_{11}}{C^{22}_{11}} &= \gamma_1+\frac{1}{\bar{\gamma}_1}+\gamma_2+\frac{1}{\bar{\gamma}_2}\\ -\frac{4C^{11}_{12}}{C^{11}_{22}} &= \bar{\gamma}_1+\frac{1}{\gamma_1}+\bar{\gamma}_2+\frac{1}{\gamma_2} \end{aligned}\right\}, \quad (7.14.31)$$

which follow from (7.14.24).

We can write (7.14.2) in the form

$$\chi_{,\alpha} = \epsilon_{\beta\alpha}(q^\beta - a^{\beta\rho}a^{\lambda\nu}R|_{\rho\nu\lambda})$$

and in complex coordinates this becomes

$$\left.\begin{aligned} \frac{\partial\chi}{\partial z} &= -\frac{i}{2}\left(\bar{\Psi}_0-8\frac{\partial^3 R}{\partial z^2\partial\bar{z}}\right)\\ \frac{\partial\chi}{\partial\bar{z}} &= \frac{i}{2}\left(\Psi_0-8\frac{\partial^3 R}{\partial z\partial\bar{z}^2}\right) \end{aligned}\right\}. \quad (7.14.32)$$

When the load on the plate is zero we see, from (7.14.29) and (7.14.32), that, since χ is real,

$$\chi = -\frac{4ih^3}{3}\{k_1\Omega''(z_1)-\bar{k}_1\bar{\Omega}''(\bar{z}_1)+k_2\omega''(z_2)-\bar{k}_2\bar{\omega}''(\bar{z}_2)\}, \quad (7.14.33)$$

apart from an arbitrary additional constant.

Again, from (7.9.11) and (7.7.33),

$$K+izF = \tfrac{1}{2}\int_A^P \{(\Lambda-z\bar{\Psi}_0)\,dz-(\Gamma-z\Psi_0)\,d\bar{z}\}+iz\chi+iz\int_A^P R|_\lambda^{\lambda\alpha} u_\alpha\,ds$$

so that, when loads on the surfaces of the plate are zero,

$$K+izF = \tfrac{1}{2}\int_A^P (\Lambda\,dz-\Gamma\,d\bar{z}+2i\chi\,dz), \tag{7.14.34}$$

if we use (7.14.32) in the form $\bar{\Psi}_0 = -2i(\partial\chi/\partial\bar{z})$. Substituting in (7.14.34) from (7.14.27)–(7.14.29) gives

$$K+izF = -\frac{2h^3}{3}\{\lambda_1\Omega'(z_1)+\mu_1\bar{\Omega}'(\bar{z}_1)+\lambda_2\,\omega'(z_2)+\mu_2\,\bar{\omega}'(\bar{z}_2)\}, \tag{7.14.35}$$

where

$$\left.\begin{aligned}\lambda_1 &= -2(C^{11}_{11}\gamma_1^2+2C^{11}_{12}\gamma_1+C^{11}_{22})/\gamma_1, \qquad \mu_1 = -2(C^{11}_{11}+2C^{11}_{12}\bar{\gamma}_1+C^{11}_{22}\bar{\gamma}_1^2)\\ \lambda_2 &= -2(C^{11}_{11}\gamma_2^2+2C^{11}_{12}\gamma_2+C^{11}_{22})/\gamma_2, \qquad \mu_2 = -2(C^{11}_{11}+2C^{11}_{12}\bar{\gamma}_2+C^{11}_{22}\bar{\gamma}_2^2)\end{aligned}\right\}. \tag{7.14.36}$$

On the other hand, the cartesian components of stress resultants and couples are

$$\left.\begin{aligned}G_x &= -\frac{2h^3}{3}\left(c^{11}_{11}\frac{\partial^2 w}{\partial x^2}+c^{11}_{22}\frac{\partial^2 w}{\partial y^2}+2c^{11}_{12}\frac{\partial^2 w}{\partial x\partial y}\right)\\ G_y &= -\frac{2h^3}{3}\left(c^{22}_{11}\frac{\partial^2 w}{\partial x^2}+c^{22}_{22}\frac{\partial^2 w}{\partial y^2}+2c^{22}_{12}\frac{\partial^2 w}{\partial x\partial y}\right)\\ H_{xy} &= -\frac{2h^3}{3}\left(c^{12}_{11}\frac{\partial^2 w}{\partial x^2}+c^{12}_{22}\frac{\partial^2 w}{\partial y^2}+2c^{12}_{12}\frac{\partial^2 w}{\partial x\partial y}\right)\\ N_x &= \frac{\partial G_x}{\partial x}+\frac{\partial H_{xy}}{\partial y}, \qquad N_y = \frac{\partial H_{xy}}{\partial x}+\frac{\partial G_y}{\partial y}\end{aligned}\right\}. \tag{7.14.37}$$

The discussion in § 7.12 on rate of work depended on the equations of equilibrium and the equations (7.12.13) and (7.12.14), and all of these equations are valid for aeolotropic plates. It follows that (7.12.18) holds for aeolotropic plates, but W is now obtained from (7.14.12) in the form

$$\left.\begin{aligned}W &= \frac{h^3}{3}E^{\alpha\beta\lambda\mu}\eta_{\alpha\beta}\,\eta_{\lambda\mu}\\ &= \frac{h^3}{3}E^{\alpha\beta\lambda\mu}w|_{\alpha\beta}\,w|_{\lambda\mu}\end{aligned}\right\}, \tag{7.14.38}$$

together with (7.12.20) which is true for both isotropic and aeolotropic plates.

The boundary condition at an edge of an aeolotropic plate at which forces are prescribed can also be expressed in the forms (7.13.7),

(7.13.10), or (7.13.18), the last form of the boundary condition reducing to (7.14.35) in complex variable notation when the loads on the plate are zero. If we compare (7.14.35) with (6.7.30) in plane strain, we see that the problem of flexure of an aeolotropic plate at which edge forces are prescribed is analytically similar to the problem of plane strain (or generalized plane stress) when displacements at boundaries are prescribed. A similar correspondence exists between problems of flexure of an aeolotropic plate which is clamped at its edges and a plane strain (or generalized plane stress) problem when the stresses applied to the boundaries are zero.

To complete the problem of flexure of an aeolotropic plate we must discuss the properties of the complex potentials, the conditions for single-valued stresses and displacements, and the form of the potentials for large values of the arguments when the plate extends to infinity. Such a discussion is, however, very similar to that carried out in Chapter 6 for plane strain in aeolotropic bodies. Also, we have already discussed the corresponding case for isotropic bodies, so we leave details to the reader. It will be found that there are some minor differences in the forms of the complex potentials at infinity for a plate under no resultant transverse force, from those in plane strain, but this makes very little difference to the type of analysis used for solving corresponding problems.

8 Plane Problems for Isotropic Bodies

In Chapter 6 we developed the theory of plane strain for isotropic bodies and in Chapter 7 we showed that a similar theory arose for generalized plane stress in a plate deformed by forces parallel to its plane. In addition, it was shown in Chapter 7 that the classical theory for the transverse flexure of a plane plate is similar in many respects to the theory of plane strain. In the present chapter special problems are discussed for plane strain and generalized plane stress for isotropic bodies, mainly with the help of complex function theory.† The analogous problems which arise in the classical theory of transverse flexure of isotropic plates may then be solved by similar methods but no further details are given here. Instead, we discuss one flexure problem using the more general theory of Reissner.

We begin by collecting in § 8.1 all the formulae which we need for applications of the theory of plane strain or generalized plane stress, and we restrict our attention to problems for which the body forces are zero. In many problems it is known from complex function theory that the solutions obtained are unique, so no general uniqueness theorems are proved.

8.1. Formulae for plane strain and generalized plane stress

If (x, y) are rectangular cartesian coordinates and $z = x+iy$, we have from Chapter 6, when body forces are zero,

$$\mu D = \mu(u_x+iu_y) = \kappa\Omega(z)-z\bar{\Omega}'(\bar{z})-\bar{\omega}'(\bar{z}), \tag{8.1.1}$$

$$\Theta = \sigma_{xx}+\sigma_{yy} = 4\{\Omega'(z)+\bar{\Omega}'(\bar{z})\}, \tag{8.1.2}$$

$$\Phi = \sigma_{xx}-\sigma_{yy}+2i\sigma_{xy} = -4\{z\bar{\Omega}''(\bar{z})+\bar{\omega}''(\bar{z})\}, \tag{8.1.3}$$

$$\sigma_{yy}-i\sigma_{xy} = 2\{\Omega'(z)+\bar{\Omega}'(\bar{z})+z\bar{\Omega}''(\bar{z})+\bar{\omega}''(\bar{z})\}, \tag{8.1.4}$$

$$P = X+iY = 2i\{z\bar{\Omega}'(\bar{z})+\Omega(z)+\bar{\omega}'(\bar{z})\}, \tag{8.1.5}$$

$$M = z\bar{z}\{\Omega'(z)+\bar{\Omega}'(\bar{z})\}+z\omega'(z)+\bar{z}\bar{\omega}'(\bar{z})-\omega(z)-\bar{\omega}(\bar{z}), \tag{8.1.6}$$

† References to original work using complex function theory in elasticity are given in the introduction to Chapter 6.

where, for plane strain, $\kappa = 3-4\eta.$ (8.1.7)

In Chapter 7 we saw that, with a different notation, similar equations hold when we consider a plate under generalized plane stress in its plane provided

$$\kappa = 3-4\sigma = \frac{3-\eta}{1+\eta}. \tag{8.1.8}$$

Here we shall use the above notation for either plane strain or generalized plane stress so that, for example, σ_{xx} in plane strain is the actual force per unit area, but σ_{xx} in generalized plane stress is force per unit length of the line $x =$ constant in the (x, y) plane.

If we have a conformal transformation of the form

$$z = z(\sigma), \qquad \sigma = e^{\zeta}, \qquad \zeta = \xi+i\eta, \tag{8.1.9}$$

where (ξ, η) are a system of orthogonal curvilinear coordinates, then, from (6.5.15)–(6.5.17),

$$\mu(u_\xi+iu_\eta) = \left(\frac{\bar\sigma}{\sigma}\frac{d\sigma}{dz}\frac{d\bar z}{d\bar\sigma}\right)^{\frac{1}{2}}\left\{\kappa f(\sigma)-z\frac{d\bar\sigma}{d\bar z}\bar f'(\bar\sigma)-\frac{d\bar\sigma}{d\bar z}\bar g'(\bar\sigma)\right\}, \tag{8.1.10}$$

$$\sigma_{\xi\xi}+\sigma_{\eta\eta} = \sigma_{xx}+\sigma_{yy} = 4\left\{\frac{d\sigma}{dz}f'(\sigma)+\frac{d\bar\sigma}{d\bar z}\bar f'(\bar\sigma)\right\}, \tag{8.1.11}$$

$$\sigma_{\xi\xi}-\sigma_{\eta\eta}+2i\sigma_{\xi\eta} = \frac{d\zeta}{dz}\frac{d\bar z}{d\bar\zeta}(\sigma_{xx}-\sigma_{yy}+2i\sigma_{xy})$$

$$= -4\frac{\bar\sigma}{\sigma}\frac{d\sigma}{dz}\frac{d}{d\bar\sigma}\left\{z\frac{d\bar\sigma}{d\bar z}\bar f'(\bar\sigma)+\frac{d\bar\sigma}{d\bar z}\bar g'(\bar\sigma)\right\}, \tag{8.1.12}$$

where $\Omega(z) = f(\sigma), \qquad \omega(z) = g(\sigma).$ (8.1.13)

The introduction of the complex variable σ as well as ζ is convenient in applications in which we transform a region to the inside, or outside, of a unit circle in the σ-plane. From (6.5.22) we also have

$$P(\eta) = i\int_A^P (\sigma_{\xi\xi}+i\sigma_{\xi\eta})\frac{dz}{d\eta}\,d\eta \tag{8.1.14}$$

on the curve $\xi =$ constant, or

$$\frac{\partial P(\eta)}{\partial\eta} = i(\sigma_{\xi\xi}+i\sigma_{\xi\eta})\frac{dz}{d\eta}. \tag{8.1.15}$$

In particular, when we have polar coordinates (r, θ) given by

$$z = \sigma = e^{\zeta} = re^{i\theta}, \qquad r = e^{\xi}, \qquad \eta = \theta, \qquad \frac{dz}{d\sigma} = 1, \tag{8.1.16}$$

then
$$\mu(u_r+iu_\theta) = e^{-i\theta}\{\kappa\Omega(z)-z\bar{\Omega}'(\bar{z})-\bar{\omega}'(\bar{z})\}, \tag{8.1.17}$$

$$\sigma_{rr}+\sigma_{\theta\theta} = 4\{\Omega'(z)+\bar{\Omega}'(\bar{z})\}, \tag{8.1.18}$$

$$\sigma_{rr}-\sigma_{\theta\theta}+2i\sigma_{r\theta} = -4\left\{\bar{z}\bar{\Omega}''(\bar{z})+\frac{\bar{z}}{z}\bar{\omega}''(\bar{z})\right\}, \tag{8.1.19}$$

$$\sigma_{rr}+i\sigma_{r\theta} = 2\left\{\Omega'(z)+\bar{\Omega}'(\bar{z})-\bar{z}\bar{\Omega}''(\bar{z})-\frac{\bar{z}}{z}\bar{\omega}''(\bar{z})\right\}, \tag{8.1.20}$$

$$P(\theta) = -\int_A^P (\sigma_{rr}+i\sigma_{r\theta})re^{i\theta}\,d\theta, \tag{8.1.21}$$

equation (8.1.21) holding on curves r = constant.

When the region S occupied by the body extends to infinity and is bounded internally by one or more non-intersecting closed curves lying entirely in the finite part of the plane, then from (6.6.4), for large $|z|$,

$$\left.\begin{aligned}\Omega'(z) &= B+iC-\frac{P}{4\pi(\kappa+1)z}+O\left(\frac{1}{z^2}\right)\\ \omega''(z) &= B'+iC'+\frac{\kappa\bar{P}}{4\pi(\kappa+1)z}+O\left(\frac{1}{z^2}\right)\end{aligned}\right\}, \tag{8.1.22}$$

when the stresses are bounded at infinity, where

$$8B = N_1+N_2, \qquad 4(B'+iC') = -(N_1-N_2)e^{-2i\alpha}. \tag{8.1.23}$$

In (8.1.23), N_1, N_2 are the values of the principal stresses at infinity and α is the angle between N_1 and the x-axis. Also $(\kappa+1)C/\mu$ is the rotation at infinity and $P = X+iY$ is the resultant force acting over all the internal boundaries.

When the stresses and rotation vanish at infinity

$$\left.\begin{aligned}\Omega'(z) &= -\frac{P}{4\pi(\kappa+1)z}+O\left(\frac{1}{z^2}\right)\\ \omega''(z) &= \frac{\kappa\bar{P}}{4\pi(\kappa+1)z}+O\left(\frac{1}{z^2}\right)\end{aligned}\right\}. \tag{8.1.24}$$

The above results do not hold when a part of the boundary of the region S extends to infinity. Suppose S is the half-plane, either $y > 0$ or $y < 0$, and $P = X+iY$ is the resultant of the external forces acting on the boundary Ox and the other boundaries which are in the finite part of the plane. Then, when stresses and rotation vanish at infinity, and any applied stresses on the boundary Ox are $O(1/x^2)$ for large $|x|$,

$$\left.\begin{aligned}\Omega'(z) &= -\frac{P}{4\pi z}+O\left(\frac{1}{z^2}\right)\\ \omega''(z) &= \frac{\bar{P}}{4\pi z}+O\left(\frac{1}{z^2}\right)\end{aligned}\right\}. \tag{8.1.25}$$

8.2. Change of origin

If the origin is changed from the point $z = 0$ to $z_1 \equiv z-c = 0$, the displacement D is given by the two expressions

$$\begin{aligned}\mu D &= \kappa\Omega(z)-z\bar{\Omega}'(\bar{z})-\bar{\omega}'(\bar{z})\\ &= \kappa\Omega_1(z_1)-z_1\bar{\Omega}_1'(\bar{z}_1)-\bar{\omega}_1'(\bar{z}_1),\end{aligned}$$

where $\Omega_1(z_1)$, $\omega_1(z_1)$ are the complex potentials corresponding to the origin $z = c$. Hence

$$\kappa\{\Omega(z)-\Omega_1(z_1)\}-z\{\bar{\Omega}'(\bar{z})-\bar{\Omega}_1'(\bar{z}_1)\}-\bar{\omega}'(\bar{z})+\bar{\omega}_1'(\bar{z}_1)-c\bar{\Omega}_1'(\bar{z}_1) = 0,$$

so that

$$\Omega(z) \equiv \Omega_1(z_1), \qquad \omega(z) \equiv \omega_1(z_1)-\bar{c}\Omega_1(z_1). \tag{8.2.1}$$

8.3. Isolated force and couple in infinite plate

Suppose an isolated force $P = X+iY$ and couple M act at the origin in an infinite plate. If the stresses and rotation due to this force and couple vanish at infinity, formulae (8.1.22) suggest that we assume for $\Omega(z)$, $\omega(z)$ the values

$$\Omega(z) = -\frac{P\ln z}{4\pi(\kappa+1)},$$

$$\omega(z) = \frac{\kappa\bar{P}z\ln z}{4\pi(\kappa+1)}+A\ln z.$$

By considering any circuit surrounding the origin we can verify at once from (8.1.5) and (8.1.6) that we have an isolated force P and couple M at the origin if

$$2\pi i(\bar{A}-A) = M.$$

We see that the term containing the real part of A does not contribute to the force or couple at the origin and we therefore take our fundamental complex potentials for a force P and couple M at $z = 0$ to be

$$\left.\begin{aligned}\Omega(z) &= -\frac{P\ln z}{4\pi(\kappa+1)}\\ \omega(z) &= \frac{\kappa\bar{P}z\ln z}{4\pi(\kappa+1)}+\frac{iM\ln z}{4\pi}\end{aligned}\right\}. \tag{8.3.1}$$

The stresses and rotation corresponding to these potentials vanish at infinity, but the displacement there is infinite. Also the stresses and displacement are infinite at the origin.

Using (8.2.1) we see that for a force $P = X+iY$ and couple M at the point $z = c$ in an infinite plate

$$\left.\begin{aligned}\Omega(z) &= -\frac{P\ln(z-c)}{4\pi(\kappa+1)}\\ \omega(z) &= \frac{\kappa\bar{P}(z-c)\ln(z-c)}{4\pi(\kappa+1)}+\frac{P\bar{c}\ln(z-c)}{4\pi(\kappa+1)}+\frac{iM\ln(z-c)}{4\pi}\end{aligned}\right\}. \tag{8.3.2}$$

8.4. General complex potentials for infinite plate

We shall suppose that the infinite plate is acted on by a finite number n of isolated forces $P_r = X_r+iY_r$ and couples M_r ($r = 1, 2,..., n$) at the points $z = z_r$. There are no other singularities of stress or displacement in the finite part of the plate, and at infinity the distribution of stress is supposed to be known. If the complex potentials are denoted by $\Omega_0(z)$, $\omega_0(z)$, we have, apart from terms giving a rigid-body displacement,

$$\Omega_0(z) = -\frac{1}{4\pi(\kappa+1)}\sum_{r=1}^{n} P_r\ln(z-z_r)+\sum_{r=1}^{m} A_r z^r, \tag{8.4.1}$$

$$\omega_0(z) = \frac{\kappa}{4\pi(\kappa+1)}\sum_{r=1}^{n} \bar{P}_r(z-z_r)\ln(z-z_r)+$$
$$+\frac{1}{4\pi(\kappa+1)}\sum_{r=1}^{n} P_r\bar{z}_r\ln(z-z_r)+\frac{i}{4\pi}\sum_{r=1}^{n} M_r\ln(z-z_r)+\sum_{r=2}^{m} B_r z^r, \tag{8.4.2}$$

where the (complex) constants A_r, B_r have known values. For example, from (8.1.23), if the state of stress at infinity is uniform and the rotation vanishes there,

$$\left.\begin{aligned}8\bar{A}_1 = 8A_1 &= N_1+N_2\\ 8B_2 &= -(N_1-N_2)e^{-2i\alpha}\end{aligned}\right\}, \tag{8.4.3}$$

the remaining constants A_r, B_r being zero.

8.5. Isolated force at boundary

Suppose an isolated force $P = X+iY$ acts at the point C ($z = c$) on a boundary of the body (see Fig. 6.2). We wish to determine the singular part of the complex potentials due to this force and therefore assume that $\Omega(z)$, $\omega(z)$ contain terms

$$\Omega(z) = \alpha\ln(z-c), \qquad \omega(z) = \beta(z-c)\ln(z-c)+\gamma\ln(z-c),$$

where α, β, γ are complex constants. We saw in § 6.4 that if the stress function ϕ is continuous at $z = c$ there is no isolated couple at this point. Therefore

$$\alpha\bar{c}+\gamma = 0.$$

The discontinuity in $\partial\phi/\partial\bar{z}$ at C must be finite and, applying (8.1.5) to a curve $C'DC''$ cut out of the material, as the curve shrinks to zero so that C lies between C' and C'', we find that

$$\alpha+\bar{\beta} = 0, \qquad P = -2\pi(\alpha-\bar{\beta}).$$

Thus the singular parts of the complex potentials are

$$\left.\begin{aligned} \Omega(z) &= -\frac{P}{4\pi}\ln(z-c) \\ \omega(z) &= \frac{\bar{P}}{4\pi}(z-c)\ln(z-c)+\frac{P\bar{c}}{4\pi}\ln(z-c) \end{aligned}\right\}. \tag{8.5.1}$$

If the isolated force P acts at the point $z = c$ on the boundary of a hole which is cut out of an infinite plate, then, in view of the forms (8.1.22) for $\Omega(z)$, $\omega'(z)$ at infinity, we see that the complex potentials must contain terms

$$\left.\begin{aligned} \Omega(z) &= -\frac{P\ln z}{4\pi(\kappa+1)}-\frac{P}{4\pi}\ln\left(\frac{z-c}{z}\right) \\ \omega(z) &= \frac{\kappa\bar{P}z\ln z}{4\pi(\kappa+1)}+\frac{\bar{P}z}{4\pi}\ln\left(\frac{z-c}{z}\right)+ \\ &\quad +\frac{P\bar{c}-\bar{P}c}{4\pi}\ln\left(\frac{z-c}{z}\right) \end{aligned}\right\}, \tag{8.5.2}$$

where the origin of coordinates is inside the hole. Further terms must, of course, be added to satisfy the complete boundary conditions over the hole.

When an isolated force P acts, at the origin, on the boundary of the semi-infinite region $y \geqslant 0$, then, putting $c = 0$ in (8.5.1), we see that these potentials must contain terms

$$\left.\begin{aligned} \Omega(z) &= -\frac{P}{4\pi}\ln z \\ \omega(z) &= \frac{\bar{P}}{4\pi}z\ln z \end{aligned}\right\}. \tag{8.5.3}$$

In this case, however, we find from (8.1.4) that

$$\sigma_{yy}-i\sigma_{xy} = \frac{P}{2\pi}\left(\frac{1}{\bar{z}}-\frac{1}{z}\right)+\frac{\bar{P}}{2\pi}\left(\frac{z}{\bar{z}^2}-\frac{1}{\bar{z}}\right),$$

and this vanishes everywhere on the boundary $y = 0$, where $\bar{z} = z$, except at the origin. The potentials (8.5.3) therefore give the distribution of stress due to an isolated force P acting at the origin on the boundary $y = 0$ of a semi-infinite region which is otherwise free from stress.

If we use polar coordinates, we find from (8.1.18) and (8.1.19) that

$$\pi r(\sigma_{rr}+\sigma_{\theta\theta}) = -2F\cos(\theta-\epsilon),$$

$$\pi r(\sigma_{rr}-\sigma_{\theta\theta}+2i\sigma_{r\theta}) = -2F\cos(\theta-\epsilon),$$

where $P = Fe^{i\epsilon}$ and is a force of magnitude F acting at an angle ϵ to the x-axis. Hence

$$\pi r\sigma_{rr} = -2F\cos(\theta-\epsilon), \qquad \sigma_{r\theta} = \sigma_{\theta\theta} = 0. \tag{8.5.4}$$

When a general distribution of stress is given along the boundary $y = 0$ of a semi-infinite region, the corresponding complex potentials can be found by integration. We will not, however, consider this method further here but will examine problems connected with the half-plane in more detail in the next sections by other methods.

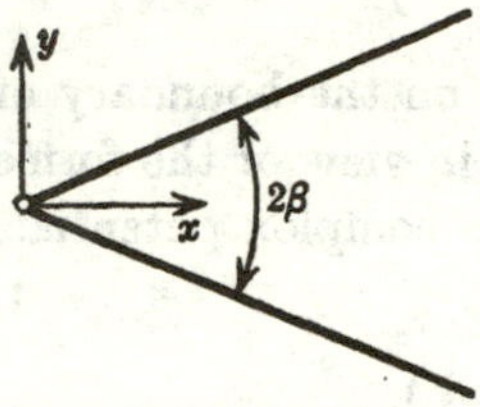

FIG. 8.1. Infinite wedge.

A slight modification of the complex potentials (8.5.3) enables us to solve the problem of an isolated force $P = Fe^{i\epsilon}$ acting at the vertex of an infinite wedge bounded by the lines $\theta = \pm\beta$ (see Fig. 8.1). Thus, assuming

$$\left.\begin{aligned} \Omega(z) &= -\frac{L}{4\pi}\ln z \\ \omega(z) &= \frac{\bar{L}}{4\pi}z\ln z \end{aligned}\right\}, \tag{8.5.5}$$

where L is a constant, we apply (8.1.5) to any circuit r = constant in the wedge from $\theta = -\beta$ to $\theta = \beta$ and obtain

$$2L\beta+\bar{L}\sin 2\beta = \pi P,$$

$$2\bar{L}\beta+L\sin 2\beta = \pi\bar{P},$$

or

$$\left.\begin{aligned} (L+\bar{L})(2\beta+\sin 2\beta) &= 2\pi F\cos\epsilon \\ (L-\bar{L})(2\beta-\sin 2\beta) &= 2\pi iF\sin\epsilon \end{aligned}\right\}. \tag{8.5.6}$$

Hence, from (8.1.18), (8.1.19), (8.5.5), and (8.5.6) we have

$$\left.\begin{aligned} \sigma_{rr} &= -\frac{2F}{r}\left(\frac{\cos\theta\cos\epsilon}{2\beta+\sin 2\beta}+\frac{\sin\theta\sin\epsilon}{2\beta-\sin 2\beta}\right) \\ \sigma_{r\theta} &= \sigma_{\theta\theta} = 0 \end{aligned}\right\}, \tag{8.5.7}$$

and the solution is therefore satisfactory for a wedge which has zero applied stresses over its edges.

8.6. Extension of complex potentials for half-plane†

Consider a body which occupies the upper half-plane $y > 0$ which we call the region S^+, the boundary $y = 0$ of the body being denoted by L. We define complex potentials $\Omega(z)$, $\omega(z)$ which are holomorphic in this half-plane except possibly at infinity. The functions

$$\bar{\Omega}(z) = \overline{\Omega(\bar{z})}, \qquad \bar{\omega}(z) = \overline{\omega(\bar{z})},$$

are then holomorphic for points z belonging to the lower half-plane S^-, since when z is in S^-, $\bar{z}$ belongs to S^+. Alternatively, if

$$\Omega(z) = \Omega_1(x, y) + i\Omega_2(x, y),$$

then

$$\bar{\Omega}(z) = \Omega_1(x, -y) - i\Omega_2(x, -y).$$

We now extend the definition of $\Omega(z)$ to the whole complex plane by putting

$$\Omega(z) = -z\bar{\Omega}'(z) - \bar{\omega}'(z) \quad (z \text{ in } S^-). \tag{8.6.1}$$

This extension of $\Omega(z)$ to the lower half-plane is chosen so that the functions $\Omega(z)$ are the analytical continuations of each other through the unloaded parts of the boundary where $P = X + iY = 0$ (see (8.1.5)).

If z belongs to S^+, then $\bar{z}$ belongs to S^- so that, from (8.6.1),

$$\Omega(\bar{z}) = -\bar{z}\bar{\Omega}'(\bar{z}) - \bar{\omega}'(\bar{z}) \quad (z \text{ in } S^+),$$

or

$$\bar{\omega}'(\bar{z}) = -\Omega(\bar{z}) - \bar{z}\bar{\Omega}'(\bar{z}) \quad (z \text{ in } S^+). \tag{8.6.2}$$

Also, differentiating this with respect to $\bar{z}$,

$$\bar{\omega}''(\bar{z}) = -\Omega'(\bar{z}) - \bar{\Omega}'(\bar{z}) - \bar{z}\bar{\Omega}''(\bar{z}) \quad (z \text{ in } S^+). \tag{8.6.3}$$

With the help of (8.6.2) and (8.6.3) the formulae (8.1.1)–(8.1.5) become, for z in S^+,

$$\mu D = \kappa\Omega(z) + \Omega(\bar{z}) + (\bar{z} - z)\bar{\Omega}'(\bar{z}), \tag{8.6.4}$$

$$\sigma_{xx} + \sigma_{yy} = 4\{\Omega'(z) + \bar{\Omega}'(\bar{z})\}, \tag{8.6.5}$$

$$\sigma_{xx} - \sigma_{yy} + 2i\sigma_{xy} = 4\{\Omega'(\bar{z}) + \bar{\Omega}'(\bar{z}) + (\bar{z} - z)\bar{\Omega}''(\bar{z})\}, \tag{8.6.6}$$

$$\sigma_{yy} - i\sigma_{xy} = 2\{\Omega'(z) - \Omega'(\bar{z}) + (z - \bar{z})\bar{\Omega}''(\bar{z})\}, \tag{8.6.7}$$

$$P = X + iY = 2i\{\Omega(z) - \Omega(\bar{z}) + (z - \bar{z})\bar{\Omega}'(\bar{z})\}. \tag{8.6.8}$$

We also observe that

$$\mu\frac{\partial D}{\partial x} = \kappa\Omega'(z) + \Omega'(\bar{z}) + (\bar{z} - z)\bar{\Omega}''(\bar{z}). \tag{8.6.9}$$

† See N. I. Muskhelishvili, *Singular Integral Equations*, loc. cit., p. 1.

When the stresses and rotation vanish at infinity in the half-plane S^+, we see from (8.1.25) and (8.6.1) that $\Omega'(z) = O(1/z)$ at infinity in S^- and hence

$$\Omega'(z) = O\left(\frac{1}{z}\right), \tag{8.6.10}$$

for large $|z|$ in the whole plane $S^+ + S^-$.

8.7. First boundary-value problem for half-plane†

In this problem the external stresses are given at all points of the boundary L of the half-plane S^+, and we can conveniently divide the problem into two parts. In the first we suppose that the whole of the boundary L is free from applied stress and that a distribution of stress with known singularities exists in S^+ such that stresses and rotation are zero at infinity. For example, we may have known distribution of isolated forces in S^+. We therefore suppose that the complex potentials $\Omega_0(z)$, $\omega_0(z)$ define a stress system which has singularities only in S^+ and which is such that

$$\left.\begin{aligned} \Omega_0'(z) &= -\frac{P}{4\pi(\kappa+1)z}+O\left(\frac{1}{z^2}\right) \\ \omega_0''(z) &= \frac{\kappa\bar{P}}{4\pi(\kappa+1)z}+O\left(\frac{1}{z^2}\right) \end{aligned}\right\}, \tag{8.7.1}$$

for large $|z|$. We now wish to find complex potentials $\Omega_1(z)$, $\omega_1(z)$ so that the combined potentials

$$\Omega(z) = \Omega_0(z)+\Omega_1(z), \qquad \omega(z) = \omega_0(z)+\omega_1(z), \tag{8.7.2}$$

give zero applied stress on L and, for large $|z|$, from (8.1.25),

$$\left.\begin{aligned} \Omega'(z) &= -\frac{P}{4\pi z}+O\left(\frac{1}{z^2}\right) \\ \omega''(z) &= \frac{\bar{P}}{4\pi z}+O\left(\frac{1}{z^2}\right) \end{aligned}\right\}. \tag{8.7.3}$$

It follows from (8.7.1) and (8.7.2) that $\Omega_1'(z)$ and $\omega_1''(z)$ are both $O(1/z)$ for large $|z|$ in S^+. If, therefore, the definition of $\Omega_1(z)$ is extended by (8.6.1) to

$$\left.\begin{aligned} \Omega_1(z) &= -z\bar{\Omega}_1'(z)-\bar{\omega}_1'(z) \\ \Omega_1'(z) &= -\bar{\Omega}_1'(z)-z\bar{\Omega}_1''(z)-\bar{\omega}_1''(z) \end{aligned}\right\} \quad (z \text{ in } S^-), \tag{8.7.4}$$

then $\Omega_1'(z)$ is defined in the whole plane and is $O(1/z)$ at infinity. Also

$$\omega_1'(z) = -\bar{\Omega}_1(z)-z\Omega_1'(z) \quad (z \text{ in } S^+). \tag{8.7.5}$$

† See I. N. Sneddon, *Fourier Transforms* (1951) for an alternative method of solution using Fourier transforms.

From (8.1.4), (8.6.7), and (8.7.2),

$$\sigma_{yy}-i\sigma_{xy} = 2\{\Omega_1'(z)-\Omega_1'(\bar{z})+(z-\bar{z})\bar{\Omega}_1''(\bar{z})\}+ \\ +2\{\Omega_0'(z)+\bar{\Omega}_0'(\bar{z})+z\bar{\Omega}_0''(\bar{z})+\bar{\omega}_0''(\bar{z})\}. \quad (8.7.6)$$

If we assume that

$$\lim_{y\to 0} y\bar{\Omega}_1''(\bar{z}) = \lim_{y\to 0} y\bar{\Omega}_1''(x-iy) = 0, \quad (8.7.7)$$

and that the limits $[\Omega_1'(x)]^+$, $[\Omega_1'(x)]^-$ exist on L, then, if the applied stress on L vanishes,

$$[\Omega_1'(x)]^+-[\Omega_1'(x)]^- = -\Omega_0'(x)-\bar{\Omega}_0'(x)-x\bar{\Omega}_0''(x)-\bar{\omega}_0''(x) \quad (8.7.8)$$

for all points x on L. From the definition of the complex potentials, $\Omega_0(z)$, $\omega_0(z)$, and from (8.7.1), we see that the right-hand side of (8.7.8) satisfies sufficient conditions (see § 1.18) for us to write the (unique) solution $\Omega_1'(z)$ of this equation, which vanishes at infinity, in the form

$$\Omega_1'(z) = -\frac{1}{2\pi i}\int\limits_{-\infty}^{\infty}\frac{\Omega_0'(t)+\bar{\Omega}_0'(t)+t\bar{\Omega}_0''(t)+\bar{\omega}_0''(t)}{t-z}\,dt. \quad (8.7.9)$$

Also, using § 1.17, it can be verified that the condition (8.7.7) is then satisfied.

From (8.7.1) we see that the integral (8.7.9), taken along the infinite semicircle either in S^+ or in S^-, vanishes, so we may evaluate (8.7.9) by residue theory. The function $\bar{\Omega}_0'(z)+z\bar{\Omega}_0''(z)+\bar{\omega}_0''(z)$ is regular in S^+ and $\Omega_0'(z)$ is regular in S^-. Hence

$$\left.\begin{aligned} \Omega_1'(z) &= -\bar{\Omega}_0'(z)-z\bar{\Omega}_0''(z)-\bar{\omega}_0''(z) && (z \text{ in } S^+) \\ \Omega_1'(z) &= \Omega_0'(z) && (z \text{ in } S^-) \end{aligned}\right\}. \quad (8.7.10)$$

This solution could also be written down by inspection from (8.7.8). It follows from (8.7.10) that

$$\left.\begin{aligned} \Omega_1(z) &= -z\bar{\Omega}_0'(z)-\bar{\omega}_0'(z) && (z \text{ in } S^+) \\ \Omega_1(z) &= \Omega_0(z) && (z \text{ in } S^-) \end{aligned}\right\}, \quad (8.7.11)$$

apart from non-essential constants.

From (8.7.11) and (8.7.5) we obtain the *complete* potentials (8.7.2) for the half-plane S^+ in the forms

$$\left.\begin{aligned} \Omega(z) &= \Omega_0(z)-z\bar{\Omega}_0'(z)-\bar{\omega}_0'(z) \\ \omega'(z) &= \omega_0'(z)-\bar{\Omega}_1(z)-z\Omega_1'(z) \\ &= \omega_0'(z)-\bar{\Omega}_0(z)+z\bar{\Omega}_0'(z)+z^2\bar{\Omega}_0''(z)+z\bar{\omega}_0''(z) \end{aligned}\right\}. \quad (8.7.12)$$

We observe from (8.7.1) that $\Omega'(z)$, $\omega''(z)$ have the correct forms for large $|z|$. The complete expressions for the displacement and stresses now follow from (8.1.1)–(8.1.3), and (8.7.12). In particular, since the applied edge stresses are zero, it follows from (8.1.2) that the value of σ_{xx} on the edge is given by

$$\sigma_{xx} = -4\{x\Omega_0''(x)+x\bar{\Omega}_0''(x)+\omega_0''(x)+\bar{\omega}_0''(x)\}. \tag{8.7.13}$$

Before considering applications of the above work, we complete the theory by solving the problem of the half-plane with a given distribution of applied stress on the boundary L and with no singularities in the half-plane S^+. We suppose that the applied stresses are $O(1/x^2)$ on L, and if the stresses and rotation vanish at infinity then, for large $|z|$, from (8.1.25), the complex potentials $\Omega(z)$, $\omega(z)$ are such that

$$\left.\begin{aligned}\Omega'(z) &= -\frac{P}{4\pi z}+O\!\left(\frac{1}{z^2}\right)\\ \omega''(z) &= \frac{\bar{P}}{4\pi z}+O\!\left(\frac{1}{z^2}\right)\end{aligned}\right\}. \tag{8.7.14}$$

On the boundary L we assume that

$$\sigma_{yy} = p(x), \qquad \sigma_{xy} = q(x), \tag{8.7.15}$$

where $p(x)$, $q(x)$ and their derivatives $p'(x)$, $q'(x)$ satisfy the H-condition on L, and x^2p, x^2q, x^3p', x^3q' satisfy the H-condition in the neighbourhood of the point at infinity.

We indicate two methods of solution. In the first method† we continue the definition of $\Omega(z)$ into S^- as in § 8.6. It then follows from (8.7.14) that $\Omega'(z) = O(1/z)$ in the whole plane, and, from (8.6.7),

$$[\Omega'(x)]^+-[\Omega'(x)]^- = \tfrac{1}{2}\{p(x)-iq(x)\}, \tag{8.7.16}$$

provided

$$\lim_{y\to 0} y\bar{\Omega}''(\bar{z}) = 0 \tag{8.7.17}$$

and provided the limits $[\Omega'(x)]^+$, $[\Omega'(x)]^-$ exist. The required (unique) solution of this boundary-value problem is, from § 1.18,

$$\Omega'(z) = \frac{1}{4\pi i}\int_{-\infty}^{\infty}\frac{p(t)-iq(t)}{t-z}\,dt, \tag{8.7.18}$$

and stresses and displacements may then be found from (8.6.4)–(8.6.6). We see, from § 1.17, that the condition (8.7.17) is satisfied.

In the second method of solution‡ we put

$$\left.\begin{aligned}V(z) &= 4\Omega'(z)+2z\Omega''(z)+2\omega''(z)\\ W(z) &= 2z\Omega''(z)+2\omega''(z)\end{aligned}\right\}, \tag{8.7.19}$$

† N. I. Muskhelishvili, loc. cit., p. 1.

‡ A. E. Green, *Proc. Camb. phil. Soc. math. phys. Sci.*, loc. cit., p. 182, and W. G Bickley, *Phil. Trans. R. Soc.* **A227** (1928) 383.

so that

$$\left.\begin{aligned}\Omega'(z) &= \tfrac{1}{4}\{V(z)-W(z)\}\\ \omega''(z) &= \tfrac{1}{2}W(z)-\tfrac{1}{4}z\{V'(z)-W'(z)\}\end{aligned}\right\}, \tag{8.7.20}$$

and we observe, from (8.1.4) and (8.7.15), that

$$\operatorname{re} V(z) = p(x), \qquad \operatorname{im} W(z) = q(x) \tag{8.7.21}$$

on the boundary L where z is real. Also, from (8.1.25),

$$V(z) = O\left(\frac{1}{z}\right), \qquad W(z) = O\left(\frac{1}{z}\right), \tag{8.7.22}$$

for large $|z|$. In (8.7.21) $\operatorname{re} V(z)$ denotes the real part of $V(z)$ and $\operatorname{im} W(z)$ denotes the imaginary part of $W(z)$. It follows by a well-known result that

$$V(z) = \frac{1}{\pi i}\int\limits_{-\infty}^{\infty}\frac{p(t)\,dt}{t-z}, \qquad W(z) = \frac{1}{\pi}\int\limits_{-\infty}^{\infty}\frac{q(t)\,dt}{t-z}. \tag{8.7.23}$$

From (8.1.2), (8.1.3), and (8.7.20),

$$\left.\begin{aligned}\sigma_{xx}+\sigma_{yy} &= V(z)-W(z)+\bar{V}(\bar{z})-\bar{W}(\bar{z})\\ \sigma_{xx}-\sigma_{yy}-2i\sigma_{xy} &= -2W(z)+(z-\bar{z})\{V'(z)-W'(z)\}\end{aligned}\right\}, \tag{8.7.24}$$

and, on the boundary $y = 0$,

$$\left.\begin{aligned}\sigma_{xx}+\sigma_{yy} &= V(x)-W(x)+\bar{V}(x)-\bar{W}(x)\\ \sigma_{xx}-\sigma_{yy}-2i\sigma_{xy} &= -2W(x)\end{aligned}\right\}. \tag{8.7.25}$$

We now consider some applications of the theory of this section.

8.8. Isolated force and couple in semi-infinite body

If an isolated force $P = X+iY = Fe^{i\epsilon}$ and an isolated couple M act at the point $z = ib$ in S^+, then, from (8.3.2),

$$\left.\begin{aligned}\Omega_0(z) &= -\frac{P\ln(z-ib)}{4\pi(\kappa+1)}\\ \omega_0(z) &= \frac{\kappa\bar{P}(z-ib)\ln(z-ib)}{4\pi(\kappa+1)}-\frac{ibP\ln(z-ib)}{4\pi(\kappa+1)}+\frac{iM}{4\pi}\ln(z-ib)\end{aligned}\right\}. \tag{8.8.1}$$

The complete distribution of stress and displacement may now be obtained from the complex potentials (8.7.12) and (8.8.1). Restricting our attention to the value of the edge stress, we have, from (8.7.13),

$$\sigma_{xx} = \frac{2}{\pi(x^2+b^2)^2}\left\{\left(\frac{3-\kappa}{1+\kappa}b^2-x^2\right)xX-\left(b^2-\frac{3-\kappa}{1+\kappa}x^2\right)bY\right\}-\frac{4Mbx}{\pi(x^2+b^2)^2}. \tag{8.8.2}$$

Results for continuous distributions of forces in S^+ may be found by integration.

8.9. Examples of given applied stresses on boundary of semi-infinite body

EXAMPLE 1. As a first example we consider the problem of a uniform distribution of normal and shear stress on $y = 0$ from $x = -a$ to $x = a$. Thus

$$\left.\begin{aligned} p(x) &= -p, \quad q(x) = q \quad (|x| \leqslant a) \\ p(x) &= 0, \quad q(x) = 0 \quad (|x| > a) \end{aligned}\right\}, \tag{8.9.1}$$

where p, q are constants, and, from (8.7.23),

$$V(z) = \frac{ip}{\pi}\ln\frac{z-a}{z+a}, \qquad W(z) = \frac{q}{\pi}\ln\frac{z-a}{z+a}. \tag{8.9.2}$$

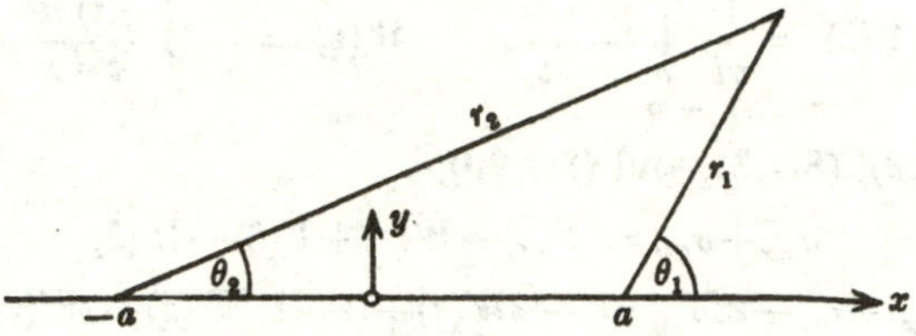

FIG. 8.2. Bipolar coordinates.

If we write $z-a = r_1 e^{i\theta_1}$, $z+a = r_2 e^{i\theta_2}$ (see Fig. 8.2), it follows immediately from (8.7.24) that

$$\left.\begin{aligned} \sigma_{xx}+\sigma_{yy} &= -\frac{2p}{\pi}(\theta_1-\theta_2)-\frac{2q}{\pi}\ln\frac{r_1}{r_2} \\ \sigma_{xx}-\sigma_{yy}+2i\sigma_{xy} &= \left(\frac{ip+q}{\pi}\right)(e^{2i\theta_1}-e^{2i\theta_2})-\frac{2q}{\pi}\left\{\ln\frac{r_1}{r_2}+i(\theta_2-\theta_1)\right\} \end{aligned}\right\}. \tag{8.9.3}$$

These results may also be obtained at once from (8.7.18), (8.6.5), and (8.6.6). In the special case when $q = 0$ equivalent results were found by Okobu† by a different method.

The limiting case of the above problem when $a \to 0$ and p, q both tend to infinity in such a way that

$$2ap \to Y, \qquad 2aq \to -X,$$

where X, Y are constants, is of some interest. From (8.9.2)

$$V(z) \to -\frac{iY}{\pi z}, \qquad W(z) \to \frac{X}{\pi z},$$

and hence (8.7.20) gives

$$\Omega'(z) = -\frac{P}{4\pi z}, \qquad \omega''(z) = \frac{\bar{P}}{4\pi z},$$

† S. Okobu, *Sci. Rep. Tôhoku Univ.* 25 (1937) 114.

which agrees with the complex potentials (8.5.3) for an isolated force $P = X+iY$ at the origin on the boundary.

EXAMPLE 2. In this example we suppose that the applied shearing stress is zero so that $q(x) = 0$, and that

$$\left.\begin{aligned} p(x) &= -\frac{Y}{\pi(a^2-x^2)^{\frac{1}{2}}} \quad (|x| \leqslant a) \\ &= 0 \quad (|x| > a) \end{aligned}\right\}, \tag{8.9.4}$$

where Y is constant and is the total normal pressure exerted on the strip. In this case $p(x)$ does not satisfy the H-condition at $x = \pm a$, but the formulae (8.7.23) are still valid, and therefore $W = 0$ and

$$V(z) = \frac{iY}{\pi^2}\int\limits_{-a}^{a} \frac{dt}{(t-z)(a^2-t^2)^{\frac{1}{2}}}. \tag{8.9.5}$$

By $(a^2-t^2)^{\frac{1}{2}}$ for $|t| < a$ we understand a positive quantity, and $(a^2-z^2)^{\frac{1}{2}}$ is that branch, holomorphic in the plane cut along $(-a,a)$, taking the positive value on the upper side. Also, for large $|z|$ this branch is $-iz+O(1)$. Now

$$\frac{1}{\pi i}\int\limits_{-a}^{a} \frac{dt}{(t-z)(a^2-t^2)^{\frac{1}{2}}} = \frac{1}{2\pi i}\oint\limits_{\Delta} \frac{dt}{(t-z)(a^2-t^2)^{\frac{1}{2}}},$$

taken clockwise round a contour Δ surrounding the cut. Since the contour integral taken round the infinite circle vanishes, it follows at once that

$$\frac{1}{\pi i}\int\limits_{-a}^{a} \frac{dt}{(t-z)(a^2-t^2)^{\frac{1}{2}}} = \frac{1}{(a^2-z^2)^{\frac{1}{2}}},$$

and therefore
$$V(z) = -\frac{Y}{\pi(a^2-z^2)^{\frac{1}{2}}} = \frac{Y}{i\pi(z^2-a^2)^{\frac{1}{2}}}. \tag{8.9.6}$$

Hence from (8.7.20) the corresponding complex potentials are

$$\left.\begin{aligned} \Omega(z) &= -\frac{iY}{4\pi}\ln\{z+(z^2-a^2)^{\frac{1}{2}}\} \\ &= -\frac{iY}{4\pi}\ln\{z+i(a^2-z^2)^{\frac{1}{2}}\} \\ \omega'(z) &= \frac{iY}{4\pi}\left[\frac{z}{(z^2-a^2)^{\frac{1}{2}}}-\ln\{z+(z^2-a^2)^{\frac{1}{2}}\}\right] \\ &= \frac{Y}{4\pi}\left[\frac{z}{(a^2-z^2)^{\frac{1}{2}}}-i\ln\{z+i(a^2-z^2)^{\frac{1}{2}}\}\right] \end{aligned}\right\}, \tag{8.9.7}$$

apart from constants, and stresses and displacements are then given by (8.1.1)–(8.1.4). In particular we may verify that the y-component of displacement, u_y, on $y = 0$, is constant when $|x| < a$ and that the stresses are

$$\left.\begin{aligned} \sigma_{xx}+\sigma_{yy} &= -\frac{2Y}{\pi(r_1 r_2)^{\frac{1}{2}}}\sin\tfrac{1}{2}(\theta_1+\theta_2) \\ \sigma_{xx}-\sigma_{yy}+2i\sigma_{xy} &= -\frac{2Yr^2\sin\theta}{\pi(r_1 r_2)^{\frac{3}{2}}}e^{-i\{\theta-\frac{3}{2}(\theta_1+\theta_2)\}} \end{aligned}\right\}. \tag{8.9.8}$$

8.10. Second boundary-value problem for half-plane

Here the values of the displacements are given at all points of the boundary L of S^+, and as in § 8.7 we consider two cases. In the first case the edge displacements are zero and we seek complex potentials (8.7.2), where $\Omega_0(z)$, $\omega_0(z)$ define a given stress system which has singularities only in S^+. If we extend the definition of $\Omega_1(z)$ to the whole plane by (8.7.4), then, from (8.1.1), (8.6.9), and (8.7.2), the derivative of the total displacement at any point of the plate is

$$\mu\frac{\partial D}{\partial x} = \kappa\Omega_1'(z)+\Omega_1'(\bar{z})+(\bar{z}-z)\bar{\Omega}_1''(\bar{z})+\kappa\Omega_0'(z)-\bar{\Omega}_0'(\bar{z})-z\bar{\Omega}_0''(\bar{z})-\bar{\omega}_0''(\bar{z}). \tag{8.10.1}$$

Hence, making assumptions similar to those in § 8.7, except that the special form (8.7.3) is deleted, it follows that, if $D = 0$ on the boundary,

$$\kappa[\Omega_1'(x)]^+ + [\Omega_1'(x)]^- = -\kappa\Omega_0'(x)+\bar{\Omega}_0'(x)+x\bar{\Omega}_0''(x)+\bar{\omega}_0''(x), \tag{8.10.2}$$

for all points x on L, where $\Omega_1'(z)$ vanishes at infinity. To solve this equation we put

$$A(z) = \kappa\Omega_1'(z) \quad (z \text{ in } S^+),$$
$$A(z) = -\Omega_1'(z) \quad (z \text{ in } S^-),$$

so that (8.10.2) becomes

$$A^+(x)-A^-(x) = -\kappa\Omega_0'(x)+\bar{\Omega}_0'(x)+x\bar{\Omega}_0''(x)+\bar{\omega}_0''(x),$$

and $A(z)$ vanishes at infinity. Hence, by § 1.18, the (unique) value of $A(z)$ is

$$A(z) = \frac{1}{2\pi i}\int\limits_{-\infty}^{\infty}\{-\kappa\Omega_0'(t)+\bar{\Omega}_0'(t)+t\bar{\Omega}_0''(t)+\bar{\omega}_0''(t)\}\frac{dt}{t-z}.$$

This integral may be evaluated by a process similar to that used in § 8.7, and we find that

$$\left.\begin{aligned} A(z) &= \kappa\Omega_1'(z) = \bar{\Omega}_0'(z)+z\bar{\Omega}_0''(z)+\bar{\omega}_0''(z) \quad &(z \text{ in } S^+) \\ A(z) &= -\Omega_1'(z) = \kappa\Omega_0'(z) \quad &(z \text{ in } S^-) \end{aligned}\right\}. \tag{8.10.3}$$

We could also obtain this result by inspection from (8.10.2). By integration of (8.10.3) we have

$$\left.\begin{aligned}\kappa\Omega_1(z) &= z\bar{\Omega}_0'(z)+\bar{\omega}_0'(z) \quad (z \text{ in } S^+)\\ \Omega_1(z) &= -\kappa\Omega_0(z) \quad (z \text{ in } S^-)\end{aligned}\right\}, \tag{8.10.4}$$

apart from non-essential constants, and hence, with the help of (8.7.5), we see that the complete complex potentials for the half-plane S^+ are

$$\left.\begin{aligned}\Omega(z) &= \Omega_0(z)+\Omega_1(z) = \Omega_0(z)+\{z\bar{\Omega}_0'(z)+\bar{\omega}_0'(z)\}/\kappa\\ \omega'(z) &= \omega_0'(z)+\omega_1'(z) = \omega_0'(z)-\bar{\Omega}_1(z)-z\Omega_1'(z)\\ &= \omega_0'(z)+\kappa\bar{\Omega}_0(z)-\{z\bar{\Omega}_0'(z)+z^2\bar{\Omega}_0''(z)+z\bar{\omega}_0''(z)\}/\kappa\end{aligned}\right\}. \tag{8.10.5}$$

Displacements and stresses now follow from (8.1.1)–(8.1.4).

When displacements take given values on the boundary L we assume that stresses and rotation vanish at infinity, and that

$$u_x = g_1(x), \qquad u_y = g_2(x). \tag{8.10.6}$$

We assume that $g_1'(x)$, $g_2'(x)$, $g_1''(x)$, $g_2''(x)$ satisfy the H-condition on L, and x^2g_1', x^2g_2', x^3g_1'', x^3g_2'' satisfy the H-condition in the neighbourhood of the point at infinity. Two methods of solution are again possible which are analogous to those used in § 8.7 for prescribed stresses, but we restrict our attention to one method here. If the definition of $\Omega(z)$ is continued into S^- by (8.6.1), then $\Omega'(z) = O(1/z)$ in the whole plane for large $|z|$, and if the boundary values of the derivatives of the displacement equal the derivatives of the boundary values, and if (8.7.17) is still satisfied, then, from (8.6.9),

$$\kappa[\Omega'(x)]^+ + [\Omega'(x)]^- = \mu\{g_1'(x)+ig_2'(x)\}. \tag{8.10.7}$$

This equation may be solved by a method similar to that used for equation (8.10.2) so that

$$\left.\begin{aligned}\Omega'(z) &= \frac{\mu}{2\pi\kappa i}\int\limits_{-\infty}^{\infty}\frac{g_1'(t)+ig_2'(t)}{t-z}\,dt \quad (z \text{ in } S^+)\\ \Omega'(z) &= -\frac{\mu}{2\pi i}\int\limits_{-\infty}^{\infty}\frac{g_1'(t)+ig_2'(t)}{t-z}\,dt \quad (z \text{ in } S^-)\end{aligned}\right\}. \tag{8.10.8}$$

The special forms (8.7.3) are not valid here and the resultant force over the boundary may be infinite. Detailed examination of particular problems is left as an exercise since it follows lines similar to those indicated for the first boundary-value problem.

8.11. Mixed boundary-value problem for half-plane†

Let $L = L_1+...+L_n$ denote a finite number of segments L_r $(r = 1, 2,..., n)$ of the boundary Ox of the half-plane S^+, and let the displacement components be given on L and the external stress components on the remaining part L' of the boundary. The problem here can be reduced to the case in which the external stress components σ_{yy}, σ_{xy}, given on L', are zero, if we have already solved the first boundary problem of § 8.7.

We assume therefore that

$$\sigma_{yy} = 0, \qquad \sigma_{xy} = 0 \quad \text{on } L', \tag{8.11.1}$$

and on the rest of the boundary L the displacement D takes prescribed values. For simplicity we restrict our discussion to the case in which L contains only one segment with end-points a, b and

$$D = u_x+iu_y = g(x) = g_1(x)+ig_2(x), \tag{8.11.2}$$

$g(x)$ being a given function on L. We shall assume that $g(x)$ has a derivative $g'(x)$ which satisfies the H-condition on L.

We adopt the notation of § 8.6, where $\Omega(z)$ is defined in the whole plane and $\Omega'(z) = O(1/z)$ for large $|z|$. We assume that the limits $[\Omega'(x)]^+$, $[\Omega'(x)]^-$ exist for all points x of the x-axis, except possibly at the points a, b, and that

$$\lim_{y\to 0} y\overline{\Omega}''(\bar{z}) = \lim_{y\to 0} y\overline{\Omega}''(x-iy) = 0 \tag{8.11.3}$$

for all x not coinciding with a, b. These assumptions may be verified when the solution of the problem is obtained. From (8.6.7), (8.11.1), and (8.11.3) it follows that

$$[\Omega'(x)]^+-[\Omega'(x)]^- = 0$$

on L', so that we may assume that $\Omega'(z)$ is holomorphic in the whole plane cut along L. From (8.6.9) the remaining boundary condition (8.11.2) on ab gives

$$\kappa[\Omega'(x)]^++[\Omega'(x)]^- = \mu g'(x) \quad \text{on } L. \tag{8.11.4}$$

The problem (8.11.4) is a particular case of the Hilbert problem which was solved in § 1.21. The general solution which admits the possibility that $\Omega'(z)$ may be infinite at the ends a, b, with degree less than 1, is

$$\Omega'(z) = \frac{\mu\{R_0(z)\}^{i\gamma}}{2\pi\kappa i\{R(z)\}^{\frac{1}{2}}}\int\limits_a^b \frac{g'(t)\{R(t)\}^{\frac{1}{2}}}{(t-z)\{R_0(t)\}^{i\gamma}}\,dt+\frac{H\{R_0(z)\}^{i\gamma}}{\{R(z)\}^{\frac{1}{2}}}, \tag{8.11.5}$$

† N. I. Muskhelishvili, loc. cit., p. 1.

where H is a complex constant,

$$2\pi\gamma = \ln(1/\kappa), \tag{8.11.6}$$

and

$$\left.\begin{aligned} R(z) &= (z-a)(b-z) \\ R_0(z) &= (z-a)/(z-b) \end{aligned}\right\}. \tag{8.11.7}$$

By $\{R(z)\}^{\frac{1}{2}}$ we understand that branch which is holomorphic in the plane cut along ab, taking positive values $\{(t-a)(b-t)\}^{\frac{1}{2}}$ on the upper side of ab. For large values of $|z|$

$$\{R(z)\}^{\frac{1}{2}} = -iz+O(1). \tag{8.11.8}$$

Also $\{R_0(z)\}^{i\gamma}$ is a branch, holomorphic in the same region and tending to 1 as $|z| \to \infty$, and $\{R_0(t)\}^{i\gamma}$ denotes the value of $\{R_0(z)\}^{i\gamma}$ on the upper side of ab.

If we assume that the total force $P = X+iY$ is given on the segment ab, then, since

$$\lim_{|z|\to\infty} z\Omega'(z) = -\frac{P}{4\pi},$$

it follows from (8.11.5) that

$$H = \frac{iP}{4\pi}. \tag{8.11.9}$$

The simplest application of the theory of this section is to the pressure of a foundation on soil in which the soil is supposed to adhere to the foundation, sliding being excluded. If the base of the foundation is straight and the body only moves vertically, then

$$g(x) = ik, \qquad g'(x) = 0,$$

where k is a real constant, so that

$$\Omega'(z) = \frac{iP\{(z-a)/(z-b)\}^{i\gamma}}{4\pi\{(z-a)(b-z)\}^{\frac{1}{2}}}. \tag{8.11.10}$$

The values of stresses and displacements may now be found from (8.6.4)–(8.6.6).

The work of this section may be generalized to apply to the problem of given displacement on more than one segment L_r of the x-axis.

8.12. Punch problems†

These problems are similar to those of the previous section but now we assume that friction is negligible and we have the boundary conditions

$$\left.\begin{aligned} \sigma_{xy} &= 0 \quad \text{everywhere on } Ox \\ \sigma_{yy} &= 0 \quad \text{on } Ox \text{ outside } ab \end{aligned}\right\}, \tag{8.12.1}$$

† For an alternative method of solving punch and crack problems see A. E. Green and A. H. England, *Proc. Camb. phil. Soc. math. phys. Sci.* **59** (1963) 489.

and on *ab* only the *normal* component of displacement is given, so that

$$u_y = f(x). \tag{8.12.2}$$

We again restrict our discussion to a single segment *ab* for clarity but the general case of *n* segments can be discussed in a similar manner. We also assume that $f(x)$ has a derivative $f'(x)$ satisfying the H-condition on *ab*. A slight adaptation of the work of the previous section could be used to solve the punch problems, but we use an alternative method which will also be available for solving 'crack' problems in later sections.

If we adopt the notation of § 8.1 we can satisfy the condition that σ_{xy} vanishes everywhere on Ox by taking

$$\omega'(z) = \Omega(z)-z\Omega'(z), \qquad \omega''(z) = -z\Omega''(z). \tag{8.12.3}$$

It then follows from (8.1.1)–(8.1.4) that the displacements and stresses are given by

$$\left.\begin{aligned} \mu D &= \kappa\Omega(z)-\bar{\Omega}(\bar{z})+(\bar{z}-z)\bar{\Omega}'(\bar{z}) \\ \mu\frac{\partial D}{\partial x} &= \kappa\Omega'(z)-\bar{\Omega}'(\bar{z})+(\bar{z}-z)\bar{\Omega}''(\bar{z}) \\ \sigma_{xx}+\sigma_{yy} &= 4\{\Omega'(z)+\bar{\Omega}'(\bar{z})\} \\ \sigma_{xx}-\sigma_{yy}+2i\sigma_{xy} &= 4(\bar{z}-z)\bar{\Omega}''(\bar{z}) \\ \sigma_{yy}-i\sigma_{xy} &= 2\{\Omega'(z)+\bar{\Omega}'(\bar{z})+(z-\bar{z})\bar{\Omega}''(\bar{z})\} \end{aligned}\right\}. \tag{8.12.4}$$

In particular, provided

$$\lim_{y\to 0} y\bar{\Omega}'(\bar{z}) = 0, \qquad \lim_{y\to 0} y\bar{\Omega}''(\bar{z}) = 0, \tag{8.12.5}$$

we have, on the boundary Ox, where $y = 0$,

$$\left.\begin{aligned} \mu D &= \kappa\Omega(z)-\bar{\Omega}(\bar{z}) \\ \mu\frac{\partial D}{\partial x} &= \kappa\Omega'(z)-\bar{\Omega}'(\bar{z}) \\ \sigma_{xx} = \sigma_{yy} &= 2\{\Omega'(z)+\bar{\Omega}'(\bar{z})\}, \quad \sigma_{xy} = 0 \end{aligned}\right\}. \tag{8.12.6}$$

We see from (8.1.25) that for large $|z|$

$$\Omega'(z) = -\frac{iY}{4\pi z}+O\left(\frac{1}{z^2}\right), \tag{8.12.7}$$

if stresses and rotation vanish at infinity, where Y is the total normal force on the edge Ox, the total tangential force X being zero.

The function $\Omega(z)$ in (8.12.4) is defined in the half-plane S^+. We now define $\Omega(z)$ for the half-plane S^- by taking

$$\left.\begin{aligned} \Omega(z) &= -\bar{\Omega}(z) \\ \Omega'(z) &= -\bar{\Omega}'(z) \end{aligned}\right\} \quad (z \text{ in } S^-), \tag{8.12.8}$$

and hence

$$\Omega(\bar{z}) = -\bar{\Omega}(\bar{z}), \qquad \Omega'(\bar{z}) = -\bar{\Omega}'(\bar{z}) \quad (z \text{ in } S^+). \tag{8.12.9}$$

It follows from (8.12.7) that $\Omega'(z) = O(1/z)$ for large $|z|$ in the whole plane, and the displacements and stresses (8.12.4) may now be written in the forms

$$\left.\begin{aligned} \mu D &= \kappa\Omega(z)+\Omega(\bar{z})+(z-\bar{z})\Omega'(\bar{z}) \\ \mu\frac{\partial D}{\partial x} &= \kappa\Omega'(z)+\Omega'(\bar{z})+(z-\bar{z})\Omega''(\bar{z}) \\ \sigma_{xx}+\sigma_{yy} &= 4\{\Omega'(z)-\Omega'(\bar{z})\} \\ \sigma_{xx}-\sigma_{yy}+2i\sigma_{xy} &= 4(z-\bar{z})\Omega''(\bar{z}) \\ \sigma_{yy}-i\sigma_{xy} &= 2\{\Omega'(z)-\Omega'(\bar{z})+(\bar{z}-z)\Omega''(\bar{z})\} \end{aligned}\right\}. \tag{8.12.10}$$

Also, from (8.12.6) and (8.12.9), provided (8.12.5) is satisfied,

$$\left.\begin{aligned} \mu D = \kappa\Omega(z)+\Omega(\bar{z}), \qquad \mu\frac{\partial D}{\partial x} = \kappa\Omega'(z)+\Omega'(\bar{z}) \\ \sigma_{xx} = \sigma_{yy} = 2\{\Omega'(z)-\Omega'(\bar{z})\}, \qquad \sigma_{xy} = 0 \end{aligned}\right\}, \tag{8.12.11}$$

on the boundary Ox. Using the second boundary condition in (8.12.1) we now see that

$$[\Omega'(x)]^+-[\Omega'(x)]^- = 0$$

on Ox outside ab, so that $\Omega'(z)$ is holomorphic in the whole plane cut along ab, and vanishes at infinity. Also, from (8.12.2) and (8.12.11),

$$[\Omega'(x)]^+ + [\Omega'(x)]^- = \frac{2i\mu f'(x)}{\kappa+1} \quad \text{on } ab. \tag{8.12.12}$$

The general solution of this problem in which we allow the possibility of infinities in $\Omega'(z)$ of order not greater than one at the points a, b is (see § 1.21)

$$\Omega'(z) = \frac{\mu}{\pi(\kappa+1)\{R(z)\}^{\frac{1}{2}}}\int_a^b \frac{f'(t)\{R(t)\}^{\frac{1}{2}}}{t-z}\,dt + \frac{H}{\{R(z)\}^{\frac{1}{2}}}, \tag{8.12.13}$$

where $R(z)$ is given by (8.11.7) and $\{R(z)\}^{\frac{1}{2}}$ is defined in § 8.11. We can verify that the condition (8.12.9) is satisfied provided H is a real constant.

The total applied force on the segment ab may be found by considering the coefficient of $1/z$ in (8.12.7) and (8.12.13). Hence

$$H = -\frac{Y}{4\pi}. \tag{8.12.14}$$

The above solution of the punch problem assumes that the segment of contact ab between the punch and the half-plane was given beforehand. This corresponds to the case in which the punch has the form shown in Fig. 8.3 (i) and the force is such that the corners A, B of the punch come into contact with the elastic body, the presence of the corners A, B of the punch explaining the infinitely large stresses at the points a, b of the half-plane coinciding with A, B.

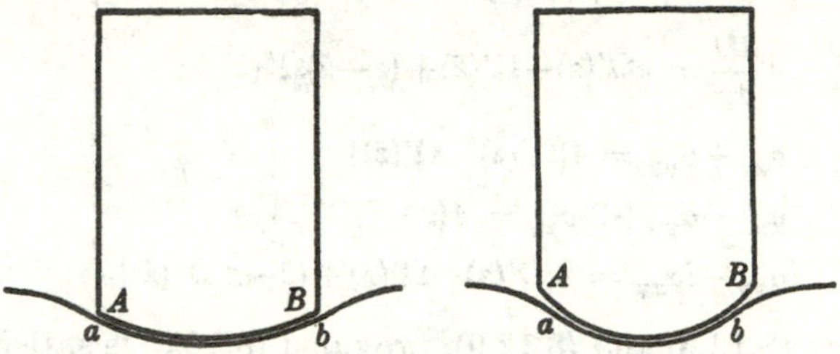

FIG. 8.3. Punch on half-plane.

When the punch has no corners or when the applied force is not sufficiently large for the corners A and B to meet the elastic half-plane, then the ends a and b of the region of contact are unknown beforehand. The problem can, however, be solved by adding the condition that the stresses at a and b are finite. In this case it is, perhaps, more convenient to write down immediately from § 1.21 the solution for $\Omega'(z)$ which is bounded at the ends a and b. Thus

$$\Omega'(z) = \frac{\mu\{R(z)\}^{\frac{1}{2}}}{\pi(\kappa+1)} \int_a^b \frac{f'(t)\,dt}{(t-z)\{R(t)\}^{\frac{1}{2}}}, \tag{8.12.15}$$

and this solution vanishes at infinity if

$$\int_a^b \frac{f'(t)\,dt}{\{(t-a)(b-t)\}^{\frac{1}{2}}} = 0. \tag{8.12.16}$$

The total normal pressure Y may be found by considering the coefficient of $1/z$ in (8.12.15), remembering (8.12.7). Thus

$$Y = -\frac{4\mu}{\kappa+1} \int_a^b \frac{tf'(t)\,dt}{\{(t-a)(b-t)\}^{\frac{1}{2}}}. \tag{8.12.17}$$

8.13. Indentation by rectangular block

We consider two examples of the theory of the previous section. The first is the indentation of a semi-infinite elastic medium by a rigid rectangular punch so that

$$f(x) = d, \tag{8.13.1}$$

where d is constant. In this case, because of symmetry, we can take $a = -l$, $b = l$, but we cannot obtain a solution which gives a finite value of $\Omega'(z)$ at $z = \pm l$. We therefore return to the more general solution (8.12.13), which, with (8.12.14) and (8.13.1), gives

$$\Omega'(z) = -\frac{Y}{4\pi(l^2-z^2)^{\frac{1}{2}}} = -\frac{iY}{4\pi(z^2-l^2)^{\frac{1}{2}}}, \tag{8.13.2}$$

and, integrating,

$$\left.\begin{aligned} \Omega(z) &= -\frac{iY}{4\pi}\ln\{z+(z^2-l^2)^{\frac{1}{2}}\} \\ &= -\frac{iY}{4\pi}\ln\{z+i(l^2-z^2)^{\frac{1}{2}}\} \end{aligned}\right\}, \tag{8.13.3}$$

apart from an arbitrary additional imaginary constant.

We observe that (8.13.3) agrees with the value given in (8.9.7) for the problem of a given pressure distribution (8.9.4) on $y = 0$ which was seen to produce constant normal displacement on $y = 0$ for $|x| \leqslant l$. The general stress components may be written down from (8.12.10) and (8.13.2) and agree with those given in (8.9.8). In particular, we notice from (8.12.11) that, on the boundary,

$$\left.\begin{aligned} \sigma_{xx} = \sigma_{yy} &= -\frac{Y}{\pi(l^2-x^2)^{\frac{1}{2}}} \quad (|x| < l) \\ \sigma_{xx} = \sigma_{yy} &= 0 \quad (|x| > l) \end{aligned}\right\}. \tag{8.13.4}$$

8.14. Indentation by circular block

Here we take

$$f(x) = d - \frac{x^2}{2R}, \tag{8.14.1}$$

which corresponds to the case of a rigid circular punch of large radius R. Because of symmetry we can again take $a = -l$, $b = l$ and we use the solution (8.12.15) for $\Omega'(z)$ which is finite at the ends a, b of the arc of contact of the punch with the elastic half-plane. The condition (8.12.16) is automatically satisfied and (8.12.17) gives

$$Y = \frac{4\mu}{(\kappa+1)R}\int\limits_{-l}^{l}\frac{t^2\,dt}{(l^2-t^2)^{\frac{1}{2}}} = \frac{2\pi\mu l^2}{(\kappa+1)R}, \tag{8.14.2}$$

and, from (8.12.15),

$$\Omega'(z) = -\frac{\mu(l^2-z^2)^{\frac{1}{2}}}{\pi R(\kappa+1)}\int\limits_{-l}^{l}\frac{t\,dt}{(t-z)(l^2-z^2)^{\frac{1}{2}}}.$$

The integral may be replaced by an integral taken clockwise along a simple contour surrounding the segment $-l, l$. Thus

$$\int_{-l}^{l} \frac{t\,dt}{(t-z)(l^2-t^2)^{\frac{1}{2}}} = \frac{1}{2}\oint \frac{t\,dt}{(t-z)(l^2-t^2)^{\frac{1}{2}}},$$

and since the contour integral taken around a large circle in an anti-clockwise direction is equal to $-\pi$, we have, by residue theory,

$$\int_{-l}^{l} \frac{t\,dt}{(t-z)(l^2-t^2)^{\frac{1}{2}}} = \frac{\pi i z}{(l^2-z^2)^{\frac{1}{2}}}+\pi,$$

so that

$$\left.\begin{aligned}\Omega'(z) &= -\frac{Y}{2\pi l^2}\{(l^2-z^2)^{\frac{1}{2}}+iz\}\\ &= \frac{iY}{2\pi l^2}\{(z^2-l^2)^{\frac{1}{2}}-z\}\end{aligned}\right\}. \tag{8.14.3}$$

From (8.12.11) and (8.14.3) we see that the stresses on the edge $y = 0$ are

$$\left.\begin{aligned}\sigma_{xx} = \sigma_{yy} &= -\frac{2Y}{\pi l}\left(1-\frac{x^2}{l^2}\right)^{\frac{1}{2}} \quad (|x| \leqslant l)\\ \sigma_{xx} = \sigma_{yy} &= 0 \qquad\qquad\qquad\quad (|x| \geqslant l)\end{aligned}\right\}. \tag{8.14.4}$$

In a similar way we can discuss the indentation of a semi-infinite medium by punches of other shapes, but we leave further examples to the reader.

8.15. Cracks

The effect of cracks on the distribution of stress inside a solid body has an important application in a theory of rupture proposed by Griffith.† He assumes that the bounding surfaces of the solid possess surface tension and that, when a crack spreads, the decrease in the strain energy is balanced by the increase in potential energy due to this surface tension.

A variety of boundary-value problems associated with cracks may be formulated, but we restrict our attention to two cases which we consider in this and the next section. In both cases we consider an infinite two-dimensional medium containing n collinear cracks situated along the segments $L_r = a_r b_r$ $(r = 1,..., n)$ of the real axis, and we shall suppose that all components of stress vanish at infinity and that there

† A. A. Griffith, *Phil. Trans. R. Soc.* **A221** (1921) 180; *Proc. Int. Congr. appl. Mech.* (Delft, 1924), p. 55.

is no resultant force over any crack. This means that

$$\Omega(z) = O\left(\frac{1}{z}\right), \qquad \omega'(z) = O\left(\frac{1}{z}\right), \tag{8.15.1}$$

at least, for large $|z|$.

We suppose that the cracks are opened symmetrically about the x-axis, by a given normal distribution of displacement along the line segments $L = L_1+L_2+...+L_n$ in the first problem, and by a given normal distribution of pressure in the second problem. Owing to the symmetry about the x-axis we can replace these problems by equivalent problems stated for the half-plane S^+ ($y > 0$). Thus

PROBLEM 1

$$\left.\begin{aligned} \sigma_{xy} &= 0 && \text{everywhere on } Ox \\ u_y &= 0 && \text{on } Ox \text{ outside } L \\ u_y &= f(x) && \text{on } L \end{aligned}\right\}, \tag{8.15.2}$$

and

PROBLEM 2

$$\left.\begin{aligned} \sigma_{xy} &= 0 && \text{everywhere on } Ox \\ \sigma_{yy} &= -p(x) && \text{on } L \\ u_y &= 0 && \text{on } Ox \text{ outside } L \end{aligned}\right\}, \tag{8.15.3}$$

where $f(x)$ and $p(x)$ satisfy the H-condition on L.

Since $\sigma_{xy} = 0$ everywhere on Ox for both problems we may use the results (8.12.4) and (8.12.6). Here, however, we define $\Omega(z)$ in the half-plane S^- by the formulae

$$\Omega(z) = \bar{\Omega}(z), \qquad \Omega'(z) = \bar{\Omega}'(z) \quad (z \text{ in } S^-), \tag{8.15.4}$$

instead of by (8.12.8), so that

$$\Omega(\bar{z}) = \bar{\Omega}(\bar{z}), \qquad \Omega'(\bar{z}) = \bar{\Omega}'(\bar{z}) \quad (z \text{ in } S^+). \tag{8.15.5}$$

It follows from (8.15.1) that $\Omega(z) = O(1/z)$ in the whole plane for large $|z|$, and, from (8.12.4) and (8.12.6), that

$$\left.\begin{aligned} \mu D &= \kappa\Omega(z)-\Omega(\bar{z})+(\bar{z}-z)\Omega'(\bar{z}) \\ \mu\frac{\partial D}{\partial x} &= \kappa\Omega'(z)-\Omega'(\bar{z})+(\bar{z}-z)\Omega''(\bar{z}) \\ \sigma_{xx}+\sigma_{yy} &= 4\{\Omega'(z)+\Omega'(\bar{z})\} \\ \sigma_{xx}-\sigma_{yy}+2i\sigma_{xy} &= 4(\bar{z}-z)\Omega''(\bar{z}) \\ \sigma_{yy}-i\sigma_{xy} &= 2\{\Omega'(z)+\Omega'(\bar{z})+(z-\bar{z})\Omega''(\bar{z})\} \end{aligned}\right\}, \tag{8.15.6}$$

and, on the boundary Ox,

$$\left.\begin{aligned} \mu D &= \kappa\Omega(z)-\Omega(\bar{z}) \\ 2i\mu u_y &= (\kappa+1)\{\Omega(z)-\Omega(\bar{z})\} \\ \sigma_{xx} &= \sigma_{yy} = 2\{\Omega'(z)+\Omega'(\bar{z})\}, \qquad \sigma_{xy} = 0 \end{aligned}\right\}, \tag{8.15.7}$$

provided the usual condition (8.12.5) still holds.

8.16. Crack of prescribed shape

We wish to determine the distribution of stress which will produce a crack of prescribed shape and we therefore take the boundary conditions of Problem 1 in § 8.15. The remaining boundary condition (8.15.2) imposed on u_y gives, using (8.15.7),

$$\left.\begin{aligned} \Omega^+(x)-\Omega^-(x) &= \frac{2i\mu f(x)}{\kappa+1} \quad \text{on } L \\ \Omega^+(x)-\Omega^-(x) &= 0 \quad \text{outside } L \end{aligned}\right\}, \tag{8.16.1}$$

and hence, since $\Omega(z)$ vanishes at infinity,

$$\Omega(z) = \frac{\mu}{\pi(\kappa+1)} \int\limits_L \frac{f(t)\,dt}{t-z}. \tag{8.16.2}$$

We consider one example in which we have one crack extending from $x = -a$ to $x = a$ and

$$f(x) = A(1-x^2/a^2) \quad (|x| \leqslant a),$$

where A is a constant. Hence, from (8.16.2),

$$\begin{aligned} \Omega(z) &= \frac{\mu A}{\pi a^2(\kappa+1)} \int\limits_{-a}^{a} \frac{a^2-t^2}{t-z}\,dt \\ &= -\frac{\mu A}{\pi a^2(\kappa+1)}\left\{2az+(z^2-a^2)\ln\frac{z-a}{z+a}\right\}. \end{aligned} \tag{8.16.3}$$

It follows from (8.15.7) that the stress applied at the crack is

$$\sigma_{yy} = -\frac{16\mu A}{\pi a(\kappa+1)}\left(1+\frac{x}{2a}\ln\frac{a-x}{a+x}\right) \quad (0 \leqslant x \leqslant a). \tag{8.16.4}$$

8.17. Cracks opened by given pressure

Problem 2 in § 8.15 gives the boundary conditions for cracks opened by a prescribed normal distribution of pressure $p(x)$ on the series of cracks L_r $(r = 1, 2,..., n)$. Hence, from (8.15.3) and (8.15.7),

$$\left.\begin{aligned} [\Omega'(x)]^+ + [\Omega'(x)]^- &= -\tfrac{1}{2}p(x) \quad \text{on } L \\ \Omega^+(x)-\Omega^-(x) &= 0 \quad \text{outside } L \end{aligned}\right\}. \tag{8.17.1}$$

It follows that $\Omega'(z)$ is holomorphic in the whole plane cut along L, of order $1/z^2$ at infinity, and satisfies the first condition in (8.17.1) on the cracks L. The solution may therefore be written down with the help of the theory of § 1.21. For simplicity we restrict further discussion to a single crack ab so that the general solution of $O(1/z^2)$ at infinity, which allows the possibility of infinities at the edges of the crack of order not greater than 1, is

$$\Omega'(z) = -\frac{1}{4\pi i\{R(z)\}^{\frac{1}{2}}} \int\limits_a^b \frac{p(t)\{R(t)\}^{\frac{1}{2}}}{t-z}\,dt, \tag{8.17.2}$$

where $\{R(z)\}^{\frac{1}{2}}$ is defined in § 8.11. It can be verified that the condition (8.15.5) is satisfied.

The simplest example is that in which a crack $-l \leqslant x \leqslant l$, $y = 0$ is opened by a constant pressure so that

$$p(x) = p,$$

where p is a constant and

$$\Omega'(z) = -\frac{p}{4\pi i(l^2-z^2)^{\frac{1}{2}}} \int\limits_{-l}^{l} \frac{(l^2-t^2)^{\frac{1}{2}}}{t-z}\,dt. \tag{8.17.3}$$

The integral in (8.17.3) may be evaluated as in § 8.14 by replacing it by a contour integral, which gives

$$\left.\begin{aligned} \Omega'(z) &= -\tfrac{1}{4}p\left\{1+\frac{iz}{(l^2-z^2)^{\frac{1}{2}}}\right\} \\ &= -\tfrac{1}{4}p\left\{1-\frac{z}{(z^2-l^2)^{\frac{1}{2}}}\right\} \\ \Omega(z) &= -\tfrac{1}{4}p\{z-i(l^2-z^2)^{\frac{1}{2}}\} \\ &= -\tfrac{1}{4}p\{z-(z^2-l^2)^{\frac{1}{2}}\} \end{aligned}\right\}. \tag{8.17.4}$$

It then follows from (8.15.6) that

$$\left.\begin{aligned} \sigma_{xx}+\sigma_{yy} &= 2p\left\{\frac{r}{(r_1 r_2)^{\frac{1}{2}}}\cos(\theta-\tfrac{1}{2}\theta_1-\tfrac{1}{2}\theta_2)-1\right\} \\ \sigma_{xx}-\sigma_{yy}+2i\sigma_{xy} &= \frac{2ipl^2 r\sin\theta}{(r_1 r_2)^{\frac{3}{2}}}\,e^{\frac{3}{2}i(\theta_1+\theta_2)} \end{aligned}\right\}, \tag{8.17.5}$$

and, from (8.15.7), on the boundary $y = 0$,

$$\left.\begin{aligned} \sigma_{xx} &= \sigma_{yy} = -p && (|x| \leqslant l) \\ \sigma_{xx} &= \sigma_{yy} = p\left\{\frac{|x|}{(x^2-l^2)^{\frac{1}{2}}}-1\right\} && (|x| > l) \\ u_y &= \frac{p(\kappa+1)}{4\mu}(l^2-x^2)^{\frac{1}{2}} && (|x| \leqslant l) \end{aligned}\right\}. \tag{8.17.6}$$

The formula for u_y in (8.17.6) shows that a uniform pressure p opens the straight crack into an elliptical crack.

If we superpose on the above solution a uniform stress

$$\sigma_{yy} = p, \quad \sigma_{xy} = 0, \quad \sigma_{xx} = 0,$$

we can find the effect of a crack on the distribution of stress in a body which is under uniform tension p at right angles to the crack. The reader is referred to Sneddon† for the application of this theory to the Griffith condition for rupture and for a more detailed numerical discussion of the stress distribution.

The general solution for a number of collinear cracks can also be obtained by the methods of this section, but we leave detailed discussion at this point. The effect of two cracks on a uniform tension has been discussed by Willmore.‡

8.18. Circular regions

The methods of solution for problems of the half-plane used in previous sections may be extended to the case when the elastic body is a circular disk or an infinite region bounded internally by a circular hole. We consider first an infinite region bounded internally by a circle of radius a whose centre is at the origin of coordinates. We denote the region $|z| > a$ by S^+ and the region $|z| < a$ by S^- so that, according to our convention, positive direction along the circle $|z| = a$ is clockwise.

Complex potentials $\Omega(z)$, $\omega(z)$ are defined for the region S^+ and we continue the definition of $\Omega(z)$ to the region S^-. If z is a point in S^- then $a^2/\bar{z}$ is in S^+ and we put

$$\left.\begin{aligned} \Omega(z) &= -z\bar{\Omega}'(a^2/z) - \bar{\omega}'(a^2/z) \\ \Omega'(z) &= -\bar{\Omega}'(a^2/z) + \frac{a^2}{z}\bar{\Omega}''(a^2/z) + \frac{a^2}{z^2}\bar{\omega}''(a^2/z) \end{aligned}\right\} \quad (z \text{ in } S^-), \qquad (8.18.1)$$

where $\quad \bar{\Omega}'(a^2/z) = \overline{\Omega'(a^2/\bar{z})}, \qquad \bar{\omega}'(a^2/z) = \overline{\omega'(a^2/\bar{z})}$

are defined for z in S^-. It will be seen later that the definition (8.18.1) has been chosen so that the function $\Omega(z)$ in S^- is the analytical continuation of $\Omega(z)$ in S^+ through the unstressed parts of the circle $|z| = a$. It follows from (8.18.1) that

$$\left.\begin{aligned} \Omega\left(\frac{a^2}{\bar{z}}\right) &= -\frac{a^2}{\bar{z}}\bar{\Omega}'(\bar{z}) - \bar{\omega}'(\bar{z}) \\ \Omega'\left(\frac{a^2}{\bar{z}}\right) &= -\bar{\Omega}'(\bar{z}) + \bar{z}\bar{\Omega}''(\bar{z}) + \frac{\bar{z}^2}{a^2}\bar{\omega}''(\bar{z}) \end{aligned}\right\} \quad (z \text{ in } S^+),$$

† I. N. Sneddon, *Fourier Transforms* (1951).
‡ T. J. Willmore, *Q. J. Mech. appl. Math.* **2** (1949) 53.

or, solving for $\bar{\omega}'(\bar{z})$ and $\bar{\omega}''(\bar{z})$,

$$\left.\begin{aligned}\bar{\omega}'(\bar{z}) &= -\Omega\left(\frac{a^2}{\bar{z}}\right)-\frac{a^2}{\bar{z}}\,\bar{\Omega}'(\bar{z})\\ \bar{\omega}''(\bar{z}) &= \frac{a^2}{\bar{z}^2}\,\Omega'\left(\frac{a^2}{\bar{z}}\right)+\frac{a^2}{\bar{z}^2}\,\bar{\Omega}'(\bar{z})-\frac{a^2}{\bar{z}}\,\bar{\Omega}''(\bar{z})\end{aligned}\right\} \quad (z \text{ in } S^+). \tag{8.18.2}$$

Hence, from (8.1.5) and (8.1.17)–(8.1.20),

$$P = X+iY = 2i\left\{\Omega(z)-\Omega\left(\frac{a^2}{\bar{z}}\right)+\left(z-\frac{a^2}{\bar{z}}\right)\bar{\Omega}'(\bar{z})\right\}, \tag{8.18.3}$$

$$\mu(u_r+iu_\theta) = e^{-i\theta}\left\{\kappa\Omega(z)+\Omega\left(\frac{a^2}{\bar{z}}\right)+\left(\frac{a^2}{\bar{z}}-z\right)\bar{\Omega}'(\bar{z})\right\}, \tag{8.18.4}$$

$$\sigma_{rr}+\sigma_{\theta\theta} = 4\{\Omega'(z)+\bar{\Omega}'(\bar{z})\}, \tag{8.18.5}$$

$$\sigma_{rr}-\sigma_{\theta\theta}+2i\sigma_{r\theta} = -\frac{4a^2}{z\bar{z}}\,\Omega'\left(\frac{a^2}{\bar{z}}\right)-\frac{4a^2}{z\bar{z}}\,\bar{\Omega}'(\bar{z})+4\left(\frac{a^2}{z}-\bar{z}\right)\bar{\Omega}''(\bar{z}), \tag{8.18.6}$$

$$\sigma_{rr}+i\sigma_{r\theta} = 2\Omega'(z)-\frac{2a^2}{z\bar{z}}\,\Omega'\left(\frac{a^2}{\bar{z}}\right)+2\left(1-\frac{a^2}{z\bar{z}}\right)\{\bar{\Omega}'(\bar{z})-\bar{z}\bar{\Omega}''(\bar{z})\}. \tag{8.18.7}$$

We also make use of a formula obtained by deriving the cartesian components of displacement with respect to θ. Thus, from (8.1.1) and (8.18.2)

$$\mu\frac{\partial D}{\partial\theta} = iz\left\{\kappa\Omega'(z)+\Omega'\left(\frac{a^2}{\bar{z}}\right)+\frac{\bar{z}}{z}\left(1-\frac{z\bar{z}}{a^2}\right)\bar{\omega}''(\bar{z})\right\}. \tag{8.18.8}$$

8.19. First boundary-value problem for circular hole

When the stresses are given at all points of the circle $|z| = a$ we can, as in § 8.7 for the half-plane, divide the problem into two parts. In the first part we suppose that complex potentials $\Omega_0(z)$, $\omega_0(z)$ define stresses and displacements which have singularities only in the region S^+ outside the circle $|z| = a$ and which are such that for large $|z|$

$$\left.\begin{aligned}\Omega_0(z) &= -\frac{P\ln z}{4\pi(\kappa+1)}+\sum_{r=0}^{k}(B_r+iC_r)z^r\\ \omega_0'(z) &= \frac{\kappa\bar{P}\ln z}{4\pi(\kappa+1)}+\sum_{r=0}^{k}(B_r'+iC_r')z^r\end{aligned}\right\}, \tag{8.19.1}$$

where $P = X+iY$ is the total resultant force on the circle at infinity due to $\Omega_0(z)$, $\omega_0(z)$. Also, when $|z|$ is small, we assume that

$$\left.\begin{aligned}\Omega_0'(z) &= c+id+O(z)\\ \omega_0'(z) &= c'+id'+O(z)\end{aligned}\right\}. \tag{8.19.2}$$

We now add a stress system, derived from complex potentials $\Omega_1(z)$, $\omega_1(z)$, which is such that the combined complex potentials

$$\Omega(z) = \Omega_0(z)+\Omega_1(z), \qquad \omega(z) = \omega_0(z)+\omega_1(z), \tag{8.19.3}$$

give zero applied stresses at the boundary $|z| = a$, and, for large $|z|$, $\Omega(z)$ and $\omega(z)$ take the forms (8.19.1). The complex potentials $\Omega_1(z)$ and $\omega_1'(z)$ therefore both vanish at infinity. If we now extend the definition of $\Omega_1(z)$ to the region S^- by formulae of the type (8.18.1) we observe that $\Omega_1(z)$ will be finite at the origin, and from (8.1.5) and (8.18.3),

$$\frac{P}{2i} = \Omega_1(z)-\Omega_1\left(\frac{a^2}{\bar{z}}\right)+\left(z-\frac{a^2}{\bar{z}}\right)\bar{\Omega}_1'(\bar{z})+\Omega_0(z)+z\bar{\Omega}_0'(\bar{z})+\bar{\omega}_0'(\bar{z}). \tag{8.19.4}$$

It follows from (8.19.4) that if the boundary $|z| = a$ is free from applied stress then

$$\Omega_1^+(t)-\Omega_1^-(t) = -\Omega_0(t)-t\bar{\Omega}_0'\left(\frac{a^2}{t}\right)-\bar{\omega}_0'\left(\frac{a^2}{t}\right), \tag{8.19.5}$$

for all points $t = ae^{i\theta}$ on the circle $|z| = a$. In deriving (8.19.5) we have, of course, made the usual assumptions about the existence of limits $\Omega_1^+(t)$, $\Omega_1^-(t)$ and we have also assumed that

$$\lim_{r\to a}\left(z-\frac{a^2}{\bar{z}}\right)\bar{\Omega}_1'(\bar{z}) = e^{i\theta}\lim_{r\to a}\left(r-\frac{a^2}{r}\right)\bar{\Omega}_1'(re^{-i\theta}) = 0.$$

Since the validity of assumptions of this type can always be tested when we have derived the solution, to avoid continual repetition we shall usually make similar assumptions in future without further statement.

The (unique) solution $\Omega_1(z)$ of (8.19.5) which is holomorphic everywhere (except on $|z| = a$) and which vanishes at infinity is

$$\Omega_1(z) = -\frac{1}{2\pi i}\int_L \left\{\Omega_0(t)+t\bar{\Omega}_0'\left(\frac{a^2}{t}\right)+\bar{\omega}_0'\left(\frac{a^2}{t}\right)\right\}\frac{dt}{t-z}, \tag{8.19.6}$$

where the integral is taken around the circle $|z| = a$ in the *clockwise* direction. To evaluate (8.19.6) we observe that $\bar{\Omega}_0'(a^2/t)$ and $\bar{\omega}_0'(a^2/t)$ are regular functions for $|t| > a$ and, from (8.19.2),

$$t\bar{\Omega}_0'\left(\frac{a^2}{t}\right)+\bar{\omega}_0'\left(\frac{a^2}{t}\right) = (c-id)t+O(1)$$

for large $|t|$, so that

$$\frac{1}{2\pi i}\int_\infty \left\{t\bar{\Omega}_0'\left(\frac{a^2}{t}\right)+\bar{\omega}_0'\left(\frac{a^2}{t}\right)\right\}\frac{dt}{t-z} = (c-id)z+\text{constant},$$

the integral being taken in an anticlockwise sense around the circle at infinity. Hence, by residue theory, (8.19.6) gives

$$\left.\begin{aligned}\Omega_1(z) &= (c-id)z - z\bar{\Omega}_0'\left(\frac{a^2}{z}\right) - \bar{\omega}_0'\left(\frac{a^2}{z}\right) \quad (z \text{ in } S^+)\\ \Omega_1(z) &= (c-id)z + \Omega_0(z) \quad (z \text{ in } S^-)\end{aligned}\right\}, \tag{8.19.7}$$

apart from an arbitrary constant which may be ignored. The results (8.19.7) could also be written down by inspection from (8.19.5) if we remember (8.19.2) and that $\Omega_1(z)$ vanishes at infinity (or is constant there). The total complex potentials for the region S^+ are therefore

$$\left.\begin{aligned}\Omega(z) &= \Omega_0(z) - z\bar{\Omega}_0'\left(\frac{a^2}{z}\right) - \bar{\omega}_0'\left(\frac{a^2}{z}\right) + (c-id)z\\ \omega'(z) &= \omega_0'(z) - \bar{\Omega}_0\left(\frac{a^2}{z}\right) + \frac{a^2}{z}\bar{\Omega}_0'\left(\frac{a^2}{z}\right) -\\ &\qquad - \frac{a^4}{z^2}\bar{\Omega}_0''\left(\frac{a^2}{z}\right) - \frac{a^4}{z^3}\bar{\omega}_0''\left(\frac{a^2}{z}\right) - \frac{2a^2c}{z}\end{aligned}\right\}, \tag{8.19.8}$$

if we use formula (8.18.2) for $\bar{\omega}_1'(\bar{z})$. Displacements and stresses can now be written down using (8.1.17)–(8.1.20). In particular, on the boundary of the circle, since $\sigma_{rr} = 0$ there, we have

$$\sigma_{\theta\theta} = 8c + 4z\Omega_0''(z) + 4\frac{a^2}{z}\bar{\Omega}_0''\left(\frac{a^2}{z}\right) + \frac{4z^2}{a^2}\omega_0''(z) + \frac{4a^2}{z^2}\bar{\omega}_0''\left(\frac{a^2}{z}\right). \tag{8.19.9}$$

Applications of these results will be given in § 8.20 after we have completed the theory by considering problems in which the applied stresses at the boundary take the prescribed forms

$$\sigma_{rr} + i\sigma_{r\theta} = p(t) + iq(t) \quad (t = ae^{i\theta}), \tag{8.19.10}$$

where $p(t)$ and $q(t)$ are real functions. To solve this problem† we suppose that the state of stress is derived from complex potentials $\Omega(z)$, $\omega(z)$, so that if stresses and rotation vanish at infinity

$$\left.\begin{aligned}\Omega'(z) &= -\frac{P}{4\pi(\kappa+1)z} + O\left(\frac{1}{z^2}\right)\\ \omega''(z) &= \frac{\kappa\bar{P}}{4\pi(\kappa+1)z} + O\left(\frac{1}{z^2}\right)\end{aligned}\right\}, \tag{8.19.11}$$

for large $|z|$, where $P = X+iY$ is the resultant force at the hole. We define $\Omega(z)$ in the region S^- by (8.18.1), so that, using (8.19.11),

$$\lim_{z\to 0} z\Omega'(z) = \frac{\kappa P}{4\pi(\kappa+1)}. \tag{8.19.12}$$

† This method is given by N. I. Muskhelishvili, loc. cit., p. 1. An alternative solution by the same author is given in *Z. angew. Math. Mech.*, loc. cit., p. 182.

Also, from (8.18.7),

$$[\Omega'(t)]^+ - [\Omega'(t)]^- = \tfrac{1}{2}\{p(t)+iq(t)\}. \tag{8.19.13}$$

Hence, provided $p(t)$, $q(t)$ satisfy suitable conditions on the circle,

$$\Omega'(z) = \frac{1}{4\pi i}\int_L \frac{p(t)+iq(t)}{t-z}\,dt + \frac{\kappa P}{4\pi(\kappa+1)z}, \tag{8.19.14}$$

where the integral is taken clockwise round the circle $|z| = a$. Displacements and stresses may now be evaluated from (8.18.3)–(8.18.7).

As in the half-plane problems in § 8.7 we can also solve the problem of prescribed non-zero stresses at the hole in terms of two potential functions $V(z)$ and $W(z)$.† We put

$$\left.\begin{aligned} V(z) &= 4\Omega'(z) - 2z\Omega''(z) - \frac{2z^2}{a^2}\,\omega''(z) \\ W(z) &= 2z\Omega''(z) + \frac{2z^2}{a^2}\,\omega''(z) \end{aligned}\right\}, \tag{8.19.15}$$

and observe from (8.1.20) that

$$\operatorname{re} V(z) = p(t), \qquad \operatorname{im} W(z) = q(t), \tag{8.19.16}$$

on the circle $|z| = a$. From (8.19.15)

$$\left.\begin{aligned} \Omega'(z) &= \tfrac{1}{4}\{V(z)+W(z)\} \\ \omega''(z) &= \frac{a^2 W(z)}{2z^2} - \frac{a^2}{4z}\{V'(z)+W'(z)\} \end{aligned}\right\}, \tag{8.19.17}$$

and using (8.19.11) we see that, for large $|z|$,

$$\left.\begin{aligned} V(z) &= -\frac{\kappa \bar{P} z}{2\pi(\kappa+1)a^2} + e + if + O\left(\frac{1}{z}\right) \\ W(z) &= \frac{\kappa \bar{P} z}{2\pi(\kappa+1)a^2} - e - if + O\left(\frac{1}{z}\right) \end{aligned}\right\}, \tag{8.19.18}$$

where e and f are real constants. From (8.1.18), (8.1.19), and (8.19.17),

$$\left.\begin{aligned} \sigma_{rr}+\sigma_{\theta\theta} &= V(z)+W(z)+\bar{V}(\bar{z})+\bar{W}(\bar{z}) \\ \sigma_{rr}-\sigma_{\theta\theta}+2i\sigma_{r\theta} &= -\frac{2a^2\bar{W}(\bar{z})}{z\bar{z}} + \left(\frac{a^2}{z}-\bar{z}\right)\{\bar{V}'(\bar{z})+\bar{W}'(\bar{z})\} \end{aligned}\right\}, \tag{8.19.19}$$

and, on the circle $|z| = a$,

$$\left.\begin{aligned} \sigma_{rr}+\sigma_{\theta\theta} &= V(z)+W(z)+\bar{V}(\bar{z})+\bar{W}(\bar{z}) \\ \sigma_{rr}-\sigma_{\theta\theta}+2i\sigma_{r\theta} &= -2\bar{W}(\bar{z}) \end{aligned}\right\}. \tag{8.19.20}$$

† W. G. Bickley, loc. cit., p. 226, and A. E. Green, *Proc. R. Soc.* A180 (1942) 173.

Expressions for $V(z)$, $W(z)$ can be written down by means of the well-known Schwarz formula. Thus, from (8.19.16) and (8.19.18),

$$\left.\begin{aligned} V(z) &= -\frac{\kappa}{2\pi(\kappa+1)}\left(\frac{\bar{P}z}{a^2}-\frac{P}{z}\right)+\frac{1}{2\pi}\int_0^{2\pi}\frac{ze^{-i\gamma}+a}{ze^{-i\gamma}-a}p(ae^{i\gamma})\,d\gamma-\frac{i}{2\pi}\int_0^{2\pi}q(ae^{i\gamma})\,d\gamma \\ W(z) &= \frac{\kappa}{2\pi(\kappa+1)}\left(\frac{\bar{P}z}{a^2}+\frac{P}{z}\right)+\frac{i}{2\pi}\int_0^{2\pi}\frac{ze^{-i\gamma}+a}{ze^{-i\gamma}-a}q(ae^{i\gamma})\,d\gamma-\frac{1}{2\pi}\int_0^{2\pi}p(ae^{i\gamma})\,d\gamma \end{aligned}\right\}. \tag{8.19.21}$$

Alternatively, for some purposes, it is more convenient to represent $V(z)$ and $W(z)$ by power series which may be obtained from (8.19.21) or written down independently. Thus

$$\left.\begin{aligned} V(z) &= -\frac{\kappa}{2\pi(\kappa+1)}\left(\frac{\bar{P}z}{a^2}-\frac{P}{z}\right)+\sum_{n=0}^{\infty}(R_n+iS_n)\frac{a^n}{z^n} \\ W(z) &= \frac{\kappa}{2\pi(\kappa+1)}\left(\frac{\bar{P}z}{a^2}+\frac{P}{z}\right)+\sum_{n=0}^{\infty}(-U_n+iT_n)\frac{a^n}{z^n} \end{aligned}\right\}, \tag{8.19.22}$$

where

$$\left.\begin{aligned} R_n &= \frac{1}{\pi}\int_0^{2\pi}p(ae^{i\theta})\cos n\theta\,d\theta, \qquad S_n = \frac{1}{\pi}\int_0^{2\pi}p(ae^{i\theta})\sin n\theta\,d\theta \\ T_n &= \frac{1}{\pi}\int_0^{2\pi}q(ae^{i\theta})\cos n\theta\,d\theta, \qquad U_n = \frac{1}{\pi}\int_0^{2\pi}q(ae^{i\theta})\sin n\theta\,d\theta \\ R_0 &= U_0 = \frac{1}{2\pi}\int_0^{2\pi}p(ae^{i\theta})\,d\theta \\ -S_0 &= T_0 = \frac{1}{2\pi}\int_0^{2\pi}q(ae^{i\theta})\,d\theta \end{aligned}\right\}, \tag{8.19.23}$$

so that S_n, R_n and U_n, T_n are respectively the Fourier coefficients in the expansions of $p(ae^{i\theta})$, $q(ae^{i\theta})$ in Fourier series. Hence

$$\left.\begin{aligned} p(t) &= p(ae^{i\theta}) = R_0+\sum_{n=1}^{\infty}(R_n\cos n\theta+S_n\sin n\theta) \\ q(t) &= q(ae^{i\theta}) = T_0+\sum_{n=1}^{\infty}(T_n\cos n\theta+U_n\sin n\theta) \end{aligned}\right\}. \tag{8.19.24}$$

The stress distributions resulting from applied surface forces of the forms (8.19.24) may be obtained from (8.19.20) and (8.19.22).

8.20. Effect of circular hole on given stress distributions

In this section we briefly indicate the solution of two problems using the first part of the theory in § 8.19. If a uniform distribution of stress is transmitted from infinity, then from (8.1.22) the values of $\Omega_0'(z)$, $\omega_0''(z)$ are

$$\left.\begin{aligned} \Omega_0'(z) &= \tfrac{1}{8}(N_1+N_2) \\ \omega_0''(z) &= -\tfrac{1}{4}(N_1-N_2)e^{-2i\alpha} \end{aligned}\right\}, \tag{8.20.1}$$

where N_1, N_2, α are defined after (8.1.23). It follows immediately from (8.19.2) that $8c = N_1+N_2$ and from (8.19.9) that the value of the hoop stress $\sigma_{\theta\theta}$ at the stress-free hole is

$$\sigma_{\theta\theta} = N_1+N_2-2(N_1-N_2)\cos 2(\theta-\alpha), \tag{8.20.2}$$

from which well-known special cases may be derived immediately. In a similar way other distributions of stress transmitted from infinity may be discussed. Here, however, we leave detailed discussion of special problems, except that we close this section by considering the distribution of stress due to an isolated force and couple in the presence of a stress-free circular hole.†

If a force $P = X+iY$ and couple M act at the point $z = -k$ where k is real and $k > a$, then, from (8.3.2) and § 8.19,

$$\left.\begin{aligned} \Omega_0(z) &= -\frac{P\ln(z+k)}{4\pi(\kappa+1)} \\ \omega_0(z) &= \frac{\kappa\bar{P}(z+k)\ln(z+k)}{4\pi(\kappa+1)}-\frac{Pk\ln(z+k)}{4\pi(\kappa+1)}+\frac{iM\ln(z+k)}{4\pi} \end{aligned}\right\}, \tag{8.20.3}$$

and, using (8.19.2),

$$c = -\frac{X}{4\pi k(\kappa+1)}. \tag{8.20.4}$$

The general stress distribution may now be obtained from (8.19.8), (8.1.18), and (8.1.19), but restricting our attention to the hoop stress on the edge of the circle we find from (8.19.9), after straightforward simplification, that

$$\begin{aligned} \pi k(\kappa+1)(a^2+k^2+2ak\cos\theta)^2\sigma_{\theta\theta}/(2F) &= \kappa\{ak^3\cos(3\theta-\epsilon)+ \\ &+k^2(k^2+2a^2)\cos(2\theta-\epsilon)+ak(a^2+2k^2)\cos(\theta-\epsilon)+a^2k^2\cos\epsilon\}+ \\ &+k^2(k^2-a^2)\cos(2\theta+\epsilon)-a^2k^2\cos(2\theta-\epsilon)+ak(k^2-2a^2)\cos(\theta+\epsilon)- \\ &-ak(a^2+2k^2)\cos(\theta-\epsilon)-(a^4+k^4+a^2k^2)\cos\epsilon+ \\ &+k(\kappa+1)(M/F)(k^2\sin 2\theta+2ak\sin\theta), \end{aligned} \tag{8.20.5}$$

where

$$P = X+iY = Fe^{i\epsilon}. \tag{8.20.6}$$

† S. Holgate, *Proc. Camb. phil. Soc. math. phys. Sci.* **40** (1944) 172.

8.21. Prescribed forces at circular hole

We have already given the solution of the problem of prescribed stresses at a circular hole which have the form (8.19.24). Here we consider one example which can be put in the general form (8.19.24), but a more compact result can be obtained by using the integral forms (8.19.21) for $V(z)$ and $W(z)$, or by using (8.19.14). We take

$$\left.\begin{array}{ll} p(t) = p, & q(t) = q \quad (\theta_1 \leqslant \theta \leqslant \theta_2) \\ p(t) = 0, & q(t) = 0 \quad \text{otherwise} \end{array}\right\}, \tag{8.21.1}$$

where p, q are constants. It follows from (8.19.14) that

$$\Omega'(z) = \frac{\kappa P}{4\pi(\kappa+1)z} - \frac{p+iq}{4\pi i}\ln\frac{z-ae^{i\theta_2}}{z-ae^{i\theta_1}}. \tag{8.21.2}$$

If stresses and rotation vanish at infinity, the form of $\Omega'(z)$ for large $|z|$ is given by (8.19.11) so that, comparing coefficients of $1/z$,

$$-\frac{P}{4\pi(\kappa+1)} = \frac{\kappa P}{4\pi(\kappa+1)} + \frac{p+iq}{4\pi i}(ae^{i\theta_2}-ae^{i\theta_1}),$$

or

$$P = X+iY = ia(p+iq)(e^{i\theta_2}-e^{i\theta_1}). \tag{8.21.3}$$

We may now write (8.21.2) in the form

$$\Omega'(z) = \frac{\kappa P}{4\pi(\kappa+1)z} + \frac{P}{4\pi a(e^{i\theta_2}-e^{i\theta_1})}\ln\frac{z-ae^{i\theta_2}}{z-ae^{i\theta_1}}, \tag{8.21.4}$$

and stresses and displacements are obtained immediately from (8.18.4)–(8.18.6). We consider in more detail a limiting case of this problem in which the applied forces are concentrated at a point $\theta = \beta$ on the circle. We assume that

$$\theta_2 \to \theta_1 = \beta, \qquad p \to \infty, \qquad q \to \infty,$$

in such a way that the resultant force P remains fixed and is then an isolated force at $\theta = \beta$. Hence, from (8.21.4), in the limit

$$\Omega'(z) = \frac{\kappa P}{4\pi(\kappa+1)z} - \frac{P}{4\pi(z-ae^{i\beta})}. \tag{8.21.5}$$

It follows from (8.18.5) by straightforward reduction that on the circle $|z| = a$

$$\sigma_{\theta\theta} = \frac{F}{\pi a}\left\{\frac{2\kappa\cos(\theta-\epsilon)}{\kappa+1} + \frac{\sin(\frac{1}{2}\theta+\frac{1}{2}\beta-\epsilon)}{\sin\frac{1}{2}(\theta-\beta)}\right\} \quad (\theta \neq \beta), \tag{8.21.6}$$

since $\sigma_{rr} = 0$ when $\theta \neq \beta$. In (8.21.6) we have put $P = Fe^{i\epsilon}$. The above results may be deduced by using the functions $V(z)$, $W(z)$ which are defined in § 8.19, but this is left as an exercise for the reader. The special result (8.21.5) may also be obtained by using the formulae (8.5.2). From these formulae the function $\Omega'(z)$ must contain terms

$$\Omega'(z) = -\frac{P}{4\pi(\kappa+1)z}-\frac{P}{4\pi}\left(\frac{1}{z-ae^{i\beta}}-\frac{1}{z}\right) = \frac{\kappa P}{4\pi(\kappa+1)z}-\frac{P}{4\pi(z-ae^{i\beta})},$$

and these terms alone satisfy the boundary condition

$$[\Omega'(t)]^+-[\Omega'(t)]^- = 0 \quad (\theta \neq \beta),$$

so that, from (8.19.13), the boundary is unstressed at all points not coinciding with $\theta = \beta$. It follows that we have, as in (8.21.5), the complete potential function $\Omega'(z)$ for an isolated force at $\theta = \beta$.

Results for a general distribution of isolated forces at the circle can now be obtained by summation.

8.22. Second boundary-value problem for circular hole

When values of the displacement are prescribed at the edge of the circular hole $|z| = a$, we again consider two cases. In the first case complex potentials $\Omega_0(z)$, $\omega_0(z)$ define a given stress distribution which has the properties defined in § 8.19 and we choose complex potentials (8.19.3) so that the total displacement at $|z| = a$ is zero. As in § 8.19 we extend the definition of the complex potential $\Omega_1(z)$ to the whole plane by formulae of the type (8.18.1). The total displacement is given by (8.18.4) and (8.1.17) in the form

$$\mu(u_r+iu_\theta) = e^{-i\theta}\left\{\kappa\Omega_1(z)+\Omega_1\left(\frac{a^2}{\bar{z}}\right)+\left(\frac{a^2}{\bar{z}}-z\right)\bar{\Omega}_1'(\bar{z})+\right.$$
$$\left.+\kappa\Omega_0(z)-z\bar{\Omega}_0'(\bar{z})-\bar{\omega}_0'(\bar{z})\right\}. \quad (8.22.1)$$

If the displacement at $|z| = a$ is zero, then, with the usual assumptions,

$$\kappa\Omega_1^+(t)+\Omega_1^-(t) = -\kappa\Omega_0(t)+t\bar{\Omega}_0'\left(\frac{a^2}{t}\right)+\bar{\omega}_0'\left(\frac{a^2}{t}\right) \quad (t = ae^{i\theta}), \quad (8.22.2)$$

and $\Omega_1(z)$ is finite at $z = 0$ and vanishes at infinity. By writing

$$A(z) = \kappa\Omega_1(z) \quad (z \text{ in } S^+),$$
$$A(z) = -\Omega_1(z) \quad (z \text{ in } S^-),$$

equation (8.22.2) becomes

$$A^+(t)-A^-(t) = -\kappa\Omega_0(t)+t\bar{\Omega}_0'\left(\frac{a^2}{t}\right)+\bar{\omega}_0'\left(\frac{a^2}{t}\right),$$

and hence

$$A(z) = \frac{1}{2\pi i}\int\limits_L \left\{-\kappa\Omega_0(t)+t\bar{\Omega}_0'\left(\frac{a^2}{t}\right)+\bar{\omega}_0'\left(\frac{a^2}{t}\right)\right\}\frac{dt}{t-z},$$

the integral being taken clockwise round the circle $|z| = a$. This integral is similar to (8.19.6) and may be evaluated in the same way, so that we have

$$\left.\begin{aligned} A(z) &= \kappa\Omega_1(z) = -(c-id)z+z\bar{\Omega}_0'\left(\frac{a^2}{z}\right)+\bar{\omega}_0'\left(\frac{a^2}{z}\right) \quad (z \text{ in } S^+) \\ A(z) &= -\Omega_1(z) = -(c-id)z+\kappa\Omega_0(z) \quad (z \text{ in } S^-) \end{aligned}\right\}. \tag{8.22.3}$$

Using the formula (8.18.2) for $\omega_1'(z)$ we see that the total complex potentials for the region S^+ are

$$\left.\begin{aligned} \Omega(z) &= \Omega_0(z)+\frac{z}{\kappa}\bar{\Omega}_0'\left(\frac{a^2}{z}\right)+\frac{1}{\kappa}\bar{\omega}_0'\left(\frac{a^2}{z}\right)-\frac{(c-id)z}{\kappa} \\ \omega'(z) &= \omega_0'(z)+\kappa\bar{\Omega}_0\left(\frac{a^2}{z}\right)-\frac{a^2}{\kappa z}\bar{\Omega}_0'\left(\frac{a^2}{z}\right)+\frac{a^4}{\kappa z^2}\bar{\Omega}_0''\left(\frac{a^2}{z}\right)+ \\ &\quad +\frac{a^4}{\kappa z^3}\bar{\omega}_0''\left(\frac{a^2}{z}\right)+\left(\frac{c-id}{\kappa}-c-id\right)\frac{a^2}{z} \end{aligned}\right\}. \tag{8.22.4}$$

Displacements and stresses may now be found from (8.1.17)–(8.1.19).

As an example we consider that a uniform distribution of stress is transmitted from infinity and that $\Omega_0(z)$, $\omega_0(z)$ are given by (8.20.1). It follows from (8.19.2) that

$$8c = N_1+N_2, \qquad d = 0, \tag{8.22.5}$$

and hence (8.22.4) gives the total complex potentials $\Omega(z)$, $\omega(z)$ as follows:

$$\left.\begin{aligned} \Omega(z) &= \frac{(N_1+N_2)z}{8}-\frac{(N_1-N_2)a^2e^{2i\alpha}}{4\kappa z} \\ \omega'(z) &= -\frac{(N_1-N_2)}{4}\left(ze^{-2i\alpha}+\frac{a^4e^{2i\alpha}}{\kappa z^3}\right)+\frac{(\kappa-1)(N_1+N_2)a^2}{8z} \end{aligned}\right\}. \tag{8.22.6}$$

The corresponding stresses and displacements can be obtained from (8.1.17)–(8.1.19). In particular, at the edge of the circular hole,

$$\left.\begin{aligned} \sigma_{rr} &= \frac{\kappa+1}{4\kappa}\{\kappa(N_1+N_2)+2(N_1-N_2)\cos 2(\theta-\alpha)\} \\ \sigma_{\theta\theta} &= \frac{3-\kappa}{4\kappa}\{\kappa(N_1+N_2)+2(N_1-N_2)\cos 2(\theta-\alpha)\} \\ \sigma_{r\theta} &= -\frac{(\kappa+1)(N_1-N_2)}{2\kappa}\sin 2(\theta-\alpha) \end{aligned}\right\}. \tag{8.22.7}$$

When the displacements take prescribed non-zero values on $|z| = a$ we assume that the cartesian components u_x, u_y take the values

$$u_x = g_1(t), \qquad u_y = g_2(t) \quad (t = ae^{i\theta}), \tag{8.22.8}$$

where $g_1(t)$, $g_2(t)$ are real functions. We also assume that $g_1(t)$, $g_2(t)$ have first derivatives $g_1'(t)$, $g_2'(t)$ satisfying the H-condition on the circle $|z| = a$. We can solve this problem by introducing two complex functions which are analogous to the functions $V(z)$ and $W(z)$ used when stresses were prescribed on the boundary, but here we restrict our attention to the first method which uses the single complex potential $\Omega(z)$ which is defined in the whole plane in § 8.19. It follows, as in § 8.19, that formulae (8.19.11) and (8.19.12) hold for large and small $|z|$ respectively, and from (8.18.8)

$$\kappa[\Omega'(t)]^+ + [\Omega'(t)]^- = \mu\{g_1'(t)+ig_2'(t)\}, \tag{8.22.9}$$

on the circle $t = ae^{i\theta}$. Hence

$$\left.\begin{aligned} \kappa\Omega'(z) &= \frac{\mu}{2\pi i}\int\limits_L \frac{g_1'(t)+ig_2'(t)}{t-z}\,dt - \frac{\kappa P}{4\pi(\kappa+1)z} \qquad (z \text{ in } S^+) \\ \Omega'(z) &= -\frac{\mu}{2\pi i}\int\limits_L \frac{g_1'(t)+ig_2'(t)}{t-z}\,dt + \frac{\kappa P}{4\pi(\kappa+1)z} \qquad (z \text{ in } S^-) \end{aligned}\right\}, \tag{8.22.10}$$

the integrals being taken clockwise round the circle $|z| = a$ on which $t = ae^{i\theta}$. Values of stresses and displacements may be found from (8.18.4)–(8.18.6), but further discussion of special problems is left to the reader since such a discussion is similar to that indicated in § 8.21 for special examples of the first boundary-value problem.

8.23. Mixed boundary-value problem for circular hole

The fundamental mixed boundary-value problem for a circular hole arises when the displacements u_x, u_y are given on the arcs $a_r b_r$ $(r = 1, 2,..., n)$ of the circle $|z| = a$ and the external stresses on the remaining part of the boundary. With the help of the solution of the first boundary-value problem in § 8.19, however, we can consider instead the slightly simpler problem in which

$$\left.\begin{aligned} D = u_x+iu_y = g(t) = g_1(t)+ig_2(t) \quad &\text{on } L \\ \sigma_{rr} = 0, \qquad \sigma_{r\theta} = 0 \quad &\text{on } L' \end{aligned}\right\}, \tag{8.23.1}$$

where L consists of the n arcs $a_r b_r$ and L' is the rest of the bounding circle.

Using the theory of § 8.18, and making the usual assumptions, we see from (8.18.7) that the second boundary condition in (8.23.1) gives

$$[\Omega'(t)]^+ - [\Omega'(t)]^- = 0 \quad \text{on } L',$$

so that $\Omega'(z)$ is holomorphic in the plane cut along L, but, as in § 8.19, when stresses and rotation vanish at infinity,

$$\Omega'(z) = -\frac{P}{4\pi(\kappa+1)z}+O\left(\frac{1}{z^2}\right), \tag{8.23.2}$$

for large $|z|$, and

$$\lim_{z\to 0} z\Omega'(z) = \frac{\kappa P}{4\pi(\kappa+1)}. \tag{8.23.3}$$

Using (8.18.8) the first boundary condition in (8.23.1) gives

$$\kappa[\Omega'(t)]^+ + [\Omega'(t)]^- = \mu g'(t) = \mu\{g_1'(t)+ig_2'(t)\}, \tag{8.23.4}$$

on L. The problem is therefore reduced to a special case of the Hilbert problem in which $\Omega'(z)$ satisfies the condition (8.23.4) on the arcs L, and also the conditions (8.23.2), (8.23.3) at infinity and at the origin respectively. We may write down a general solution with the help of § 1.21, but for simplicity we consider the case of a single arc ab. If we admit solutions in which $\Omega'(z)$ is infinite at a, b with degree less than 1, then

$$\Omega'(z) = \frac{\mu\{R_0(z)\}^{i\gamma}}{2\pi\kappa i\{R(z)\}^{\frac{1}{2}}}\int\limits_a^b \frac{g'(t)\{R(t)\}^{\frac{1}{2}}}{(t-z)\{R_0(t)\}^{i\gamma}}\,dt+\frac{N\{R_0(z)\}^{i\gamma}}{\{R(z)\}^{\frac{1}{2}}}+\frac{\kappa P}{4\pi(\kappa+1)z}, \tag{8.23.5}$$

where the integration is clockwise along the arc ab of the circle, and where N is a constant,

$$2\pi\gamma = \log(1/\kappa), \tag{8.23.6}$$

and

$$\left.\begin{aligned} R(z) &= (z-a)(b-z) \\ R_0(z) &= \frac{z-a}{z-b} \end{aligned}\right\}. \tag{8.23.7}$$

By $\{R(z)\}^{\frac{1}{2}}$ we understand that branch which is holomorphic in the plane cut along ab, such that

$$\lim_{|z|\to\infty} \{R(z)\}^{\frac{1}{2}}/z = -i, \tag{8.23.8}$$

and by $\{R_0(z)\}^{i\gamma}$ a branch, holomorphic in the same region, which tends to 1 as $|z|\to\infty$. Also $\{R(t)\}^{\frac{1}{2}}$, $\{R_0(t)\}^{i\gamma}$ denote the values of the specified branches as $z\to t$ from S^+ (i.e. outside the circle).

If the total force P exerted on the segment ab is given, then, in view of (8.23.2), we have

$$-\frac{P}{4\pi(\kappa+1)} = \frac{\kappa P}{4\pi(\kappa+1)}-\frac{N}{i},$$

or

$$N = \frac{iP}{4\pi}. \tag{8.23.9}$$

Other types of boundary conditions may also be considered at the edge of the circular hole, but we leave the discussion at this point. Problems concerned with circular disks may be solved by methods similar to those used for the circular hole. We indicate, briefly, in the next section, the solution of the first boundary-value problem for a disk.

8.24. First boundary-value problem for a circular disk

The region $|z| < a$ is now denoted by S^+ and the region $|z| > a$ by S^-, so that the positive direction along the circle $|z| = a$ is anti-clockwise. With this notation for S^+ and S^-, the definitions of $\Omega(z)$, $\omega(z)$ follow exactly as in § 8.18 and the displacements and stresses are given by (8.18.4) to (8.18.8).

Following the plan of § 8.19, we first suppose that complex potentials $\Omega_0(z)$, $\omega_0(z)$ define stresses and displacements which have singularities only in the region S^+ inside the circular disk $|z| = a$. We also assume that these potentials produce a self-equilibrating system of stress so that, for large $|z|$,

$$\left.\begin{aligned} \Omega_0(z) &= O\left(\frac{1}{z}\right) \\ \omega_0'(z) &= \frac{B_1'}{z}+\frac{B_2'+iC_2'}{z^2}+O\left(\frac{1}{z^3}\right) \end{aligned}\right\}, \tag{8.24.1}$$

where B_1', B_2', C_2' are real constants. The coefficient B_1' of $1/z$ is real since there is no resultant couple over any large circle. We add a stress system derived from complex potentials $\Omega_1(z)$, $\omega_1(z)$ which is such that the combined potentials

$$\Omega(z) = \Omega_0(z)+\Omega_1(z), \qquad \omega(z) = \omega_0(z)+\omega_1(z), \tag{8.24.2}$$

give zero applied stresses at the boundary $|z| = a$ of the disk. The complex potentials $\Omega_1(z)$, $\omega_1(z)$ also give rise to a self-equilibrating system of stresses, and we suppose that, for small $|z|$,

$$\left.\begin{aligned} \Omega_1'(z) &= \alpha_0+\alpha_1 z+O(z^2) \\ \omega_1'(z) &= O(z) \end{aligned}\right\}. \tag{8.24.3}$$

By a convenient choice of the arbitrary constants α', β' in § 6.6 we may always take $\omega_1'(0) = 0$ and this has been done in (8.24.3).

We now extend the definition of $\Omega_1(z)$ to the region S^- by formulae of the type (8.18.1) so that

$$\Omega_1(z) = -z\bar{\Omega}_1'\left(\frac{a^2}{z}\right)-\bar{\omega}_1'\left(\frac{a^2}{z}\right) \quad (z \text{ in } S^-), \tag{8.24.4}$$

and hence

$$\bar{\omega}_1'(\bar{z}) = -\Omega_1\left(\frac{a^2}{\bar{z}}\right)-\frac{a^2}{\bar{z}}\bar{\Omega}_1'(\bar{z}) \quad (z \text{ in } S^+). \tag{8.24.5}$$

It follows from (8.18.3) and (8.1.5) that

$$\frac{P}{2i} = \Omega_1(z) - \Omega_1\left(\frac{a^2}{\bar{z}}\right) + \left(z - \frac{a^2}{\bar{z}}\right)\bar{\Omega}_1'(\bar{z}) + \Omega_0(z) + z\bar{\Omega}_0'(\bar{z}) + \bar{\omega}_0'(\bar{z}).$$

Hence, if the boundary of the disk is free from applied stress, we have, with the usual assumptions,

$$\Omega_1^+(t) - \Omega_1^-(t) = -\Omega_0(t) - t\bar{\Omega}_0'\left(\frac{a^2}{t}\right) - \bar{\omega}_0'\left(\frac{a^2}{t}\right), \tag{8.24.6}$$

for all points t on $|z| = a$. Also, from (8.24.3) and (8.24.4), we see that, for large $|z|$,

$$\Omega_1(z) = -\bar{\alpha}_0 z - \bar{\alpha}_1 a^2 + O\left(\frac{1}{z}\right). \tag{8.24.7}$$

The solution $\Omega_1(z)$ of (8.24.6) which is holomorphic everywhere, except on $|z| = a$ and at infinity where it behaves like (8.24.7), is

$$\Omega_1(z) = -\bar{\alpha}_0 z - \bar{\alpha}_1 a^2 - \frac{1}{2\pi i}\int \left\{\Omega_0(t) + t\bar{\Omega}_0'\left(\frac{a^2}{t}\right) + \bar{\omega}_0'\left(\frac{a^2}{t}\right)\right\}\frac{dt}{t-z}, \tag{8.24.8}$$

where the integral is taken anti-clockwise around the circle $|z| = a$. To evaluate this integral we observe that $\bar{\Omega}_0'(a^2/t)$ and $\bar{\omega}_0'(a^2/t)$ are regular for $|t| < a$, $\Omega_0(t)$ is regular for $|t| > a$, and $\Omega_0(t) = O(1/t)$ for large $|t|$. Hence

$$\left.\begin{aligned} \Omega_1(z) &= -z\bar{\Omega}_0'\left(\frac{a^2}{z}\right) - \bar{\omega}_0'\left(\frac{a^2}{z}\right) - \bar{\alpha}_0 z - \bar{\alpha}_1 a^2 \quad (z \text{ in } S^+) \\ \Omega_1(z) &= \Omega_0(z) - \bar{\alpha}_0 z - \bar{\alpha}_1 a^2 \quad (z \text{ in } S^-) \end{aligned}\right\}, \tag{8.24.9}$$

and the total complex potentials for the region S^+ are

$$\left.\begin{aligned} \Omega(z) &= \Omega_0(z) - z\bar{\Omega}_0'\left(\frac{a^2}{z}\right) - \bar{\omega}_0'\left(\frac{a^2}{z}\right) - \bar{\alpha}_0 z - \bar{\alpha}_1 a^2 \\ \omega'(z) &= \omega_0'(z) - \bar{\Omega}_0\left(\frac{a^2}{z}\right) + \\ &\quad + \frac{a^2}{z}\bar{\Omega}_0'\left(\frac{a^2}{z}\right) - \frac{a^4}{z^2}\bar{\Omega}_0''\left(\frac{a^2}{z}\right) - \frac{a^4}{z^3}\bar{\omega}_0''\left(\frac{a^2}{z}\right) + \frac{(\alpha_0 + \bar{\alpha}_0)a^2}{z} + \alpha_1 a^2 \end{aligned}\right\}. \tag{8.24.10}$$

From (8.24.1) and (8.24.9), when $|z|$ is small,

$$\Omega_1'(z) = -\bar{\alpha}_0 - \frac{B_1'}{a^2} - \frac{2(B_2' - iC_2')z}{a^4} + \dots$$

and comparison with (8.24.3) gives

$$\alpha_0 + \bar{\alpha}_0 = -\frac{B_1'}{a^2}, \qquad \alpha_1 = -\frac{2(B_2' - iC_2')}{a^4}.$$

Since the imaginary part of the constant α_0 only gives rise to a rigid-body displacement we may put

$$\alpha_0 = -\frac{B_1'}{2a^2}, \qquad \alpha_1 = -\frac{2(B_2'-iC_2')}{a^4}, \tag{8.24.11}$$

and ignore the imaginary part of α_0.

Displacements and stresses may now be written down using (8.24.10), (8.24.11), and (8.1.17)–(8.1.20). The stress on the edge of the disk $r = a$ reduces to

$$\sigma_{\theta\theta} = \frac{4B_1'}{a^2}+4z\Omega_0''(z)+\frac{4a^2}{z}\bar{\Omega}_0''\left(\frac{a^2}{z}\right)+\frac{4z^2}{a^2}\omega_0''(z)+\frac{4a^2}{z^2}\bar{\omega}_0''\left(\frac{a^2}{z}\right), \tag{8.24.12}$$

since $\sigma_{rr} = 0$ on $r = a$. It is now a simple matter to write down the edge stress in a given problem. For example, if we have a force $P = X+iY$ at the point $z = z_0$ ($|z_0| < a$), balanced by a force $-P$ and a couple $-M$ at the centre $z = 0$ of the disk, where

$$M = \tfrac{1}{2}i(z_0\bar{P}-\bar{z}_0 P),$$

then, from (8.3.2),

$$\left.\begin{aligned}\Omega_0(z) &= -\frac{P}{4\pi(\kappa+1)}\{\ln(z-z_0)-\ln z\}\\ \omega_0'(z) &= \frac{\kappa\bar{P}}{4\pi(\kappa+1)}\{\ln(z-z_0)-\ln z\}+\frac{P\bar{z}_0}{4\pi(\kappa+1)(z-z_0)}-\frac{iM}{4\pi z}\end{aligned}\right\}. \tag{8.24.13}$$

Also, from (8.24.1),

$$B_1' = -\frac{(\kappa-1)(\bar{z}_0 P+z_0\bar{P})}{8\pi(\kappa+1)}. \tag{8.24.14}$$

The evaluation of the edge stress $\sigma_{\theta\theta}$, using (8.24.12), is left as an exercise.

To complete the first boundary-value problem for the circular disk we consider a disk under the action of prescribed edge forces

$$\sigma_{rr}+i\sigma_{r\theta} = p(t)+iq(t) \quad (t = ae^{i\theta}), \tag{8.24.15}$$

which are self-equilibrating. Here $p(t)$, $q(t)$ are real functions. We suppose that the stress in the disk is derived from complex potentials $\Omega(z)$, $\omega(z)$ such that, for small $|z|$,

$$\left.\begin{aligned}\Omega'(z) &= \beta_0+\beta_1 z+\ldots\\ \omega'(z) &= O(z)\end{aligned}\right\}, \tag{8.24.16}$$

where β_0 is a real constant and β_1 a complex constant. We define $\Omega(z)$ in the region S^- ($|z| > a$) by (8.18.1) so that, when $|z|$ is large,

$$\Omega(z) = -\beta_0 z - \bar{\beta}_1 a^2 + O\left(\frac{1}{z}\right). \tag{8.24.17}$$

Also, from (8.1.21), (8.18.3), and (8.24.15),

$$\Omega^+(t) - \Omega^-(t) = S(t) + iT(t) \quad (t = ae^{i\theta}), \tag{8.24.18}$$

where
$$S(t) + iT(t) = \tfrac{1}{2} \int^t \{p(t) + iq(t)\}\, dt. \tag{8.24.19}$$

Hence, provided $S(t)$, $T(t)$ satisfy suitable conditions on the circle,

$$\Omega(z) = \frac{1}{2\pi i} \int \frac{S(t) + iT(t)}{t - z}\, dt - \beta_0 z - \bar{\beta}_1 a^2, \tag{8.24.20}$$

where the integral is taken anti-clockwise around the circle $|z| = a$.

An alternative solution of the problem of prescribed stresses at the boundary of a disk $|z| = a$ can be given in terms of two potential functions $V(z)$, $W(z)$. As in (8.19.15) and (8.19.16) we put

$$\left.\begin{aligned} V(z) &= 4\Omega'(z) - 2z\Omega''(z) - \frac{2z^2}{a^2}\omega''(z) \\ W(z) &= 2z\Omega''(z) + \frac{2z^2}{a^2}\omega''(z) \end{aligned}\right\}, \tag{8.24.21}$$

so that, on the circle $|z| = a$,

$$\operatorname{re} V(z) = p(t), \qquad \operatorname{im} W(z) = q(t). \tag{8.24.22}$$

Also

$$\left.\begin{aligned} \Omega'(z) &= \tfrac{1}{4}\{V(z) + W(z)\} \\ \omega''(z) &= \frac{a^2 W(z)}{2z^2} - \frac{a^2}{4z}\{V'(z) + W'(z)\} \end{aligned}\right\}, \tag{8.24.23}$$

and, from (8.24.16), for small values of z,

$$\left.\begin{aligned} V(z) &= 4\beta_0 + 2\beta_1 z + O(z^2) \\ W(z) &= 2\beta_1 z + O(z^2) \end{aligned}\right\}. \tag{8.24.24}$$

From (8.1.18), (8.1.19), and (8.24.23),

$$\left.\begin{aligned} \sigma_{rr} + \sigma_{\theta\theta} &= V(z) + W(z) + \bar{V}(\bar{z}) + \bar{W}(\bar{z}) \\ \sigma_{rr} - \sigma_{\theta\theta} + 2i\sigma_{r\theta} &= -\frac{2a^2\bar{W}(\bar{z})}{z\bar{z}} + \left(\frac{a^2}{z} - \bar{z}\right)\{\bar{V}'(\bar{z}) + \bar{W}'(\bar{z})\} \end{aligned}\right\}, \tag{8.24.25}$$

and, on the circle $r = a$,

$$\left.\begin{aligned} \sigma_{rr} + \sigma_{\theta\theta} &= V(z) + W(z) + \bar{V}(\bar{z}) + \bar{W}(\bar{z}) \\ \sigma_{rr} - \sigma_{\theta\theta} + 2i\sigma_{r\theta} &= -2\bar{W}(\bar{z}) \end{aligned}\right\}. \tag{8.24.26}$$

Expressions for $V(z)$, $W(z)$ can be found by using the Schwarz formulae

$$\left.\begin{aligned} V(z) &= \frac{1}{2\pi}\int\limits_0^{2\pi} \frac{ae^{i\gamma}+z}{ae^{i\gamma}-z}\,p(ae^{i\gamma})\,d\gamma \\ W(z) &= \frac{i}{2\pi}\int\limits_0^{2\pi} \frac{ae^{i\gamma}+z}{ae^{i\gamma}-z}\,q(ae^{i\gamma})\,d\gamma \end{aligned}\right\}. \tag{8.24.27}$$

The functions $V(z)$ and $W(z)$ may also be represented as power series

$$\left.\begin{aligned} V(z) &= R_0+\sum_{n=1}^{\infty}(R_n-iS_n)\frac{z^n}{a^n} \\ W(z) &= \sum_{n=1}^{\infty}(U_n+iT_n)\frac{z^n}{a^n} \end{aligned}\right\}, \tag{8.24.28}$$

where

$$\left.\begin{aligned} R_n &= \frac{1}{\pi}\int\limits_0^{2\pi} p(ae^{i\theta})\cos n\theta\,d\theta, & S_n &= \frac{1}{\pi}\int\limits_0^{2\pi} p(ae^{i\theta})\sin n\theta\,d\theta \\ T_n &= \frac{1}{\pi}\int\limits_0^{2\pi} q(ae^{i\theta})\cos n\theta\,d\theta, & U_n &= \frac{1}{\pi}\int\limits_0^{2\pi} q(ae^{i\theta})\sin n\theta\,d\theta \\ R_0 &= \frac{1}{2\pi}\int\limits_0^{2\pi} p(ae^{i\theta})\,d\theta, & & \int\limits_0^{2\pi} q(ae^{i\theta})\,d\theta = 0 \end{aligned}\right\}, \tag{8.24.29}$$

and, in view of (8.24.24),

$$R_1 = U_1, \qquad S_1 = -T_1. \tag{8.24.30}$$

The coefficients R_n, S_n and U_n, T_n are respectively the Fourier coefficients in the expansions of $p(ae^{i\theta})$ and $q(ae^{i\theta})$ as Fourier series. Hence

$$\left.\begin{aligned} p(ae^{i\theta}) &= R_0+\sum_{n=1}^{\infty}(R_n\cos n\theta+S_n\sin n\theta) \\ q(ae^{i\theta}) &= \sum_{n=1}^{\infty}(T_n\cos n\theta+U_n\sin n\theta) \end{aligned}\right\}. \tag{8.24.31}$$

8.25. Curvilinear polygonal hole

We now turn our attention to some problems connected with stress distributions in an infinite plate containing a hole of general curvilinear polygonal shape† given by $|\sigma| = 1$ (or $\xi = 0$) in the transformation

$$z = e^{i\beta}(na\sigma+b\sigma^{-n}), \qquad \sigma = e^{\zeta}, \qquad \zeta = \xi+i\eta, \tag{8.25.1}$$

where a and b are real constants, $a > b$, and β is an angle which determines the orientation of the axes. The subsidiary transformation $\sigma = e^{\zeta}$

† See A. E. Green, *Proc. R. Soc.* A**184** (1945) 231; A. C. Stevenson, *Phil. Mag.* **34** (1943) 766.

is useful as it provides us with curvilinear coordinates (ξ, η) such that the boundary of the hole in the z-plane is given by the circle $\sigma\bar{\sigma} = 1$ in the σ-plane, i.e. by $\xi = 0$. We observe that $z'(\sigma) = ne^{i\beta}(a-b/\sigma^{n+1})$ has zeros only in the region $|\sigma| < 1$, and $\bar{z}'(1/\sigma)$ has zeros only in the region $|\sigma| > 1$. We consider briefly special cases of (8.25.1).

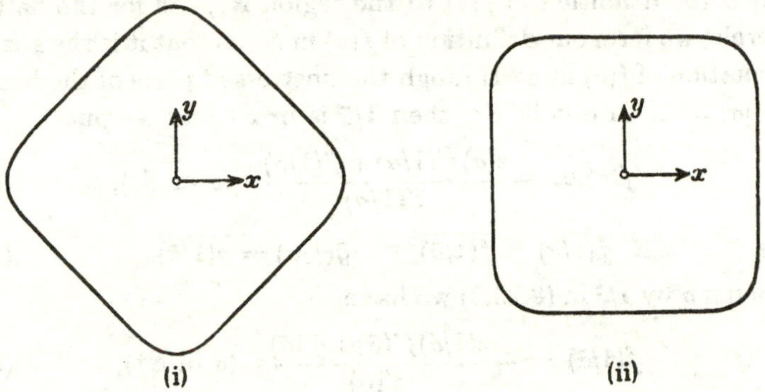

FIG. 8.4. Polygonal hole: (i) $n = 3$, $a = 3b$, $\beta = 0$; (ii) $n = 3$, $a = 3b$, $\beta = \frac{1}{4}\pi$.

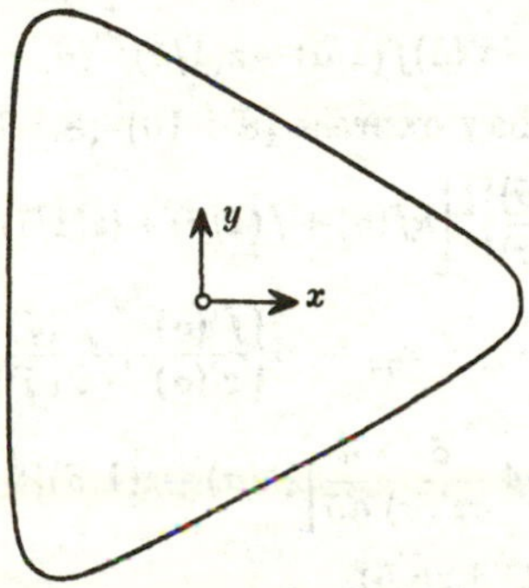

FIG. 8.5. Polygonal hole: $n = 2$, $a = 2b$, $\beta = 0$.

When the hole is circular and of radius a the transformation is given by $n = 1$, $b = \beta = 0$.

For an elliptical hole whose semi-axes are a', b', where

$$a' = a+b = c\cosh\epsilon, \qquad b' = a-b = c\sinh\epsilon,$$

we have $n = 1$, the major axis being at an angle β with the x-axis.

The curve for which $n = 3$, $a = 3b$ represents approximately a square with rounded corners. When $\beta = 0$ the 'sides' of the square are inclined at angles of $\pm\frac{1}{4}\pi$ to the coordinate axes, and when $\beta = \frac{1}{4}\pi$ the 'sides' are parallel to the axes (see Fig. 8.4).

An equilateral triangle with rounded corners is represented approximately by $n = 2$, $a = 2b$, the orientation of the triangle for $\beta = 0$ being shown in Fig. 8.5.

We make use here of the formulae (8.1.9)–(8.1.15). We denote the region $|\sigma| > 1$ in the σ-plane, which corresponds to the region outside the polygonal hole in the z-plane, by S^+ and the region $|\sigma| < 1$ by S^-, so that positive direction along the circle $|\sigma| = 1$ is clockwise. Complex potentials $f(\sigma)$, $g(\sigma)$ are defined for the region S^+ and we continue the definition of $f(\sigma)$ to the region S^-. As for the half-plane and circle, we form our definition of $f(\sigma)$ in S^- so that it is the analytical continuation of $f(\sigma)$ in S^+ through the unstressed parts of the boundary circle $|\sigma| = 1$. If σ is in S^-, then $1/\bar{\sigma}$ is in S^+ and we put

$$f(\sigma) = -\frac{z(\sigma)\bar{f}'(1/\sigma)+\bar{g}'(1/\sigma)}{\bar{z}'(1/\sigma)} \quad (\sigma \text{ in } S^-), \tag{8.25.2}$$

where $$\bar{f}(1/\sigma) = \overline{f(1/\bar{\sigma})}, \qquad \bar{g}(1/\sigma) = \overline{g(1/\bar{\sigma})}. \tag{8.25.3}$$

Replacing σ by $1/\bar{\sigma}$ in (8.25.2) we have

$$f(1/\bar{\sigma}) = -\frac{z(1/\bar{\sigma})\bar{f}'(\bar{\sigma})+\bar{g}'(\bar{\sigma})}{\bar{z}'(\bar{\sigma})} \quad (\sigma \text{ in } S^+), \tag{8.25.4}$$

or, solving for $\bar{g}'(\bar{\sigma})$,

$$\bar{g}'(\bar{\sigma}) = -\bar{z}'(\bar{\sigma})f(1/\bar{\sigma})-z(1/\bar{\sigma})\bar{f}'(\bar{\sigma}) \quad (\sigma \text{ in } S^+). \tag{8.25.5}$$

Using this result we may express (8.1.10)–(8.1.12) in the forms

$$\mu(u_\xi+iu_\eta) = \left\{\frac{\bar{\sigma}\,\bar{z}'(\bar{\sigma})}{\sigma\,z'(\sigma)}\right\}^{\frac{1}{2}}\left[\kappa f(\sigma)+f(1/\bar{\sigma})+\{z(1/\bar{\sigma})-z(\sigma)\}\frac{\bar{f}'(\bar{\sigma})}{\bar{z}'(\bar{\sigma})}\right], \tag{8.25.6}$$

$$\sigma_{\xi\xi}+\sigma_{\eta\eta} = 4\left\{\frac{f'(\sigma)}{z'(\sigma)}+\frac{\bar{f}'(\bar{\sigma})}{\bar{z}'(\bar{\sigma})}\right\}, \tag{8.25.7}$$

$$\sigma_{\xi\xi}-\sigma_{\eta\eta}+2i\sigma_{\xi\eta} = -4\frac{\bar{\sigma}}{\sigma z'(\sigma)}\frac{d}{d\bar{\sigma}}\left[\{z(\sigma)-z(1/\bar{\sigma})\}\frac{\bar{f}'(\bar{\sigma})}{\bar{z}'(\bar{\sigma})}-f(1/\bar{\sigma})\right]. \tag{8.25.8}$$

Also, from (8.1.5) and (8.25.5),

$$\frac{P}{2i} = f(\sigma)-f(1/\bar{\sigma})+\{z(\sigma)-z(1/\bar{\sigma})\}\frac{\bar{f}'(\bar{\sigma})}{\bar{z}'(\bar{\sigma})}. \tag{8.25.9}$$

8.26. First boundary-value problem for polygonal hole

As in previous cases, when stresses are prescribed at a boundary we divide the problem into two parts. In the first part we suppose that complex potentials $f_0(\sigma)$, $g_0(\sigma)$ define stresses and displacements for a plate which does not contain a polygonal hole. We now add a stress system derived from complex potentials $f_1(\sigma)$, $g_1(\sigma)$ which is such that the combined complex potentials

$$f(\sigma) = f_0(\sigma)+f_1(\sigma), \qquad g(\sigma) = g_0(\sigma)+g_1(\sigma) \tag{8.26.1}$$

give zero applied stresses at the boundary $|\sigma| = 1$ of the hole. Also,

$f_1(\sigma)$ and $g_1(\sigma)$ are potentials which produce zero stresses and rotation at infinity and zero resultant force over the hole, so that

$$f_1'(\sigma) = O\left(\frac{1}{\sigma^2}\right), \qquad g_1''(\sigma) = O\left(\frac{1}{\sigma^2}\right), \tag{8.26.2}$$

for large $|\sigma|$. We put

$$f_1'(\sigma) = \sum_{r=0}^{\infty} A_r \sigma^{-r-2}, \tag{8.26.3}$$

for large $|\sigma|$, and observe that

$$\bar{f}_1'(1/\sigma) = \sum_{r=0}^{\infty} \bar{A}_r \sigma^{r+2}, \tag{8.26.4}$$

for small $|\sigma|$.

We now extend the definition of $f_1(\sigma)$ to the region S^- ($|\sigma| < 1$) by a formula of the type (8.25.2), so that

$$f_1(\sigma) = -\frac{z(\sigma)\bar{f}_1'(1/\sigma)+\bar{g}_1'(1/\sigma)}{\bar{z}'(1/\sigma)} \quad (\sigma \text{ in } S^-). \tag{8.26.5}$$

From (8.26.4) and (8.25.1) we see that this definition of $f_1(\sigma)$ implies that $f_1(\sigma)$ has a pole at the origin of σ of the form

$$f_1(\sigma) = -\frac{e^{2i\beta}(na\sigma+b\sigma^{-n})}{n(a-b\sigma^{n+1})} \sum_{r=0}^{\infty} \bar{A}_r \sigma^{r+2},$$

so that the principal part of $f_1(\sigma)$ at the pole $\sigma = 0$ is

$$f_1(\sigma) = -\frac{be^{2i\beta}}{na} \sum_{r=0}^{n-3} \bar{A}_r \sigma^{r+2-n} \quad (n \geqslant 3). \tag{8.26.6}$$

The resultant force P acting on any arc of the hole $|\sigma| = 1$ may be obtained from (8.25.9), (8.1.5), and (8.26.1) in the form

$$\frac{P}{2i} = f_1(\sigma) - f_1(1/\bar{\sigma}) + \{z(\sigma) - z(1/\bar{\sigma})\}\frac{\bar{f}_1'(\bar{\sigma})}{\bar{z}'(\bar{\sigma})} + f_0(\sigma) + \frac{z(\sigma)\bar{f}_0'(\bar{\sigma})}{\bar{z}'(\bar{\sigma})} + \frac{\bar{g}_0'(\bar{\sigma})}{\bar{z}'(\bar{\sigma})}. \tag{8.26.7}$$

Hence, making the usual assumptions about the existence of various limits, we see from (8.26.7) that if the boundary of the hole is to be free from applied stress then

$$f_1^+(t) - f_1^-(t) = -f_0(t) - \frac{z(t)\bar{f}_0'(1/t)}{\bar{z}'(1/t)} - \frac{\bar{g}_0'(1/t)}{\bar{z}'(1/t)}, \tag{8.26.8}$$

where t denotes a point on the circle $|\sigma| = 1$. The solution of (8.26.8) which is sectionally holomorphic, except at the origin where it takes the form (8.26.6), and which vanishes at infinity, is

$$f_1(\sigma) = -\frac{1}{2\pi i}\int\left\{f_0(t)+\frac{z(t)\bar{f}'_0(1/t)}{\bar{z}'(1/t)}+\frac{\bar{g}'_0(1/t)}{\bar{z}'(1/t)}\right\}\frac{dt}{t-\sigma}-\frac{be^{2i\beta}}{na}\sum_{r=0}^{n-3}\bar{A}_r\sigma^{r+2-n}, \tag{8.26.9}$$

the integral being clockwise around the unit circle $|\sigma| = 1$. The second term in (8.26.9) is replaced by zero if $n \leqslant 2$.

Owing to the presence of the factor $\bar{z}'(1/t)$ in the integrand of (8.26.9) it is not very convenient to attempt to evaluate the integral in terms of the general functions $f_0(\sigma)$, $g_0(\sigma)$ as was done in the case of the circular hole. We therefore leave the evaluation of $f_1(\sigma)$ to special problems in § 8.27 and consider next the second part of the first boundary-value problem for a curvilinear polygonal hole in which stresses take prescribed non-zero values at the boundary of the hole.

We suppose that the prescribed boundary stresses take the forms

$$\sigma_{\xi\xi}+i\sigma_{\xi\eta} = p(t)+iq(t) \quad (t = e^{i\eta}), \tag{8.26.10}$$

where $p(t)$, $q(t)$ are real functions, and we suppose that the state of stress is derived from complex potentials $f(\sigma)$, $g(\sigma)$ so that if stresses and rotation vanish at infinity

$$\left.\begin{aligned} f'(\sigma) &= -\frac{P}{4\pi(\kappa+1)\sigma}+\sum_{r=0}^{\infty}B_r\sigma^{-r-2} \\ g''(\sigma) &= O\left(\frac{1}{\sigma}\right) \end{aligned}\right\} \tag{8.26.11}$$

for large $|\sigma|$, where P is the resultant force at the hole. We extend the definition of $f(\sigma)$ to the region S^- by (8.25.2) and see from (8.26.11) that $f(\sigma)$ has a pole at $\sigma = 0$ with principal part

$$\left.\begin{aligned} f(\sigma) &= \frac{be^{2i\beta}\bar{P}}{4\pi(\kappa+1)na\sigma^{n-1}}-\frac{be^{2i\beta}}{na}\sum_{r=0}^{n-3}\bar{B}_r\sigma^{r+2-n} \quad (n \geqslant 3) \\ f(\sigma) &= \frac{be^{2i\beta}\bar{P}}{8\pi(\kappa+1)\sigma a} \quad (n = 2) \end{aligned}\right\}. \tag{8.26.12}$$

From (8.26.10), (8.1.14) and (8.25.1),

$$P(t) = i\int^{t}\{p(t)+iq(t)\}ne^{i\beta}(a-bt^{-n-1})\,dt \quad (t = e^{i\eta}), \tag{8.26.13}$$

and, with the usual assumptions about the existence of various limits, we have from (8.25.9)

$$f^+(t)-f^-(t) = -\tfrac{1}{2}iP(t). \tag{8.26.14}$$

The solution of this boundary-value problem can only conveniently be written down when the prescribed forces at the hole are self-equilibrating so that $P = 0$ and $f(\sigma)$ vanishes at infinity. If for the present we consider this case, then, remembering (8.26.12), we have

$$f(\sigma) = -\frac{1}{4\pi}\int\frac{P(t)\,dt}{t-\sigma}-\frac{be^{2i\beta}}{na}\sum_{r=0}^{n-3}\bar{B}_r\sigma^{r+2-n}, \qquad (8.26.15)$$

the integral being clockwise around the unit circle $t = e^{i\eta}$. The second term in (8.26.15) is replaced by zero when $n \leqslant 2$.

If the prescribed stresses at the hole are not self-equilibrating and have a resultant force P, then $f(\sigma)$ does not vanish at infinity. In this case, from (8.26.13), we write

$$\frac{\partial P(t)}{\partial\eta} = -ne^{i\beta}(at-bt^{-n})\{p(t)+iq(t)\}$$

on the circle $t = e^{i\eta}$. Also, using the result

$$\frac{\partial}{\partial\eta} = i\sigma\frac{\partial}{\partial\sigma}-i\bar{\sigma}\frac{\partial}{\partial\bar{\sigma}} \quad (\sigma = e^{\xi+i\eta}),$$

we have, from (8.25.9),

$$-\frac{1}{2}\frac{\partial P(t)}{\partial\eta} = \sigma f'(\sigma)-\frac{1}{\bar{\sigma}}f'(1/\bar{\sigma})+\left\{\sigma z'(\sigma)-\frac{z'(1/\bar{\sigma})}{\bar{\sigma}}\right\}\frac{\bar{f}'(\bar{\sigma})}{\bar{z}'(\bar{\sigma})}-$$
$$-\{z(\sigma)-z(1/\bar{\sigma})\}\bar{\sigma}\frac{\partial}{\partial\bar{\sigma}}\left\{\frac{\bar{f}'(\bar{\sigma})}{\bar{z}'(\bar{\sigma})}\right\},$$

and hence, with the usual assumptions,

$$[f'(t)]^+-[f'(t)]^- = R(t), \qquad (8.26.16)$$

where $$R(t) = \tfrac{1}{2}ne^{i\beta}(a-bt^{-n-1})\{p(t)+iq(t)\}. \qquad (8.26.17)$$

The function $f'(\sigma)$ vanishes at infinity and is sectionally holomorphic, with discontinuity given by (8.26.16), except at the origin, where, from (8.26.12), $f'(\sigma)$ has a pole with principal part

$$\left.\begin{aligned} f'(\sigma) &= \frac{(1-n)be^{2i\beta}\bar{P}}{4\pi(\kappa+1)na\sigma^n}-\frac{be^{2i\beta}}{na}\sum_{r=0}^{n-3}(r+2-n)\bar{B}_r\sigma^{r+1-n} \quad (n \geqslant 3) \\ f'(\sigma) &= -\frac{be^{2i\beta}\bar{P}}{8\pi(\kappa+1)a\sigma^2} \quad (n = 2) \end{aligned}\right\}. \qquad (8.26.18)$$

The function $f'(\sigma)$ is therefore given by

$$f'(\sigma) = \frac{1}{2\pi i}\int\frac{R(t)\,dt}{t-\sigma}+\frac{(1-n)be^{2i\beta}\bar{P}}{4\pi(\kappa+1)na\sigma^n}-\frac{be^{2i\beta}}{na}\sum_{r=0}^{n-3}(r+2-n)\bar{B}_r\sigma^{r+1-n}, \qquad (8.26.19)$$

the integral being clockwise around the unit circle $t = e^{i\eta}$. When $n = 1$

the second and third terms in (8.26.19) are replaced by zero. When $n = 2$ the third group of terms is replaced by zero.

8.27. Effect of polygonal hole on given stress distributions

Using the notation of the previous two sections we suppose that an infinite plate containing a stress-free polygonal hole is in a uniform state of stress at infinity, so that

$$\left.\begin{aligned} 8f_0(\sigma) &= (N_1+N_2)z \\ 8g_0(\sigma) &= -(N_1-N_2)e^{-2i\alpha}z^2 \end{aligned}\right\}, \tag{8.27.1}$$

where N_1, N_2, α are defined after (8.1.23). Since we only require the stress system corresponding to the potentials (8.27.1) to be uniform for large values of $|z|$, and hence for large $|\sigma|$, we could take the fundamental potentials to be

$$8f_0(\sigma) = (N_1+N_2)nae^{i\beta}\sigma,$$

$$8g_0(\sigma) = -(N_1-N_2)n^2a^2e^{2i\beta-2i\alpha}\sigma^2,$$

but it is found that the form (8.27.1) is more suitable for our purpose. Using (8.27.1), it follows from (8.26.9) that

$$f_1(\sigma) = -\frac{1}{2\pi i}\int \{(N_1+N_2)z(t)-(N_1-N_2)e^{2i\alpha}\bar{z}(1/t)\}\frac{dt}{4(t-\sigma)}-$$

$$-\frac{be^{2i\beta}}{na}\sum_{r=0}^{n-3}\bar{A}_r\sigma^{r+2-n}, \tag{8.27.2}$$

the integral being clockwise around the unit circle $t = e^{i\eta}$ in the σ-plane. The terms apart from the integral in (8.27.2) are replaced by zero when $n \leqslant 2$. The various terms in the integral in (8.27.2) may be evaluated either by residues inside the unit circle or by residues outside the circle together with an integral over the infinite circle. Thus

$$4f_1(\sigma) = -(N_1+N_2)z(\sigma)+(N_1+N_2)nae^{i\beta}\sigma+(N_1-N_2)na\sigma^{-1}e^{2i\alpha-i\beta}-$$

$$-\frac{4be^{2i\beta}}{na}\sum_{r=0}^{n-3}\bar{A}_r\sigma^{r+2-n} \quad (\sigma \text{ in } S^+), \tag{8.27.3}$$

$$4f_1(\sigma) = (N_1+N_2)nae^{i\beta}\sigma-(N_1-N_2)be^{2i\alpha-i\beta}\sigma^n-$$

$$-\frac{4be^{2i\beta}}{na}\sum_{r=0}^{n-3}\bar{A}_r\sigma^{r+2-n} \quad (\sigma \text{ in } S^-). \tag{8.27.4}$$

If we now compare (8.27.3) with the expansion (8.26.3) for $f_1'(\sigma)$, valid for large $|\sigma|$, we have

$$\frac{be^{2i\beta}}{na}\bar{A}_{n-3}-\frac{(N_1-N_2)nae^{2i\alpha-i\beta}}{4}=A_0\quad(n\geqslant 3),$$

$$A_{n-1}=\tfrac{1}{4}(N_1+N_2)nbe^{i\beta},\qquad A_r=0\quad\text{otherwise},$$

$$\frac{b(n-2)e^{2i\beta}\bar{A}_0}{na}=A_{n-3}\quad(n\geqslant 4),$$

and hence

$$\left.\begin{aligned}A_0\left\{1-\frac{b^2(n-2)}{n^2a^2}\right\}&=-\tfrac{1}{4}na(N_1-N_2)e^{2i\alpha-i\beta}\\ A_{n-3}\left\{1-\frac{b^2(n-2)}{n^2a^2}\right\}&=-\tfrac{1}{4}b(n-2)(N_1-N_2)e^{-2i\alpha+3i\beta}\end{aligned}\right\}\quad(n\geqslant 3).\qquad(8.27.5)$$

When $n=3$, A_0 is the *sum* of values in (8.27.5).

We may now obtain all the components of stress and displacement from (8.25.6)–(8.25.8), (8.27.3), (8.27.4), and (8.1.10)–(8.1.12), together with (8.27.1). We restrict our attention, however, to the hoop stress at the boundary of the hole which is given by

$$\sigma_{\eta\eta}=4\left\{\frac{f_1'(t)}{z'(t)}+\frac{\bar{f}_1'(\bar{t})}{\bar{z}'(\bar{t})}+\frac{f_0'(t)}{z'(t)}+\frac{\bar{f}_0'(\bar{t})}{\bar{z}'(\bar{t})}\right\}\qquad(8.27.6)$$

evaluated at $t=e^{i\eta}$, since $\sigma_{\xi\xi}=0$. Consider first the case $n\leqslant 2$, so that, from (8.27.1) and (8.27.3),

$$8\{f_0'(\sigma)+f_1'(\sigma)\}=(N_1+N_2)nae^{i\beta}+(N_1+N_2)nbe^{i\beta}\sigma^{-n-1}-\\-2(N_1-N_2)nae^{2i\alpha-i\beta}\sigma^{-2}.$$

Therefore, a straightforward calculation gives

$$\{a^2+b^2-2ab\cos(n+1)\eta\}\sigma_{\eta\eta}=(N_1+N_2)(a^2-b^2)-\\-2(N_1-N_2)[a^2\cos(2\eta-2\alpha+2\beta)-ab\cos\{(n-1)\eta+2\alpha-2\beta\}].\qquad(8.27.7)$$

When $n\geqslant 3$ additional terms containing the constants A_r must be used in (8.27.3) where the constants are given by (8.27.5). Hence, the terms in the edge stress $\sigma_{\eta\eta}$ which must be *added* to those in (8.27.7) are given by

$$\left\{1-\frac{b^2(n-2)}{n^2a^2}\right\}\{a^2+b^2-2ab\cos(n+1)\eta\}\sigma_{\eta\eta}\\=-\frac{2(N_1-N_2)(n-2)}{n}\left[ab\left(1-\frac{b^2}{na^2}\right)\cos\{(n-1)\eta+2\alpha-2\beta\}-\right.\\\left.-b^2\left(1-\frac{1}{n}\right)\cos(2\eta-2\alpha+2\beta)\right],\qquad(8.27.8)$$

so that the *total* edge stress $\sigma_{\eta\eta}$ for $n \geqslant 3$, obtained by adding contributions from (8.27.7) and (8.27.8), is

$$\{n-(n-2)b^2/(na^2)\}\{a^2+b^2-2ab\cos(n+1)\eta\}\sigma_{\eta\eta}$$
$$= (N_1+N_2)(a^2-b^2)\{n-(n-2)b^2/(na^2)\}-$$
$$-2(N_1-N_2)[\{na^2-(n-2)b^2\}\cos(2\eta-2\alpha+2\beta)-$$
$$-2ab\cos\{(n-1)\eta+2\alpha-2\beta\}]. \tag{8.27.9}$$

From the general formulae (8.27.7) and (8.27.9) we can at once deduce various special cases. For an elliptical hole we put $a+b = c\cosh\epsilon$, $a-b = c\sinh\epsilon$, and $n = 1$ in (8.27.7) so that

$$(\cosh 2\epsilon-\cos 2\eta)\sigma_{\eta\eta} = (N_1+N_2)\sinh 2\epsilon-$$
$$-(N_1-N_2)\{e^{2\epsilon}\cos(2\eta-2\alpha+2\beta)-\cos(2\alpha-2\beta)\}. \tag{8.27.10}$$

If we have a tension T parallel to the major axis, then putting $\beta = \alpha = N_2 = 0$, $N_1 = T$ in (8.27.10), the edge stress becomes

$$\sigma_{\eta\eta} = T\frac{\sinh 2\epsilon+1-e^{2\epsilon}\cos 2\eta}{\cosh 2\epsilon-\cos 2\eta}. \tag{8.27.11}$$

For a tension T parallel to the minor axis, $\beta = \alpha = N_1 = 0$, $N_2 = T$, and

$$\sigma_{\eta\eta} = T\frac{\sinh 2\epsilon-1+e^{2\epsilon}\cos 2\eta}{\cosh 2\epsilon-\cos 2\eta}. \tag{8.27.12}$$

For a shearing force S parallel to the axis of the ellipse, $\beta = 0$, $\alpha = \frac{1}{4}\pi$, $N_1 = -N_2 = S$, and

$$\sigma_{\eta\eta} = -\frac{2Se^{2\epsilon}\sin 2\eta}{\cosh 2\epsilon-\cos 2\eta}. \tag{8.27.13}$$

An approximately equilateral triangle with rounded corners is given by $n = 2$, $a = 2b$, $\beta = 0$ and is orientated as shown in Fig. 8.5. If a tension T is then applied parallel to the x-axis, the edge stress is obtained from (8.27.7) by putting $N_1 = T$, $N_2 = \alpha = 0$. Hence

$$\sigma_{\eta\eta} = \frac{T(3-8\cos 2\eta+4\cos\eta)}{5-4\cos 3\eta}. \tag{8.27.14}$$

For a tension T parallel to the y-axis, $N_1 = \alpha = 0$, $N_2 = T$, and

$$\sigma_{\eta\eta} = \frac{T(3+8\cos 2\eta-4\cos\eta)}{5-4\cos 3\eta}. \tag{8.27.15}$$

If the hole is approximately a square with rounded corners we must use formula (8.27.9) with $n = 3$, $a = 3b$. If $\beta = \frac{1}{4}\pi$ the sides of the square are parallel to the axis as shown in Fig. 8.4 (ii) and for a tension T parallel to the x-axis $N_1 = T$, $N_2 = \alpha = 0$. Thus the edge stress is

$$\sigma_{\eta\eta} = \frac{4T(1+2{\cdot}7\sin 2\eta)}{5-3\cos 4\eta}. \tag{8.27.16}$$

When the force applied at infinity is a shearing stress S then $\alpha = \frac{1}{4}\pi$, $N_1 = -N_2 = S$ and the edge stress becomes

$$\sigma_{\eta\eta} = \frac{-27S\cos 2\eta}{2(5-3\cos 4\eta)}. \tag{8.27.17}$$

All the formulae (8.27.11) to (8.27.17) for edge stresses are comparatively simple to evaluate numerically.†

8.28. Prescribed forces at polygonal hole

We consider briefly one example in which the prescribed forces at the hole are self-equilibrating so that we may use (8.26.15). Suppose two isolated forces $\pm X$ act parallel to the x-axis at points $\sigma = \pm 1$. Then

$$\left.\begin{aligned} P(t) &= X \quad (0 \leqslant \eta \leqslant \pi) \\ P(t) &= 0 \quad (\pi < \eta < 2\pi) \end{aligned}\right\} \quad (t = e^{i\eta}), \tag{8.28.1}$$

and, from (8.26.15),

$$\left.\begin{aligned} f(\sigma) &= \frac{X}{4\pi}\ln\frac{\sigma+1}{\sigma-1} - \frac{be^{2i\beta}}{na}\sum_{r=0}^{n-3} \bar{B}_r \sigma^{r+2-n} \\ f'(\sigma) &= -\frac{X}{2\pi(\sigma^2-1)} - \frac{be^{2i\beta}}{na}\sum_{r=0}^{n-3} (r+2-n)\bar{B}_r \sigma^{r+1-n} \end{aligned}\right\}, \tag{8.28.2}$$

the terms containing the constants $\bar{B}_r$ being added when $n \geqslant 3$. If we compare (8.28.2) with (8.26.11) for large $|\sigma|$, we have

$$\left.\begin{aligned} B_r - \frac{(r+1)be^{2i\beta}}{na}\bar{B}_{n-r-3} &= -\frac{X}{2\pi} \quad (r \text{ even}) \\ &= 0 \quad (r \text{ odd}) \\ \bar{B}_{n-r-3} - \frac{(n-r-2)be^{-2i\beta}}{na} B_r &= -\frac{X}{2\pi} \quad (n-r \text{ odd}) \\ &= 0 \quad (n-r \text{ even}) \end{aligned}\right\} \quad \begin{aligned} (r = 0, 1,\ldots, n-3;& \\ n \geqslant 3),& \end{aligned} \tag{8.28.3}$$

$$\left.\begin{aligned} B_r &= -\frac{X}{2\pi} \quad (r \text{ even}) \\ &= 0 \quad (r \text{ odd}) \end{aligned}\right\} \quad (r > n-3). \tag{8.28.4}$$

† See A. E. Green, loc. cit., p. 294.

Since $\sigma_{\xi\xi} = 0$ at the hole, except at the isolated forces, the value of the hoop stress at the hole is given by

$$\sigma_{\eta\eta} = 4\left\{\frac{f'(t)}{z'(t)}+\frac{\bar{f}'(t)}{\bar{z}'(t)}\right\} \quad (t = e^{i\eta}), \tag{8.28.5}$$

except at $t = \pm 1$. Hence, from (8.28.2),

$$\sigma_{\eta\eta} = \frac{2X}{\pi n \sin\eta}\left\{\frac{a\sin(\eta+\beta)+b\sin(n\eta-\beta)}{a^2+b^2-2ab\cos(n+1)\eta}\right\} \quad (n \leqslant 2). \tag{8.28.6}$$

When $n \geqslant 3$ additional terms containing the constants

$$B_r \quad (r = 0, 1, ..., n-3)$$

must be used in (8.28.2), these constants being given by (8.28.3). The evaluation of the extra terms required in the hoop stress is left as an exercise.

For an ellipse whose major axis is along the x-axis, the hoop stress for a pair of equal and opposite isolated forces at opposite ends of the major axis is obtained by putting $\beta = 0$, $n = 1$ in (8.28.6). Thus

$$\sigma_{\eta\eta} = \frac{2X(a+b)}{\pi(a^2+b^2-2ab\cos 2\eta)} = \frac{4X\cosh\epsilon}{\pi c(\cosh 2\epsilon-\cos 2\eta)} \quad (\eta \neq 0, \pi). \tag{8.28.7}$$

If equal and opposite forces Y act at opposite ends of the minor axis of an elliptical hole, we can evaluate the hoop stress in a similar manner.† Thus

$$\sigma_{\eta\eta} = \frac{2Y(a-b)}{\pi(a^2+b^2-2ab\cos 2\eta)} = \frac{4Y\sinh\epsilon}{\pi c(\cosh 2\epsilon-\cos 2\eta)} \quad (\eta \neq \pm\tfrac{1}{2}\pi). \tag{8.28.8}$$

8.29. Distribution of stress in infinite wedge‡

We consider the distribution of stress in an infinite wedge (Fig. 8.6) which is bounded by the lines $\theta = \pm\alpha$ ($\alpha \leqslant \frac{1}{2}\pi$), the origin of coordinates being at the vertex of the wedge. It is convenient to use a conformal transformation

$$z = ae^{\zeta}, \qquad \xi = \log\frac{r}{a}, \qquad \eta = \theta; \qquad z\frac{d}{dz} = \frac{d}{d\zeta}, \tag{8.29.1}$$

which transforms the boundary lines $\theta = \pm\alpha$ of the wedge into the lines $\eta = \pm\alpha$ in the ζ-plane, and the interior of the wedge into the region (Fig. 8.6 (ii)) between the lines $\eta = \pm\alpha$. The constant a denotes

† A. E. Green, *J. appl. Mech.* **14** (1947) 246.

‡ An alternative solution of wedge problems using Mellin transforms is given by C. J. Tranter, *Q. J. Mech. appl. Math.* **1** (1948) 125.

the distance from the vertex O of two points B, C on the boundary lines of the wedge, and these transform into the points $(0, \alpha)$ and $(0, -\alpha)$ respectively in the ζ-plane. From (8.1.18)–(8.1.20) we have

$$\sigma_{rr}+\sigma_{\theta\theta} = 4\{\Omega'(z)+\bar{\Omega}'(\bar{z})\}, \tag{8.29.2}$$

$$\sigma_{rr}-\sigma_{\theta\theta}+2i\sigma_{r\theta} = -4\left\{\bar{z}\bar{\Omega}''(\bar{z})+\frac{\bar{z}}{z}\bar{\omega}''(\bar{z})\right\}, \tag{8.29.3}$$

$$\sigma_{\theta\theta}-i\sigma_{r\theta} = 2\left\{\Omega'(z)+\bar{\Omega}'(\bar{z})+\bar{z}\bar{\Omega}''(\bar{z})+\frac{\bar{z}}{z}\bar{\omega}''(\bar{z})\right\}. \tag{8.29.4}$$

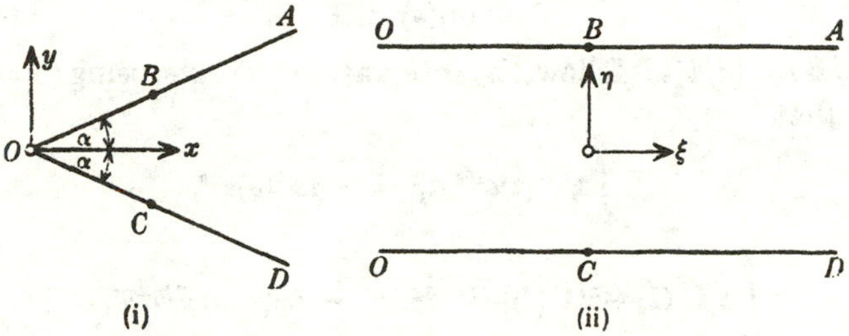

FIG. 8.6. Infinite wedge: (i) z-plane; (ii) ζ-plane.

We suppose that the surface stresses are prescribed on the boundary lines of the wedge so that

$$\left.\begin{aligned} \sigma_{\theta\theta} &= p_1(\xi), \qquad \sigma_{r\theta} = q_1(\xi) \quad (\theta = \alpha) \\ \sigma_{\theta\theta} &= p_2(\xi), \qquad \sigma_{r\theta} = q_2(\xi) \quad (\theta = -\alpha) \end{aligned}\right\}. \tag{8.29.5}$$

If the stress distribution on the faces of the wedge is such that the stresses and rotation vanish at infinity in the wedge, then, for large $|z|$ or large positive ξ,

$$\left.\begin{aligned} \Omega(z) = O(\zeta), \qquad \Omega'(z) = O(e^{-\zeta}), \qquad \Omega''(z) = O(e^{-2\zeta}) \\ \bar{\omega}'(\bar{z}) = O(\bar{\zeta}), \qquad \bar{\omega}''(\bar{z}) = O(e^{-\bar{\zeta}}) \end{aligned}\right\}. \tag{8.29.6}$$

We also assume that, for small $|z|$ or large negative ξ,

$$\left.\begin{aligned} \Omega(z) = O(z), \qquad \Omega'(z) = O(1), \qquad \Omega''(z) = O(1) \\ \bar{\omega}'(\bar{z}) = O(\bar{z}), \qquad \bar{\omega}''(\bar{z}) = O(1) \end{aligned}\right\}. \tag{8.29.7}$$

To solve the wedge problem we make use of Fourier integral theorems for holomorphic functions and we assume that the various functions used satisfy sufficient conditions for the validity of our solution (see § 1.22). The correctness of such assumptions can be checked when we have obtained the solution of a given problem.

Let

$$\left.\begin{aligned}\phi(s) &= \int_{-\infty}^{\infty} \Omega(z)e^{i\zeta s}\,d\xi\\ \psi(s) &= \int_{-\infty}^{\infty} \bar{\omega}'(\bar{z})e^{i\zeta s}\,d\xi\end{aligned}\right\}, \tag{8.29.8}$$

then

$$\bar{\phi}(-s) = \int_{-\infty}^{\infty} \bar{\Omega}(\bar{z})e^{i\zeta s}\,d\xi, \tag{8.29.9}$$

and, in view of (8.29.6) and (8.29.7), these integrals converge if

$$0 < \mathrm{im}(s) < 1. \tag{8.29.10}$$

Since $z(d/dz) = (d/d\zeta)$ it follows, by integration by parts, using (8.29.6) and (8.29.7), that

$$\int_{-\infty}^{\infty} z\Omega'(z)e^{i\xi s}\,d\xi = -is\phi(s)e^{\eta s},$$

$$\int_{-\infty}^{\infty} z\{\bar{\Omega}'(\bar{z})+\bar{z}\bar{\Omega}''(\bar{z})\}e^{i\xi s}\,d\xi = -s^2\bar{\phi}(-s)e^{2i\eta-\eta s},$$

$$\int_{-\infty}^{\infty} \bar{z}\bar{\omega}''(\bar{z})e^{i\xi s}\,d\xi = -is\psi(s)e^{-\eta s}.$$

Hence, multiplying (8.29.4) by $ae^{\xi+i\xi s}$ and integrating with respect to ξ from $-\infty$ to ∞ gives

$$is\phi(s)e^{\eta s-i\eta}+s^2\bar{\phi}(-s)e^{-\eta s+i\eta}+is\psi(s)e^{-\eta s-i\eta} = -\tfrac{1}{2}a\int_{-\infty}^{\infty} (\sigma_{\theta\theta}-i\sigma_{r\theta})e^{\xi+i\xi s}\,d\xi. \tag{8.29.11}$$

In particular, if this equation holds for $\eta = \pm\alpha$, from (8.29.5) it follows that

$$\left.\begin{aligned}s\phi(s)-is^2e^{2i\alpha-2\alpha s}\bar{\phi}(-s)+se^{-2\alpha s}\psi(s) &= \frac{iae^{i\alpha-\alpha s}}{2}\{f_1(s)-ig_1(s)\}\\ s\phi(s)-is^2e^{-2i\alpha+2\alpha s}\bar{\phi}(-s)+se^{2\alpha s}\psi(s) &= \frac{iae^{-i\alpha+\alpha s}}{2}\{f_2(s)-ig_2(s)\}\end{aligned}\right\}, \tag{8.29.12}$$

where

$$f_\alpha(s) = \int_{-\infty}^{\infty} p_\alpha(\xi)e^{(1+is)\xi}\,d\xi, \qquad g_\alpha(s) = \int_{-\infty}^{\infty} q_\alpha(\xi)e^{(1+is)\xi}\,d\xi \quad (\alpha = 1,2). \tag{8.29.13}$$

Equations (8.29.12), together with their complex conjugates, are sufficient to determine the functions $\phi(s)$, $\psi(s)$. In order to avoid undue complications in the algebra it is convenient to consider separately the cases for which the applied stress system is symmetrical or anti-symmetrical about the line $\theta = 0$, and is such that either the shearing stresses or the normal stresses are zero. For symmetrical systems in which the shearing stresses vanish

$$f_2(s) = f_1(s) = f(s), \qquad g_2(s) = g_1(s) = 0, \tag{8.29.14}$$

and eliminating $\psi(s)$ from (8.29.12) gives

$$4s\{\phi(s)\sinh 2s\alpha+\bar{\phi}(-s)s\sin 2\alpha\} = ia(e^{i\alpha+\alpha s}-e^{-i\alpha-\alpha s})f(s).$$

Taking the conjugate of this equation and changing s into $-s$ gives

$$4s\{\bar{\phi}(-s)\sinh 2s\alpha+\phi(s)s\sin 2\alpha\} = ia(e^{i\alpha+\alpha s}-e^{-i\alpha-\alpha s})f(s),$$

and hence
$$\phi(s) = \bar{\phi}(-s) = \frac{ia(e^{i\alpha+\alpha s}-e^{-i\alpha-\alpha s})f(s)}{4s(\sinh 2s\alpha+s\sin 2\alpha)}. \tag{8.29.15}$$

Also, from (8.29.12),

$$\psi(s) = -\frac{ia(e^{i\alpha-\alpha s}-e^{-i\alpha+\alpha s})(1+is)f(s)}{4s(\sinh 2s\alpha+s\sin 2\alpha)}. \tag{8.29.16}$$

To obtain the stresses at any point we observe from § 1.22 that the formulae inverse to (8.29.8) and (8.29.9) are

$$\Omega(z) = \frac{1}{2\pi}\int\limits_{-\infty+ic}^{\infty+ic} \phi(s)e^{-i\zeta s}\,ds,$$

$$\bar{\Omega}(\bar{z}) = \frac{1}{2\pi}\int\limits_{-\infty+ic}^{\infty+ic} \bar{\phi}(-s)e^{-i\bar{\zeta}s}\,ds,$$

$$\bar{\omega}'(\bar{z}) = \frac{1}{2\pi}\int\limits_{-\infty+ic}^{\infty+ic} \psi(s)e^{-i\bar{\zeta}s}\,ds,$$

where $0 < c < 1$. Moreover, we may differentiate these integrals under the integral sign. Hence, from (8.29.3) and (8.29.4),

$$r(\sigma_{rr}+\sigma_{\theta\theta}) = -\frac{2i}{\pi}\int\limits_{-\infty+ic}^{\infty+ic} \{s\phi(s)e^{\theta s-i\theta}+s\bar{\phi}(-s)e^{-\theta s+i\theta}\}e^{-i\xi s}\,ds,$$

$$r(\sigma_{rr}-\sigma_{\theta\theta}+2i\sigma_{r\theta}) = -\frac{2i}{\pi}\int\limits_{-\infty+ic}^{\infty+ic} \{s(1+is)\bar{\phi}(-s)e^{-\theta s+i\theta}-s\psi(s)e^{-\theta s-i\theta}\}e^{-i\xi s}\,ds,$$

and using (8.29.15) and (8.29.16), these become

$$\frac{\pi r}{a}(\sigma_{rr}+\sigma_{\theta\theta}) = \int\limits_{-\infty+ic}^{\infty+ic} \{Q(s)+iP(s)\}f(s)e^{-i\xi s}\,ds, \qquad (8.29.17)$$

$$\frac{\pi r}{a}(\sigma_{rr}-\sigma_{\theta\theta}+2i\sigma_{r\theta}) = i\int\limits_{-\infty+ic}^{\infty+ic} (1+is)\{P(s)-R(s)\}f(s)e^{-i\xi s}\,ds, \qquad (8.29.18)$$

where

$$\left.\begin{aligned}(\sinh 2\alpha s+s\sin 2\alpha)P(s) &= \sin(\alpha-\theta)\cosh(\alpha+\theta)s+\sin(\alpha+\theta)\cosh(\alpha-\theta)s\\ (\sinh 2\alpha s+s\sin 2\alpha)Q(s) &= \cos(\alpha-\theta)\sinh(\alpha+\theta)s+\cos(\alpha+\theta)\sinh(\alpha-\theta)s\\ (\sinh 2\alpha s+s\sin 2\alpha)R(s) &= \sin(\alpha-\theta)\sinh(\alpha+\theta)s-\sin(\alpha+\theta)\sinh(\alpha-\theta)s\end{aligned}\right\}. \qquad (8.29.19)$$

The symmetrical case for shearing forces and zero normal forces and the unsymmetrical cases can be dealt with in a similar way.

As an example of the above theory we consider the problem of a constant symmetrical normal distribution of stress on each boundary of the wedge acting over a length b of each face so that

$$\left.\begin{aligned} p_1(\xi) &= p_2(\xi) = -p \quad (-\beta \leqslant \xi \leqslant 0)\\ p_1(\xi) &= p_2(\xi) = 0 \quad \begin{pmatrix}\xi < -\beta\\ \xi > 0\end{pmatrix}\\ q_1(\xi) &= q_2(\xi) = 0\end{aligned}\right\}, \qquad (8.29.20)$$

and $$b = a(1-e^{-\beta}). \qquad (8.29.21)$$

From (8.29.13) and (8.29.14)

$$f(s) = -p\int\limits_{-\beta}^{0} e^{(1+is)\xi}\,d\xi = -\frac{p(1-e^{-(1+is)\beta})}{1+is}. \qquad (8.29.22)$$

Formula (8.29.18) now becomes

$$\frac{\pi r}{2ap}(\sigma_{rr}-\sigma_{\theta\theta}+2i\sigma_{r\theta}) = -\tfrac{1}{2}i\int\limits_{-\infty+ic}^{\infty+ic} \{P(s)-R(s)\}(1-e^{-(1+is)\beta})e^{-i\xi s}\,ds \quad (0 < c < 1).$$

The integrand has a pole at $s = 0$ and no other poles in the region $0 < \mathrm{im}(s) < 1$, so we may replace the integral by one along the real axis from $-\infty$ to ∞, less πi times the residue at $s = 0$. Hence, after straightforward reduction,

$$\frac{\pi r}{2ap}(\sigma_{rr}-\sigma_{\theta\theta}) = -\frac{\pi(1-e^{-\beta})\sin\alpha\cos\theta}{2\alpha+\sin 2\alpha} - \\ -\int_0^\infty \{\sin\xi s - e^{-\beta}\sin(\xi+\beta)s\}P(s)\,ds, \quad (8.29.23)$$

$$\frac{\pi r\sigma_{r\theta}}{ap} = \int_0^\infty \{\cos\xi s - e^{-\beta}\cos(\xi+\beta)s\}R(s)\,ds. \quad (8.29.24)$$

Similarly, we find from (8.29.17) and (8.29.22),

$$\frac{\pi r}{2ap}(\sigma_{rr}+\sigma_{\theta\theta}) = -\frac{\pi(1-e^{-\beta})\sin\alpha\cos\theta}{2\alpha+\sin 2\alpha} - \\ -\int_0^\infty \{\cos\xi s - e^{-\beta}\cos(\xi+\beta)s\}\left(\frac{Q+sP}{1+s^2}\right)ds - \\ -\int_0^\infty \{\sin\xi s - e^{-\beta}\sin(\xi+\beta)s\}\left(\frac{P-sQ}{1+s^2}\right)ds. \quad (8.29.25)$$

8.30. Overlapped circular holes

A problem of considerable interest is that of two overlapped circular holes† in an infinite medium in which the stresses at infinity are uniform. For this problem we use the transformation

$$z = ia\coth\tfrac{1}{2}\zeta, \qquad \zeta = \xi+i\eta, \quad (8.30.1)$$

which maps one circular arc AXB on to the line $\eta = \alpha$ and the other circular arc AYB on to the line $\eta = -\beta$ in the ζ-plane, the region outside the circles being transformed into the inside of the strip in the ζ-plane between these lines (see Fig. 8.7). From (8.30.1) we can verify the following results:

$$\left.\begin{aligned} z = J(\sin\eta+i\sinh\xi), &\qquad \frac{a}{J} = \cosh\xi-\cos\eta \\ z+\bar{z} = 2J\sin\eta, &\qquad z\bar{z}-a^2 = 2aJ\cos\eta \\ \frac{z\bar{z}}{a^2} &= \frac{\cosh\xi+\cos\eta}{\cosh\xi-\cos\eta} \\ z \to \frac{2ia}{\zeta} &\quad \text{as} \quad \zeta\to 0 \end{aligned}\right\}. \quad (8.30.2)$$

† This section is a modification of a paper by C. Ling, *J. appl. Phys.* **19** (1948) 405. See also E. Weinel, *Z. angew. Math. Mech.* **21** (1941) 228.

If the radii of the circles are each equal to b then $\beta = \alpha$ and

$$AB = 2a = 2b \sin \alpha. \tag{8.30.3}$$

We saw in § 6.4 that, in the absence of body forces, Airy's stress function ϕ satisfies the biharmonic equation and we expressed ϕ in terms

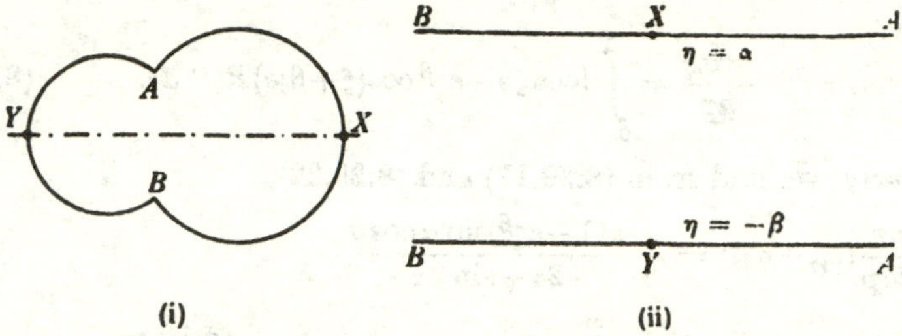

FIG. 8.7. Overlapped circular holes: (i) z-plane; (ii) ζ-plane.

of two harmonic functions $\Omega(z)$, $\omega(z)$. Here we make a change of notation and put ϕ in the form

$$2\phi = a(z+\bar{z})\{\Omega(z)+\bar{\Omega}(\bar{z})\}+(z\bar{z}-a^2)\{\omega(z)+\bar{\omega}(\bar{z})\}.$$

Hence

$$\chi = \frac{\phi}{aJ} = \{\Omega(z)+\bar{\Omega}(\bar{z})\}\sin\eta+\{\omega(z)+\bar{\omega}(\bar{z})\}\cos\eta = f\sin\eta+g\cos\eta, \tag{8.30.4}$$

where f and g are real harmonic functions satisfying the equations

$$\left(\frac{\partial^2}{\partial x^2}+\frac{\partial^2}{\partial y^2}\right)\begin{matrix} f \\ g \end{matrix} = 0 \quad \text{or} \quad \left(\frac{\partial^2}{\partial \xi^2}+\frac{\partial^2}{\partial \eta^2}\right)\begin{matrix} f \\ g \end{matrix} = 0. \tag{8.30.5}$$

If the stress at infinity consists of a uniform tension N_1 parallel to the x-axis and N_2 parallel to the y-axis, then the corresponding stress function ϕ_0 is

$$\phi_0 = \tfrac{1}{2}N_1(z\bar{z}-a^2)-\tfrac{1}{8}(N_1-N_2)(z+\bar{z})^2,$$

and

$$\chi_0 = \frac{\phi_0}{aJ} = N_1\cos\eta-\frac{(N_1-N_2)\sin^2\eta}{2(\cosh\xi-\cos\eta)}. \tag{8.30.6}$$

We observe that when $\eta \to 0$ ($\xi \neq 0$) then $\chi_0 \to N_1$.

The function $\phi = \ln z\bar{z}$, which represents an all-round pressure at the origin, is a possible stress function, and therefore a possible value of χ is

$$\chi = \frac{\phi}{aJ} = (\cosh\xi-\cos\eta)\ln\frac{\cosh\xi-\cos\eta}{\cosh\xi+\cos\eta}. \tag{8.30.7}$$

We now take the complete stress function for the problem of overlapped circular holes in an infinite plane under uniform tensions parallel and perpendicular to the line of centres of the holes to be

$$\chi = \frac{\phi}{aJ} = -\frac{(N_1-N_2)\sin^2\eta}{2(\cosh\xi-\cos\eta)}+$$
$$+K\left\{2\cos\eta+(\cosh\xi-\cos\eta)\ln\frac{\cosh\xi-\cos\eta}{\cosh\xi+\cos\eta}\right\}+$$
$$+f(\xi,\eta)\sin\eta+g(\xi,\eta)\cos\eta. \quad (8.30.8)$$

The term $N_1\cos\eta$ in (8.30.6) has been combined with a term of the type (8.30.7) so that the resulting contribution to χ in (8.30.8) vanishes as $\xi \to \pm\infty$. The coefficient of this term is also replaced by a constant K.

We shall now restrict our attention to the case when the circular arcs are equal so that the stress system is symmetrical about the x- and y-axes, and therefore about the ξ- and η-axes. We shall complete the solution by expressing χ in (8.30.8) as a Fourier cosine integral. We proceed formally and then we can check from our final solution that it is completely satisfactory from the mathematical point of view. Suitable forms for f and g, since they are harmonic, are

$$f(\xi,\eta) = \int_0^\infty f(s)\sinh s\eta\cos s\xi\,ds,$$

$$g(\xi,\eta) = \int_0^\infty g(s)\cosh s\eta\cos s\xi\,ds.$$

We must now express the remaining terms in (8.30.8) as Fourier cosine integrals and for this purpose we observe that

$$\int_0^\infty \frac{\cos s\xi\,d\xi}{\cosh\xi+\cos\eta} = \frac{\pi\sinh s\eta}{\sinh s\pi\sin\eta} \quad (0 \leqslant \eta < \pi), \qquad (8.30.9)$$

a result which may be obtained by contour integration. From this we have also

$$\int_0^\infty \frac{\cos s\xi\,d\xi}{\cosh\xi-\cos\eta} = \frac{\pi\sinh(\pi-\eta)s}{\sinh s\pi\sin\eta} \quad (0 < \eta \leqslant \pi), \qquad (8.30.10)$$

and therefore

$$\int_0^\infty \left(\frac{\sin\eta}{\cosh\xi-\cos\eta}+\frac{\sin\eta}{\cosh\xi+\cos\eta}\right)\cos s\xi\,d\xi = \frac{\pi\cosh(\frac{1}{2}\pi-\eta)s}{\cosh\frac{1}{2}s\pi}$$
$$(0 < \eta < \pi). \quad (8.30.11)$$

Integrating this with respect to η gives

$$\int_0^\infty \ln\frac{\cosh\xi-\cos\eta}{\cosh\xi+\cos\eta}\cos s\xi\,d\xi = -\frac{\pi\sinh(\frac{1}{2}\pi-\eta)s}{s\cosh\frac{1}{2}s\pi} \quad (0 \leqslant \eta \leqslant \pi). \tag{8.30.12}$$

Combining this with (8.30.9) we have

$$\int_0^\infty \left\{\ln\frac{\cosh\xi-\cos\eta}{\cosh\xi+\cos\eta}-\frac{2\cos\eta}{\cosh\xi+\cos\eta}\right\}\cos s\xi\,d\xi$$
$$= -\frac{\pi\sinh(\frac{1}{2}\pi-\eta)s}{s\cosh\frac{1}{2}s\pi}-\frac{2\pi\sinh s\eta\cos\eta}{\sinh s\pi\sin\eta} \quad (0 \leqslant \eta < \pi), \tag{8.30.13}$$

and integrating this with respect to η gives

$$\int_0^\infty \left\{2\cos\eta+(\cosh\xi-\cos\eta)\ln\frac{\cosh\xi-\cos\eta}{\cosh\xi+\cos\eta}\right\}\cos s\xi\,d\xi$$
$$= -\frac{2\pi s\cos\eta\cosh s\eta}{(s^2+1)\sinh s\pi}+\frac{\pi\sinh(\frac{1}{2}\pi-\eta)s}{s(s^2+1)}\left\{\frac{\cos\eta}{\cosh\frac{1}{2}\pi s}+\frac{s\sin\eta}{\sinh\frac{1}{2}\pi s}\right\}$$
$$(0 \leqslant \eta \leqslant \pi). \tag{8.30.14}$$

If we now invert this by Fourier's inversion formula (1.22.4), we obtain

$$2\cos\eta+(\cosh\xi-\cos\eta)\ln\frac{\cosh\xi-\cos\eta}{\cosh\xi+\cos\eta}$$
$$= 2\int_0^\infty \left[\left\{\frac{s\sin\eta}{\sinh\frac{1}{2}\pi s}+\frac{\cos\eta}{\cosh\frac{1}{2}\pi s}\right\}\frac{\sinh(\frac{1}{2}\pi-\eta)s}{s(s^2+1)}-\frac{2s\cos\eta\cosh s\eta}{(s^2+1)\sinh s\pi}\right]\cos s\xi\,ds$$
$$(0 \leqslant \eta < \pi). \tag{8.30.15}$$

Also, inverting (8.30.10) gives

$$2\int_0^\infty \frac{\sinh(\pi-\eta)s\cos s\xi}{\sinh s\pi}\,ds = \frac{\sin\eta}{\cosh\xi-\cos\eta} \quad (0 < \eta \leqslant \pi). \tag{8.30.16}$$

Hence (8.30.8) may be written

$$\chi = 2K\int_0^\infty \left[\left\{\frac{s\sin\eta}{\sinh\frac{1}{2}\pi s}+\frac{\cos\eta}{\cosh\frac{1}{2}\pi s}\right\}\frac{\sinh(\frac{1}{2}\pi-\eta)s}{s(s^2+1)}-\frac{2s\cos\eta\cosh s\eta}{(s^2+1)\sinh s\pi}\right]\cos s\xi\,ds-$$
$$-(N_1-N_2)\sin\eta\int_0^\infty \frac{\sinh(\pi-\eta)s\cos s\xi}{\sinh s\pi}\,ds+$$
$$+\int_0^\infty \{f(s)\sin\eta\sinh s\eta+g(s)\cos\eta\cosh s\eta\}\cos s\xi\,ds, \tag{8.30.17}$$

and we assume that $f(s)$, $g(s)$ are such that this expression for χ is valid for $0 \leqslant \eta \leqslant \alpha \leqslant \pi$.

By grouping together various terms in (8.30.17) we may write χ in the simpler form

$$\chi = \int_0^\infty \Big\{F(s)\sin\eta\sinh s\eta + G(s)\cos\eta\cosh s\eta - (N_1-N_2)\sin\eta\cosh s\eta + \\ + \frac{2K\sin\eta\cosh s\eta}{s^2+1} - \frac{2K\cos\eta\sinh s\eta}{s(s^2+1)}\Big\}\cos s\xi\,ds, \quad (8.30.18)$$

where $F(s)$, $G(s)$ are arbitrary functions to be determined.

We must now satisfy the condition that the hole is free from applied stress and owing to the symmetry we need only satisfy this condition for the hole $\eta = \alpha$. At a stress-free hole $\eta = \alpha$,

$$\phi = \frac{\partial\phi}{\partial\eta} = 0,$$

or, alternatively,
$$\chi = \frac{\partial\chi}{\partial\eta} = 0 \quad (\eta = \alpha). \quad (8.30.19)$$

These conditions can be satisfied if

$$F(s)\sin\alpha\sinh s\alpha + G(s)\cos\alpha\cosh s\alpha - (N_1-N_2)\sin\alpha\cosh s\alpha + \\ + \frac{2K\sin\alpha\cosh s\alpha}{s^2+1} - \frac{2K\cos\alpha\sinh s\alpha}{s(s^2+1)} = 0,$$

$$(s\sin\alpha\cosh s\alpha + \cos\alpha\sinh s\alpha)F(s) + (s\cos\alpha\sinh s\alpha - \sin\alpha\cosh s\alpha)G(s) - \\ -(N_1-N_2)(s\sin\alpha\sinh s\alpha + \cos\alpha\cosh s\alpha) + 2(K/s)\sin\alpha\sinh s\alpha = 0.$$

If we solve these equations for $F(s)$, $G(s)$, and substitute in (8.30.18) we find that χ takes the form

$$\chi = 4K\int_0^\infty \frac{s\cosh(\alpha-\eta)s\sin\eta + \sinh(\alpha-\eta)s\cos\eta}{s(s^2+1)(\sinh 2s\alpha + s\sin 2\alpha)}\sinh s\alpha\cos s\xi\,ds - \\ -4K\sin\alpha\int_0^\infty \frac{s\cosh s\eta\sin(\alpha-\eta) + \sinh s\eta\cos(\alpha-\eta)}{(s^2+1)(\sinh 2s\alpha + s\sin 2\alpha)}\cos s\xi\,ds + \\ +2(N_1-N_2)\int_0^\infty \frac{s\cosh s\eta\sin(\alpha-\eta)\sin\alpha - \cosh s\alpha\sinh(\alpha-\eta)s\sin\eta}{\sinh 2s\alpha + s\sin 2\alpha}\times \\ \times\cos s\xi\,ds \quad (0 \leqslant \eta \leqslant \alpha \leqslant \pi). \quad (8.30.20)$$

The unknown constant K must be determined by the condition at infinity that if $\eta \to 0$ and subsequently $\xi \to 0$, the value of χ tends to N_1 (see 8.30.6). Hence

$$4K\int\limits_0^\infty \frac{(\sinh^2 s\alpha - s^2\sin^2\alpha)\,ds}{s(s^2+1)(\sinh 2s\alpha + s\sin 2\alpha)} + 2(N_1-N_2)\int\limits_0^\infty \frac{s\sin^2\alpha\,ds}{\sinh 2s\alpha + s\sin 2\alpha} = N_1. \tag{8.30.21}$$

It is not difficult to verify that the form (8.30.20) for χ, valid for $0 \leqslant \eta \leqslant \alpha \leqslant \pi$, satisfies sufficient conditions for the validity of the various differentiations under the integral sign which are required.

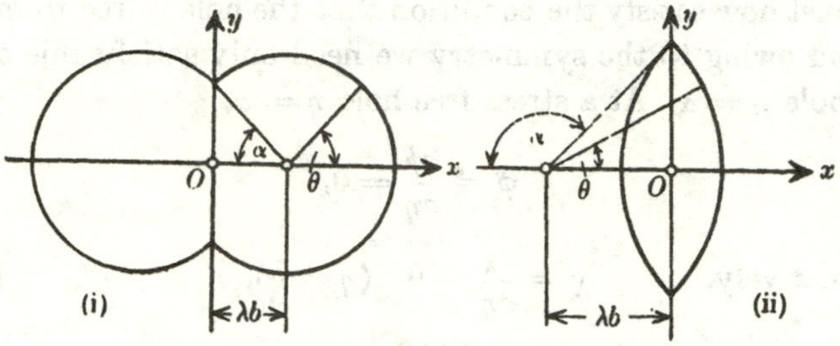

FIG. 8.8. Overlapped equal circular holes: (i) $\alpha < \frac{1}{2}\pi$, $\lambda > 0$; (ii) $\alpha > \frac{1}{2}\pi$, $\lambda < 0$.

To evaluate the stress at the edge of the hole we observe from § 6.5 that

$$\sigma_{\xi\xi}+\sigma_{\eta\eta} = \sigma_{xx}+\sigma_{yy} = 4\frac{\partial^2\phi}{\partial\zeta\partial\bar\zeta}\frac{d\zeta}{dz}\frac{d\bar\zeta}{d\bar z} = \frac{1}{J^2}\left(\frac{\partial^2\phi}{\partial\xi^2}+\frac{\partial^2\phi}{\partial\eta^2}\right),$$

and since $\sigma_{\eta\eta} = 0$ and $\phi = \dfrac{\partial\phi}{\partial\eta} = \dfrac{\partial^2\phi}{\partial\xi^2} = 0$ at $\eta = \alpha$,

$$(\sigma_{\xi\xi})_{\eta=\alpha} = \frac{a}{J}\left(\frac{\partial^2\chi}{\partial\eta^2}\right)_{\eta=\alpha},$$

or

$$(\sigma_{\xi\xi})_{\eta=\alpha} = 4(\cosh\xi-\cos\alpha)\sin\alpha\int\limits_0^\infty \frac{2K-(N_1-N_2)s(s-\cot\alpha\coth s\alpha)}{\sinh 2s\alpha+s\sin 2\alpha}\times$$
$$\times\sinh s\alpha\cos s\xi\,ds. \tag{8.30.22}$$

For convenience, the points on the rim of the hole may also be denoted by λ and θ, where (see Fig. 8.8)

$$\lambda = \cos\alpha, \qquad \cosh\xi = \frac{1+\lambda\cos\theta}{\lambda+\cos\theta}. \tag{8.30.23}$$

The limiting cases of the above results $\alpha = \pi$, $\alpha = \frac{1}{2}\pi$ correspond respectively to a slit in a plane and to a single circular hole. The limiting case $\alpha = 0$ corresponds to two complete circular holes tangential to each other. From (8.30.23), when $\alpha \to 0$ we have $\lambda \to 1$ and $\xi \to 0$ and

$$\xi/\alpha \to \tan \tfrac{1}{2}\theta.$$

Hence, putting $s\alpha = u$ and $4\alpha^2 K = K'$ in (8.30.21) and (8.30.22) we have, for holes tangential to each other,

$$K' \int_0^\infty \frac{(\sinh^2 u - u^2)\, du}{u^3(\sinh 2u + 2u)} + 2(N_1 - N_2) \int_0^\infty \frac{u\, du}{\sinh 2u + 2u} = N_1, \tag{8.30.24}$$

$$(\sigma_{\xi\xi})_{\eta=\alpha} = \sec^2 \tfrac{1}{2}\theta \int_0^\infty \frac{K' - 2(N_1 - N_2)u(u - \coth u)}{\sinh 2u + 2u} \sinh u \cos\{u \tan \tfrac{1}{2}\theta\}\, du. \tag{8.30.25}$$

The stress at the edge of the hole may now be evaluated from equations (8.30.21) and (8.30.22) for the three cases of all-round tension, longitudinal tension, transverse tension, when $\alpha \neq 0$, the corresponding values for $\alpha = 0$ being obtained from (8.30.24) and (8.30.25). The reader is referred to a paper by Ling† for numerical results but some values are quoted from his paper, for unit stress at infinity.

λ	α	*All-rouhd tension. Stress at* $\theta = 0$	*Longitudinal tension. Stress at* $\theta = \frac{1}{2}\pi$	*Transverse tension. Stress at* $\theta = 0$	
1	0°	2·894	2·569	3·869	Tangential holes
0·5	60°	2·524	2·630	3·493	
0	90°	2·000	3·000	3·000	Single hole
−0·5	120°	..	..	2·101	
−1	180°	..	..	1·000	Slit

8.31. Notch or mound and straight boundary

By a slight adaptation of the work of the previous section we can find the effect of a circular arc notch (Fig. 8.9) cut in the straight edge of a semi-infinite body under tension T parallel to the straight edge.‡ Using the notation of § 8.30 the straight edge is $\eta = 0$ and the circular arc notch is $\eta = \alpha$ $(0 \leqslant \alpha \leqslant \pi)$. A circular arc mound on a straight edge is given by $\eta = \alpha$ $(\pi \leqslant \alpha \leqslant 2\pi)$. Although the boundary curve of the body consists of two parts, the straight edge and the circular arc, it is still a single boundary and since it is free from applied stress we may

† loc. cit., p. 309.
‡ C. B. Ling, *J. Math. Phys.* **26** (1947) 284. See also E. Weinel, *Z. angew. Math. Mech.* **21** (1941) 228.

take the boundary conditions to be

$$\chi = 0, \qquad \frac{\partial \chi}{\partial \eta} = 0 \quad (\eta = 0, \alpha). \tag{8.31.1}$$

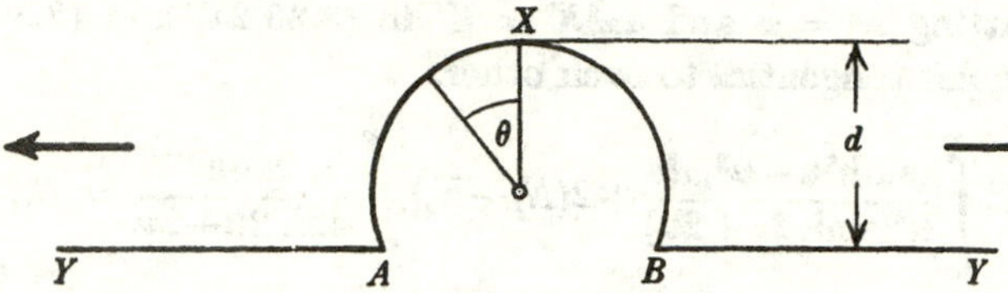

FIG. 8.9. Circular arc notch in straight edge.

Putting $N_1 = K = 0$, $N_2 = T$ in (8.30.8) we see that a suitable stress function χ is

$$\chi = \frac{T \sin^2\eta}{2(\cosh\xi - \cos\eta)} + \int\limits_0^\infty [\sin\eta\{F_1(s)\sinh s\eta + F_2(s)\cosh s\eta\} +$$
$$+ \cos\eta\{f(s)\sinh s\eta + G_2(s)\cosh s\eta\}]\cos s\xi \, ds, \tag{8.31.2}$$

where $F_1(s)$, $F_2(s)$, $f(s)$, $G_2(s)$ are functions to be determined. The condition (8.31.1) on $\eta = 0$ is satisfied if

$$G_2(s) = 0, \qquad F_2(s) + sf(s) = 0, \tag{8.31.3}$$

provided we may differentiate under the integral sign. Then, using (8.30.16) and (8.31.3), the expression (8.31.2) may be written

$$\chi = T \sin\eta \int\limits_0^\infty \frac{\sinh(\pi - \eta)s \cos s\xi}{\sinh s\pi} \, ds +$$
$$+ \int\limits_0^\infty [\sin\eta\{F_1(s)\sinh s\eta - sf(s)\cosh s\eta\} + f(s)\cos\eta \sinh s\eta]\cos s\xi \, ds,$$

or, more compactly,

$$\chi = \int\limits_0^\infty [F(s)\sin\eta \sinh s\eta + f(s)\cos\eta \sinh s\eta +$$
$$+ \{T - sf(s)\}\sin\eta \cosh s\eta]\cos s\xi \, ds, \tag{8.31.4}$$

where the original arbitrary functions have been reduced to arbitrary functions $F(s)$, $f(s)$ which are to be found by the boundary conditions on $\eta = \alpha$. We observe that the form (8.31.4) cannot be used for the boundary condition $\partial\chi/\partial\eta = 0$ on $\eta = 0$ since the derivative of the Fourier integral of the first term in (8.31.2) is not valid for $\eta = 0$. If we now assume that we may differentiate (8.31.4) with respect to η at

$\eta = \alpha$ we may determine $F(s)$, $f(s)$ from the boundary conditions on $\eta = \alpha$, and if these values are substituted in (8.31.4) the final result is

$$\chi = T\int\limits_0^\infty \frac{\sinh(\alpha-\eta)s\sinh s\alpha\sin\eta - s\sin\alpha\sin(\alpha-\eta)\sinh s\eta}{\sinh^2 s\alpha - s^2\sin^2\alpha}\cos s\xi\,ds \tag{8.31.5}$$

for χ.

The stress at the edge of the notch is now

$$(\sigma_{\xi\xi})_{\eta=\alpha} = \left(\frac{a}{J}\frac{\partial^2\chi}{\partial\eta^2}\right)_{\eta=\alpha}$$
$$= 2T(\cosh\xi-\cos\alpha)\int\limits_0^\infty \frac{s(s\cosh s\alpha\sin\alpha - \sinh s\alpha\cos\alpha)}{\sinh^2 s\alpha - s^2\sin^2\alpha}\cos s\xi\,ds. \tag{8.31.6}$$

The limiting case of a circular notch which touches the straight edge is given by $\alpha\to 0$, and in this case, by an argument similar to that used in the previous section,

$$(\sigma_{\xi\xi})_{\eta=\alpha} = T\sec^2\tfrac{1}{2}\theta\int\limits_0^\infty \frac{u(u\cosh u - \sinh u)}{\sinh^2 u - u^2}\cos(u\tan\tfrac{1}{2}\theta)\,du. \tag{8.31.7}$$

The stress at the lowest point of the notch or mound, i.e. at $\xi = 0$ or $\theta = 0$, is

$$\left.\begin{aligned}(\sigma_{\xi\xi})_{\xi=0} &= 2T(1-\cos\alpha)\int\limits_0^\infty \frac{s(s\cosh s\alpha\sin\alpha-\sinh s\alpha\cos\alpha)}{\sinh^2 s\alpha - s^2\sin^2\alpha}\,ds \quad (\alpha\neq 0)\\ &= T\int\limits_0^\infty \frac{u(u\cosh u-\sinh u)}{\sinh^2 u - u^2}\,du \quad (\alpha = 0)\end{aligned}\right\}. \tag{8.31.8}$$

If d is the depth of the notch or height of the mound, then the stress concentration factors (s.c. factors) at the point $\theta = 0$ are recorded in the accompanying table. These results are quoted from Ling.

Using bipolar coordinates a number of other problems have been solved for a region bounded by non-concentric circles, for an infinite plate containing two circular holes, and for a semi-infinite plate containing a single circular hole.†

† See e.g. G. B. Jeffery, *Phil. Trans. R. Soc.* **A221** (1921) 265; C. B. Ling, *J. appl. Phys.* **19** (1948) 77; A. C. Stevenson, *Proc. R. Soc.* **A184** (1945) 129. For other problems of stress distributions in plates containing circular holes see R. C. J. Howland, *Proc. R. Soc.* **A124** (1929) 89; **A148** (1935) 471; *Phil. Trans. R. Soc.* **A229** (1930) 49. R. C. J. Howland and R. C. Knight, *Phil. Trans. R. Soc.* **A238** (1939) 357. R. C. J. Howland and A. C. Stevenson, *Phil. Trans. R. Soc.* **A232** (1933) 155. A. E. Green, *Proc. R. Soc* **A176** (1940) 121. K. J. Schulz, 'Doctor's Thesis' (Delft, 1941), *Proc. K. ned. Akad. Wet.*, Series B, **45** (1942); **48** (1945).

$\alpha°$	d/a	s.c. factor	
0	∞	3·999	Tangential hole
30	3·732	3·882	
60	1·732	3·568	
90	1·000	3·065	Semicircular notch
120	0·577	2·424	
150	0·268	1·707	
180	0·000	1·000	Unnotched plate
210	−0·268	0·411	
240	−0·577	0·039	
270	−1·000	−0·096	Semicircular mound
300	−1·732	−0·076	
330	−3·732	−0·021	
360	$-\infty$	−0·000	Tangential mound

As already indicated in the introduction to this chapter all the problems solved here have analogues in the classical theory of transverse flexure of plates but further details are left to the reader. We close this chapter by considering one problem which can be solved using Reissner's theory and we begin by recapitulating the main results of § 7.9.

8.32. Reissner's theory of flexure: circular hole

Using the notation of § 7.9 the displacements, stress resultants, and couples are given by

$$\left.\begin{aligned} w &= -R/D - z\bar{\Omega}(\bar{z}) - \bar{z}\Omega(z) - \omega(z) - \bar{\omega}(\bar{z}) \\ \bar{u}_x + i\bar{u}_y &= -2\frac{\partial w}{\partial \bar{z}} - \frac{6i}{5\mu h}\frac{\partial \psi}{\partial \bar{z}} \\ \Lambda &= (1+\eta)\nabla_1^2 R + 4D(1+\eta)\{\Omega'(z) + \bar{\Omega}'(\bar{z})\} \\ \Gamma &= 4(1-\eta)\frac{\partial^2 R}{\partial \bar{z}^2} + 4D(1-\eta)\{z\bar{\Omega}''(\bar{z}) + \bar{\omega}''(\bar{z})\} - \frac{16ih^2}{5}\frac{\partial^2 \psi}{\partial \bar{z}^2} \\ \Psi_0 &= 2\frac{\partial \nabla_1^2 R}{\partial \bar{z}} + 8D\bar{\Omega}''(\bar{z}) - 2i\frac{\partial \psi}{\partial \bar{z}} \end{aligned}\right\}, \tag{8.32.1}$$

where $$\psi - \frac{8h^2}{5}\frac{\partial^2 \psi}{\partial z \partial \bar{z}} = 0. \tag{8.32.2}$$

Also, from (7.9.18), if a single boundary is free from applied stress and the plate is unloaded on its faces, then

$$\left.\begin{aligned} \psi + 4iD\{\bar{\Omega}'(\bar{z}) - \Omega'(z)\} &= 0 \\ \frac{4ih^2}{5}\frac{\partial \psi}{\partial \bar{z}} + D[(3+\eta)\Omega(z) - (1-\eta)\{z\bar{\Omega}'(\bar{z}) + \bar{\omega}'(\bar{z})\}] &= 0 \end{aligned}\right\}, \tag{8.32.3}$$

on that boundary.

We consider the special problem of an infinite plate containing an unstressed circular hole $|z| = a$ bent by uniform couples at infinity so that, from (7.10.10) and (7.10.11), for large $|z|$,

$$\left.\begin{aligned} \Omega(z) &= \frac{(M_1+M_2)z}{8(1+\eta)D}+O(1) \\ \omega'(z) &= \frac{(M_1-M_2)e^{-2i\alpha}z}{4(1-\eta)D}+\frac{A}{z}+O\left(\frac{1}{z^2}\right) \end{aligned}\right\}, \tag{8.32.4}$$

provided ψ and its derivatives vanish at infinity, where M_1, M_2 are the principal stress couples at infinity, α is the angle between M_1 and the axis Ox, and A is a real constant. We adopt the semi-inverse method of solution and assume that suitable forms for the complex potentials and the function ψ are

$$\left.\begin{aligned} \Omega(z) &= \frac{(M_1+M_2)z}{8(1+\eta)D}+\frac{B_1+iB_2}{Dz} \\ \omega'(z) &= \frac{(M_1-M_2)e^{-2i\alpha}z}{4(1-\eta)D}+\frac{A}{z}+\frac{A_1+iA_2}{Dz^3} \\ \psi &= \{(C_1+iC_2)e^{2i\theta}+(C_1-iC_2)e^{-2i\theta}\}K_2(\xi) \\ \xi &= \frac{r\sqrt{10}}{2h} \end{aligned}\right\}. \tag{8.32.5}$$

The function chosen for ψ satisfies equation (8.32.2) and vanishes at infinity if $K_2(\xi)$ is the Bessel function of the second kind which vanishes for large positive ξ. The arbitrary constants B_1, B_2, A_1, A_2, C_1, C_2, A in (8.32.5) are all real. If the boundary $r = a$ of the circular hole is free from applied stress, we may apply the boundary conditions (8.32.3) using also the results

$$\left.\begin{aligned} 2e^{-i\theta}\frac{\partial\psi}{\partial\bar{z}} &= \frac{\partial\psi}{\partial r}+\frac{i}{r}\frac{\partial\psi}{\partial\theta} \\ \xi K_2'(\xi) &= -2K_2(\xi)-\tfrac{1}{2}\xi^2\{K_2(\xi)-K_0(\xi)\} \end{aligned}\right\}. \tag{8.32.6}$$

Thus, adopting the notation

$$\mu = \frac{a\sqrt{10}}{2h} \tag{8.32.7}$$

and understanding that all Bessel functions have argument μ, we find, on equating the coefficients of $e^{2i\theta}$ and $e^{-2i\theta}$ and the constant term to zero, that

$$(C_1+iC_2)K_2-4i(B_1-iB_2)/a^2 = 0,$$
$$(C_1-iC_2)K_2+4i(B_1+iB_2)/a^2 = 0,$$
$$D(1-\eta)A-(M_1+M_2)a^2/4 = 0,$$
$$\{8K_2/\mu^2+K_2-K_0\}(C_1+iC_2)+2i(1-\eta)(B_1-iB_2)/a^2- \\ -2i(1-\eta)(A_1-iA_2)/a^4 = 0,$$
$$2(K_2-K_0)(C_1-iC_2)+4i(3+\eta)(B_1+iB_2)/a^2-i(M_1-M_2)e^{2i\alpha} = 0.$$

Solving these equations we have

$$\left.\begin{aligned} B_1+iB_2 &= \frac{(M_1-M_2)a^2K_2\, e^{2i\alpha}}{4\{(1+\eta)K_2+2K_0\}} \\ C_1+iC_2 &= \frac{i(M_1-M_2)e^{-2i\alpha}}{(1+\eta)K_2+2K_0} \\ A_1+iA_2 &= \frac{(M_1-M_2)\{(3-\eta)K_2-2K_0+16\mu^{-2}K_2\}a^4e^{2i\alpha}}{4\{(1+\eta)K_2+2K_0\}(1-\eta)} \end{aligned}\right\}. \qquad (8.32.8)$$

If we now transfer to polar coordinates with the help of § 7.8 we see that since

$$G_r+G_\theta = \Lambda' = \Lambda,$$

and $G_r = 0$ at $r = a$, the stress couple G_θ at the hole is

$$G_\theta = 4D(1+\eta)\{\Omega'(z)+\bar{\Omega}'(\bar{z})\},$$

and with the help of (8.32.5) and (8.32.8) this becomes

$$G_\theta = M_1+M_2-\frac{2(1+\eta)(M_1-M_2)K_2\cos 2(\theta-\alpha)}{(1+\eta)K_2+2K_0}. \qquad (8.32.9)$$

When $\mu = a\surd 10/(2h)$ is large the following asymptotic expansions for K_0, K_2 may be used:

$$K_0(\mu) = \sqrt{\left(\frac{\pi}{2\mu}\right)}e^{-\mu}\left(1-\frac{1}{8\mu}\right),$$

$$K_2(\mu) = \sqrt{\left(\frac{\pi}{2\mu}\right)}e^{-\mu}\left(1+\frac{15}{8\mu}\right).$$

Hence

$$\lim_{a/h\to\infty} G_\theta = M_1+M_2-2\{(1+\eta)/(3+\eta)\}(M_1-M_2)\cos 2(\theta-\alpha), \qquad (8.32.10)$$

which is the result for classical thin plate theory.

On the other hand, for small values of μ,

$$\lim_{\mu\to 0} K_2/K_0 \to \infty,$$

and consequently

$$\lim_{a/h\to 0} G_\theta = M_1+M_2-2(M_1-M_2)\cos 2(\theta-\alpha). \qquad (8.32.11)$$

For further discussion of these results the reader should refer to the original paper by Reissner.† The problem of a rigid circular inclusion in a plate bent by uniform couples at infinity may be solved by a method similar to that used here, taking boundary conditions

$$w = \bar{u}_x = \bar{u}_y = 0 \quad (r = a).$$

One aspect of this problem has been solved by Hirsch‡ and a numerical discussion has also been given by him.

† E. Reissner, *J. appl. Mech.* **12** (1945) 68.
‡ R. A. Hirsch, ibid. **19** (1952) 28.

9 Plane Problems for Aeolotropic Bodies

IN this chapter we consider some problems of plane strain or generalized plane stress in aeolotropic bodies. As indicated in Chapter 7 many of the problems have analogues in the classical theory of bending of thin aeolotropic plates, but such analogues are not discussed here since they may be obtained with the help of the theory of Chapter 7. Complex variable methods are used throughout this chapter.†

9.1. Formulae for plane strain and generalized plane stress

We first summarize from Chapter 6 the formulae which are needed for plane strain, assuming that body forces are zero. Let (x, y) be rectangular cartesian coordinates and let

$$\left.\begin{aligned} z &= x+iy, & \bar{z} &= x-iy \\ z_1 &= z+\gamma_1\bar{z}, & \bar{z}_1 &= \bar{z}+\bar{\gamma}_1 z \\ z_2 &= z+\gamma_2\bar{z}, & \bar{z}_2 &= \bar{z}+\bar{\gamma}_2 z \end{aligned}\right\}, \tag{9.1.1}$$

where

$$|\gamma_1| < 1, \qquad |\gamma_2| < 1, \tag{9.1.2}$$

and γ_1, γ_2 are roots of equation (6.7.20). Airy's stress function ϕ is then

$$\phi = \Omega(z_1)+\bar{\Omega}(\bar{z}_1)+\omega(z_2)+\bar{\omega}(\bar{z}_2), \tag{9.1.3}$$

and the displacements and stresses are

$$D = u_x+iu_y = \delta_1\Omega'(z_1)+\rho_1\bar{\Omega}'(\bar{z}_1)+\delta_2\omega'(z_2)+\rho_2\bar{\omega}'(\bar{z}_2), \tag{9.1.4}$$

$$\Theta = \sigma_{xx}+\sigma_{yy} = 4\gamma_1\Omega''(z_1)+4\bar{\gamma}_1\bar{\Omega}''(\bar{z}_1)+4\gamma_2\omega''(z_2)+4\bar{\gamma}_2\bar{\omega}''(\bar{z}_2), \tag{9.1.5}$$

$$\Phi = \sigma_{xx}-\sigma_{yy}+2i\sigma_{xy} = -4\gamma_1^2\Omega''(z_1)-4\bar{\Omega}''(\bar{z}_1)-4\gamma_2^2\omega''(z_2)-4\bar{\omega}''(\bar{z}_2), \tag{9.1.6}$$

$$\begin{aligned} \sigma_{yy}-i\sigma_{xy} = {} & 2\gamma_1(1+\gamma_1)\Omega''(z_1)+2(1+\bar{\gamma}_1)\bar{\Omega}''(\bar{z}_1)+ \\ & +2\gamma_2(1+\gamma_2)\omega''(z_2)+2(1+\bar{\gamma}_2)\bar{\omega}''(\bar{z}_2), \end{aligned} \tag{9.1.7}$$

where the complex constants δ_1, δ_2, ρ_1, ρ_2 are given in (6.7.31). Also the resultant force and couple (about the origin) exerted across an arc

† References to original papers are given in the introduction to Chapter 6.

AP of a curve are

$$P = X+iY = 2i\frac{\partial\phi}{\partial\bar{z}} = 2i\{\gamma_1\Omega'(z_1)+\bar{\Omega}'(\bar{z}_1)+\gamma_2\omega'(z_2)+\bar{\omega}'(\bar{z}_2)\}, \tag{9.1.8}$$

$$\begin{aligned} M &= z\frac{\partial\phi}{\partial z}+\bar{z}\frac{\partial\phi}{\partial\bar{z}}-\phi \\ &= z_1\Omega'(z_1)+\bar{z}_1\bar{\Omega}'(\bar{z}_1)-\Omega(z_1)-\bar{\Omega}(\bar{z}_1)+z_2\omega'(z_2)+\bar{z}_2\bar{\omega}'(\bar{z}_2)-\omega(z_2)-\bar{\omega}(\bar{z}_2). \end{aligned} \tag{9.1.9}$$

The above formulae are suitable for plane strain provided the constants $\gamma_1, \gamma_2, \delta_1, \delta_2, \rho_1, \rho_2$ are defined in terms of elastic coefficients $c_{\lambda\mu}^{\alpha\beta}$ by equations (6.7.9), (6.7.17), (6.7.20), and (6.7.31).

For a plate under generalized plane stress in its plane similar formulae are valid provided, for example, stress is interpreted per unit length of a line in the middle surface of the plate and provided the constants $\gamma_1, \gamma_2, \delta_1, \delta_2, \rho_1, \rho_2$ are now defined in terms of elastic coefficients $s_{\lambda\mu}^{\alpha\beta}$ (equation (7.6.1)) through equations (6.7.17), (6.7.20), and (6.7.31).

If we have a conformal transformation of the form

$$z = z(\sigma), \qquad \sigma = e^{\zeta}, \qquad \zeta = \xi+i\eta,$$

where (ξ, η) are a system of orthogonal curvilinear coordinates, then displacements and stresses are obtained from (6.5.14) and (6.5.12) in the forms

$$\left.\begin{aligned} u_\xi+iu_\eta &= \left(\frac{\bar{\sigma}}{\sigma}\frac{d\sigma}{dz}\frac{d\bar{z}}{d\bar{\sigma}}\right)^{\frac{1}{2}}(u_x+iu_y) \\ \sigma_{\xi\xi}+\sigma_{\eta\eta} &= \sigma_{xx}+\sigma_{yy} \\ \sigma_{\xi\xi}-\sigma_{\eta\eta}+2i\sigma_{\xi\eta} &= \frac{\bar{\sigma}}{\sigma}\frac{d\sigma}{dz}\frac{d\bar{z}}{d\bar{\sigma}}(\sigma_{xx}-\sigma_{yy}+2i\sigma_{xy}) \end{aligned}\right\}. \tag{9.1.10}$$

Also, from (6.5.22),

$$P(\eta) = i\int_A^P (\sigma_{\xi\xi}+i\sigma_{\xi\eta})\frac{dz}{d\eta}\,d\eta, \qquad \frac{\partial P(\eta)}{\partial\eta} = i(\sigma_{\xi\xi}+i\sigma_{\xi\eta})\frac{dz}{d\eta}, \tag{9.1.11}$$

on the curve ξ = constant.

In particular, for polar coordinates (r, θ),

$$\left.\begin{aligned} z = \sigma = e^{\zeta} = re^{i\theta}, \qquad r = e^{\xi}, \qquad \theta = \eta \\ u_r+iu_\theta = e^{-i\theta}(u_x+iu_y) \\ \sigma_{rr}+\sigma_{\theta\theta} = \sigma_{xx}+\sigma_{yy} \\ \sigma_{rr}-\sigma_{\theta\theta}+2i\sigma_{r\theta} = e^{-2i\theta}(\sigma_{xx}-\sigma_{yy}+2i\sigma_{xy}) \end{aligned}\right\}, \tag{9.1.12}$$

and, on the curve $r =$ constant,

$$P(\theta) = -\int_A^P (\sigma_{rr}+i\sigma_{r\theta})re^{i\theta}\,d\theta, \qquad \frac{\partial P(\theta)}{\partial\theta} = -(\sigma_{rr}+i\sigma_{r\theta})re^{i\theta}. \tag{9.1.13}$$

When the region occupied by the body is bounded internally by one or more non-intersecting closed curves lying entirely in the finite part of the plane, then, from (6.8.8),

$$\left.\begin{aligned}\Omega''(z_1) &= B+iC+\frac{H}{2\pi z_1}+O\left(\frac{1}{z_1^2}\right)\\ \omega''(z_2) &= B'+iC'+\frac{K}{2\pi z_2}+O\left(\frac{1}{z_2^2}\right)\end{aligned}\right\}, \tag{9.1.14}$$

when stresses are bounded at infinity and rotation vanishes there, where, from (6.7.39), (6.8.6), (6.8.7), (6.8.9), and (6.8.10),

$$\left.\begin{aligned}P = X+iY &= 2(\bar{H}-\gamma_1 H+\bar{K}-\gamma_2 K)\\ \delta_1 H-\rho_1\bar{H}+\delta_2 K-\rho_2\bar{K} &= 0\end{aligned}\right\}, \tag{9.1.15}$$

$$N_1+N_2 = 4\gamma_1(B+iC)+4\bar{\gamma}_1(B-iC)+4\gamma_2(B'+iC')+4\bar{\gamma}_2(B'-iC'), \tag{9.1.16}$$

$$(N_1-N_2)e^{2i\alpha} = -4\gamma_1^2(B+iC)-4(B-iC)-4\gamma_2^2(B'+iC')-4(B'-iC'), \tag{9.1.17}$$

$$\begin{aligned}(1-\gamma_1\bar{\gamma}_1)\{(\gamma_1-\gamma_2)(1-\gamma_1\bar{\gamma}_2)(B+iC)-(\bar{\gamma}_1-\bar{\gamma}_2)(1-\bar{\gamma}_1\gamma_2)(B-iC)\}-&\\ -(1-\gamma_2\bar{\gamma}_2)\{(\gamma_1-\gamma_2)(1-\bar{\gamma}_1\gamma_2)(B'+iC')-&\\ -(\bar{\gamma}_1-\bar{\gamma}_2)(1-\gamma_1\bar{\gamma}_2)(B'-iC')\} = 0.&\end{aligned} \tag{9.1.18}$$

In equations (9.1.14) to (9.1.18) P is the resultant force acting over all the internal boundaries, N_1, N_2 are the values of the principal stresses at infinity, and α is the angle between N_1 and the x-axis. Equation (9.1.18) holds when the rotation vanishes at infinity. When stresses as well as rotation vanish at infinity, then, for large $|z_1|$, $|z_2|$,

$$\left.\begin{aligned}\Omega''(z_1) &= \frac{H}{2\pi z_1}+O\left(\frac{1}{z_1^2}\right)\\ \omega''(z_2) &= \frac{K}{2\pi z_2}+O\left(\frac{1}{z_2^2}\right)\end{aligned}\right\}, \tag{9.1.19}$$

where H, K are still given by equations (9.1.15).

When the region S occupied by the body is the half-plane, either $y \geqslant 0$ or $y \leqslant 0$, and $P = X+iY$ is the resultant of the external forces acting on the boundary $y = 0$ and the other boundaries which are in the finite part of the plane, the above results are altered. Thus, if stresses and rotation vanish at infinity, and any applied stresses on the boundary $y = 0$ are $O(1/x^2)$ for large $|x|$, we have, from (6.8.17),

$$\left.\begin{aligned} \Omega''(z_1) &= -\frac{P+\gamma_2\bar{P}}{4\pi(\gamma_1-\gamma_2)z_1}+O\left(\frac{1}{z_1^2}\right) \\ \omega''(z_2) &= \frac{P+\gamma_1\bar{P}}{4\pi(\gamma_1-\gamma_2)z_2}+O\left(\frac{1}{z_2^2}\right) \end{aligned}\right\}, \tag{9.1.20}$$

for large $|z_1|$, $|z_2|$.

9.2. Orthotropic plates: numerical calculations

In this chapter we shall see that many results for aeolotropic bodies are strikingly different from those for isotropic bodies. In order to illustrate this difference we shall give a few graphical illustrations of results for some of the problems which are discussed, and we shall restrict such illustrations to problems of generalized plane stress in planks made of either oak or spruce wood. These materials are chosen because they possess a very high degree of aeolotropy. The planks are cut parallel to the grain of the wood. The direction of the fibres (which are parallel to the core of the original tree trunk) is taken as the axis of y. With this orientation of the saw cuts, the axis of x is in the direction which is perpendicular to the fibres and also to a plane which passes through the central core of the tree. The appropriate elastic constants s_{rs}^{ij} are defined in (7.6.1)–(7.6.3) and their values, taken from a paper by Hörig, are reproduced† in Table 1.

TABLE 1

	s_{11}^{11}	s_{22}^{22}	s_{22}^{11}	s_{12}^{12}	α_1	α_2	γ_1	γ_2
Oak	10·15	1·72	−0·87	3·2	2·3067	1·053	0·395	0·026
Spruce	15·5	0·587	−0·33	2·875	4·112	1·249	0·608	0·111

The corresponding values of α_1, α_2, γ_1, γ_2 may be found from (6.9.2) and (6.9.1) and are recorded in Table 1, and it is seen that they are all real. It follows that these specimens of oak and spruce woods come under the analysis of case (*b*) of § 6.9 and that β_1, β_2, δ_1, δ_2, ρ_1, ρ_2 are

† H. Hörig, *Ing.-Arch.* **6** (1935) 8. We remind the reader that the elastic constants s_{rs}^{ij} may be expressed in terms of the more usual elastic constants s_{mn} (m, n = 1, 2,..., 6) with the help of § 5.11.

also real. We recall, therefore, from § 6.9, that

$$\left.\begin{aligned} \gamma_1 &= \frac{\alpha_1-1}{\alpha_1+1}, \quad \gamma_2 = \frac{\alpha_2-1}{\alpha_2+1}, \quad \alpha_1^2\alpha_2^2 = \frac{\lambda}{\mu} \\ \beta_1 &= \nu-\alpha_1^2\mu, \quad \beta_2 = \nu-\alpha_2^2\mu \\ \delta_1 &= (1+\gamma_1)\beta_2-(1-\gamma_1)\beta_1, \quad \delta_2 = (1+\gamma_2)\beta_1-(1-\gamma_2)\beta_2 \\ \rho_1 &= (1+\gamma_1)\beta_2+(1-\gamma_1)\beta_1, \quad \rho_2 = (1+\gamma_2)\beta_1+(1-\gamma_2)\beta_2 \end{aligned}\right\}, \tag{9.2.1}$$

and

$$\left.\begin{aligned} 4(\alpha_1^2-\alpha_2^2)\mu H &= (1+\alpha_1)(\beta_2 X+i\beta_1 Y/\alpha_1) \\ 4(\alpha_1^2-\alpha_2^2)\mu K &= -(1+\alpha_2)(\beta_1 X+i\beta_2 Y/\alpha_2) \end{aligned}\right\}, \tag{9.2.2}$$

where, to avoid continual repetition of indices, we have put

$$s_{11}^{11} = \lambda, \qquad s_{22}^{22} = \mu, \qquad s_{22}^{11} = \nu. \tag{9.2.3}$$

Also, equations (6.9.8) for the constants B, B', C, C' may be solved to give

$$\left.\begin{aligned} 8(\gamma_1-\gamma_2)(1-\gamma_1\gamma_2)B &= (1+\gamma_2^2)(N_1+N_2)+2\gamma_2(N_1-N_2)\cos 2\alpha \\ 8(\gamma_1-\gamma_2)(1-\gamma_1\gamma_2)B' &= -(1+\gamma_1^2)(N_1+N_2)-2\gamma_1(N_1-N_2)\cos 2\alpha \\ 8(1-\gamma_1^2)C = 8(1-\gamma_2^2)C' &= (N_1-N_2)\sin 2\alpha \end{aligned}\right\}, \tag{9.2.4}$$

where here α denotes the angle between N_1 and the x-axis.

9.3. Isolated force in infinite plate†

The required complex potentials for an isolated force at the origin in an infinite plate are suggested by (9.1.14). Thus, if the stresses and rotation due to this force vanish at infinity we take

$$\left.\begin{aligned} \Omega(z_1) &= \frac{H}{2\pi}z_1\ln z_1 \\ \omega(z_2) &= \frac{K}{2\pi}z_2\ln z_2 \end{aligned}\right\}, \tag{9.3.1}$$

where H, K are given by (9.1.15). By applying (9.1.8) and (9.1.9) to any circuit surrounding the origin we can at once verify that (9.3.1) is a satisfactory form for the complex potentials corresponding to an isolated force at the origin. The displacements and stresses may be obtained from (9.1.4)–(9.1.7) but here we record only the results for the special case of the orthotropic plank described in § 9.2, using values

† See A. E. Green and G. I. Taylor, *Proc. R. Soc.* A173 (1939) 162, and A. E. Green, *Phil. Mag.* 34 (1943) 416.

for H, K given in (9.2.2). Moreover, the stresses are best represented in polar coordinates (r, θ) with the help of (9.1.12). Thus

$$\sigma_{rr}+\sigma_{\theta\theta} = \frac{2}{\pi}\left(\frac{\gamma_1 H}{z_1}+\frac{\bar{\gamma}_1 \bar{H}}{\bar{z}_1}+\frac{\gamma_2 K}{z_2}+\frac{\bar{\gamma}_2 \bar{K}}{\bar{z}_2}\right),$$

$$\sigma_{rr}-\sigma_{\theta\theta}+2i\sigma_{r\theta} = -\frac{2e^{-2i\theta}}{\pi}\left(\frac{\gamma_1^2 H}{z_1}+\frac{\bar{H}}{\bar{z}_1}+\frac{\gamma_2^2 K}{z_2}+\frac{\bar{K}}{\bar{z}_2}\right),$$

and using the special values (9.2.2) for H and K for an orthotropic plank we find, after some reduction,

$$2\pi(\alpha_1^2-\alpha_2^2)\mu r\sigma_{rr}$$
$$= X\cos\theta\left[\frac{\alpha_1\beta_2\{\alpha_1^2-3-(\alpha_1^2-1)\cos 2\theta\}}{\alpha_1^2+1+(\alpha_1^2-1)\cos 2\theta}-\frac{\alpha_2\beta_1\{\alpha_2^2-3-(\alpha_2^2-1)\cos 2\theta\}}{\alpha_2^2+1+(\alpha_2^2-1)\cos 2\theta}\right]+$$
$$+Y\sin\theta\left[\frac{\beta_1\{3\alpha_1^2-1+(\alpha_1^2-1)\cos 2\theta\}}{\alpha_1\{\alpha_1^2+1+(\alpha_1^2-1)\cos 2\theta\}}-\frac{\beta_2\{3\alpha_2^2-1+(\alpha_2^2-1)\cos 2\theta\}}{\alpha_2\{\alpha_2^2+1+(\alpha_2^2-1)\cos 2\theta\}}\right], \tag{9.3.2}$$

$$2\pi(\alpha_1+\alpha_2)\mu r\sigma_{\theta\theta} = (\nu+\mu\alpha_1\alpha_2)\left(X\cos\theta+\frac{Y\sin\theta}{\alpha_1\alpha_2}\right), \tag{9.3.3}$$

$$2\pi(\alpha_1+\alpha_2)\mu r\sigma_{r\theta} = (\nu+\mu\alpha_1\alpha_2)\left(X\sin\theta-\frac{Y\cos\theta}{\alpha_1\alpha_2}\right). \tag{9.3.4}$$

The stresses due to an isolated force in an isotropic plate can be obtained from the above results by taking the limit as $\alpha_1 \to \alpha_2 \to 1$.

Radial diagrams representing the values of $(r/P)\sigma_{rr}$ and $(r/P)\sigma_{r\theta}$ for planks of oak and spruce are shown in Fig. 9.1 for the cases when the resultant force P acts along and perpendicular to the grain.† The corresponding results for an isotropic plate whose Poisson's ratio is 0·25 are included in these diagrams for comparison. A force acting at a point of a sheet of spruce, in the direction of the grain, produces a high stress along the radii which are nearly parallel to the grain but very little radial stress in directions which differ from this by more than about 30°. At the same time the radial-tangential stress $\sigma_{r\theta}$ is extremely small, its maximum value being only 0·0368 of the maximum radial stress. This may be compared with the corresponding ratio 0·231 for an isotropic material whose Poisson's ratio is 0·25. When the force is applied perpendicularly to the grain the radial stress σ_{rr} is not a maximum in the direction of the applied force, being greatest at an angle of approximately 75° with this direction and therefore more nearly in the direction of the grain. In this case the ratio of the maximum shear stress to the maximum radial stress is very much greater than

† These diagrams were given by A. E. Green and G. I. Taylor, loc. cit., p. 326.

in the previous case, being about 0·764. The results for oak are of a similar nature but as oak is not so highly aeolotropic as spruce the effect of the grain is less striking.

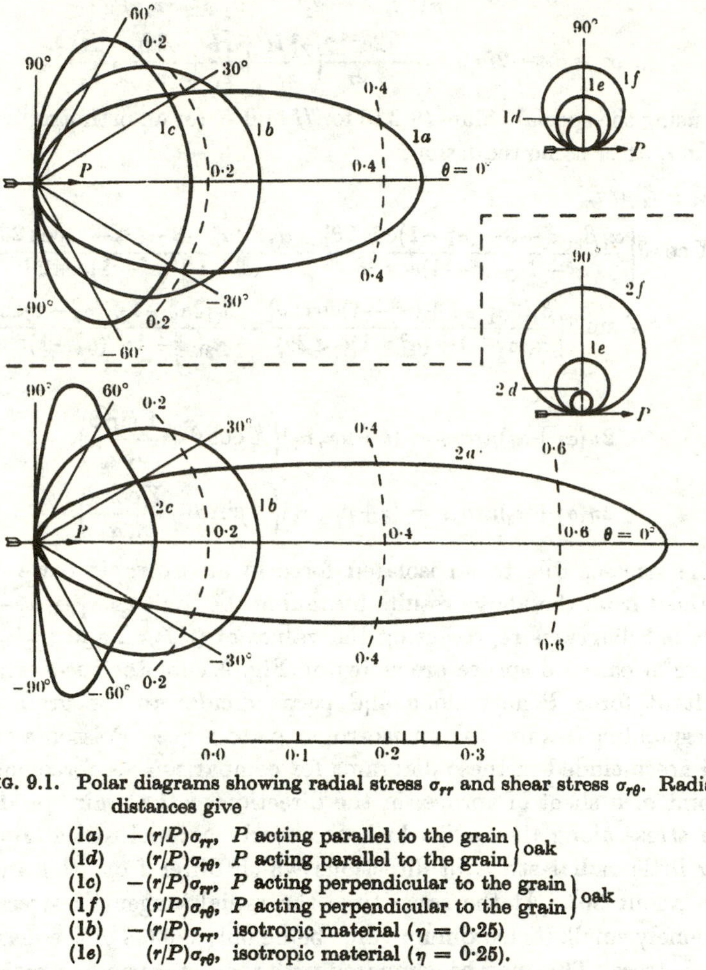

FIG. 9.1. Polar diagrams showing radial stress σ_{rr} and shear stress $\sigma_{r\theta}$. Radial distances give

(1*a*)	$-(r/P)\sigma_{rr}$, P acting parallel to the grain	oak
(1*d*)	$(r/P)\sigma_{r\theta}$, P acting parallel to the grain	oak
(1*c*)	$-(r/P)\sigma_{rr}$, P acting perpendicular to the grain	oak
(1*f*)	$(r/P)\sigma_{r\theta}$, P acting perpendicular to the grain	oak
(1*b*)	$-(r/P)\sigma_{rr}$, isotropic material ($\eta = 0{\cdot}25$)	
(1*e*)	$(r/P)\sigma_{r\theta}$, isotropic material ($\eta = 0{\cdot}25$).	

The diagrams 2*a*, 2*d*, 2*c*, 2*f* represent the corresponding results for spruce.

9.4. Isolated force at boundary

Suppose an isolated force $P = X+iY$ acts at the point C ($z = c$) on a boundary of the body (see Fig. 6.2). We know that at C the stress function ϕ is continuous, and that if we apply (9.1.8) to a curve $C'DC''$ cut out of the material, as the curve shrinks to zero so that C lies between C' and C'', we must obtain the isolated force P which acts at C. It is not difficult to verify that the necessary singular parts of the complex potentials are

$$\left.\begin{aligned}\Omega(z_1) &= -\frac{P+\gamma_2\bar{P}}{4\pi(\gamma_1-\gamma_2)}(z_1-c-\gamma_1\bar{c})\ln(z_1-c-\gamma_1\bar{c})\\ \omega(z_2) &= \frac{P+\gamma_1\bar{P}}{4\pi(\gamma_1-\gamma_2)}(z_2-c-\gamma_2\bar{c})\ln(z_2-c-\gamma_2\bar{c})\end{aligned}\right\}. \tag{9.4.1}$$

If the isolated force P acts at the point $z = c$ on the boundary of a hole which is cut in an infinite plate, then, in view of the forms (9.1.19) for $\Omega(z_1)$, $\omega(z_2)$ at infinity, we see that the complex potentials must contain terms

$$\left.\begin{aligned}\Omega(z_1) &= \frac{H}{2\pi}z_1\ln z_1-\frac{P+\gamma_2\bar{P}}{4\pi(\gamma_1-\gamma_2)}(z_1-c-\gamma_1\bar{c})\ln\left(\frac{z_1-c-\gamma_1\bar{c}}{z_1}\right)\\ \omega(z_2) &= \frac{K}{2\pi}z_2\ln z_2+\frac{P+\gamma_1\bar{P}}{4\pi(\gamma_1-\gamma_2)}(z_2-c-\gamma_2\bar{c})\ln\left(\frac{z_2-c-\gamma_2\bar{c}}{z_2}\right)\end{aligned}\right\}, \tag{9.4.2}$$

where the origin of coordinates is inside the hole and where H, K are given by (9.1.15). Further terms must be added to satisfy the boundary conditions over the hole.

When an isolated force P acts at the origin on the boundary $y = 0$ of the semi-infinite region $y \geqslant 0$ then from (9.4.1) we see that the complex potentials must contain the terms

$$\left.\begin{aligned}\Omega(z_1) &= -\frac{P+\gamma_2\bar{P}}{4\pi(\gamma_1-\gamma_2)}z_1\ln z_1\\ \omega(z_2) &= \frac{P+\gamma_1\bar{P}}{4\pi(\gamma_1-\gamma_2)}z_2\ln z_2\end{aligned}\right\}. \tag{9.4.3}$$

From (9.1.7) we find that the corresponding value of $\sigma_{yy}-i\sigma_{xy}$ is zero everywhere on the boundary $y = 0$ except at the origin. The potentials (9.4.3) therefore give the distribution of stress due to an isolated force P acting at the origin on the boundary $y = 0$ of a semi-infinite region $y \geqslant 0$ which is otherwise free from applied stress.

Using polar coordinates we find from (9.1.5), (9.1.6), (9.1.12), and (9.4.3) that

$$\sigma_{rr}+\sigma_{\theta\theta} = -\frac{\gamma_1(P+\gamma_2\bar{P})}{\pi(\gamma_1-\gamma_2)z_1}-\frac{\bar{\gamma}_1(\bar{P}+\bar{\gamma}_2P)}{\pi(\bar{\gamma}_1-\bar{\gamma}_2)\bar{z}_1}+\frac{\gamma_2(P+\gamma_1\bar{P})}{\pi(\gamma_1-\gamma_2)z_2}+\frac{\bar{\gamma}_2(\bar{P}+\bar{\gamma}_1P)}{\pi(\bar{\gamma}_1-\bar{\gamma}_2)\bar{z}_2},$$

$$\begin{aligned}\sigma_{rr}-\sigma_{\theta\theta}+2i\sigma_{r\theta} &= \frac{e^{-2i\theta}}{\pi}\left\{\frac{\gamma_1^2(P+\gamma_2\bar{P})}{(\gamma_1-\gamma_2)z_1}+\frac{\bar{P}+\bar{\gamma}_2P}{(\bar{\gamma}_1-\bar{\gamma}_2)\bar{z}_1}-\frac{\gamma_2^2(P+\gamma_1\bar{P})}{(\gamma_1-\gamma_2)z_2}-\frac{\bar{P}+\bar{\gamma}_1P}{(\bar{\gamma}_1-\bar{\gamma}_2)\bar{z}_2}\right\}\\ &= -\frac{\gamma_1(P+\gamma_2\bar{P})}{\pi(\gamma_1-\gamma_2)z_1}-\frac{\bar{\gamma}_1(\bar{P}+\bar{\gamma}_2P)}{\pi(\bar{\gamma}_1-\bar{\gamma}_2)\bar{z}_1}+\frac{\gamma_2(P+\gamma_1\bar{P})}{\pi(\gamma_1-\gamma_2)z_2}+\frac{\bar{\gamma}_2(\bar{P}+\bar{\gamma}_1P)}{\pi(\bar{\gamma}_1-\bar{\gamma}_2)\bar{z}_2},\end{aligned}$$

and hence $\sigma_{\theta\theta} = \sigma_{r\theta} = 0$ and

$$\pi r\{1+\gamma_1\bar{\gamma}_1+(\gamma_1+\bar{\gamma}_1)\cos 2\theta+i(\bar{\gamma}_1-\gamma_1)\sin 2\theta\}\times$$
$$\times\{1+\gamma_2\bar{\gamma}_2+(\gamma_2+\bar{\gamma}_2)\cos 2\theta+i(\bar{\gamma}_2-\gamma_2)\sin 2\theta\}\sigma_{rr}$$
$$= -X\{2(1-\gamma_1\bar{\gamma}_1\gamma_2\bar{\gamma}_2)+(\gamma_1+\gamma_2)(1-\bar{\gamma}_1\bar{\gamma}_2)+(\bar{\gamma}_1+\bar{\gamma}_2)(1-\gamma_1\gamma_2)\}\cos\theta-$$
$$-iX\{(\bar{\gamma}_1+\bar{\gamma}_2)(1+\gamma_1\gamma_2)-(\gamma_1+\gamma_2)(1+\bar{\gamma}_1\bar{\gamma}_2)\}\sin\theta+$$
$$+Y\{-2(1-\gamma_1\bar{\gamma}_1\gamma_2\bar{\gamma}_2)+(\gamma_1+\gamma_2)(1-\bar{\gamma}_1\bar{\gamma}_2)+(\bar{\gamma}_1+\bar{\gamma}_2)(1-\gamma_1\gamma_2)\}\sin\theta-$$
$$-iY\{(\bar{\gamma}_1+\bar{\gamma}_2)(1+\gamma_1\gamma_2)-(\gamma_1+\gamma_2)(1+\bar{\gamma}_1\bar{\gamma}_2)\}\cos\theta. \quad (9.4.4)$$

In the special case of orthotropic plates, discussed in § 9.2, when γ_1, γ_2 are real, this formula reduces to

$$\pi r(1+\gamma_1^2+2\gamma_1\cos 2\theta)(1+\gamma_2^2+2\gamma_2\cos 2\theta)\sigma_{rr}$$
$$= -2(1-\gamma_1\gamma_2)(1+\gamma_1)(1+\gamma_2)X\cos\theta-$$
$$-2(1-\gamma_1\gamma_2)(1-\gamma_1)(1-\gamma_2)Y\sin\theta. \quad (9.4.5)$$

The result $$\pi r\sigma_{rr} = -2X\cos\theta-2Y\sin\theta$$
for an isotropic plate may be obtained from (9.4.5) by putting $\gamma_1 = 0$, $\gamma_2 = 0$.

Fig. 9.2 shows a radial diagram for $r\sigma_{rr}/P$ when an isolated force P acts normally to the straight edge of a semi-infinite spruce plank in the two cases when the edge is parallel and perpendicular to the grain.† The diagram for an isotropic plank is included for comparison.

9.5. The half-plane

We suppose that the body occupies the half-plane $y \geqslant 0$, the boundary being $y = 0$. We define two complex variables ζ_1, ζ_2 by the affine transformations

$$\left.\begin{aligned}\zeta_1 &= \frac{z_1}{1+\gamma_1} = x+i\left(\frac{1-\gamma_1}{1+\gamma_1}\right)y\\ \zeta_2 &= \frac{z_2}{1+\gamma_2} = x+i\left(\frac{1-\gamma_2}{1+\gamma_2}\right)y\end{aligned}\right\}, \quad (9.5.1)$$

and corresponding to each point (x, y) in the half-plane $y \geqslant 0$ we have one point in each ζ_α-plane ($\alpha = 1, 2$). Moreover, since $|\gamma_1| < 1$, $|\gamma_2| < 1$ we always have

$$\left.\begin{aligned}\operatorname{im}\zeta_1 &= \frac{1-\gamma_1\bar{\gamma}_1}{(1+\gamma_1)(1+\bar{\gamma}_1)}y \geqslant 0\\ \operatorname{im}\zeta_2 &= \frac{1-\gamma_2\bar{\gamma}_2}{(1+\gamma_2)(1+\bar{\gamma}_2)}y \geqslant 0\end{aligned}\right\}, \quad (9.5.2)$$

† See A. E. Green, *Proc. R. Soc.* A**173** (1939) 173, and *Phil. Mag.* **34** (1943) 420.

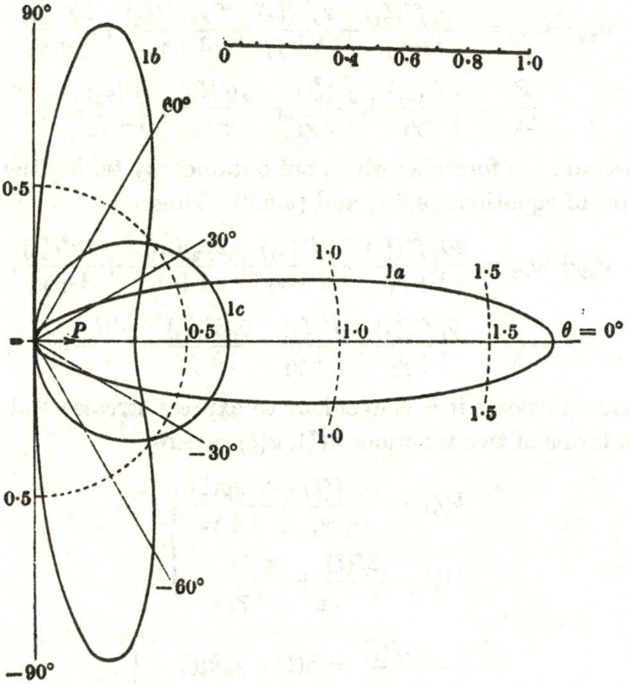

FIG. 9.2. Polar diagrams showing radial stress σ_{rr} for isolated force at edge of semi-infinite plate. Radial distances give

(1a) $-(r/P)\sigma_{rr}$, P acting parallel to the grain for spruce,
(1b) $-(r/P)\sigma_{rr}$, P acting perpendicular to the grain for spruce,
(1c) $-(r/P)\sigma_{rr}$, isotropic material ($\eta = 0{\cdot}25$).

where im ζ_α is the imaginary part of ζ_α, so that the points which correspond to (x,y) always lie in the half-planes given by (9.5.2). Also, when $y = 0$, $\zeta_1 = \zeta_2 = x$, so that the real axes of the z-, ζ_1-, ζ_2-planes coincide, and coordinates along these axes may be denoted by x.

If we put $$\Omega(z_1) = f(\zeta_1), \qquad \omega(z_2) = g(\zeta_2), \tag{9.5.3}$$

then the formulae (9.1.4) to (9.1.8) become

$$D = \frac{\delta_1 f'(\zeta_1)}{1+\gamma_1} + \frac{\rho_1 \bar{f}'(\bar{\zeta}_1)}{1+\bar{\gamma}_1} + \frac{\delta_2 g'(\zeta_2)}{1+\gamma_2} + \frac{\rho_2 \bar{g}'(\bar{\zeta}_2)}{1+\bar{\gamma}_2}, \tag{9.5.4}$$

$$\Theta = \frac{4\gamma_1 f''(\zeta_1)}{(1+\gamma_1)^2} + \frac{4\bar{\gamma}_1 \bar{f}''(\bar{\zeta}_1)}{(1+\bar{\gamma}_1)^2} + \frac{4\gamma_2 g''(\zeta_2)}{(1+\gamma_2)^2} + \frac{4\bar{\gamma}_2 \bar{g}''(\bar{\zeta}_2)}{(1+\bar{\gamma}_2)^2}, \tag{9.5.5}$$

$$\Phi = -\frac{4\gamma_1^2 f''(\zeta_1)}{(1+\gamma_1)^2} - \frac{4\bar{f}''(\bar{\zeta}_1)}{(1+\bar{\gamma}_1)^2} - \frac{4\gamma_2^2 g''(\zeta_2)}{(1+\gamma_2)^2} - \frac{4\bar{g}''(\bar{\zeta}_2)}{(1+\bar{\gamma}_2)^2}, \tag{9.5.6}$$

$$\sigma_{yy}-i\sigma_{xy} = \frac{2\gamma_1 f''(\zeta_1)}{1+\gamma_1}+\frac{2\bar{f}''(\bar{\zeta}_1)}{1+\bar{\gamma}_1}+\frac{2\gamma_2 g''(\zeta_2)}{1+\gamma_2}+\frac{2\bar{g}''(\bar{\zeta}_2)}{1+\bar{\gamma}_2}, \tag{9.5.7}$$

$$\frac{P}{2i} = \frac{\gamma_1 f'(\zeta_1)}{1+\gamma_1}+\frac{\bar{f}'(\bar{\zeta}_1)}{1+\bar{\gamma}_1}+\frac{\gamma_2 g'(\zeta_2)}{1+\gamma_2}+\frac{\bar{g}'(\bar{\zeta}_2)}{1+\bar{\gamma}_2}. \tag{9.5.8}$$

We also add the formulae which are obtained by taking the complex conjugates of equations (9.5.7) and (9.5.8). Thus

$$\sigma_{yy}+i\sigma_{xy} = \frac{2\bar{\gamma}_1 \bar{f}''(\bar{\zeta}_1)}{1+\bar{\gamma}_1}+\frac{2f''(\zeta_1)}{1+\gamma_1}+\frac{2\bar{\gamma}_2 \bar{g}''(\bar{\zeta}_2)}{1+\bar{\gamma}_2}+\frac{2g''(\zeta_2)}{1+\gamma_2}, \tag{9.5.9}$$

$$-\frac{\bar{P}}{2i} = \frac{\bar{\gamma}_1 \bar{f}'(\bar{\zeta}_1)}{1+\bar{\gamma}_1}+\frac{f'(\zeta_1)}{1+\gamma_1}+\frac{\bar{\gamma}_2 \bar{g}'(\bar{\zeta}_2)}{1+\bar{\gamma}_2}+\frac{g'(\zeta_2)}{1+\gamma_2}. \tag{9.5.10}$$

For some purposes it is convenient to express stresses and displacements in terms of two functions $h(\zeta)$, $k(\zeta)$, where

$$\left.\begin{aligned} h(\zeta) &= \frac{2\gamma_1 f(\zeta)}{1+\gamma_1}+\frac{2\gamma_2 g(\zeta)}{1+\gamma_2} \\ k(\zeta) &= \frac{2f(\zeta)}{1+\gamma_1}+\frac{2g(\zeta)}{1+\gamma_2} \end{aligned}\right\}, \tag{9.5.11}$$

so that

$$\left.\begin{aligned} \frac{2(\gamma_1-\gamma_2)f(\zeta)}{1+\gamma_1} &= h(\zeta)-\gamma_2 k(\zeta) \\ \frac{2(\gamma_1-\gamma_2)g(\zeta)}{1+\gamma_2} &= -h(\zeta)+\gamma_1 k(\zeta) \end{aligned}\right\}. \tag{9.5.12}$$

Hence, from (9.5.7) and (9.5.9),

$$\begin{aligned} \sigma_{yy}-i\sigma_{xy} = {} & [\gamma_1 h''(\zeta_1)-\gamma_2 h''(\zeta_2)-\gamma_1\gamma_2\{k''(\zeta_1)-k''(\zeta_2)\}]/(\gamma_1-\gamma_2)+ \\ & +\{\bar{h}''(\bar{\zeta}_1)-\bar{h}''(\bar{\zeta}_2)+\bar{\gamma}_1 \bar{k}''(\bar{\zeta}_2)-\bar{\gamma}_2 \bar{k}''(\bar{\zeta}_1)\}/(\bar{\gamma}_1-\bar{\gamma}_2), \end{aligned} \tag{9.5.13}$$

$$\begin{aligned} \sigma_{yy}+i\sigma_{xy} = {} & \{h''(\zeta_1)-h''(\zeta_2)+\gamma_1 k''(\zeta_2)-\gamma_2 k''(\zeta_1)\}/(\gamma_1-\gamma_2)+ \\ & +[\bar{\gamma}_1 \bar{h}''(\bar{\zeta}_1)-\bar{\gamma}_2 \bar{h}''(\bar{\zeta}_2)-\bar{\gamma}_1\bar{\gamma}_2\{\bar{k}''(\bar{\zeta}_1)-\bar{k}''(\bar{\zeta}_2)\}]/(\bar{\gamma}_1-\bar{\gamma}_2). \end{aligned} \tag{9.5.14}$$

The functions $f(\zeta_1)$, $g(\zeta_1)$ are defined and holomorphic at every point of the half-planes $\mathrm{im}(\zeta_1) > 0$, $\mathrm{im}(\zeta_2) > 0$ respectively, except possibly at infinity, so that $h(\zeta)$, $k(\zeta)$ are defined and holomorphic at every point in $\mathrm{im}(\zeta) > 0$. We now extend the definition of these functions to the whole of the $\zeta = \xi+i\eta$-plane by the formulae

$$\left.\begin{aligned} h(\zeta) &= -\bar{k}(\zeta) \\ k(\zeta) &= -\bar{h}(\zeta) \end{aligned}\right\} \quad (\eta < 0), \tag{9.5.15}$$

where, as in § 8.6,

$$\bar{h}(\zeta) = \overline{h(\bar{\zeta})}, \qquad \bar{k}(\zeta) = \overline{k(\bar{\zeta})},$$

and these functions are defined for $\eta < 0$, since when ζ is in the lower half-plane, $\bar{\zeta}$ belongs to the upper half-plane. It will be seen later that these definitions of $h(\zeta)$, $k(\zeta)$ for $\eta < 0$ are chosen so that they are respectively the analytical continuations of $h(\zeta)$, $k(\zeta)$, for $\eta > 0$, through the unloaded part of the boundary. If ζ belongs to $\eta > 0$, then $\bar{\zeta}$ belongs to $\eta < 0$, so, from (9.5.15),

$$\left.\begin{aligned} h(\bar{\zeta}) &= -\bar{k}(\zeta) \\ k(\bar{\zeta}) &= -\bar{h}(\zeta) \end{aligned}\right\} \quad (\eta > 0), \tag{9.5.16}$$

and (9.5.13), (9.5.14) become

$$\sigma_{yy}-i\sigma_{xy} = [\gamma_1 h''(\zeta_1)-\gamma_2 h''(\zeta_2)-\gamma_1\gamma_2\{k''(\zeta_1)-k''(\zeta_2)\}]/(\gamma_1-\gamma_2)-$$
$$-\{k''(\bar{\zeta}_1)-k''(\bar{\zeta}_2)+\bar{\gamma}_1 h''(\bar{\zeta}_2)-\bar{\gamma}_2 h''(\bar{\zeta}_1)\}/(\bar{\gamma}_1-\bar{\gamma}_2), \tag{9.5.17}$$

$$\sigma_{yy}+i\sigma_{xy} = \{h''(\zeta_1)-h''(\zeta_2)+\gamma_1 k''(\zeta_2)-\gamma_2 k''(\zeta_1)\}/(\gamma_1-\gamma_2)-$$
$$-[\bar{\gamma}_1 k''(\bar{\zeta}_1)-\bar{\gamma}_2 k''(\bar{\zeta}_2)-\bar{\gamma}_1\bar{\gamma}_2\{h''(\bar{\zeta}_1)-h''(\bar{\zeta}_2)\}]/(\bar{\gamma}_1-\bar{\gamma}_2). \tag{9.5.18}$$

When the stresses and rotation in the half-plane $y > 0$ vanish at infinity we see from (9.5.3) and (9.5.11) that $h''(\zeta)$, $k''(\zeta)$ are $O(1/\zeta)$ at infinity in the half-plane $\eta > 0$ and hence, from (9.5.15),

$$h''(\zeta) = O\left(\frac{1}{\zeta}\right), \qquad k''(\zeta) = O\left(\frac{1}{\zeta}\right), \tag{9.5.19}$$

for all large $|\zeta|$.

From (9.5.4), (9.5.11), and (9.5.16) we also have

$$2D = [\delta_1\{h'(\zeta_1)-\gamma_2 k'(\zeta_1)\}-\delta_2\{h'(\zeta_2)-\gamma_1 k'(\zeta_2)\}]/(\gamma_1-\gamma_2)-$$
$$-[\rho_1\{k'(\bar{\zeta}_1)-\bar{\gamma}_2 h'(\bar{\zeta}_1)\}-\rho_2\{k'(\bar{\zeta}_2)-\bar{\gamma}_1 h'(\bar{\zeta}_2)\}]/(\bar{\gamma}_1-\bar{\gamma}_2), \tag{9.5.20}$$

and

$$2\bar{D} = [\bar{\rho}_1\{h'(\zeta_1)-\gamma_2 k'(\zeta_1)\}-\bar{\rho}_2\{h'(\zeta_2)-\gamma_1 k'(\zeta_2)\}]/(\gamma_1-\gamma_2)-$$
$$-[\bar{\delta}_1\{k'(\bar{\zeta}_1)-\bar{\gamma}_2 h'(\bar{\zeta}_1)\}-\bar{\delta}_2\{k'(\bar{\zeta}_2)-\bar{\gamma}_1 h'(\bar{\zeta}_2)\}]/(\bar{\gamma}_1-\bar{\gamma}_2). \tag{9.5.21}$$

9.6. First boundary-value problem for half-plane: part 1

As in § 8.7 for isotropic bodies it is convenient to divide the first boundary-value problem for the half-plane $y \geqslant 0$ into two parts. In the first part, which we consider here, we suppose that the whole of the boundary $y = 0$ is free from applied stress and that a distribution of stress with known singularities exists in $y > 0$ such that the stresses and rotation are zero at infinity. We therefore suppose that the complex potentials $f_0(\zeta_1)$, $g_0(\zeta_2)$ define a stress system which has singularities only in $y > 0$ and which is such that

$$\left.\begin{aligned} f_0''(\zeta_1) &= \frac{(1+\gamma_1)H}{2\pi\zeta_1}+O\left(\frac{1}{\zeta_1^2}\right) \\ g_0''(\zeta_2) &= \frac{(1+\gamma_2)K}{2\pi\zeta_2}+O\left(\frac{1}{\zeta_2^2}\right) \end{aligned}\right\}, \tag{9.6.1}$$

for large $|\zeta_1|$ and $|\zeta_2|$, where H, K are given by (9.1.15) in terms of the resultant force P acting on the plate. We now wish to find complex potentials $f_1(\zeta_1)$, $g_1(\zeta_2)$ which are such that the combined potentials

$$f(\zeta_1) = f_0(\zeta_1)+f_1(\zeta_1). \qquad g(\zeta_2) = g_0(\zeta_2)+g_1(\zeta_2) \tag{9.6.2}$$

give zero applied stress on $y = 0$, and, from (9.1.20), for large $|\zeta_1|$, $|\zeta_2|$,

$$\left.\begin{aligned} f''(\zeta_1) &= f_0''(\zeta_1)+f_1''(\zeta_1) = -\frac{(1+\gamma_1)(P+\gamma_2\bar{P})}{4\pi(\gamma_1-\gamma_2)\zeta_1}+O\left(\frac{1}{\zeta_1^2}\right) \\ g''(\zeta_2) &= g_0''(\zeta_2)+g_1''(\zeta_2) = \frac{(1+\gamma_2)(P+\gamma_1\bar{P})}{4\pi(\gamma_1-\gamma_2)\zeta_2}+O\left(\frac{1}{\zeta_2^2}\right) \end{aligned}\right\}. \tag{9.6.3}$$

Corresponding to $f_1(\zeta)$, $g_1(\zeta)$ we define $h_1(\zeta)$, $k_1(\zeta)$ by formulae of the type (9.5.11) so that

$$\left.\begin{aligned} h_1(\zeta) &= \frac{2\gamma_1 f_1(\zeta)}{1+\gamma_1}+\frac{2\gamma_2 g_1(\zeta)}{1+\gamma_2} \\ k_1(\zeta) &= \frac{2f_1(\zeta)}{1+\gamma_1}+\frac{2g_1(\zeta)}{1+\gamma_2} \end{aligned}\right\}. \tag{9.6.4}$$

It follows from (9.6.1), (9.6.2), and (9.6.3) that $h_1''(\zeta)$ and $k_1''(\zeta)$ are both $O(1/\zeta)$ for large $|\zeta|$ in $\eta > 0$. If the definitions of $h_1(\zeta)$, $k_1(\zeta)$ are extended by formulae of the type (9.5.15) to

$$\left.\begin{aligned} h_1(\zeta) &= -\bar{k}_1(\zeta) \\ k_1(\zeta) &= -\bar{h}_1(\zeta) \end{aligned}\right\} \quad (\eta < 0), \tag{9.6.5}$$

then $h_1''(\zeta)$, $k_1''(\zeta)$ are sectionally holomorphic in the whole plane, with line of discontinuity $\eta = 0$, and $O(1/\zeta)$ for large $|\zeta|$.

From (9.5.7), (9.5.17), and (9.6.2),

$$\begin{aligned} \sigma_{yy}-i\sigma_{xy} = {} & \frac{2\gamma_1 f_0''(\zeta_1)}{1+\gamma_1}+\frac{2\bar{f}_0''(\bar{\zeta}_1)}{1+\bar{\gamma}_1}+\frac{2\gamma_2 g_0''(\zeta_2)}{1+\gamma_2}+\frac{2\bar{g}_0''(\bar{\zeta}_2)}{1+\bar{\gamma}_2}+ \\ & +[\gamma_1 h_1''(\zeta_1)-\gamma_2 h_1''(\zeta_2)-\gamma_1\gamma_2\{k_1''(\zeta_1)-k_1''(\zeta_2)\}]/(\gamma_1-\gamma_1)- \\ & -\{k_1''(\bar{\zeta}_1)-k_1''(\bar{\zeta}_2)+\bar{\gamma}_1 h_1''(\bar{\zeta}_2)-\bar{\gamma}_2 h_1''(\bar{\zeta}_1)\}/(\bar{\gamma}_1-\bar{\gamma}_2). \end{aligned} \tag{9.6.6}$$

Similarly, from (9.5.9), (9.5.18), and (9.6.2),

$$\sigma_{yy}+i\sigma_{xy} = \frac{2\bar{\gamma}_1\bar{f}''_0(\bar{\zeta}_1)}{1+\bar{\gamma}_1}+\frac{2f''_0(\zeta_1)}{1+\gamma_1}+\frac{2\bar{\gamma}_2\bar{g}''_0(\bar{\zeta}_2)}{1+\bar{\gamma}_2}+\frac{2g''_0(\zeta_2)}{1+\gamma_2}+$$
$$+\{h''_1(\zeta_1)-h''_1(\zeta_2)+\gamma_1 k''_1(\zeta_2)-\gamma_2 k''_1(\zeta_1)\}/(\gamma_1-\gamma_2)-$$
$$-[\bar{\gamma}_1 k''_1(\bar{\zeta}_1)-\bar{\gamma}_2 k''_1(\bar{\zeta}_2)-\bar{\gamma}_1\bar{\gamma}_2\{h''_1(\bar{\zeta}_1)-h''_1(\bar{\zeta}_2)\}]/(\bar{\gamma}_1-\bar{\gamma}_2). \quad (9.6.7)$$

When $y \to 0$, $\zeta_1 \to x$, $\zeta_2 \to x$, and if the limits $\{h''_1(x)\}^+$, $\{h''_1(x)\}^-$, $\{k''_1(x)\}^+$, $\{k''_1(x)\}^-$ exist and the applied stress on $y = 0$ vanishes, then (9.6.6) and (9.6.7) give

$$\{h''_1(x)\}^+-\{h''_1(x)\}^- = -\frac{2\gamma_1 f''_0(x)}{1+\gamma_1}-\frac{2\bar{f}''_0(x)}{1+\bar{\gamma}_1}-\frac{2\gamma_2 g''_0(x)}{1+\gamma_2}-\frac{2\bar{g}''_0(x)}{1+\bar{\gamma}_2}, \quad (9.6.8)$$

$$\{k''_1(x)\}^+-\{k''_1(x)\}^- = -\frac{2f''_0(x)}{1+\gamma_1}-\frac{2\bar{\gamma}_1\bar{f}''_0(x)}{1+\bar{\gamma}_1}-\frac{2g''_0(x)}{1+\gamma_2}-\frac{2\bar{\gamma}_2\bar{g}''_0(x)}{1+\bar{\gamma}_2}, \quad (9.6.9)$$

where $\{\ \}^+$, $\{\ \}^-$ denote the limits as $y \to +0$ and $y \to -0$ respectively. Since $h''_1(\zeta)$, $k''_1(\zeta)$ are sectionally holomorphic functions in the whole ζ-plane, vanishing at infinity, and with discontinuities given by (9.6.8) and (9.6.9), we may write, uniquely,

$$h''_1(\zeta) = -\frac{1}{\pi i}\int\limits_{-\infty}^{\infty}\left\{\frac{\gamma_1 f''_0(t)}{1+\gamma_1}+\frac{\bar{f}''_0(t)}{1+\bar{\gamma}_1}+\frac{\gamma_2 g''_0(t)}{1+\gamma_2}+\frac{\bar{g}''_0(t)}{1+\bar{\gamma}_2}\right\}\frac{dt}{t-\zeta}, \quad (9.6.10)$$

$$k''_1(\zeta) = -\frac{1}{\pi i}\int\limits_{-\infty}^{\infty}\left\{\frac{f''_0(t)}{1+\gamma_1}+\frac{\bar{\gamma}_1\bar{f}''_0(t)}{1+\bar{\gamma}_1}+\frac{g''_0(t)}{1+\gamma_2}+\frac{\bar{\gamma}_2\bar{g}''_0(t)}{1+\bar{\gamma}_2}\right\}\frac{dt}{t-\zeta}. \quad (9.6.11)$$

From (9.6.1) we see that the integrals (9.6.10), (9.6.11) taken along the infinite semicircle in either the upper or the lower half of a t-plane vanish so we may evaluate the integrals by residue theory. The functions $f''_0(\zeta)$, $g''_0(\zeta)$ are regular for $\eta < 0$, and $\bar{f}''_0(\zeta)$, $\bar{g}''_0(\zeta)$ are regular for $\eta > 0$, hence

$$\left.\begin{aligned} h''_1(\zeta) &= -\frac{2\bar{f}''_0(\zeta)}{1+\bar{\gamma}_1}-\frac{2\bar{g}''_0(\zeta)}{1+\bar{\gamma}_2} \quad (\eta \geqslant 0) \\ h''_1(\zeta) &= \frac{2\gamma_1 f''_0(\zeta)}{1+\gamma_1}+\frac{2\gamma_2 g''_0(\zeta)}{1+\gamma_2} \quad (\eta \leqslant 0) \end{aligned}\right\}, \quad (9.6.12)$$

and

$$\left.\begin{aligned} k''_1(\zeta) &= -\frac{2\bar{\gamma}_1\bar{f}''_0(\zeta)}{1+\bar{\gamma}_1}-\frac{2\bar{\gamma}_2\bar{g}''_0(\zeta)}{1+\bar{\gamma}_2} \quad (\eta \geqslant 0) \\ k''_1(\zeta) &= \frac{2f''_0(\zeta)}{1+\gamma_1}+\frac{2g''_0(\zeta)}{1+\gamma_2} \quad (\eta \leqslant 0) \end{aligned}\right\}. \quad (9.6.13)$$

These solutions could also be written down by inspection from (9.6.8) and (9.6.9). From the first results in (9.6.12) and (9.6.13), we may, with the help of (9.6.4), solve for $f_1''(\zeta)$, $g_1''(\zeta)$ to get

$$\left.\begin{aligned}\frac{(\gamma_1-\gamma_2)f_1''(\zeta)}{1+\gamma_1} &= -\frac{(1-\bar{\gamma}_1\gamma_2)\bar{f}_0''(\zeta)}{1+\bar{\gamma}_1}-\frac{(1-\gamma_2\bar{\gamma}_2)\bar{g}_0''(\zeta)}{1+\bar{\gamma}_2}\\ \frac{(\gamma_1-\gamma_2)g_1''(\zeta)}{1+\gamma_2} &= \frac{(1-\gamma_1\bar{\gamma}_1)\bar{f}_0''(\zeta)}{1+\bar{\gamma}_1}+\frac{(1-\gamma_1\bar{\gamma}_2)\bar{g}_0''(\zeta)}{1+\bar{\gamma}_2}\end{aligned}\right\}, \tag{9.6.14}$$

for $\eta \geqslant 0$. The complete complex potentials $f(\zeta)$, $g(\zeta)$ may now be written down from (9.6.2) and (9.6.14) and we may verify, with the help of (9.1.15), that they have the correct form (9.6.3) at infinity. Thus

$$\left.\begin{aligned}f(\zeta) &= f_0(\zeta)-\frac{1+\gamma_1}{\gamma_1-\gamma_2}\left\{\frac{(1-\bar{\gamma}_1\gamma_2)\bar{f}_0(\zeta)}{1+\bar{\gamma}_1}+\frac{(1-\gamma_2\bar{\gamma}_2)\bar{g}_0(\zeta)}{1+\bar{\gamma}_2}\right\}\\ g(\zeta) &= g_0(\zeta)+\frac{1+\gamma_2}{\gamma_1-\gamma_2}\left\{\frac{(1-\gamma_1\bar{\gamma}_1)\bar{f}_0(\zeta)}{1+\bar{\gamma}_1}+\frac{(1-\gamma_1\bar{\gamma}_2)\bar{g}_0(\zeta)}{1+\bar{\gamma}_2}\right\}\end{aligned}\right\}, \tag{9.6.15}$$

for $\eta \geqslant 0$, and the corresponding displacements and stresses may be found from (9.5.3)–(9.5.7). In particular, from (9.5.5), the value of σ_{xx} on the edge $y = 0$ of the plate is given by

$$\begin{aligned}\sigma_{xx} = &-\frac{4(1-\gamma_1\bar{\gamma}_1)(1-\gamma_1\bar{\gamma}_2)f_0''(x)}{(1+\gamma_1)^2(1+\bar{\gamma}_1)(1+\bar{\gamma}_2)}-\frac{4(1-\gamma_1\bar{\gamma}_1)(1-\bar{\gamma}_1\gamma_2)\bar{f}_0''(x)}{(1+\bar{\gamma}_1)^2(1+\gamma_1)(1+\gamma_2)}-\\ &-\frac{4(1-\gamma_2\bar{\gamma}_2)(1-\bar{\gamma}_1\gamma_2)g_0''(x)}{(1+\gamma_2)^2(1+\bar{\gamma}_1)(1+\bar{\gamma}_2)}-\frac{4(1-\gamma_2\bar{\gamma}_2)(1-\gamma_1\bar{\gamma}_2)\bar{g}_0''(x)}{(1+\bar{\gamma}_2)^2(1+\gamma_1)(1+\gamma_2)},\end{aligned} \tag{9.6.16}$$

since $\sigma_{yy} = 0$ when $y = 0$.

We consider one example of the previous theory. If an isolated force P acts at the point $z = ib$, then, from § 9.3,

$$\left.\begin{aligned}f_0(\zeta) &= \frac{H}{2\pi}\{(1+\gamma_1)\zeta-i(1-\gamma_1)b\}\ln\{(1+\gamma_1)\zeta-i(1-\gamma_1)b\}\\ g_0(\zeta) &= \frac{K}{2\pi}\{(1+\gamma_2)\zeta-i(1-\gamma_2)b\}\ln\{(1+\gamma_2)\zeta-i(1-\gamma_2)b\}\end{aligned}\right\}. \tag{9.6.17}$$

The value of the edge stress may now be obtained from (9.6.16), but we record here only the results for the special case of orthotropy described in § 9.2 for which γ_1, γ_2 are real and H, K are given by (9.2.2). Thus

$$\begin{aligned}&\pi\alpha_1\alpha_2(\alpha_2-\alpha_1)\mu\sigma_{xx}\\ &\quad = xX\left(\frac{\beta_2}{x^2+b^2/\alpha_1^2}-\frac{\beta_1}{x^2+b^2/\alpha_2^2}\right)+bY\left(\frac{\beta_2}{\alpha_2^2x^2+b^2}-\frac{\beta_1}{\alpha_1^2x^2+b^2}\right).\end{aligned} \tag{9.6.18}$$

If we consider the limit of this formula as $\alpha_1 \to \alpha_2 \to 1$, we recover the result (8.8.2) for an isotropic body.

Numerical results have been given by Green† for a spruce plank whose edge is parallel or perpendicular to the grain.

9.7. First boundary-value problem for half-plane: part 2

To complete the first boundary-value problem we consider the problem of the stress distribution due to given stresses, of $O(1/x^2)$ for large $|x|$, applied to the boundary $y = 0$. We suppose that the stresses and rotation in the body vanish at infinity so that, from (9.1.20), the complex potentials $f(\zeta_1)$, $g(\zeta_1)$ are such that

$$\left.\begin{aligned} f''(\zeta) &= -\frac{(1+\gamma_1)(P+\gamma_2\bar{P})}{4\pi(\gamma_1-\gamma_2)\zeta}+O\left(\frac{1}{\zeta^2}\right) \\ g''(\zeta) &= \frac{(1+\gamma_2)(P+\gamma_1\bar{P})}{4\pi(\gamma_1-\gamma_2)\zeta}+O\left(\frac{1}{\zeta^2}\right) \end{aligned}\right\}, \tag{9.7.1}$$

for large $|\zeta|$. On the boundary $y = 0$ we assume that

$$\sigma_{yy} = p(x), \qquad \sigma_{xy} = q(x), \tag{9.7.2}$$

where $p(x)$, $q(x)$ satisfy the H-condition and x^2p, x^2q satisfy the H-condition in the neighbourhood of the point at infinity. As in the corresponding isotropic problem we indicate two methods of solution.

In the first method we define $h(\zeta)$, $k(\zeta)$ in the region $\eta > 0$ by (9.5.11) and then extend their definitions by (9.5.15) to the region $\eta < 0$. It then follows that $h''(\zeta)$, $k''(\zeta)$ are both $O(1/\zeta)$ for all large $|\zeta|$, and, from (9.5.17) and (9.5.18),

$$\{h''(x)\}^+-\{h''(x)\}^- = p(x)-iq(x), \tag{9.7.3}$$

$$\{k''(x)\}^+-\{k''(x)\}^- = p(x)+iq(x). \tag{9.7.4}$$

Hence

$$\left.\begin{aligned} \frac{2\gamma_1 f''(\zeta)}{1+\gamma_1}+\frac{2\gamma_2 g''(\zeta)}{1+\gamma_2} &= h''(\zeta) = \frac{1}{2\pi i}\int_{-\infty}^{\infty}\frac{p(t)-iq(t)}{t-\zeta}\,dt \\ \frac{2f''(\zeta)}{1+\gamma_1}+\frac{2g''(\zeta)}{1+\gamma_2} &= k''(\zeta) = \frac{1}{2\pi i}\int_{-\infty}^{\infty}\frac{p(t)+iq(t)}{t-\zeta}\,dt \end{aligned}\right\}. \tag{9.7.5}$$

From equations (9.7.5) we may find $f(\zeta)$, $g(\zeta)$, and then the stresses and displacements are given by (9.5.3)–(9.5.6).

† loc. cit., p. 330.

In the second method of solution† we put

$$\left.\begin{aligned} 2f''(\zeta)+2g''(\zeta) &= V(\zeta) \\ \frac{2(1-\gamma_1)f''(\zeta)}{1+\gamma_1}+\frac{2(1-\gamma_2)g''(\zeta)}{1+\gamma_2} &= W(\zeta) \end{aligned}\right\}, \tag{9.7.6}$$

so that, on the boundary $y = 0$, from (9.5.7),

$$\operatorname{re} V(z) = p(x), \qquad \operatorname{im} W(z) = q(x). \tag{9.7.7}$$

Also, from (9.7.1),

$$V(\zeta) = O\left(\frac{1}{\zeta}\right), \qquad W(\zeta) = O\left(\frac{1}{\zeta}\right), \tag{9.7.8}$$

for large $|\zeta|$. It follows that

$$V(\zeta) = \frac{1}{\pi i}\int\limits_{-\infty}^{\infty}\frac{p(t)\,dt}{t-\zeta}, \qquad W(\zeta) = \frac{1}{\pi}\int\limits_{-\infty}^{\infty}\frac{q(t)\,dt}{t-\zeta}, \tag{9.7.9}$$

and we now see that equations (9.7.6) and (9.7.9) are equivalent to (9.7.5) which were obtained by the first method.

If we solve equations (9.7.6) for $f''(\zeta)$ and $g''(\zeta)$, we have

$$\left.\begin{aligned} \frac{4(\gamma_1-\gamma_2)f''(\zeta)}{(1+\gamma_1)(1+\gamma_2)} &= \frac{1-\gamma_2}{1+\gamma_2}V(\zeta)-W(\zeta) \\ \frac{4(\gamma_1-\gamma_2)g''(\zeta)}{(1+\gamma_1)(1+\gamma_2)} &= -\frac{1-\gamma_1}{1+\gamma_1}V(\zeta)+W(\zeta) \end{aligned}\right\}, \tag{9.7.10}$$

and hence, from (9.5.5) and (9.5.6), we can obtain expressions for the components of stress. In particular, on the edge $y = 0$ of the body.

$$\sigma_{xx} = \frac{(1-\gamma_1)(1-\gamma_2)V(x)-2(1-\gamma_1\gamma_2)W(x)}{2(1+\gamma_1)(1+\gamma_2)}+ \\ +\frac{(1-\bar{\gamma}_1)(1-\bar{\gamma}_2)\bar{V}(x)-2(1-\bar{\gamma}_1\bar{\gamma}_2)\bar{W}(x)}{2(1+\bar{\gamma}_1)(1+\bar{\gamma}_2)}. \tag{9.7.11}$$

For the special orthotropic material of § 9.2 this formula for the edge stress reduces further to

$$\sigma_{xx} = [V(x)+\bar{V}(x)-(\alpha_1+\alpha_2)\{W(x)+\bar{W}(x)\}]/(2\alpha_1\alpha_2). \tag{9.7.12}$$

In § 8.9 two special problems of given applied stresses on the boundary of a semi-infinite isotropic body were solved. These problems may now be solved for aeolotropic bodies since the appropriate expressions for

† See A. E. Green, *Proc. Camb. phil. Soc. math. phys. Sci.* **41** (1945) 224.

$V(z)$, $W(z)$ are the same. Further details are, however, left as an exercise.

9.8. Second boundary-value problem for half-plane

Two cases arise when values of the displacements are given at all points of the boundary $y = 0$. In the first case the edge displacements are zero and we wish to find complex potentials (9.6.2), where $f_0(\zeta)$, $g_0(\zeta)$ define a given stress system which has singularities in the upper half-plane $\eta > 0$. The special forms (9.6.3) for $f(\zeta)$, $g(\zeta)$ are omitted. As in § 9.6 we define $h_1(\zeta)$, $k_1(\zeta)$ by (9.6.4) and we extend the definitions of $h_1(\zeta)$, $k_1(\zeta)$ to the whole plane so that $h_1''(\zeta)$ and $k_1''(\zeta)$ are $O(1/\zeta)$ for all large $|\zeta|$. Then, from (9.5.4), (9.5.20), and (9.6.2) we find that the derivative of the total displacement D is given by

$$\frac{\partial D}{\partial x} = \frac{\delta_1 f_0''(\zeta_1)}{1+\gamma_1}+\frac{\rho_1 \bar{f}_0''(\bar{\zeta}_1)}{1+\bar{\gamma}_1}+\frac{\delta_2 g_0''(\zeta_2)}{1+\gamma_2}+\frac{\rho_2 \bar{g}_0''(\bar{\zeta}_2)}{1+\bar{\gamma}_2}+$$
$$+[\delta_1\{h_1''(\zeta_1)-\gamma_2 k_1''(\zeta_1)\}-\delta_2\{h_1''(\zeta_2)-\gamma_1 k_1''(\zeta_2)\}]/\{2(\gamma_1-\gamma_2)\}-$$
$$-[\rho_1\{k_1''(\bar{\zeta}_1)-\bar{\gamma}_2 h_1''(\bar{\zeta}_1)\}-\rho_2\{k_1''(\bar{\zeta}_2)-\bar{\gamma}_1 h_1''(\bar{\zeta}_2)\}]/\{2(\bar{\gamma}_1-\bar{\gamma}_2)\}. \quad (9.8.1)$$

Similarly, using (9.5.21),

$$\frac{\partial \bar{D}}{\partial x} = \frac{\bar{\delta}_1 \bar{f}_0''(\bar{\zeta}_1)}{1+\bar{\gamma}_1}+\frac{\bar{\rho}_1 f_0''(\zeta_1)}{1+\gamma_1}+\frac{\bar{\delta}_2 \bar{g}_0''(\bar{\zeta}_2)}{1+\bar{\gamma}_2}+\frac{\bar{\rho}_2 g_0''(\zeta_2)}{1+\gamma_2}+$$
$$+[\bar{\rho}_1\{h_1''(\zeta_1)-\gamma_2 k_1''(\zeta_1)\}-\bar{\rho}_2\{h_1''(\zeta_2)-\gamma_1 k_1''(\zeta_2)\}]/\{2(\gamma_1-\gamma_2)\}-$$
$$-[\bar{\delta}_1\{k_1''(\bar{\zeta}_1)-\bar{\gamma}_2 h_1''(\bar{\zeta}_1)\}-\bar{\delta}_2\{k_1''(\bar{\zeta}_2)-\bar{\gamma}_1 h_1''(\bar{\zeta}_2)\}]/\{2(\bar{\gamma}_1-\bar{\gamma}_2)\}. \quad (9.8.2)$$

If D vanishes everywhere on the boundary $y = 0$, then $\partial D/\partial x$ and $\partial \bar{D}/\partial x$ are also zero and hence, with the usual notation,

$$\{(\delta_1-\delta_2)h_1''(x)-(\gamma_2\delta_1-\gamma_1\delta_2)k_1''(x)\}^+/\{2(\gamma_1-\gamma_2)\}-$$
$$-\{(\rho_1-\rho_2)k_1''(x)-(\rho_1\bar{\gamma}_2-\rho_2\bar{\gamma}_1)h_1''(x)\}^-/\{2(\bar{\gamma}_1-\bar{\gamma}_2)\}$$
$$= -\frac{\delta_1 f_0''(x)}{1+\gamma_1}-\frac{\rho_1 \bar{f}_0''(x)}{1+\bar{\gamma}_1}-\frac{\delta_2 g_0''(x)}{1+\gamma_2}-\frac{\rho_2 \bar{g}_0''(x)}{1+\bar{\gamma}_2}, \quad (9.8.3)$$

and

$$\{(\bar{\rho}_1-\bar{\rho}_2)h_1''(x)-(\gamma_2\bar{\rho}_1-\gamma_1\bar{\rho}_2)k_1''(x)\}^+/\{2(\gamma_1-\gamma_2)\}-$$
$$-\{(\bar{\delta}_1-\bar{\delta}_2)k_1''(x)-(\bar{\gamma}_2\bar{\delta}_1-\bar{\gamma}_1\bar{\delta}_2)h_1''(x)\}^-/\{2(\bar{\gamma}_1-\bar{\gamma}_2)\}$$
$$= -\frac{\bar{\delta}_1 \bar{f}_0''(x)}{1+\bar{\gamma}_1}-\frac{\bar{\rho}_1 f_0''(x)}{1+\gamma_1}-\frac{\bar{\delta}_2 \bar{g}_0''(x)}{1+\bar{\gamma}_2}-\frac{\bar{\rho}_2 g_0''(x)}{1+\gamma_2}. \quad (9.8.4)$$

These equations may be solved by integral formulae as in previous sections, or by inspection. Hence, for $\eta \geqslant 0$, using (9.6.4),

$$\{(\delta_1-\delta_2)h_1''(\zeta)-(\gamma_2\delta_1-\gamma_1\delta_2)k_1''(\zeta)\}/\{2(\gamma_1-\gamma_2)\}$$
$$=\frac{\delta_1 f_1''(\zeta)}{1+\gamma_1}+\frac{\delta_2 g_1''(\zeta)}{1+\gamma_2}=-\frac{\rho_1\bar{f}_0''(\zeta)}{1+\bar{\gamma}_1}-\frac{\rho_2\bar{g}_0''(\zeta)}{1+\bar{\gamma}_2}, \quad (9.8.5)$$

$$\{(\bar{\rho}_1-\bar{\rho}_2)h_1''(\zeta)-(\gamma_2\bar{\rho}_1-\gamma_1\bar{\rho}_2)k_1''(\zeta)\}/\{2(\gamma_1-\gamma_2)\}$$
$$=\frac{\bar{\rho}_1 f_1''(\zeta)}{1+\gamma_1}+\frac{\bar{\rho}_2 g_1''(\zeta)}{1+\gamma_2}=-\frac{\bar{\delta}_1\bar{f}_0''(\zeta)}{1+\bar{\gamma}_1}-\frac{\bar{\delta}_2\bar{g}_0''(\zeta)}{1+\bar{\gamma}_2}, \quad (9.8.6)$$

and solving these for $f_1''(\zeta)$, $g_1''(\zeta)$ gives, for $\eta \geqslant 0$,

$$\left.\begin{aligned}\frac{(\delta_1\bar{\rho}_2-\delta_2\bar{\rho}_1)f_1''(\zeta)}{1+\gamma_1}&=\frac{(\delta_2\bar{\delta}_1-\rho_1\bar{\rho}_2)\bar{f}_0''(\zeta)}{1+\bar{\gamma}_1}+\frac{(\delta_2\bar{\delta}_2-\rho_2\bar{\rho}_2)\bar{g}_0''(\zeta)}{1+\bar{\gamma}_2}\\ \frac{(\delta_1\bar{\rho}_2-\delta_2\bar{\rho}_1)g_1''(\zeta)}{1+\gamma_2}&=\frac{(\rho_1\bar{\rho}_1-\delta_1\bar{\delta}_1)\bar{f}_0''(\zeta)}{1+\bar{\gamma}_1}+\frac{(\rho_2\bar{\rho}_1-\delta_1\bar{\delta}_2)\bar{g}_0''(\zeta)}{1+\bar{\gamma}_2}\end{aligned}\right\}. \quad (9.8.7)$$

The complete stress potentials are now given by (9.6.2) and (9.8.7) for $\eta \geqslant 0$, and the corresponding values of the displacement and stresses may be obtained from (9.5.4)–(9.5.7).

The formulae (9.8.7) simplify when the material is orthotropic (§ 9.2) and become

$$\left.\begin{aligned}f_1(\zeta)&=-\frac{(\alpha_1\beta_2^2+\alpha_2\beta_1^2)\bar{f}_0(\zeta)+2\alpha_1\beta_1\beta_2\bar{g}_0(\zeta)}{\alpha_1\beta_2^2-\alpha_2\beta_1^2}\\ g_1(\zeta)&=\frac{2\alpha_2\beta_1\beta_2\bar{f}_0(\zeta)+(\alpha_1\beta_2^2+\alpha_2\beta_1^2)\bar{g}_0(\zeta)}{\alpha_1\beta_2^2-\alpha_2\beta_1^2}\end{aligned}\right\}. \quad (9.8.8)$$

If we add to these the original stress functions $f_0(\zeta)$, $g_0(\zeta)$, we find, from (9.5.5) and (9.5.7), that the stress components at the edge $y=0$ of an orthotropic plate are

$$\left.\begin{aligned}\sigma_{xx}&=-\frac{2\nu(\alpha_1^2-\alpha_2^2)}{\alpha_1^2\alpha_2^2(\alpha_1\beta_2^2-\alpha_2\beta_1^2)}[\alpha_2\beta_1\{f_0''(x)+\bar{f}_0''(x)\}+\alpha_1\beta_2\{g_0''(x)+\bar{g}_0''(x)\}]\\ \sigma_{yy}&=\frac{2\mu(\alpha_1^2-\alpha_2^2)}{\alpha_1\beta_2^2-\alpha_2\beta_1^2}[\alpha_2\beta_1\{f_0''(x)+\bar{f}_0''(x)\}+\alpha_1\beta_2\{g_0''(x)+\bar{g}_0''(x)\}]\\ \sigma_{xy}&=-\frac{2i\mu(\alpha_1^2-\alpha_2^2)}{\alpha_1\beta_2^2-\alpha_2\beta_1^2}[\beta_2\{f_0''(x)-\bar{f}_0''(x)\}+\beta_1\{g_0''(x)-\bar{g}_0''(x)\}]\end{aligned}\right\}.$$
(9.8.9)

The second boundary-value problem is completed by considering the case when non-zero values of the displacement are given at the edge of the body in the form

$$u_x=r(x), \qquad u_y=s(x). \quad (9.8.10)$$

We assume that $r(x)$, $s(x)$ have first derivatives $r'(x)$, $s'(x)$ satisfying the H-condition and x^2r', x^2s' satisfy the H-condition in the neighbourhood of the point at infinity. We also assume that stresses and rotation vanish at infinity. Two methods of solution are again possible which are analogous to those used in § 9.7 for the corresponding part of the first boundary-value problem but we restrict our attention to one method here. We define $h(\zeta)$, $k(\zeta)$ in the region $\eta < 0$ as in § 9.5 and it follows from (9.5.20) and (9.8.10) that

$$\{(\delta_1-\delta_2)h''(x)-(\delta_1\gamma_2-\delta_2\gamma_1)k''(x)\}^+/\{2(\gamma_1-\gamma_2)\}-$$
$$-\{(\rho_1-\rho_2)k''(x)-(\rho_1\bar{\gamma}_2-\rho_2\bar{\gamma}_1)h''(x)\}^-/\{2(\bar{\gamma}_1-\bar{\gamma}_2)\} = r'(x)+is'(x),$$

and hence, by an argument similar to that used in the previous section,

$$\{(\delta_1-\delta_2)h''(\zeta)-(\delta_1\gamma_2-\delta_2\gamma_1)k''(\zeta)\}/\{2(\gamma_1-\gamma_2)\}$$
$$= \frac{\delta_1 f''(\zeta)}{1+\gamma_1}+\frac{\delta_2 g''(\zeta)}{1+\gamma_2} = \frac{1}{2\pi i}\int_{-\infty}^{\infty}\frac{r'(t)+is'(t)}{t-\zeta}\,dt \quad (\eta \geqslant 0), \tag{9.8.11}$$

if we use (9.5.11) and (9.5.19) (but not (9.6.3)). Similarly, from (9.5.11), (9.5.19), (9.5.21), and (9.8.10),

$$\frac{\bar{\rho}_1 f''(\zeta)}{1+\gamma_1}+\frac{\bar{\rho}_2 g''(\zeta)}{1+\gamma_2} = \frac{1}{2\pi i}\int_{-\infty}^{\infty}\frac{r'(t)-is'(t)}{t-\zeta}\,dt \quad (\eta \geqslant 0), \tag{9.8.12}$$

and then $f''(\zeta)$, $g''(\zeta)$ may be found from (9.8.11) and (9.8.12), the corresponding stresses being calculated from (9.5.4)–(9.5.6).

As for isotropic bodies various problems arise of a mixed type. Here we omit problems corresponding to those of § 8.11 and consider only 'punch' and 'crack' problems corresponding to §§ 8.12 to 8.15 for isotropic bodies.

9.9. Punch problems

In this section and in § 9.10 we consider problems for which the shearing stress is zero along the whole of the boundary $y = 0$. We take

$$k(\zeta) = h(\zeta) \tag{9.9.1}$$

so that, from (9.5.12),

$$\left.\begin{aligned}\frac{2(\gamma_1-\gamma_2)f(\zeta)}{1+\gamma_1} &= (1-\gamma_2)h(\zeta)\\ \frac{2(\gamma_1-\gamma_2)g(\zeta)}{1+\gamma_2} &= -(1-\gamma_1)h(\zeta)\end{aligned}\right\}, \tag{9.9.2}$$

and hence, from (9.5.4)–(9.5.7),

$$2D = \{\delta_1(1-\gamma_2)h'(\zeta_1)-\delta_2(1-\gamma_1)h'(\zeta_2)\}/(\gamma_1-\gamma_2)+$$
$$+\{\rho_1(1-\bar{\gamma}_2)\bar{h}'(\bar{\zeta}_1)-\rho_2(1-\bar{\gamma}_1)\bar{h}'(\bar{\zeta}_2)\}/(\bar{\gamma}_1-\bar{\gamma}_2), \quad (9.9.3)$$

$$\Theta = \frac{2}{\gamma_1-\gamma_2}\left\{\frac{\gamma_1(1-\gamma_2)}{1+\gamma_1}h''(\zeta_1)-\frac{\gamma_2(1-\gamma_1)}{1+\gamma_2}h''(\zeta_2)\right\}+$$
$$+\frac{2}{\bar{\gamma}_1-\bar{\gamma}_2}\left\{\frac{\bar{\gamma}_1(1-\bar{\gamma}_2)}{1+\bar{\gamma}_1}\bar{h}''(\bar{\zeta}_1)-\frac{\bar{\gamma}_2(1-\bar{\gamma}_1)}{1+\bar{\gamma}_2}\bar{h}''(\bar{\zeta}_2)\right\}, \quad (9.9.4)$$

$$\sigma_{yy}-i\sigma_{xy} = \{\gamma_1(1-\gamma_2)h''(\zeta_1)-\gamma_2(1-\gamma_1)h''(\zeta_2)\}/(\gamma_1-\gamma_2)+$$
$$+\{(1-\bar{\gamma}_2)\bar{h}''(\bar{\zeta}_1)-(1-\bar{\gamma}_1)\bar{h}''(\bar{\zeta}_2)\}/(\bar{\gamma}_1-\bar{\gamma}_2). \quad (9.9.5)$$

Also, on the boundary $y = 0$ where $\zeta_1 = \zeta_2 = \zeta = x$,

$$D = \frac{\delta_1-\delta_2+\gamma_1\delta_2-\gamma_2\delta_1}{2(\gamma_1-\gamma_2)}h'(\zeta)+\frac{\rho_1-\rho_2+\bar{\gamma}_1\rho_2-\bar{\gamma}_2\rho_1}{2(\bar{\gamma}_1-\bar{\gamma}_2)}\bar{h}'(\bar{\zeta}), \quad (9.9.6)$$

$$\sigma_{yy} = h''(\zeta)+\bar{h}''(\bar{\zeta}), \quad (9.9.7)$$

$$\sigma_{xx} = \frac{(1-\gamma_1)(1-\gamma_2)h''(\zeta)}{(1+\gamma_1)(1+\gamma_2)}+\frac{(1-\bar{\gamma}_1)(1-\bar{\gamma}_2)\bar{h}''(\bar{\zeta})}{(1+\bar{\gamma}_1)(1+\bar{\gamma}_2)}, \quad (9.9.8)$$

and the shear stress σ_{xy} vanishes.

We now examine the distribution of stress in a semi-infinite elastic medium $y \geqslant 0$ when the surface is deformed by the pressure against it of a smooth rigid body so that the boundary conditions on $y = 0$ are

$$\sigma_{xy} = 0 \quad (-\infty < x < \infty), \quad (9.9.9)$$

$$\left.\begin{aligned} \sigma_{yy} &= 0 \quad (x < a,\ x > b) \\ u_y &= f(x) \quad (a < x < b) \end{aligned}\right\}, \quad (9.9.10)$$

where ab is a single segment of the x-axis. The discussion for n segments (i.e. n punches) follows similarly. We assume that $f(x)$ has a derivative $f'(x)$ satisfying the H-condition on ab. Condition (9.9.9) is satisfied by the system of stresses defined earlier in this section. If the total normal force on the edge $y = 0$ is Y (the total tangential traction being zero), and if stresses and rotation vanish at infinity, then, from (9.1.20), (9.5.1), (9.5.3), and (9.9.2),

$$h''(\zeta) = -\frac{iY}{2\pi\zeta}+O\left(\frac{1}{\zeta^2}\right) \quad (9.9.11)$$

for large $|\zeta|$ in $\eta \geqslant 0$. The function $h(\zeta)$ is defined for $\eta > 0$ and we now define $h(\zeta)$ for the half-plane $\eta < 0$ by the equation

$$h(\zeta) = -\bar{h}(\zeta) \quad (\eta < 0), \tag{9.9.12}$$

and hence

$$h(\bar{\zeta}) = -\bar{h}(\bar{\zeta}) \quad (\eta > 0). \tag{9.9.13}$$

It follows from (9.9.11) that $h''(\zeta) = O(1/\zeta)$ for large $|\zeta|$ in the whole plane and, from (9.9.6)–(9.9.8), on the boundary $\eta = 0$,

$$\frac{i}{S}\frac{\partial u_y}{\partial x} = \{h''(x)\}^+ + \{h''(x)\}^-, \tag{9.9.14}$$

$$\sigma_{yy} = \{h''(x)\}^+ - \{h''(x)\}^-, \tag{9.9.15}$$

$$\sigma_{xx} = \frac{(1-\gamma_1)(1-\gamma_2)\{h''(x)\}^+}{(1+\gamma_1)(1+\gamma_2)} - \frac{(1-\bar{\gamma}_1)(1-\bar{\gamma}_2)\{h''(x)\}^-}{(1+\bar{\gamma}_1)(1+\bar{\gamma}_2)}, \tag{9.9.16}$$

where

$$\bar{S} = S$$
$$= \frac{S_{22}^{11}}{4\gamma_1\gamma_2}\{2-\gamma_1-\gamma_2-\bar{\gamma}_1-\bar{\gamma}_2+\gamma_1\gamma_2(\bar{\gamma}_1+\bar{\gamma}_2)+\bar{\gamma}_1\bar{\gamma}_2(\gamma_1+\gamma_2)-$$
$$-2\gamma_1\gamma_2\bar{\gamma}_1\bar{\gamma}_2\}. \tag{9.9.17}$$

In deriving formula (9.9.14) we have used the results (6.7.37) and (6.7.38). The first boundary condition in (9.9.10) now gives

$$\{h''(x)\}^+ - \{h''(x)\}^- = 0,$$

on $y = 0$ outside the segment ab, so that $h''(\zeta)$ is holomorphic in the whole plane cut along ab. Also, from (9.9.10) and (9.9.14),

$$\{h''(x)\}^+ + \{h''(x)\}^- = \frac{if'(x)}{S}, \tag{9.9.18}$$

on ab. The general solution of this equation, remembering the conditions satisfied by $h''(\zeta)$, and allowing for the possibility of infinities in $h''(\zeta)$ of order not greater than 1 at the ends a, b, is

$$h''(\zeta) = \frac{1}{2\pi S\sqrt{\{R(\zeta)\}}}\int_a^b \frac{f'(t)\sqrt{\{R(t)\}}}{t-\zeta}\,dt + \frac{H'}{\sqrt{\{R(\zeta)\}}}, \tag{9.9.19}$$

where $R(\zeta)$ is given by (8.11.7) and $\sqrt{\{R(\zeta)\}}$ is defined in § 8.11. The condition (9.9.13) is satisfied if H' is a real constant. By considering the coefficient of $1/\zeta$ in (9.9.11) and (9.9.19) we have

$$Y = -2\pi H'. \tag{9.9.20}$$

Alternatively, the solution of (9.9.18) which is finite everywhere and which vanishes at infinity is

$$h''(\zeta) = \frac{\sqrt{\{R(\zeta)\}}}{2\pi S}\int\limits_a^b \frac{f'(t)\,dt}{(t-\zeta)\sqrt{\{R(t)\}}}, \tag{9.9.21}$$

provided

$$\int\limits_a^b \frac{f'(t)\,dt}{\sqrt{\{(t-a)(b-t)\}}} = 0, \tag{9.9.22}$$

and

$$Y = -\frac{1}{S}\int\limits_a^b \frac{tf'(t)\,dt}{\sqrt{\{(t-a)(b-t)\}}}. \tag{9.9.23}$$

The solution of the punch problem is now in a form similar to that given for isotropic bodies in § 8.12, except for the differences in expressions for the stresses and displacements. We therefore leave the reader to complete detailed examinations of special problems such as those discussed in §§ 8.13, 8.14, and we now turn our attention to crack problems.

9.10. Cracks

As in § 8.15 for isotropic bodies, we consider an infinite two-dimensional body containing n collinear cracks situated along the segments $L_r = a_r b_r$ ($r = 1, 2,..., n$) of the real axis and we suppose that all components of stress and both components of displacement vanish at infinity, the resultant force over the entire surface of each crack being zero. Then, for large $|\zeta|$,

$$h'(\zeta) = O\left(\frac{1}{\zeta}\right). \tag{9.10.1}$$

If the cracks are opened symmetrically about the x-axis by a given normal distribution of displacement or pressure along the line segments $L = L_1+L_2+...+L_n$, we can consider equivalent problems in the half-plane $y \geqslant 0$. Thus:

PROBLEM 1.

$$\left.\begin{aligned} \sigma_{xy} &= 0 \qquad (-\infty < x < \infty) \\ u_y &= 0 \qquad \text{on } y = 0 \text{ outside } L \\ u_y &= s(x) \quad \text{on } L \end{aligned}\right\}. \tag{9.10.2}$$

PROBLEM 2.

$$\left.\begin{aligned} \sigma_{xy} &= 0 \qquad (-\infty < x < \infty) \\ \sigma_{yy} &= -p(x) \quad \text{on } L \\ u_y &= 0 \qquad \text{on } y = 0 \text{ outside } L \end{aligned}\right\}, \tag{9.10.3}$$

where $p(x)$, $s(x)$ satisfy the H-condition on L.

Since $\sigma_{xy} = 0$ on $y = 0$ for both problems we may use the formulae (9.9.3) to (9.9.8) but now we define $h(\zeta)$ in the region $\eta < 0$ by the formula

$$h(\zeta) = \bar{h}(\zeta) \quad (\eta < 0), \tag{9.10.4}$$

so that

$$h(\bar{\zeta}) = \bar{h}(\bar{\zeta}) \quad (\eta > 0), \tag{9.10.5}$$

and, from (9.10.1), $h'(\zeta)$ vanishes for all large $|\zeta|$. Hence, using (9.9.6)–(9.9.8), it follows that, when $\eta = 0$,

$$\left.\begin{aligned} \frac{iu_y}{S} &= \{h'(x)\}^+ - \{h'(x)\}^- \\ \sigma_{yy} &= \{h''(x)\}^+ + \{h''(x)\}^- \\ \sigma_{xx} &= \frac{(1-\gamma_1)(1-\gamma_2)\{h''(x)\}^+}{(1+\gamma_1)(1+\gamma_2)} + \frac{(1-\bar{\gamma}_1)(1-\bar{\gamma}_2)\{h''(x)\}^-}{(1+\bar{\gamma}_1)(1+\bar{\gamma}_2)} \end{aligned}\right\}. \tag{9.10.6}$$

Problem 1. The second and third boundary conditions in (9.10.2) give, on using (9.10.6),

$$\left.\begin{aligned} \{h'(x)\}^+ - \{h'(x)\}^- &= 0 \quad \text{outside } L \\ \{h'(x)\}^+ - \{h'(x)\}^- &= \frac{is(x)}{S} \quad \text{on } L \end{aligned}\right\}, \tag{9.10.7}$$

and hence, since $h'(\zeta)$ vanishes at infinity,

$$h'(\zeta) = \frac{1}{2\pi S} \int\limits_L \frac{s(t)\,dt}{t-\zeta}. \tag{9.10.8}$$

Problem 2. Here the conditions (9.10.3), together with (9.10.6), give

$$\left.\begin{aligned} \{h''(x)\}^+ + \{h''(x)\}^- &= -p(x) \quad \text{on } L \\ \{h'(x)\}^+ - \{h'(x)\}^- &= 0 \quad \text{outside } L \end{aligned}\right\}. \tag{9.10.9}$$

It follows that $h''(\zeta)$ is holomorphic in the whole plane cut along L, of $O(1/\zeta^2)$ at infinity, and satisfying the first condition in (9.10.9) on L. Restricting the discussion for simplicity to a single crack ab, the general solution which allows the possibility of infinities at the edges of the crack of order not greater than 1 is

$$h''(\zeta) = -\frac{1}{2\pi i\sqrt{\{R(\zeta)\}}} \int\limits_a^b \frac{p(t)\sqrt{\{R(t)\}}}{t-\zeta}\,dt, \tag{9.10.10}$$

where $\sqrt{\{R(\zeta)\}}$ is defined in § 8.11. It can be verified that condition (9.10.5) is satisfied.

The solutions of both Problems 1 and 2 are in forms similar to those given in §§ 8.16, 8.17 for isotropic materials, except that the stresses and displacements are now given by the formulae of § 9.9. We therefore leave further discussion of special problems to the reader.

9.11. Circular hole in infinite plate

Consider an infinite region bounded internally by the single circular hole $|z| = a$. We define two complex variables ζ_1, ζ_2 by the transformations

$$\left.\begin{aligned} z_1 &= z+\gamma_1\bar{z} = \zeta_1+\gamma_1 a^2/\zeta_1 \\ z_2 &= z+\gamma_2\bar{z} = \zeta_2+\gamma_2 a^2/\zeta_2 \end{aligned}\right\}, \tag{9.11.1}$$

and the ζ_1-, ζ_2-planes are chosen so that the circles $|\zeta_1| = a$, $|\zeta_2| = a$ correspond to the circle $|z| = a$ in the z-plane, and $|\zeta_1| \to \infty$, $|\zeta_2| \to \infty$ when $|z| \to \infty$. Also, since $|\gamma_1| < 1$, $|\gamma_2| < 1$, the singularities of the transformation from the z_1- to the ζ_1-plane, and from the z_2- to the ζ_2-plane,† are inside the circles $|\zeta_1| = a$, $|\zeta_2| = a$ respectively, For convenience we may imagine that the three circles $|z| = a$, $|\zeta_1| = a$, and $|\zeta_2| = a$ coincide and are given by

$$\zeta_1 = \zeta_2 = z = ae^{i\theta}, \qquad \bar{z} = a^2/z. \tag{9.11.2}$$

To every point in the ζ_1-plane, on or outside the circle (9.11.2), corresponds one point in the (x, y) plane on or outside this circle, but the transformation between these planes is not conformal. A similar correspondence exists between the ζ_2-plane and the (x, y) plane. If we put

$$\Omega'(z_1) = f(\zeta_1), \qquad \omega'(z_2) = g(\zeta_2), \tag{9.11.3}$$

then, from the formulae (9.1.4)–(9.1.8), and (9.1.12),

$$P = X+iY = 2i\{\gamma_1 f(\zeta_1)+\bar{f}(\bar{\zeta}_1)+\gamma_2 g(\zeta_2)+\bar{g}(\bar{\zeta}_2)\}, \tag{9.11.4}$$

$$\sigma_{rr}+\sigma_{\theta\theta} = \frac{4\gamma_1 f'(\zeta_1)}{1-\gamma_1 a^2/\zeta_1^2}+\frac{4\bar{\gamma}_1\bar{f}'(\bar{\zeta}_1)}{1-\bar{\gamma}_1 a^2/\bar{\zeta}_1^2}+\frac{4\gamma_2 g'(\zeta_2)}{1-\gamma_2 a^2/\zeta_2^2}+\frac{4\bar{\gamma}_2\bar{g}'(\bar{\zeta}_2)}{1-\bar{\gamma}_2 a^2/\bar{\zeta}_2^2}, \tag{9.11.5}$$

$$\sigma_{rr}-\sigma_{\theta\theta}+2i\sigma_{r\theta} = -\frac{4\bar{z}}{z}\left\{\frac{\gamma_1^2 f'(\zeta_1)}{1-\gamma_1 a^2/\zeta_1^2}+\frac{\bar{f}'(\bar{\zeta}_1)}{1-\bar{\gamma}_1 a^2/\bar{\zeta}_1^2}+\frac{\gamma_2^2 g'(\zeta_2)}{1-\gamma_2 a^2/\zeta_2^2}+\frac{\bar{g}'(\bar{\zeta}_2)}{1-\bar{\gamma}_2 a^2/\bar{\zeta}_2^2}\right\}, \tag{9.11.6}$$

$$\sigma_{rr}+i\sigma_{r\theta} = \frac{2\gamma_1(1-\gamma_1\bar{z}/z)f'(\zeta_1)}{1-\gamma_1 a^2/\zeta_1^2}+\frac{2(\bar{\gamma}_1-\bar{z}/z)\bar{f}'(\bar{\zeta}_1)}{1-\bar{\gamma}_1 a^2/\bar{\zeta}_1^2}+ \\ +\frac{2\gamma_2(1-\gamma_2\bar{z}/z)g'(\zeta_2)}{1-\gamma_2 a^2/\zeta_2^2}+\frac{2(\bar{\gamma}_2-\bar{z}/z)\bar{g}'(\bar{\zeta}_2)}{1-\bar{\gamma}_2 a^2/\bar{\zeta}_2^2}, \tag{9.11.7}$$

$$D = \delta_1 f(\zeta_1)+\rho_1\bar{f}(\bar{\zeta}_1)+\delta_2 g(\zeta_2)+\rho_2\bar{g}(\bar{\zeta}_2). \tag{9.11.8}$$

† i.e. points at which $z_1'(\zeta_1) = 0$, $z_2'(\zeta_2) = 0$.

If we now define $h(z)$, $k(z)$ by the formulae

$$\left.\begin{aligned} h(z) &= 2\gamma_1 f(z)+2\gamma_2 g(z) \\ k(z) &= 2f(z)+2g(z) \end{aligned}\right\}, \tag{9.11.9}$$

so that

$$\left.\begin{aligned} 2(\gamma_1-\gamma_2)f(z) &= h(z)-\gamma_2 k(z) \\ 2(\gamma_1-\gamma_2)g(z) &= -h(z)+\gamma_1 k(z) \end{aligned}\right\}, \tag{9.11.10}$$

then stresses and displacements may be expressed in terms of $h(\zeta_1)$, $h(\zeta_2)$, $k(\zeta_1)$, $k(\zeta_2)$. In particular,

$$P = \frac{i}{\gamma_1-\gamma_2}\{\gamma_1 h(\zeta_1)-\gamma_1\gamma_2 k(\zeta_1)-\gamma_2 h(\zeta_2)+\gamma_1\gamma_2 k(\zeta_2)\}+ \\ +\frac{i}{\bar{\gamma}_1-\bar{\gamma}_2}\{\bar{h}(\bar{\zeta}_1)-\bar{\gamma}_2\bar{k}(\bar{\zeta}_1)-\bar{h}(\bar{\zeta}_2)+\bar{\gamma}_1\bar{k}(\bar{\zeta}_2)\}, \tag{9.11.11}$$

$$\bar{P} = -\frac{i}{\gamma_1-\gamma_2}\{h(\zeta_1)-\gamma_2 k(\zeta_1)-h(\zeta_2)+\gamma_1 k(\zeta_2)\}- \\ -\frac{i}{\bar{\gamma}_1-\bar{\gamma}_2}\{\bar{\gamma}_1\bar{h}(\bar{\zeta}_1)-\bar{\gamma}_1\bar{\gamma}_2\bar{k}(\bar{\zeta}_1)-\bar{\gamma}_2\bar{h}(\bar{\zeta}_2)+\bar{\gamma}_1\bar{\gamma}_2\bar{k}(\bar{\zeta}_2)\}. \tag{9.11.12}$$

The functions $f(\zeta_1)$, $g(\zeta_2)$ are defined and holomorphic at every point outside the circles $\zeta_1 = \zeta_2 = ae^{i\theta}$, except possibly at infinity, so that $h(z)$, $k(z)$ are defined and holomorphic at every point outside the circle $z = ae^{i\theta}$ except possibly at infinity. We now extend the definitions of these functions to the whole z-plane by the formulae

$$\left.\begin{aligned} h(z) &= -\bar{k}(a^2/z) \\ k(z) &= -\bar{h}(a^2/z) \end{aligned}\right\} \quad (|z| < a), \tag{9.11.13}$$

where, as in § 8.18,

$$\bar{h}(a^2/z) = \overline{h(a^2/\bar{z})}, \qquad \bar{k}(a^2/z) = \overline{k(a^2/\bar{z})},$$

are defined for $|z| < a$. If z belongs to $|z| > a$, then $a^2/\bar{z}$ belongs to $|z| < a$, so, from (9.11.13),

$$\left.\begin{aligned} h(a^2/\bar{z}) &= -\bar{k}(\bar{z}) \\ k(a^2/\bar{z}) &= -\bar{h}(\bar{z}) \end{aligned}\right\} \quad (|z| > a). \tag{9.11.14}$$

With the help of (9.11.14) the expressions (9.11.11) and (9.11.12) become

$$P = \frac{i}{\gamma_1-\gamma_2}[\gamma_1 h(\zeta_1)-\gamma_2 h(\zeta_2)+\gamma_1\gamma_2\{k(\zeta_2)-k(\zeta_1)\}]+ \\ +\frac{i}{\bar{\gamma}_1-\bar{\gamma}_2}\left\{k\left(\frac{a^2}{\bar{\zeta}_2}\right)-k\left(\frac{a^2}{\bar{\zeta}_1}\right)+\bar{\gamma}_2 h\left(\frac{a^2}{\bar{\zeta}_1}\right)-\bar{\gamma}_1 h\left(\frac{a^2}{\bar{\zeta}_2}\right)\right\}, \tag{9.11.15}$$

$$\bar{P} = -\frac{i}{\gamma_1-\gamma_2}\{h(\zeta_1)-h(\zeta_2)+\gamma_1 k(\zeta_2)-\gamma_2 k(\zeta_1)\}- \\ -\frac{i}{\bar{\gamma}_1-\bar{\gamma}_2}\left[\bar{\gamma}_2 k\left(\frac{a^2}{\bar{\zeta}_2}\right)-\bar{\gamma}_1 k\left(\frac{a^2}{\bar{\zeta}_1}\right)+\bar{\gamma}_1\bar{\gamma}_2\left\{h\left(\frac{a^2}{\bar{\zeta}_1}\right)-h\left(\frac{a^2}{\bar{\zeta}_2}\right)\right\}\right]. \tag{9.11.16}$$

Similarly, we find from (9.11.8), (9.11.10), and (9.11.14), that

$$D = \frac{1}{2(\gamma_1-\gamma_2)}[\delta_1\{h(\zeta_1)-\gamma_2 k(\zeta_1)\}-\delta_2\{h(\zeta_2)-\gamma_1 k(\zeta_2)\}]+$$
$$+\frac{1}{2(\bar\gamma_1-\bar\gamma_2)}\left[\rho_1\left\{\bar\gamma_2 h\left(\frac{a^2}{\bar\zeta_1}\right)-k\left(\frac{a^2}{\bar\zeta_1}\right)\right\}-\rho_2\left\{\bar\gamma_1 h\left(\frac{a^2}{\bar\zeta_2}\right)-k\left(\frac{a^2}{\bar\zeta_2}\right)\right\}\right], \quad (9.11.17)$$

$$\bar D = \frac{1}{2(\gamma_1-\gamma_2)}[\bar\rho_1\{h(\zeta_1)-\gamma_2 k(\zeta_1)\}-\bar\rho_2\{h(\zeta_2)-\gamma_1 k(\zeta_2)\}]+$$
$$+\frac{1}{2(\bar\gamma_1-\bar\gamma_2)}\left[\delta_1\left\{\bar\gamma_2 h\left(\frac{a^2}{\bar\zeta_1}\right)-k\left(\frac{a^2}{\bar\zeta_1}\right)\right\}-\delta_2\left\{\bar\gamma_1 h\left(\frac{a^2}{\bar\zeta_2}\right)-k\left(\frac{a^2}{\bar\zeta_2}\right)\right\}\right]. \quad (9.11.18)$$

9.12. First boundary-value problem for circular hole: part 1

We suppose that the circular hole $|z| = a$ is free from applied stress and that a distribution of stress exists, with known singularities only in $|z| > a$, which is derived from complex potentials $f_0(\zeta_1)$, $g_0(\zeta_2)$. The distribution of stress is such that, for large $|z|$,

$$\left.\begin{aligned} f_0(z) &= \frac{H\ln z}{2\pi}+\sum_{r=0}^{n}(B_r+iC_r)z^r \\ g_0(z) &= \frac{K\ln z}{2\pi}+\sum_{r=0}^{n}(B'_r+iC'_r)z^r \end{aligned}\right\}, \quad (9.12.1)$$

where H, K are given in terms of the resultant force $P = X+iY$ due to $f_0(z)$, $g_0(z)$ by (9.1.15). We now add a stress system derived from potentials $f_1(\zeta_1)$, $g_1(\zeta_2)$ which is such that the combined potentials

$$f(\zeta_1) = f_0(\zeta_1)+f_1(\zeta_1), \qquad g(\zeta_2) = g_0(\zeta_2)+g_1(\zeta_2), \quad (9.12.2)$$

give zero applied stresses at the boundary $|z| = a$, and, for large $|z|$, $f(z)$, $g(z)$ take the forms (9.12.1). The complex potentials $f_1(z)$, $g_1(z)$ therefore both vanish at infinity. Corresponding to $f_1(z)$, $g_1(z)$ we define $h_1(z)$, $k_1(z)$ by formulae of the type (9.11.9) so that

$$\left.\begin{aligned} h_1(z) &= 2\gamma_1 f_1(z)+2\gamma_2 g_1(z) \\ k_1(z) &= 2f_1(z)+2g_1(z) \end{aligned}\right\}. \quad (9.12.3)$$

It follows that $h_1(z)$, $k_1(z)$ are both $O(1/z)$ for large $|z|$. If the definitions of $h_1(z)$, $k_1(z)$ are extended by formulae of the type (9.11.13) to

$$\left.\begin{aligned} h_1(z) &= -\bar k_1(a^2/z) \\ k_1(z) &= -\bar h_1(a^2/z) \end{aligned}\right\} \quad (|z| < a), \quad (9.12.4)$$

then $h_1(z)$, $k_1(z)$ are sectionally holomorphic in the whole plane, of $O(1/z)$ for large $|z|$, with curve of discontinuity $|z| = a$. Also, from (9.12.4) we see that these functions are finite at the origin.

If the total applied stress at the edge of the circle $|z| = a$ is zero, it follows that the sum of the values of P obtained from (9.12.2), (9.11.4), and (9.11.15) is zero, where, in (9.11.15), $h_1(z)$, $k_1(z)$ replace $h(z)$, $k(z)$ respectively. Similarly, the value of $\bar{P}$ is zero at $|z| = a$. Hence

$$\left.\begin{aligned} h_1^+(t)-h_1^-(t) &= -2\gamma_1 f_0(t)-2\gamma_2 g_0(t)-2\bar{f}_0(a^2/t)-2\bar{g}_0(a^2/t) \\ k_1^+(t)-k_1^-(t) &= -2\bar{\gamma}_1 \bar{f}_0(a^2/t)-2\bar{\gamma}_2 \bar{g}_0(a^2/t)-2f_0(t)-2g_0(t) \end{aligned}\right\}, \tag{9.12.5}$$

where t denotes any point on the circle $|z| = a$. As usual $h_1^+(t)$, $h_1^-(t)$ denote the limits of $h_1(z)$ as we approach the point t on the circle from the outside and inside respectively. The unique solution of these equations is

$$h_1(z) = -\frac{1}{\pi i}\int\limits_L \{\gamma_1 f_0(t)+\gamma_2 g_0(t)+\bar{f}_0(a^2/t)+\bar{g}_0(a^2/t)\}\frac{dt}{t-z},$$

$$k_1(z) = -\frac{1}{\pi i}\int\limits_L \{f_0(t)+g_0(t)+\bar{\gamma}_1 \bar{f}_0(a^2/t)+\bar{\gamma}_2 \bar{g}_0(a^2/t)\}\frac{dt}{t-z},$$

the integrals being taken around the circle $|t| = a$ in a clockwise direction. Consider the integral for $h_1(z)$. We observe that $\bar{f}_0(a^2/t)$, $\bar{g}_0(a^2/t)$ are regular for $|t| > a$ and finite at infinity. Hence

$$\frac{1}{2\pi i}\int\limits_\infty \{\bar{f}_0(a^2/t)+\bar{g}_0(a^2/t)\}\frac{dt}{t-z} = \bar{f}_0(0)+\bar{g}_0(0),$$

the integral being taken in the anti-clockwise sense round the infinite circle, and therefore

$$\begin{aligned} -\frac{1}{2\pi i}\int\limits_L & \{\bar{f}_0(a^2/t)+\bar{g}_0(a^2/t)\}\frac{dt}{t-z} \\ &= -\bar{f}_0(a^2/z)-\bar{g}_0(a^2/z)+\bar{f}_0(0)+\bar{g}_0(0) \quad (|z| \geqslant a), \\ &= \bar{f}_0(0)+\bar{g}_0(0) \quad (|z| \leqslant a). \end{aligned}$$

Also $f_0(t)$, $g_0(t)$ are regular for $|t| < a$ so that

$$\begin{aligned} -\frac{1}{2\pi i}\int\limits_L \{\gamma_1 f_0(t)+\gamma_2 g_0(t)\}\frac{dt}{t-z} &= 0 \quad (|z| \geqslant a), \\ &= \gamma_1 f_0(z)+\gamma_2 g_0(z) \quad (|z| \leqslant a). \end{aligned}$$

Combining these results we have

$$\left.\begin{aligned} h_1(z) &= -2\bar{f}_0(a^2/z)-2\bar{g}_0(a^2/z)+2\bar{f}_0(0)+2\bar{g}_0(0) \quad (|z| \geqslant a) \\ h_1(z) &= 2\gamma_1 f_0(z)+2\gamma_2 g_0(z)+2\bar{f}_0(0)+2\bar{g}_0(0) \quad (|z| \leqslant a) \end{aligned}\right\}. \tag{9.12.6}$$

Similarly

$$\left.\begin{aligned} k_1(z) &= -2\bar{\gamma}_1\bar{f}_0(a^2/z)-2\bar{\gamma}_2\bar{g}_0(a^2/z)+2\bar{\gamma}_1\bar{f}_0(0)+2\bar{\gamma}_2\bar{g}_0(0) \quad (|z| \geqslant a) \\ k_1(z) &= 2f_0(z)+2g_0(z)+2\bar{\gamma}_1\bar{f}_0(0)+2\bar{\gamma}_2\bar{g}_0(0) \quad (|z| \leqslant a) \end{aligned}\right\}. \tag{9.12.7}$$

The total complex potentials for the region $|z| \geqslant a$, given by (9.12.2), (9.12.3), (9.12.6), and (9.12.7), are

$$\left.\begin{aligned} f(z) &= f_0(z)-\{(1-\bar{\gamma}_1\gamma_2)\bar{f}_0(a^2/z)+(1-\gamma_2\bar{\gamma}_2)\bar{g}_0(a^2/z)\}/(\gamma_1-\gamma_2) \\ g(z) &= g_0(z)+\{(1-\gamma_1\bar{\gamma}_1)\bar{f}_0(a^2/z)+(1-\gamma_1\bar{\gamma}_2)\bar{g}_0(a^2/z)\}/(\gamma_1-\gamma_2) \end{aligned}\right\}, \tag{9.12.8}$$

apart from non-essential constants. The stresses and displacements now follow from these formulae and from (9.11.5)–(9.11.8). In particular, if we remember that $\sigma_{rr} = 0$ at the edge of the hole, from (9.11.5) and (9.12.8), after a little reduction,

$$\begin{aligned} S_1 S_2 \sigma_{\theta\theta} & \\ = 4(1-\gamma_1\bar{\gamma}_1)&\{(1-\gamma_1\bar{\gamma}_2)(e^{2i\theta}-\gamma_2)f_0'(z)+(1-\bar{\gamma}_1\gamma_2)(e^{-2i\theta}-\bar{\gamma}_2)\bar{f}_0'(\bar{z})\}+ \\ &+4(1-\gamma_2\bar{\gamma}_2)\{(1-\bar{\gamma}_1\gamma_2)(e^{2i\theta}-\gamma_1)g_0'(z)+(1-\gamma_1\bar{\gamma}_2)(e^{-2i\theta}-\bar{\gamma}_1)\bar{g}_0'(\bar{z})\} \\ &\qquad (z = ae^{i\theta}), \end{aligned} \tag{9.12.9}$$

where

$$\left.\begin{aligned} S_1 &= 1+\gamma_1\bar{\gamma}_1-(\gamma_1+\bar{\gamma}_1)\cos 2\theta+i(\gamma_1-\bar{\gamma}_1)\sin 2\theta \\ S_2 &= 1+\gamma_2\bar{\gamma}_2-(\gamma_2+\bar{\gamma}_2)\cos 2\theta+i(\gamma_2-\bar{\gamma}_2)\sin 2\theta \end{aligned}\right\}. \tag{9.12.10}$$

In the case of orthotropic materials considered in § 9.2, for which γ_1, γ_2 are real, the formula for the edge stress becomes

$$\begin{aligned} (1+\gamma_1^2-2\gamma_1\cos 2\theta)&(1+\gamma_2^2-2\gamma_2\cos 2\theta)\sigma_{\theta\theta} \\ = 4(1-\gamma_1\gamma_2)&(1-\gamma_1^2)\{(e^{2i\theta}-\gamma_2)f_0'(z)+(e^{-2i\theta}-\gamma_2)\bar{f}_0'(\bar{z})\}+ \\ &+4(1-\gamma_1\gamma_2)(1-\gamma_2^2)\{(e^{2i\theta}-\gamma_1)g_0'(z)+(e^{-2i\theta}-\gamma_1)\bar{g}_0'(\bar{z})\} \quad (z = ae^{i\theta}). \end{aligned} \tag{9.12.11}$$

We now consider a few special examples.†

9.13. Circular hole: uniform stress at infinity

If a plate containing a circular hole is subject to uniform stress at infinity, defined, as in § 9.1, by N_1, N_2, and α, then, from (9.1.14), for large $|z_1|$, $|z_2|$,

$$\Omega''(z_1) = B+iC, \qquad \omega''(z_2) = B'+iC',$$

where B, B', C, C' are given by equations (9.1.15)–(9.1.18). Hence, from (9.11.3), for large $|z|$,

$$f_0'(z) = B+iC, \qquad g_0'(z) = B'+iC', \tag{9.13.1}$$

† These examples are taken from A. E. Green, *Proc. R. Soc.* A180 (1942) 173. See also loc. cit., p. 326; C. B. Smith, loc. cit., p. 183.

and we see that these potentials have no singularities inside the hole, and are suitable values to take for $f'_0(z)$, $g'_0(z)$. The edge stress $\sigma_{\theta\theta}$ at the hole is now given immediately by (9.12.9) and (9.13.1) and may be reduced to a form suitable for calculations. We confine our attention here, however, to the orthotropic material of § 9.2 so that we may use the simpler formulae (9.2.4) for B, C, B', C' and the expression (9.12.11) for $\sigma_{\theta\theta}$. Thus, after some reduction,

$$\begin{aligned}(1+\gamma_1^2-2\gamma_1\cos 2\theta)&(1+\gamma_2^2-2\gamma_2\cos 2\theta)\sigma_{\theta\theta}\\ &= (N_1+N_2)(1+\gamma_1)(1+\gamma_2)(1+\gamma_1+\gamma_2-\gamma_1\gamma_2-2\cos 2\theta)-\\ &\quad -4\{\gamma_1+\gamma_2-(1+\gamma_1\gamma_2)\cos 2\theta\}(N_1\sin^2\alpha+N_2\cos^2\alpha)+\\ &\quad +2(\gamma_1\gamma_2-1)(N_1-N_2)\sin 2\alpha\sin 2\theta. \qquad (9.13.2)\end{aligned}$$

In the special case of a uniform tension T parallel to the x-axis, $N_1 = T$, $N_2 = \alpha = 0$, and the edge stress becomes

$$\sigma_{\theta\theta} = \frac{T(1+\gamma_1)(1+\gamma_2)(1+\gamma_1+\gamma_2-\gamma_1\gamma_2-2\cos 2\theta)}{(1+\gamma_1^2-2\gamma_1\cos 2\theta)(1+\gamma_2^2-2\gamma_2\cos 2\theta)}. \qquad (9.13.3)$$

For an isotropic material $\gamma_1 = \gamma_2 = 0$ and we recover the formula

$$\sigma_{\theta\theta} = T(1-2\cos 2\theta).$$

The distribution of $\sigma_{\theta\theta}$ over one quadrant of the hole is shown in Fig. 9.3 for the case when the tension is parallel to the grain and in Fig. 9.4 for the case when the tension is perpendicular to the grain, for both oak and spruce. The curve for an isotropic plank is included for comparison. In these figures the sheet is supposed to be in a state of tension in the direction $\theta = 0$. The reader is referred to a paper by Green and Taylor† for more detailed numerical results and for a discussion on the type of failure which might be expected to arise at the edge of a hole in a spruce plank under tension.

Another special case which can be derived from (9.13.2) is when the plate is acted on by a uniform shearing stress S at infinity, parallel to the axes. In this case $N_2 = -S$, $N_1 = S$, $\alpha = \frac{1}{4}\pi$, and‡

$$\sigma_{\theta\theta} = \frac{4S(\gamma_1\gamma_2-1)\sin 2\theta}{(1+\gamma_1^2-2\gamma_1\cos 2\theta)(1+\gamma_2^2-2\gamma_2\cos 2\theta)}. \qquad (9.13.4)$$

9.14. Circular hole: uniform couple at infinity

Here we suppose that the plate is bounded by two straight edges $y = \pm b$, where b is large compared with the radius a of the hole. The

† *Proc. R. Soc.* A184 (1945) 181.

‡ See A. E. Green, loc. cit., p. 350, for a numerical discussion of this formula and of the formula (9.14.3) of the next section.

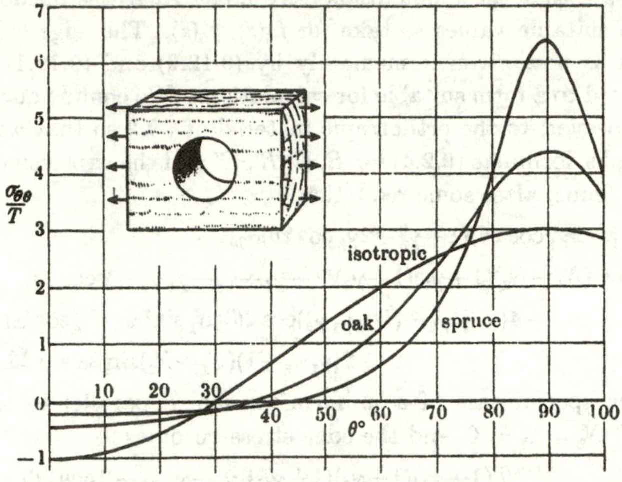

Fig. 9.3. Stress distribution round the edge of a circular hole in a plank under tension. Stress applied parallel to the grain, parallel to $\theta = 0$.

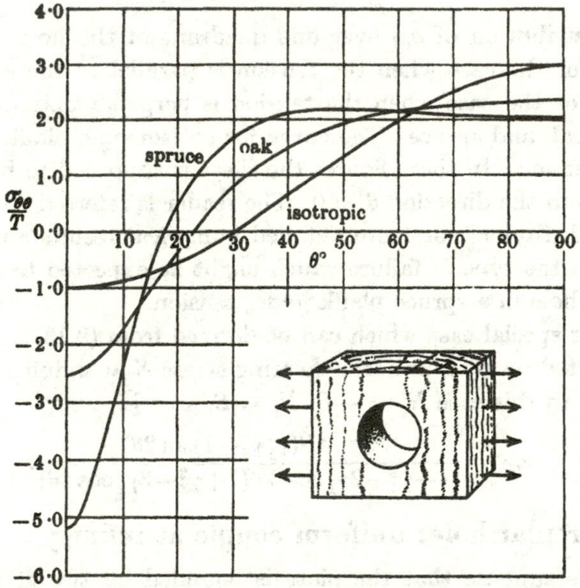

Fig. 9.4. Stress distribution round the edge of a hole in a plank under tension. Stress applied perpendicular to the grain, parallel to $\theta = 0$.

plate is acted on by a bending moment which is such that for large values of x

$$\sigma_{xx} = \frac{3My}{2b^3}, \qquad \sigma_{xy} = \sigma_{yy} = 0 \tag{9.14.1}$$

so that across every section of the plate parallel to the y-axis there is a pure bending moment of amount M. Restricting consideration to the orthotropic materials of § 9.2 we may verify from (9.1.5) and (9.1.6) that the appropriate complex potentials corresponding to the stress distribution (9.14.1) are

$$\Omega''(z_1) = -\frac{3Mi(1+\gamma_2)^2 z_1}{16b^3(\gamma_1-\gamma_2)(1-\gamma_1)(1-\gamma_1\gamma_2)},$$

$$\omega''(z_2) = \frac{3Mi(1+\gamma_1)^2 z_2}{16b^3(\gamma_1-\gamma_2)(1-\gamma_2)(1-\gamma_1\gamma_2)}.$$

It follows from (9.11.3) that the values of $f_0'(z)$, $g_0'(z)$ which must have no singularities inside the circular hole and which must produce the stress system (9.14.1) at infinity, are

$$\left.\begin{aligned} f_0'(z) &= -\frac{3Mi(1+\gamma_2)^2 z}{16b^3(\gamma_1-\gamma_2)(1-\gamma_1)(1-\gamma_1\gamma_2)} \\ g_0'(z) &= \frac{3Mi(1+\gamma_1)^2 z}{16b^3(\gamma_1-\gamma_2)(1-\gamma_2)(1-\gamma_1\gamma_2)} \end{aligned}\right\}. \tag{9.14.2}$$

The edge stress may now be obtained from (9.12.11) and, after straightforward reduction, we have

$$\frac{b^3\sigma_{\theta\theta}}{aM} = \frac{3(1+\gamma_1)(1+\gamma_2)\{(1+\gamma_1+\gamma_2)\sin\theta-\sin 3\theta\}}{2(1+\gamma_1^2-2\gamma_1\cos 2\theta)(1+\gamma_2^2-2\gamma_2\cos 2\theta)}. \tag{9.14.3}$$

When $\gamma_1 = \gamma_2 = 0$ we obtain the formula

$$\frac{b^3\sigma_{\theta\theta}}{aM} = \tfrac{3}{2}(\sin\theta-\sin 3\theta) \tag{9.14.4}$$

for an isotropic material.

9.15. Isolated force near circular hole

We now consider, as a final example of the theory of § 9.12, the problem of the influence of a circular hole on the stress distribution due to an isolated force.† Suppose a force $P = X+iY = Fe^{i\epsilon}$ acts at the point $z = b$ in the z-plane where, in general, b is complex and $|b| > a$. If we denote the corresponding points in the ζ_1- and ζ_2-planes

† See S. Holgate, *Proc. Camb. phil. Soc. math. phys. Sci.* **40** (1944) 172.

by b_1 and b_2 respectively, then, from (9.11.1),

$$\left.\begin{aligned} 2b_1 &= b+\gamma_1\bar{b}+\sqrt{\{(b+\gamma_1\bar{b})^2-4\gamma_1 a^2\}} \\ 2b_2 &= b+\gamma_2\bar{b}+\sqrt{\{(b+\gamma_2\bar{b})^2-4\gamma_2 a^2\}} \\ b+\gamma_1\bar{b} = b_1+\gamma_1 a^2/b_1, &\quad b+\gamma_2\bar{b} = b_2+\gamma_2 a^2/b_2 \end{aligned}\right\}, \tag{9.15.1}$$

the root signs being chosen so that $|b_1|$, $|b_2|$ both tend to infinity with $|b|$ and so that $|b_1| > a$, $|b_2| > a$. The appropriate complex potentials $\Omega(z_1)$, $\omega(z_2)$ for an isolated force at $z = b$ are, from (9.3.1), for an infinite plate, given by

$$\left.\begin{aligned} \Omega''(z_1) &= \frac{H}{2\pi(z_1-b-\gamma_1\bar{b})} \\ \omega''(z_2) &= \frac{K}{2\pi(z_2-b-\gamma_2\bar{b})} \end{aligned}\right\}. \tag{9.15.2}$$

The required values of the potentials $f_0(z)$, $g_0(z)$ must have singularities which give rise to an isolated force at the point $z = b$ in the plate and must be regular inside the circle $|z| = a$. From (9.11.1) and (9.15.2), since $f'(\zeta_1) = \Omega''(z_1)(dz_1/d\zeta_1)$, $g'(\zeta_2) = \omega''(z_2)(dz_2/d\zeta_2)$, we see that

$$\left.\begin{aligned} f_0'(\zeta_1) &= \frac{H}{2\pi(\zeta_1-b_1)} \\ g_0'(\zeta_2) &= \frac{K}{2\pi(\zeta_2-b_2)} \end{aligned}\right\}. \tag{9.15.3}$$

These complex potentials have the correct singularity at points corresponding to the isolated force in the z-plane and they have no singularities inside the circles $|\zeta_1| = a$, $|\zeta_2| = a$. The edge stress may therefore be evaluated at once from (9.15.3) and (9.12.9), where H, K are given by (9.1.15). For the simpler case of orthotropy we use (9.12.11) for $\sigma_{\theta\theta}$, and (9.2.2) for H, K but further discussion is left as an exercise.

9.16. First boundary-value problem for circular hole: part 2

We assume that the applied stresses at the edge of the circular hole take the prescribed forms

$$\sigma_{rr}+i\sigma_{r\theta} = p(t)+iq(t), \tag{9.16.1}$$

where $t = ae^{i\theta}$ is a point on the boundary of the hole and $p(t)$, $q(t)$ are real functions satisfying the H-condition. In the isotropic case we indicated two methods of solution for this problem. Here we give the solution in terms of two potential functions $V(z)$ and $W(z)$. An equivalent solution can be derived by the method of extending the definitions of the complex potentials $f(\zeta_1)$, $g(\zeta_2)$ to the whole plane.†

† This method is used in § 9.20 for an elliptical hole.

If the stresses and rotation vanish at infinity, then, from (9.11.1), (9.11.3), and (9.1.19)

$$\left.\begin{aligned} f'(\zeta_1) &= \frac{H}{2\pi\zeta_1}+O\left(\frac{1}{\zeta_1^2}\right) \\ g'(\zeta_2) &= \frac{K}{2\pi\zeta_2}+O\left(\frac{1}{\zeta_2^2}\right) \end{aligned}\right\}, \tag{9.16.2}$$

for large $|\zeta_1|$, $|\zeta_2|$. We now put

$$\left.\begin{aligned} V(z) &= -\frac{2z^2}{a^2}\left(1-\frac{\gamma_1 a^2}{z^2}\right)f'(z)-\frac{2z^2}{a^2}\left(1-\frac{\gamma_2 a^2}{z^2}\right)g'(z) \\ W(z) &= \frac{2z^2}{a^2}\left(1+\frac{\gamma_1 a^2}{z^2}\right)f'(z)+\frac{2z^2}{a^2}\left(1+\frac{\gamma_2 a^2}{z^2}\right)g'(z) \end{aligned}\right\}, \tag{9.16.3}$$

so that, from (9.11.7),

$$\operatorname{re} V(z) = p(t), \qquad \operatorname{im} W(z) = q(t), \tag{9.16.4}$$

on the circle $|z| = a$. If we solve equations (9.16.3) for $f'(z)$, $g'(z)$, we obtain

$$\left.\begin{aligned} 4(\gamma_1-\gamma_2)f'(z) &= (1+\gamma_2 a^2/z^2)V(z)+(1-\gamma_2 a^2/z^2)W(z) \\ 4(\gamma_1-\gamma_2)g'(z) &= -(1+\gamma_1 a^2/z^2)V(z)-(1-\gamma_1 a^2/z^2)W(z) \end{aligned}\right\}, \tag{9.16.5}$$

and using (9.16.2) we see that, for large $|z|$,

$$\left.\begin{aligned} V(z) &= -\frac{(H+K)z}{\pi a^2}+A+iA'+O\left(\frac{1}{z}\right) \\ W(z) &= \frac{(H+K)z}{\pi a^2}-A-iA'+O\left(\frac{1}{z}\right) \end{aligned}\right\}, \tag{9.16.6}$$

where A and A' are real constants. From (9.11.5), (9.11.7), and (9.16.5) we have

$$\begin{aligned} \sigma_{rr}+\sigma_{\theta\theta} = {} & \frac{\gamma_1\{(1+\gamma_2 a^2/\zeta_1^2)V(\zeta_1)+(1-\gamma_2 a^2/\zeta_1^2)W(\zeta_1)\}}{(\gamma_1-\gamma_2)(1-\gamma_1 a^2/\zeta_1^2)}+ \\ & +\frac{\bar{\gamma}_1\{(1+\bar{\gamma}_2 a^2/\bar{\zeta}_1^2)\overline{V}(\bar{\zeta}_1)+(1-\bar{\gamma}_2 a^2/\bar{\zeta}_1^2)\overline{W}(\bar{\zeta}_1)\}}{(\bar{\gamma}_1-\bar{\gamma}_2)(1-\bar{\gamma}_1 a^2/\bar{\zeta}_1^2)}- \\ & -\frac{\gamma_2\{(1+\gamma_1 a^2/\zeta_2^2)V(\zeta_2)+(1-\gamma_1 a^2/\zeta_2^2)W(\zeta_2)\}}{(\gamma_1-\gamma_2)(1-\gamma_2 a^2/\zeta_2^2)}- \\ & -\frac{\bar{\gamma}_2\{(1+\bar{\gamma}_1 a^2/\bar{\zeta}_2^2)\overline{V}(\bar{\zeta}_2)+(1-\bar{\gamma}_1 a^2/\bar{\zeta}_2^2)\overline{W}(\bar{\zeta}_2)\}}{(\bar{\gamma}_1-\bar{\gamma}_2)(1-\bar{\gamma}_2 a^2/\bar{\zeta}_2^2)}, \end{aligned} \tag{9.16.7}$$

$$\sigma_{rr}+i\sigma_{r\theta} = \frac{\gamma_1(1-\gamma_1\bar{z}/z)\{(1+\gamma_2 a^2/\zeta_1^2)V(\zeta_1)+(1-\gamma_2 a^2/\zeta_1^2)W(\zeta_1)\}}{2(\gamma_1-\gamma_2)(1-\gamma_1 a^2/\zeta_1^2)}+$$
$$+\frac{(\bar{\gamma}_1-\bar{z}/z)\{(1+\bar{\gamma}_2 a^2/\bar{\zeta}_1^2)\bar{V}(\bar{\zeta}_1)+(1-\bar{\gamma}_2 a^2/\bar{\zeta}_1^2)\bar{W}(\bar{\zeta}_1)\}}{2(\bar{\gamma}_1-\bar{\gamma}_2)(1-\bar{\gamma}_1 a^2/\bar{\zeta}_1^2)}-$$
$$-\frac{\gamma_2(1-\gamma_2\bar{z}/z)\{(1+\gamma_1 a^2/\zeta_2^2)V(\zeta_2)+(1-\gamma_1 a^2/\zeta_2^2)W(\zeta_2)\}}{2(\gamma_1-\gamma_2)(1-\gamma_2 a^2/\zeta_2^2)}-$$
$$-\frac{(\bar{\gamma}_2-\bar{z}/z)\{(1+\bar{\gamma}_1 a^2/\bar{\zeta}_2^2)\bar{V}(\bar{\zeta}_2)+(1-\bar{\gamma}_1 a^2/\bar{\zeta}_2^2)\bar{W}(\bar{\zeta}_2)\}}{2(\bar{\gamma}_1-\bar{\gamma}_2)(1-\bar{\gamma}_2 a^2/\bar{\zeta}_2^2)}. \quad (9.16.8)$$

At the edge of the circle $|z| = a$ these expressions for the stresses reduce to

$$\sigma_{\theta\theta} = \frac{(1+\gamma_1 e^{-2i\theta})(1+\gamma_2 e^{-2i\theta})V(z)+2(1-\gamma_1\gamma_2 e^{-4i\theta})W(z)}{2(1-\gamma_1 e^{-2i\theta})(1-\gamma_2 e^{-2i\theta})}+$$
$$+\frac{(1+\bar{\gamma}_1 e^{2i\theta})(1+\bar{\gamma}_2 e^{2i\theta})\bar{V}(\bar{z})+2(1-\bar{\gamma}_1\bar{\gamma}_2 e^{4i\theta})\bar{W}(\bar{z})}{2(1-\bar{\gamma}_1 e^{2i\theta})(1-\bar{\gamma}_2 e^{2i\theta})}, \quad (9.16.9)$$

$$\left.\begin{aligned}\sigma_{rr} &= \tfrac{1}{2}\{V(z)+\bar{V}(\bar{z})\}\\ \sigma_{r\theta} &= \frac{1}{2i}\{W(z)-\bar{W}(\bar{z})\}\end{aligned}\right\}. \quad (9.16.10)$$

When the material is the orthotropic material of § 9.2 equation (9.16.9) becomes

$$\sigma_{\theta\theta} = \frac{(1+\gamma_1 e^{-2i\theta})(1+\gamma_2 e^{-2i\theta})V(z)+2(1-\gamma_1\gamma_2 e^{-4i\theta})W(z)}{2(1-\gamma_1 e^{-2i\theta})(1-\gamma_2 e^{-2i\theta})}+$$
$$+\frac{(1+\gamma_1 e^{2i\theta})(1+\gamma_2 e^{2i\theta})\bar{V}(\bar{z})+2(1-\gamma_1\gamma_2 e^{4i\theta})\bar{W}(\bar{z})}{2(1-\gamma_1 e^{2i\theta})(1-\gamma_2 e^{2i\theta})}. \quad (9.16.11)$$

Expressions for $V(z)$, $W(z)$ can now be written down in a way similar to that used in § 8.19 for isotropic materials. Thus, remembering (9.16.4) and (9.16.6),

$$\left.\begin{aligned}V(z) &= -\frac{(H+K)z}{\pi a^2}+\frac{\bar{H}+\bar{K}}{\pi z}+\\ &\quad+\frac{1}{2\pi}\int_0^{2\pi}\frac{ze^{-i\gamma}+a}{ze^{-i\gamma}-a}p(ae^{i\gamma})\,d\gamma-\frac{i}{2\pi}\int_0^{2\pi}q(ae^{i\gamma})\,d\gamma\\ W(z) &= \frac{(H+K)z}{\pi a^2}+\frac{\bar{H}+\bar{K}}{\pi z}+\\ &\quad+\frac{i}{2\pi}\int_0^{2\pi}\frac{ze^{-i\gamma}+a}{ze^{-i\gamma}-a}q(ae^{i\gamma})\,d\gamma-\frac{1}{2\pi}\int_0^{2\pi}p(ae^{i\gamma})\,d\gamma\end{aligned}\right\}, \quad (9.16.12)$$

or, alternatively,

$$\left.\begin{aligned} V(z) &= -\frac{(H+K)z}{\pi a^2}+\frac{\bar{H}+\bar{K}}{\pi z}+\sum_{n=0}^{\infty}(R_n+iS_n)\frac{a^n}{z^n} \\ W(z) &= \frac{(H+K)z}{\pi a^2}+\frac{\bar{H}+\bar{K}}{\pi z}+\sum_{n=0}^{\infty}(-U_n+iT_n)\frac{a^n}{z^n} \end{aligned}\right\}, \qquad (9.16.13)$$

where R_n, S_n, U_n, T_n are defined in (8.19.23). We also recall that the applied stresses $p(t)$, $q(t)$ have the Fourier expansions (8.19.24), when they exist, so that

$$\left.\begin{aligned} p(ae^{i\theta}) &= R_0+\sum_{n=1}^{\infty}(R_n\cos n\theta+S_n\sin n\theta \\ q(ae^{i\theta}) &= T_0+\sum_{n=1}^{\infty}(T_n\cos n\theta+U_n\sin n\theta) \end{aligned}\right\}. \qquad (9.16.14)$$

We add the solution of some special problems in the next section.

9.17. Examples of applied stress at a circular hole

EXAMPLE 1. If the circular hole is acted on by uniform normal pressure p, then

$$R_0 = U_0 = -p, \qquad H = K = 0,$$

all other coefficients R_n, S_n, U_n, T_n in (9.16.14) being zero. Hence, from (9.16.13),

$$V(z) = -p, \qquad W(z) = p, \qquad (9.17.1)$$

and, when the material is orthotropic, the edge stress $\sigma_{\theta\theta}$ at the hole, which is given by (9.16.11), is

$$\frac{\sigma_{\theta\theta}}{p} = \frac{1+(\gamma_1+\gamma_2)^2-3\gamma_1^2\gamma_2^2-2(\gamma_1+\gamma_2)(1-\gamma_1\gamma_2)\cos 2\theta-2\gamma_1\gamma_2\cos 4\theta}{(1+\gamma_1^2-2\gamma_1\cos 2\theta)(1+\gamma_2^2-2\gamma_2\cos 2\theta)}. \qquad (9.17.2)$$

For an isotropic material, $\gamma_1 = \gamma_2 = 0$, and this reduces to the simple result $\sigma_{\theta\theta} = p$.

EXAMPLE 2. If a uniform shear stress is applied at the hole so that it produces a couple G in an anticlockwise direction at the edge of the hole, then

$$S_0 = -T_0 = \frac{G}{2\pi a^2}, \qquad H = K = 0,$$

all other coefficients R_n, S_n, U_n, T_n in (9.16.14) being zero. It follows from (9.16.13) that

$$V(z) = \frac{iG}{2\pi a^2}, \qquad W(z) = -\frac{iG}{2\pi a^2}, \qquad (9.17.3)$$

and, using (9.16.11), for an orthotropic material, the edge stress $\sigma_{\theta\theta}$ at the hole becomes

$$2\pi a^2\sigma_{\theta\theta} = \frac{4G\gamma_1\gamma_2\{\sin 4\theta-(\gamma_1+\gamma_2)\sin 2\theta\}}{(1+\gamma_1^2-2\gamma_1\cos 2\theta)(1+\gamma_2^2-2\gamma_2\cos 2\theta)}. \tag{9.17.4}$$

A numerical discussion of the formulae (9.17.2) and (9.17.4) has been given in a paper by Green.† This paper also contains other examples of stress distributions due to prescribed applied stresses at a circular hole.

9.18. Second boundary-value problem for circular hole

We first suppose that complex potentials $f_0(z)$, $g_0(z)$ define a given stress distribution which has the properties defined in § 9.12, and we wish to find complex potentials (9.12.2) so that total displacement at $|z| = a$ is zero. As in § 9.12 we define complex potentials $h_1(z)$, $k_1(z)$ by (9.12.3) and then extend the definitions of these functions to the whole plane by (9.12.4). These functions vanish at infinity and are finite at the origin. If D vanishes at the edge of the circle, it follows from (9.11.8), (9.11.17) and (9.12.2) that

$$\begin{aligned}\{(\delta_1-\delta_2)h_1(t)-(\gamma_2\delta_1-\gamma_1\delta_2)k_1(t)\}^+/\{2(\gamma_1-\gamma_2)\}-\\ -\{(\rho_1-\rho_2)k_1(t)-(\bar\gamma_2\rho_1-\bar\gamma_1\rho_2)h_1(t)\}^-/\{2(\bar\gamma_1-\bar\gamma_2)\}\\ = -\delta_1 f_0(t)-\rho_1\bar f_0(a^2/t)-\delta_2 g_0(t)-\rho_2\bar g_0(a^2/t).\end{aligned} \tag{9.18.1}$$

Similarly, since $\bar D$ also vanishes at the hole,

$$\begin{aligned}\{(\bar\rho_1-\bar\rho_2)h_1(t)-(\gamma_2\bar\rho_1-\gamma_1\bar\rho_2)k_1(t)\}^+/\{2(\gamma_1-\gamma_2)\}-\\ -\{(\bar\delta_1-\bar\delta_2)k_1(t)-(\bar\gamma_2\bar\delta_1-\bar\gamma_1\bar\delta_2)h_1(t)\}^-/\{2(\bar\gamma_1-\bar\gamma_2)\}\\ = -\bar\delta_1\bar f_0(a^2/t)-\bar\rho_1 f_0(t)-\bar\delta_2\bar g_0(a^2/t)-\bar\rho_2 g_0(t),\end{aligned} \tag{9.18.2}$$

where $t = ae^{i\theta}$ denotes any point on the circle $|z| = a$.

These equations may be solved in the usual manner to give, with the help of (9.12.3), for the region $|z| \geqslant a$,

$$\left.\begin{aligned}\{(\delta_1-\delta_2)h_1(z)-(\gamma_2\delta_1-\gamma_1\delta_2)k_1(z)\}/\{2(\gamma_1-\gamma_2)\}\\ = \delta_1 f_1(z)+\delta_2 g_1(z) = -\rho_1\bar f_0(a^2/z)-\rho_2\bar g_0(a^2/z)\\ \{(\bar\rho_1-\bar\rho_2)h_1(z)-(\gamma_2\bar\rho_1-\gamma_1\bar\rho_2)k_1(z)\}/\{2(\gamma_1-\gamma_2)\}\\ = \bar\rho_1 f_1(z)+\bar\rho_2 g_1(z) = -\bar\delta_1\bar f_0(a^2/z)-\bar\delta_2\bar g_0(a^2/z)\end{aligned}\right\}, \tag{9.18.3}$$

apart from additional constants. Hence,

$$\left.\begin{aligned}(\delta_1\bar\rho_2-\delta_2\bar\rho_1)f_1(z) = (\delta_2\bar\delta_1-\rho_1\bar\rho_2)\bar f_0(a^2/z)+(\delta_2\bar\delta_2-\rho_2\bar\rho_2)\bar g_0(a^2/z)\\ (\delta_1\bar\rho_2-\delta_2\bar\rho_1)g_1(z) = (\rho_1\bar\rho_1-\delta_1\bar\delta_1)\bar f_0(a^2/z)+(\rho_2\bar\rho_1-\delta_1\bar\delta_2)\bar g_0(a^2/z)\end{aligned}\right\}. \tag{9.18.4}$$

† loc. cit., p. 350.

The complete stress potentials are now given by (9.18.4) and (9.12.2) for $|z| \geqslant a$ and the corresponding stresses and displacements follow from (9.11.5)–(9.11.8).

For an orthotropic material the formulae (9.18.4) simplify with the help of (9.2.1), so that, adding the original complex potentials $f_0(z)$, $g_0(z)$, we see that the total complex potentials (9.12.2) become

$$\left.\begin{aligned} f(z) &= f_0(z) - \frac{\alpha_1\beta_2^2+\alpha_2\beta_1^2}{\alpha_1\beta_2^2-\alpha_2\beta_1^2}\bar{f}_0\left(\frac{a^2}{z}\right) - \frac{(1+\gamma_2)(\alpha_1+1)\beta_1\beta_2}{\alpha_1\beta_2^2-\alpha_2\beta_1^2}\bar{g}_0\left(\frac{a^2}{z}\right) \\ g(z) &= g_0(z) + \frac{(1+\gamma_1)(\alpha_2+1)\beta_1\beta_2}{\alpha_1\beta_2^2-\alpha_2\beta_1^2}\bar{f}_0\left(\frac{a^2}{z}\right) + \frac{\alpha_1\beta_2^2+\alpha_2\beta_1^2}{\alpha_1\beta_2^2-\alpha_2\beta_1^2}\bar{g}_0\left(\frac{a^2}{z}\right) \end{aligned}\right\}, \tag{9.18.5}$$

for $|z| \geqslant a$. Again, the corresponding stresses and displacements follow from (9.11.5)–(9.11.8), remembering now that γ_1, γ_2, δ_1, δ_2, ρ_1, ρ_2 are real.

To complete the second boundary-value problem for a circle we suppose that the cartesian components of displacement take the prescribed values

$$u_x = r(t), \qquad u_y = s(t) \quad (t = ae^{i\theta}), \tag{9.18.6}$$

at the edge of the hole, where $r(t)$, $s(t)$ are real functions. We assume that $r(t)$, $s(t)$ have first derivatives $r'(t)$, $s'(t)$ satisfying the H-condition on the circle $|z| = a$. The stresses and displacements are derived from complex potentials $f(\zeta_1)$, $g(\zeta_2)$, so that, from (9.11.1), (9.11.3), and (9.1.19),

$$\left.\begin{aligned} f'(\zeta_1) &= \frac{H}{2\pi\zeta_1} + O\left(\frac{1}{\zeta_1^2}\right) \\ g'(\zeta_2) &= \frac{K}{2\pi\zeta_2} + O\left(\frac{1}{\zeta_2^2}\right) \end{aligned}\right\}, \tag{9.18.7}$$

for large $|\zeta_1|$, $|\zeta_2|$, if stresses and rotation vanish at infinity. If we define complex potentials $h(z)$, $k(z)$ by (9.11.9) for $|z| \geqslant a$, then it follows that

$$\left.\begin{aligned} h'(z) &= \frac{\gamma_1 H + \gamma_2 K}{\pi z} + O\left(\frac{1}{z^2}\right) \\ k'(z) &= \frac{H+K}{\pi z} + O\left(\frac{1}{z^2}\right) \end{aligned}\right\}, \tag{9.18.8}$$

for large $|z|$. We now define $h(z)$, $k(z)$ for the region $|z| < a$ by (9.11.13) so that

$$\left.\begin{aligned} h'(z) &= \frac{a^2}{z^2}\bar{k}'\left(\frac{a^2}{z}\right) \\ k'(z) &= \frac{a^2}{z^2}\bar{h}'\left(\frac{a^2}{z}\right) \end{aligned}\right\} \quad (|z| < a), \tag{9.18.9}$$

and therefore, when $|z|$ is small,

$$\left.\begin{aligned} h'(z) &= \frac{\bar{H}+\bar{K}}{\pi z}+O(1) \\ k'(z) &= \frac{\bar{\gamma}_1\bar{H}+\bar{\gamma}_2\bar{K}}{\pi z}+O(1) \end{aligned}\right\}. \tag{9.18.10}$$

If we now obtain $\partial D/\partial\theta$ and $\partial\bar{D}/\partial\theta$ from the formulae (9.11.17) and (9.11.18) and use the boundary condition (9.18.6), we find, under conditions suitable for the existence of various limits, that

$$\begin{aligned} &\{(\delta_1-\delta_2)h'(t)-(\gamma_2\delta_1-\gamma_1\delta_2)k'(t)\}^+/\{2(\gamma_1-\gamma_2)\}- \\ &\quad -\{(\rho_1-\rho_2)k'(t)-(\rho_1\bar{\gamma}_2-\rho_2\bar{\gamma}_1)h'(t)\}^-/\{2(\bar{\gamma}_1-\bar{\gamma}_2)\} \\ &\qquad = r'(t)+is'(t), \end{aligned} \tag{9.18.11}$$

$$\begin{aligned} &\{(\bar{\rho}_1-\bar{\rho}_2)h'(t)-(\gamma_2\bar{\rho}_1-\gamma_1\bar{\rho}_2)k'(t)\}^+/\{2(\gamma_1-\gamma_2)\}- \\ &\quad -\{(\bar{\delta}_1-\bar{\delta}_2)k'(t)-(\bar{\gamma}_2\bar{\delta}_1-\bar{\gamma}_1\bar{\delta}_2)h'(t)\}^-/\{2(\bar{\gamma}_1-\bar{\gamma}_2)\} \\ &\qquad = r'(t)-is'(t). \end{aligned} \tag{9.18.12}$$

Remembering (9.18.10), we can solve these equations by the usual methods, and we obtain, for the region $|z| \geqslant a$,

$$\begin{aligned} &\{(\delta_1-\delta_2)h'(z)-(\gamma_2\delta_1-\gamma_1\delta_2)k'(z)\}/\{2(\gamma_1-\gamma_2)\} \\ &\qquad = \delta_1 f'(z)+\delta_2 g'(z) \\ &\qquad = \frac{\rho_1\bar{H}+\rho_2\bar{K}}{2\pi z}+\frac{1}{2\pi i}\int\limits_L \frac{r'(t)+is'(t)}{t-z}\,dt, \end{aligned} \tag{9.18.13}$$

$$\begin{aligned} &\{(\bar{\rho}_1-\bar{\rho}_2)h'(z)-(\gamma_2\bar{\rho}_1-\gamma_1\bar{\rho}_2)k'(z)\}/\{2(\gamma_1-\gamma_2)\} \\ &\qquad = \bar{\rho}_1 f'(z)+\bar{\rho}_2 g'(z) \\ &\qquad = \frac{\bar{\delta}_1\bar{H}+\bar{\delta}_2\bar{K}}{2\pi z}+\frac{1}{2\pi i}\int\limits_L \frac{r'(t)-is'(t)}{t-z}\,dt, \end{aligned} \tag{9.18.14}$$

the integrals being taken clockwise around the circle $|z| = a$. Hence, solving these equations for $f'(z)$, $g'(z)$ we have, for $|z| \geqslant a$,

$$\begin{aligned} (\delta_1\bar{\rho}_2-\delta_2\bar{\rho}_1)f'(z) &= \frac{(\rho_1\bar{\rho}_2-\delta_2\bar{\delta}_1)\bar{H}+(\rho_2\bar{\rho}_2-\delta_2\bar{\delta}_2)\bar{K}}{2\pi z}+ \\ &\quad +\frac{1}{2\pi i}\int\limits_L [\bar{\rho}_2\{r'(t)+is'(t)\}-\delta_2\{r'(t)-is'(t)\}]\frac{dt}{t-z}, \end{aligned} \tag{9.18.15}$$

$$\begin{aligned} (\delta_1\bar{\rho}_2-\delta_2\bar{\rho}_1)g'(z) &= \frac{(\delta_1\bar{\delta}_1-\rho_1\bar{\rho}_1)\bar{H}+(\delta_1\bar{\delta}_2-\rho_2\bar{\rho}_1)\bar{K}}{2\pi z}+ \\ &\quad +\frac{1}{2\pi i}\int\limits_L [\delta_1\{r'(t)-is'(t)\}-\bar{\rho}_1\{r'(t)+is'(t)\}]\frac{dt}{t-z}. \end{aligned} \tag{9.18.16}$$

Using (9.18.15) and (9.18.16) the stresses and displacements can be evaluated from (9.11.5)–(9.11.8).

We consider, briefly, one example. If a uniform radial displacement c is prescribed at the hole $|z| = a$, then

$$r(t)+is(t) = ce^{i\theta} = ct/a,$$

$$r(t)-is(t) = ce^{-i\theta} = ca/t,$$

and

$$r'(t)+is'(t) = c/a,$$

$$r'(t)-is'(t) = -ca/t^2.$$

Hence

$$\left.\begin{aligned} \int\limits_L \frac{r'(t)+is'(t)}{t-z}\,dt &= 0 \\ \frac{1}{2\pi i}\int\limits_L \frac{r'(t)-is'(t)}{t-z}\,dt &= -\frac{ca}{z^2} \end{aligned}\right\} \quad (|z| \geqslant a),$$

and since the resultant force at the hole is zero, $H = K = 0$. It follows that

$$\left.\begin{aligned} f'(z) &= \frac{ca\delta_2}{(\delta_1\bar{\rho}_2-\delta_2\bar{\rho}_1)z^2} \\ g'(z) &= -\frac{ca\delta_1}{(\delta_1\bar{\rho}_2-\delta_2\bar{\rho}_1)z^2} \end{aligned}\right\}. \tag{9.18.17}$$

For the special case of orthotropic material these potentials reduce, with the help of § 9.2, to

$$\left.\begin{aligned} f'(z) &= \frac{ca(\alpha_2\beta_1-\beta_2)(\alpha_1+1)}{4(\alpha_1\beta_2^2-\alpha_2\beta_1^2)z^2} \\ g'(z) &= -\frac{ca(\alpha_1\beta_2-\beta_1)(\alpha_2+1)}{4(\alpha_1\beta_2^2-\alpha_2\beta_1^2)z^2} \end{aligned}\right\}. \tag{9.18.18}$$

From (9.11.7) and (9.18.18) we find that at the edge of the circular hole

$$\left.\begin{aligned} \sigma_{rr} &= \frac{c}{2a(\alpha_1\beta_2^2-\alpha_2\beta_1^2)}\{\beta_2(\alpha_1\alpha_2+2\alpha_1+1)-\beta_1(\alpha_1\alpha_2+2\alpha_2+1)+ \\ &\qquad\qquad +(\alpha_1\alpha_2-1)(\beta_1-\beta_2)\cos 2\theta\} \\ \sigma_{r\theta} &= \frac{c(1-\alpha_1\alpha_2)(\beta_1-\beta_2)\sin 2\theta}{2a(\alpha_1\beta_2^2-\alpha_2\beta_1^2)} \end{aligned}\right\}, \tag{9.18.19}$$

and, from (9.11.5),

$$\sigma_{rr}+\sigma_{\theta\theta} = \frac{2c}{a(\alpha_1\beta_2^2-\alpha_2\beta_1^2)}\times$$
$$\times\left\{\frac{(\alpha_1-1)(\alpha_2\beta_1-\beta_2)(\cos 2\theta-\gamma_1)}{1+\gamma_1^2-2\gamma_1\cos 2\theta}-\frac{(\alpha_2-1)(\alpha_1\beta_2-\beta_1)(\cos 2\theta-\gamma_2)}{1+\gamma_2^2-2\gamma_2\cos 2\theta}\right\}. \tag{9.18.20}$$

These results have been discussed numerically in a paper by Green.† This paper also contains the solution of a variety of other problems concerned with displacements and stresses at a circular hole in an orthotropic plank.

9.19. Elliptical hole in infinite plate

If an infinite region is bounded internally by an elliptical hole, we consider the transformation

$$z = c\sigma+d/\sigma, \qquad \sigma = e^{\zeta}, \qquad \zeta = \xi+i\eta, \tag{9.19.1}$$

where c and d are real and positive and $c > d$. This transforms an ellipse with semi-axes (a, b) along the axes of coordinates in the z-plane, into a circle $|\sigma| = 1$ (or $\xi = 0$) in the σ-plane, where

$$\left.\begin{aligned} a = c+d = k\cosh\epsilon, \qquad b = c-d = k\sinh\epsilon \\ 2c = ke^{\epsilon}, \qquad 2d = ke^{-\epsilon} \end{aligned}\right\}. \tag{9.19.2}$$

Also the outside of the circle in the σ-plane transforms into the outside of the ellipse in the z-plane. We define two other complex variables σ_1, σ_2 by the transformations

$$\left.\begin{aligned} z_1 = z+\gamma_1\bar{z} = (c+\gamma_1 d)\sigma_1+(d+\gamma_1 c)/\sigma_1 \\ z_2 = z+\gamma_2\bar{z} = (c+\gamma_2 d)\sigma_2+(d+\gamma_2 c)/\sigma_2 \end{aligned}\right\}, \tag{9.19.3}$$

and the σ_1-, σ_2-planes are chosen so that the circles $|\sigma_1| = 1$, $|\sigma_2| = 1$ correspond to the circle $|\sigma| = 1$ in the σ-plane, and $|\sigma_1| \to \infty$, $|\sigma_2| \to \infty$ when $|z| \to \infty$. Also, since $|\gamma_1| < 1$, $|\gamma_2| < 1$, and $c > d$, the singularities of the transformation‡ from the z_1- to the σ_1-plane, and from the z_2- to the σ_2-plane, are inside the circles $|\sigma_1| = 1$, $|\sigma_2| = 1$ respectively. For convenience we may imagine that the three circles $|\sigma| = 1$, $|\sigma_1| = 1$, and $|\sigma_2| = 1$ coincide and are given by

$$\sigma_1 = \sigma_2 = \sigma = e^{i\eta}, \qquad \bar{\sigma} = 1/\sigma. \tag{9.19.4}$$

† *Proc. R. Soc.* A184 (1945) 301.
‡ i.e. points at which $z_1'(\sigma_1) = 0$, $z_2'(\sigma_2) = 0$.

To every point in the σ_1-plane, on or outside the circle (9.19.4), corresponds one point in the (x,y) plane, but the transformation between these planes is not conformal. A similar correspondence exists between the σ_2-plane and the (x,y) plane. If we put

$$\Omega'(z_1) = f(\sigma_1), \qquad \omega'(z_2) = g(\sigma_2), \tag{9.19.5}$$

then, from the formulae (9.1.4)–(9.1.8), and (9.1.10),

$$P = 2i\{\gamma_1 f(\sigma_1)+\bar{f}(\bar{\sigma}_1)+\gamma_2 g(\sigma_2)+\bar{g}(\bar{\sigma}_2)\}, \tag{9.19.6}$$

$$\sigma_{\xi\xi}+\sigma_{\eta\eta} = \frac{4\gamma_1 f'(\sigma_1)}{c+\gamma_1 d-(d+\gamma_1 c)/\sigma_1^2}+\frac{4\bar{\gamma}_1\bar{f}'(\bar{\sigma}_1)}{c+\bar{\gamma}_1 d-(d+\bar{\gamma}_1 c)/\bar{\sigma}_1^2}+$$
$$+\frac{4\gamma_2 g'(\sigma_2)}{c+\gamma_2 d-(d+\gamma_2 c)/\sigma_2^2}+\frac{4\bar{\gamma}_2\bar{g}'(\bar{\sigma}_2)}{c+\bar{\gamma}_2 d-(d+\bar{\gamma}_2 c)/\bar{\sigma}_2^2}, \tag{9.19.7}$$

$$D = \delta_1 f(\sigma_1)+\rho_1\bar{f}(\bar{\sigma}_1)+\delta_2 g(\sigma_2)+\rho_2\bar{g}(\bar{\sigma}_2). \tag{9.19.8}$$

Following the plan of § 9.11 we define $h(\sigma)$, $k(\sigma)$ by the formulae

$$\left.\begin{aligned} h(\sigma) &= 2\gamma_1 f(\sigma)+2\gamma_2 g(\sigma)\\ k(\sigma) &= 2f(\sigma)+2g(\sigma)\end{aligned}\right\}, \tag{9.19.9}$$

so that

$$\left.\begin{aligned} 2(\gamma_1-\gamma_2)f(\sigma) &= h(\sigma)-\gamma_2 k(\sigma)\\ 2(\gamma_1-\gamma_2)g(\sigma) &= -h(\sigma)+\gamma_1 k(\sigma)\end{aligned}\right\}. \tag{9.19.10}$$

Then, from (9.19.6) and (9.19.9),

$$P = i\{\gamma_1 h(\sigma_1)-\gamma_1\gamma_2 k(\sigma_1)-\gamma_2 h(\sigma_2)+\gamma_1\gamma_2 k(\sigma_2)\}/(\gamma_1-\gamma_2)+$$
$$+i\{\bar{h}(\bar{\sigma}_1)-\bar{\gamma}_2\bar{k}(\bar{\sigma}_1)-\bar{h}(\bar{\sigma}_2)+\bar{\gamma}_1\bar{k}(\bar{\sigma}_2)\}/(\bar{\gamma}_1-\bar{\gamma}_2), \tag{9.19.11}$$

$$\bar{P} = -i\{h(\sigma_1)-\gamma_2 k(\sigma_1)-h(\sigma_2)+\gamma_1 k(\sigma_2)\}/(\gamma_1-\gamma_2)-$$
$$-i\{\bar{\gamma}_1\bar{h}(\bar{\sigma}_1)-\bar{\gamma}_1\bar{\gamma}_2\bar{k}(\bar{\sigma}_1)-\bar{\gamma}_2\bar{h}(\bar{\sigma}_2)+\bar{\gamma}_1\bar{\gamma}_2\bar{k}(\bar{\sigma}_2)\}/(\bar{\gamma}_1-\bar{\gamma}_2). \tag{9.19.12}$$

The functions $f(\sigma_1)$, $g(\sigma_2)$ are defined and holomorphic at every point outside the circles $\sigma_1 = \sigma_2 = e^{i\eta}$ except possibly at infinity, so that $h(\sigma)$, $k(\sigma)$ are defined and holomorphic at every point outside the circle $\sigma = e^{i\eta}$, except possibly at infinity. We now define

$$\left.\begin{aligned} h(\sigma) &= -\bar{k}(1/\sigma)\\ k(\sigma) &= -\bar{h}(1/\sigma)\end{aligned}\right\} \quad (|\sigma|<1), \tag{9.19.13}$$

and, hence,

$$\left.\begin{aligned} h(1/\bar{\sigma}) &= -\bar{k}(\bar{\sigma})\\ k(1/\bar{\sigma}) &= -\bar{h}(\bar{\sigma})\end{aligned}\right\} \quad (|\sigma|>1). \tag{9.19.14}$$

With the help of (9.19.14) the formulae (9.19.11) and (9.19.12) become

$$P = i[\gamma_1 h(\sigma_1)-\gamma_2 h(\sigma_2)+\gamma_1\gamma_2\{k(\sigma_2)-k(\sigma_1)\}]/(\gamma_1-\gamma_2)+$$
$$+i\left\{k\left(\frac{1}{\bar{\sigma}_2}\right)-k\left(\frac{1}{\bar{\sigma}_1}\right)+\bar{\gamma}_2 h\left(\frac{1}{\bar{\sigma}_1}\right)-\bar{\gamma}_1 h\left(\frac{1}{\bar{\sigma}_2}\right)\right\}\Big/(\bar{\gamma}_1-\bar{\gamma}_2), \quad (9.19.15)$$

$$\bar{P} = -i\{h(\sigma_1)-h(\sigma_2)+\gamma_1 k(\sigma_2)-\gamma_2 k(\sigma_1)\}/(\gamma_1-\gamma_2)-$$
$$-i\left[\bar{\gamma}_2 k\left(\frac{1}{\bar{\sigma}_2}\right)-\bar{\gamma}_1 k\left(\frac{1}{\bar{\sigma}_1}\right)+\bar{\gamma}_1\bar{\gamma}_2\left\{h\left(\frac{1}{\bar{\sigma}_1}\right)-h\left(\frac{1}{\bar{\sigma}_2}\right)\right\}\right]\Big/(\bar{\gamma}_1-\bar{\gamma}_2). \quad (9.19.16)$$

9.20. First boundary-value problem for elliptical hole

We first suppose that complex potentials $f_0(\sigma)$, $g_0(\sigma)$ define stresses and displacements for a plate which does not contain an elliptical hole, the singularities of $f_0(\sigma)$, $g_0(\sigma)$ being only in the region $|\sigma| > 1$. We add a stress system derived from potentials $f_1(\sigma)$, $g_1(\sigma)$ which is such that the combined potentials

$$f(\sigma) = f_0(\sigma)+f_1(\sigma), \qquad g(\sigma) = g_0(\sigma)+g_1(\sigma), \quad (9.20.1)$$

give zero applied stresses at the boundary $|\sigma| = 1$, and, for large $|\sigma|$, $f_1(\sigma)$, $g_1(\sigma)$ both vanish at infinity. The evaluation of $f_1(\sigma)$, $g_1(\sigma)$ follows exactly the analysis of § 9.12 if we replace z, a in § 9.12 by σ, 1 respectively. Hence, from (9.12.8), the total complex potentials for the region $|\sigma| \geqslant 1$ are

$$\left.\begin{aligned} f(\sigma) &= f_0(\sigma)-\{(1-\bar{\gamma}_1\gamma_2)\bar{f}_0(1/\sigma)+(1-\gamma_2\bar{\gamma}_2)\bar{g}_0(1/\sigma)\}/(\gamma_1-\gamma_2)\\ g(\sigma) &= g_0(\sigma)+\{(1-\gamma_1\bar{\gamma}_1)\bar{f}_0(1/\sigma)+(1-\gamma_1\bar{\gamma}_2)\bar{g}_0(1/\sigma)\}/(\gamma_1-\gamma_2)\end{aligned}\right\}. \quad (9.20.2)$$

The stress at the edge of the elliptical hole is obtained from (9.19.7) and (9.20.2). Thus, since $\sigma_{\xi\xi} = 0$ when $\xi = 0$,

$$\begin{aligned} &T_1 T_2\,\sigma_{\eta\eta}/(c^2+d^2-2cd\cos 2\eta)\\ &= 4(1-\gamma_1\bar{\gamma}_1)[(1-\gamma_1\bar{\gamma}_2)\{(c+\gamma_2 d)e^{2i\eta}-(d+\gamma_2 c)\}f_0'(\sigma)+\\ &\quad +(1-\bar{\gamma}_1\gamma_2)\{(c+\bar{\gamma}_2 d)e^{-2i\eta}-(d+\bar{\gamma}_2 c)\}\bar{f}_0'(\bar{\sigma})]+\\ &\quad +4(1-\gamma_2\bar{\gamma}_2)[(1-\bar{\gamma}_1\gamma_2)\{(c+\gamma_1 d)e^{2i\eta}-(d+\gamma_1 c)\}g_0'(\sigma)+\\ &\quad +(1-\gamma_1\bar{\gamma}_2)\{(c+\bar{\gamma}_1 d)e^{-2i\eta}-(d+\bar{\gamma}_1 c)\}\bar{g}_0'(\bar{\sigma})], \end{aligned} \quad (9.20.3)$$

where $\sigma = e^{i\eta}$ and

$$\left.\begin{aligned} T_1 &= (1+\gamma_1\bar{\gamma}_1)(c^2+d^2)+2(\gamma_1+\bar{\gamma}_1)cd-(\gamma_1+\bar{\gamma}_1)(c^2+d^2)\cos 2\eta-\\ &\quad -2(1+\gamma_1\bar{\gamma}_1)cd\cos 2\eta+i(\gamma_1-\bar{\gamma}_1)(c^2-d^2)\sin 2\eta\\ T_2 &= (1+\gamma_2\bar{\gamma}_2)(c^2+d^2)+2(\gamma_2+\bar{\gamma}_2)cd-(\gamma_2+\bar{\gamma}_2)(c^2+d^2)\cos 2\eta-\\ &\quad -2(1+\gamma_2\bar{\gamma}_2)cd\cos 2\eta+i(\gamma_2-\bar{\gamma}_2)(c^2-d^2)\sin 2\eta \end{aligned}\right\}, \quad (9.20.4)$$

When the material is orthotropic these formulae reduce to

$$[(1+\gamma_1^2)(c^2+d^2)+4\gamma_1 cd-2\{\gamma_1(c^2+d^2)+(1+\gamma_1^2)cd\}\cos 2\eta]\times$$
$$\times[(1+\gamma_2^2)(c^2+d^2)+4\gamma_2 cd-2\{\gamma_2(c^2+d^2)+(1+\gamma_2^2)cd\}\cos 2\eta]\times$$
$$\times\sigma_{\eta\eta}/(c^2+d^2-2cd\cos 2\eta)$$
$$= 4(1-\gamma_1\gamma_2)(1-\gamma_1^2)[\{(c+\gamma_2 d)e^{2i\eta}-(d+\gamma_2 c)\}f_0'(\sigma)+$$
$$+\{(c+\gamma_2 d)e^{-2i\eta}-(d+\gamma_2 c)\}\bar{f}_0'(\bar{\sigma})]+$$
$$+4(1-\gamma_1\gamma_2)(1-\gamma_2^2)[\{(c+\gamma_1 d)e^{2i\eta}-(d+\gamma_1 c)\}g_0'(\sigma)+$$
$$+\{(c+\gamma_1 d)e^{-2i\eta}-(d+\gamma_1 c)\}\bar{g}_0'(\bar{\sigma})]. \quad (9.20.5)$$

If a plate containing an elliptical hole is subject to uniform stress at infinity, defined, as in § 9.1, by N_1, N_2, and α, then, from (9.1.14), for large $|z_1|$, $|z_2|$,

$$\Omega''(z_1) = B+iC, \qquad \omega''(z_2) = B'+iC',$$

where B, C, B', C' are given by equations (9.1.16)–(9.1.18). Hence, from (9.19.3) and (9.19.5), we see that the appropriate values for $f_0'(\sigma)$, $g_0'(\sigma)$ are

$$f_0'(\sigma) = (B+iC)(c+\gamma_1 d), \qquad g_0'(\sigma) = (B'+iC')(c+\gamma_2 d), \quad (9.20.6)$$

since these functions have no singularities in $|\sigma| < 1$ and take the correct values for large $|\sigma|$. The edge stress at the elliptical hole is now given by (9.20.6) and (9.20.3). In the case of orthotropy we may use (9.20.5), where B, C, B', C' are now given by (9.2.4). Hence

$$[(1+\gamma_1^2)(c^2+d^2)+4\gamma_1 cd-2\{\gamma_1(c^2+d^2)+(1+\gamma_1^2)cd\}\cos 2\eta]\times$$
$$\times[(1+\gamma_2^2)(c^2+d^2)+4\gamma_2 cd-2\{\gamma_2(c^2+d^2)+(1+\gamma_2^2)cd\}\cos 2\eta]\times$$
$$\times\sigma_{\eta\eta}/(c^2+d^2-2cd\cos 2\eta)$$
$$= (N_1+N_2)(1+\gamma_1)(1+\gamma_2)\{c^2(1+\gamma_1+\gamma_2-\gamma_1\gamma_2)+2cd(1+\gamma_1\gamma_2)-$$
$$-d^2(1-\gamma_1-\gamma_2-\gamma_1\gamma_2)-2(c+\gamma_1 d)(c+\gamma_2 d)\cos 2\eta\}-$$
$$-4\{(\gamma_1+\gamma_2)(c^2+\gamma_1\gamma_2 d^2)+cd(1+\gamma_1\gamma_2)^2-$$
$$-(1+\gamma_1\gamma_2)(c+\gamma_1 d)(c+\gamma_2 d)\cos 2\eta\}(N_1\sin^2\alpha+N_2\cos^2\alpha)+$$
$$+2(\gamma_1\gamma_2-1)(N_1-N_2)(c+\gamma_1 d)(c+\gamma_2 d)\sin 2\alpha\sin 2\eta. \quad (9.20.7)$$

We observe that when $d = 0$ this formula reduces to (9.13.2) for a circular hole. Also, when $\gamma_1 = \gamma_2 = 0$ we recover the formula (8.27.10) for an elliptical hole in an isotropic plate, if $\beta = 0$ and we remember

(9.19.2). Numerical discussion of special cases of (9.20.7) has been given by A. E. Green† and C. B. Smith.‡

The first boundary-value problem for an elliptical hole is completed by considering the case of prescribed stresses at the hole of the form

$$\sigma_{\xi\xi}+i\sigma_{\xi\eta} = p(t)+iq(t) \quad (t = e^{i\eta}), \tag{9.20.8}$$

where $p(t)$, $q(t)$ are real functions. We may find the resulting stress distribution by introducing two potential functions as in § 9.16 for the corresponding problem for the circular hole. Here, however, we use the second method and suppose that the stresses are derived from complex potentials $f(\sigma)$, $g(\sigma)$ defined in § 9.19. If stresses and rotation vanish at infinity, it follows from (9.19.1), (9.19.3), (9.19.5), and (9.1.19) that, for large $|\sigma|$,

$$\left.\begin{aligned} f'(\sigma) &= \frac{H}{2\pi\sigma}+O\left(\frac{1}{\sigma^2}\right) \\ g'(\sigma) &= \frac{K}{2\pi\sigma}+O\left(\frac{1}{\sigma^2}\right) \end{aligned}\right\}, \tag{9.20.9}$$

and, hence, from (9.19.9),

$$\left.\begin{aligned} h'(\sigma) &= \frac{\gamma_1 H+\gamma_2 K}{\pi\sigma}+O\left(\frac{1}{\sigma^2}\right) \\ k'(\sigma) &= \frac{H+K}{\pi\sigma}+O\left(\frac{1}{\sigma^2}\right) \end{aligned}\right\}, \tag{9.20.10}$$

for large $|\sigma|$. If we define $h(\sigma)$, $k(\sigma)$ for the region $|\sigma| < 1$ by (9.19.13), it follows that

$$\left.\begin{aligned} h'(\sigma) &= \frac{\bar{k}'(1/\sigma)}{\sigma^2} \\ k'(\sigma) &= \frac{\bar{h}'(1/\sigma)}{\sigma^2} \end{aligned}\right\} \quad (|\sigma| < 1), \tag{9.20.11}$$

and therefore, when $|\sigma|$ is small,

$$\left.\begin{aligned} h'(\sigma) &= \frac{\bar{H}+\bar{K}}{\pi\sigma}+O(1) \\ k'(\sigma) &= \frac{\bar{\gamma}_1\bar{H}+\bar{\gamma}_2\bar{K}}{\pi\sigma}+O(1) \end{aligned}\right\}. \tag{9.20.12}$$

From (9.1.11), (9.19.1), and (9.20.8),

$$\left.\begin{aligned} P(t) &= i\int^{t} \{p(t)+iq(t)\}(c-d/t^2)\,dt \\ \bar{P}(t) &= -i\int^{t} \{p(t)-iq(t)\}(d-c/t^2)\,dt \end{aligned}\right\} \quad (t = e^{i\eta}), \tag{9.20.13}$$

† *Proc. R. Soc.* **A184** (1945) 231.

‡ *U.S. Dept. Agric., For. Prod. Lab.* Mimeo. **1510** (1944) 40.

and hence, from (9.19.15) and (9.19.16), under suitable conditions,

$$\left.\begin{aligned} h^+(t)-h^-(t) &= -iP(t)\\ k^+(t)-k^-(t) &= i\bar{P}(t)\end{aligned}\right\}. \tag{9.20.14}$$

The solution of this boundary-value problem can be obtained in a convenient form when the prescribed forces at the hole are self-equilibrating so that $H = K = 0$, and $h(\sigma)$, $k(\sigma)$ vanish at infinity and are finite at the origin. For this case

$$\left.\begin{aligned} h(\sigma) &= -\frac{1}{2\pi}\int\limits_L \frac{P(t)\,dt}{t-\sigma}\\ k(\sigma) &= \frac{1}{2\pi}\int\limits_L \frac{\bar{P}(t)\,dt}{t-\sigma}\end{aligned}\right\}, \tag{9.20.15}$$

the integrals being taken clockwise around the unit circle $t = e^{i\eta}$.

If the prescribed forces at the edge of the hole are not self-equilibrating, then H, K are not zero and we must use the results (9.20.10) and (9.20.12). Also

$$\frac{\partial P(t)}{\partial \eta} = -(ct-d/t)\{p(t)+iq(t)\},$$

$$\frac{\partial \bar{P}(t)}{\partial \eta} = (dt-c/t)\{p(t)-iq(t)\},$$

on the circle $t = e^{i\eta}$, and from (9.19.15) and (9.19.16) we now have, under suitable conditions,

$$\left.\begin{aligned} [h'(t)]^+-[h'(t)]^- &= (c-d/t^2)\{p(t)+iq(t)\}\\ [k'(t)]^+-[k'(t)]^- &= (d-c/t^2)\{p(t)-iq(t)\}\end{aligned}\right\}. \tag{9.20.16}$$

Hence, since $h'(\sigma)$ and $k'(\sigma)$ vanish at infinity and satisfy the conditions (9.20.12) for small $|\sigma|$, we have

$$\left.\begin{aligned} h'(\sigma) &= \frac{\bar{H}+\bar{K}}{\pi\sigma}+\frac{1}{2\pi i}\int\limits_L \frac{(c-d/t^2)\{p(t)+iq(t)\}}{t-\sigma}\,dt\\ k'(\sigma) &= \frac{\bar{\gamma}_1\bar{H}+\bar{\gamma}_2\bar{K}}{\pi\sigma}+\frac{1}{2\pi i}\int\limits_L \frac{(d-c/t^2)\{p(t)-iq(t)\}}{t-\sigma}\,dt\end{aligned}\right\}, \tag{9.20.17}$$

the integrals being taken clockwise around the unit circle $t = e^{i\eta}$.

One of the simplest examples is that of a constant normal pressure applied at the hole so that

$$p(t) = -p, \qquad q(t) = 0, \tag{9.20.18}$$

where p is a constant. In this problem the resultant force at the hole is zero so we may use (9.20.15), where, from (9.20.13),

$$\left.\begin{aligned} P(t) &= -ip(ct+d/t) \\ \bar{P}(t) &= ip(dt+c/t) \end{aligned}\right\}. \tag{9.20.19}$$

Hence

$$\left.\begin{aligned} h(\sigma) &= \frac{ip}{2\pi}\int\limits_L \frac{ct+d/t}{t-\sigma}\,dt \\ k(\sigma) &= \frac{ip}{2\pi}\int\limits_L \frac{td+c/t}{t-\sigma}\,dt \end{aligned}\right\}, \tag{9.20.20}$$

the integrals being taken clockwise around the unit circle $t = e^{i\eta}$. When σ is in the region $|\sigma| > 1$ these integrals are best evaluated by residues inside the contour, so that

$$h(\sigma) = -pd/\sigma, \qquad k(\sigma) = -pc/\sigma. \tag{9.20.21}$$

Therefore, from (9.19.10),

$$\left.\begin{aligned} 2(\gamma_1-\gamma_2)f(\sigma) &= p(\gamma_2 c-d)/\sigma \\ 2(\gamma_1-\gamma_2)g(\sigma) &= p(d-\gamma_1 c)/\sigma \end{aligned}\right\}. \tag{9.20.22}$$

The evaluation of stresses is left as an exercise.

The second boundary-value problem for an elliptical hole may be studied by a method similar to that used for the circular hole in § 9.18, but details of this work are omitted.

9.21. Hyperbolic notch

In this section we examine two simple problems associated with a wide plate which contains two hyperbolic notches (Fig. 9.5).† We use the transformation

$$z = c\sinh\zeta, \qquad \zeta = \xi+i\eta, \tag{9.21.1}$$

which maps the plate in the z-plane which is bounded by two hyperbolas, into a strip in the ζ-plane bounded by the lines $\eta = \pm\alpha$. If $2a$ is the width of the plate and ρ the radius of curvature of the hyperbolas at the bottom of the notches, then

$$a = c\sin\alpha, \qquad a/\rho = \tan^2\alpha. \tag{9.21.2}$$

On the boundary $\eta = \alpha$

$$z_1 = z+\gamma_1\bar{z} = c\sinh(\xi+i\alpha)+c\gamma_1\sinh(\xi-i\alpha),$$

so that

$$(z+\gamma_1\bar{z})^2 = c^2\sinh^2(\xi+i\alpha)+c^2\gamma_1^2\sinh^2(\xi-i\alpha)+c^2\gamma_1(\cosh 2\xi-\cos 2\alpha),$$

† See A. E. Green, *Proc. R. Soc.* A184 (1945) 289. Also C. B. Smith, *Q. appl. Math.* 6 (1949) 452; H. Okubo, *Phil. Mag.* 40 (1949) 913.

and

$$(z+\gamma_1\bar{z})^2+(1+\gamma_1^2)c^2+2\gamma_1 c^2\cos 2\alpha$$
$$= c^2\cosh^2(\xi+i\alpha)+c^2\gamma_1^2\cosh^2(\xi-i\alpha)+c^2\gamma_1(\cosh 2\xi+\cos 2\alpha)$$
$$= c^2\{\cosh(\xi+i\alpha)+\gamma_1\cosh(\xi-i\alpha)\}^2,$$

$$\sqrt{\{c^2(1+2\gamma_1\cos 2\alpha+\gamma_1^2)+(z+\gamma_1\bar{z})^2\}} = c\{\cosh(\xi+i\alpha)+\gamma_1\cosh(\xi-i\alpha)\}. \tag{9.21.3}$$

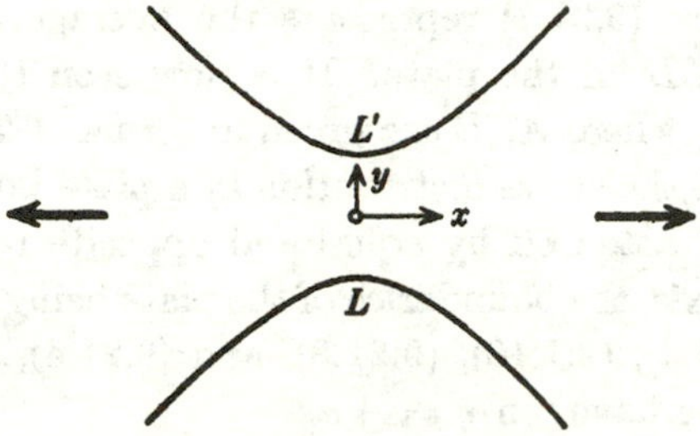

FIG. 9.5. Hyperbolic notch.

We now consider the stress functions

$$\left.\begin{aligned}\Omega'(z_1) &= A\ln[z_1+\sqrt{\{c^2(1+2\gamma_1\cos 2\alpha+\gamma_1^2)+z_1^2\}}]\\ \omega'(z_2) &= B\ln[z_2+\sqrt{\{c^2(1+2\gamma_2\cos 2\alpha+\gamma_2^2)+z_2^2\}}]\end{aligned}\right\}, \tag{9.21.4}$$

where A, B are constants. By $\sqrt{\{c^2(1+2\gamma_1\cos 2\alpha+\gamma_1^2)+z_1^2\}}$ and $\sqrt{\{c^2(1+2\gamma_2\cos 2\alpha+\gamma_2^2)+z_2^2\}}$ we understand those branches which are respectively $z_1+O(1/z_1)$ and $z_2+O(1/z_2)$ for large $|z|$. When $\eta = \pm\alpha$, we see from (9.21.3) that

$$\left.\begin{aligned}\Omega'(z_1) &= A\ln\{ce^{\xi}(e^{\pm i\alpha}+\gamma_1 e^{\mp i\alpha})\}\\ \omega'(z_2) &= B\ln\{ce^{\xi}(e^{\pm i\alpha}+\gamma_2 e^{\mp i\alpha})\}\end{aligned}\right\}, \tag{9.21.5}$$

so that (9.1.8) gives

$$P = 2i\xi(\gamma_1 A+\gamma_2 B+\bar{A}+\bar{B})+\text{constant}.$$

It follows that the applied stress on $\eta = \pm\alpha$ is zero if

$$\gamma_1 A+\gamma_2 B+\bar{A}+\bar{B} = 0.$$

We may therefore put

$$A = \frac{K(1+\gamma_2)}{\gamma_1-\gamma_2}, \qquad B = -\frac{K(1+\gamma_1)}{\gamma_1-\gamma_2}, \tag{9.21.6}$$

as one solution of this equation, where K is a real constant.

Restricting the remainder of the discussion to the orthotropic materials of § 9.2 for which γ_1, γ_2 are real, we see, from (9.21.5) and

(9.1.8), that across any section $x =$ constant of the plate there is a resultant force X given by

$$\left.\begin{aligned} 2aT = X &= \frac{4K}{\gamma_1-\gamma_2}\{(1+\gamma_1)(1-\gamma_2)\phi_2-(1+\gamma_2)(1-\gamma_1)\phi_1\} \\ \tan\phi_1 &= \frac{1-\gamma_1}{1+\gamma_1}\tan\alpha, \qquad \tan\phi_2 = \frac{1-\gamma_2}{1+\gamma_2}\tan\alpha \end{aligned}\right\}. \tag{9.21.7}$$

The constant T in (9.21.7) represents the average stress across the smallest section LL' of the plate. It is now seen that the potential functions (9.21.4), where A, B are given in terms of T by (9.21.6) and (9.21.7), represent the stress distribution in a plate bounded by hyperbolic notches and acted on by equal and opposite resultant forces X parallel to the x-axis, the boundaries of the plate being free from applied stress. From (9.1.5), (9.1.10), (9.21.3), and (9.21.4), since $\sigma_{\eta\eta} = 0$ on each boundary, we have, on $\eta = \pm\alpha$,

$$\begin{aligned} \sigma_{\xi\xi}\{(1+\gamma_1^2)(\cosh 2\xi+\cos 2\alpha)+2\gamma_1(1+\cosh 2\xi\cos 2\alpha)\}\times \\ \times\{(1+\gamma_2^2)(\cosh 2\xi+\cos 2\alpha)+2\gamma_2(1+\cosh 2\xi\cos 2\alpha)\} \\ = \lambda T\cosh\xi(\cosh 2\xi+\cos 2\alpha), \end{aligned} \tag{9.21.8}$$

where

$$\begin{aligned} \lambda &= 8K(1-\gamma_1\gamma_2)(1+\gamma_1)(1+\gamma_2)\sin 2\alpha/(aT) \\ &= \frac{4(\gamma_1-\gamma_2)(1-\gamma_1\gamma_2)(1+\gamma_1)(1+\gamma_2)\sin 2\alpha}{(1+\gamma_1)(1-\gamma_2)\phi_2-(1+\gamma_2)(1-\gamma_1)\phi_1}. \end{aligned} \tag{9.21.9}$$

Results for an isotropic material† may be deduced as a limiting case from the above results by letting $\gamma_1 \to \gamma_2 \to 0$. In particular the edge stress on $\eta = \pm\alpha$ is given by

$$\sigma_{\xi\xi} = \frac{\lambda T\cosh\xi}{\cosh 2\xi+\cos 2\alpha}, \qquad \lambda = \frac{4\tan\alpha}{\tan\alpha+\alpha(1+\tan^2\alpha)}. \tag{9.21.10}$$

A numerical discussion of the tension problem has been given by Green for a spruce plank.‡

We now consider the complex potentials

$$\left.\begin{aligned} \Omega'(z_1) &= A'\sqrt{\{c^2(1+2\gamma_1\cos 2\alpha+\gamma_1^2)+z_1^2\}} \\ \omega'(z_2) &= B'\sqrt{\{c^2(1+2\gamma_2\cos 2\alpha+\gamma_2^2)+z_2^2\}} \end{aligned}\right\}, \tag{9.21.11}$$

† Given originally by A. A. Griffith, *Rep. Memo. aeronaut. Res. Comm.* (*Coun.*) 1928, no. 1152. See also H. Neuber, *Z. angew. Math. Mech.* **13** (1933) 439.

‡ loc. cit., p. 368.

where A', B' are complex constants. From (9.1.8) and (9.21.3) we see that on the edges $\eta = \pm\alpha$ of the plate

$$\frac{P}{2ic} = A'\gamma_1\{\cosh(\xi\pm i\alpha)+\gamma_1\cosh(\xi\mp i\alpha)\}+ +\bar{A}'\{\cosh(\xi\mp i\alpha)+\bar{\gamma}_1\cosh(\xi\pm i\alpha)\}+ +B'\gamma_2\{\cosh(\xi\pm i\alpha)+\gamma_2\cosh(\xi\mp i\alpha)\}+ +\bar{B}'\{\cosh(\xi\mp i\alpha)+\bar{\gamma}_2\cosh(\xi\pm i\alpha)\},$$

and this vanishes if

$$\gamma_1 A'+\bar{\gamma}_1\bar{A}'+\gamma_2 B'+\bar{\gamma}_2\bar{B}' = 0,$$
$$\gamma_1^2 A'+\bar{A}'+\gamma_2^2 B'+\bar{B}' = 0,$$

so that the edges $\eta = \pm\alpha$ of the plate are then free from applied stress. The solution of these equations may be put in the form

$$\left.\begin{aligned} A' &= \frac{iK'(1-\gamma_2\bar{\gamma}_2)(1-\bar{\gamma}_1\gamma_2)}{\gamma_1-\gamma_2} \\ B' &= -\frac{iK'(1-\gamma_1\bar{\gamma}_1)(1-\gamma_1\bar{\gamma}_2)}{\gamma_1-\gamma_2} \end{aligned}\right\}, \tag{9.21.12}$$

where K' is a real constant. It will be observed that the displacements corresponding to (9.21.11) become infinite at infinity, but, in view of (6.8.1), (6.8.2), and (9.21.12), the infinite part of the displacement represents a rigid-body rotation and may be removed without affecting the stresses.

For the case of orthotropy, the constants in (9.21.12) become

$$\left.\begin{aligned} A' &= \frac{iK''(1-\gamma_2^2)}{\gamma_1-\gamma_2} \\ B' &= -\frac{iK''(1-\gamma_1^2)}{\gamma_1-\gamma_2} \end{aligned}\right\}, \tag{9.21.13}$$

where $K'' = K'(1-\gamma_1\gamma_2)$ is a real constant, and from (9.1.9) we see that the potentials (9.21.11) then give rise to a constant couple M in the plane of the plate, across any section $x =$ constant, where

$$M = \frac{2K''a^2(1-\gamma_1)(1-\gamma_2)}{\gamma_1-\gamma_2}\left\{\frac{(1-\gamma_1)(1+\gamma_2)\phi_1}{\sin^2\phi_1}-\frac{(1-\gamma_2)(1+\gamma_1)\phi_2}{\sin^2\phi_2}\right\}. \tag{9.21.14}$$

The edge stress $\sigma_{\xi\xi}$ on the boundaries $\eta = \pm\alpha$ may now be evaluated from (9.1.5), (9.1.10), (9.21.11), and (9.21.13), where K'' is given by

(9.21.14). Thus, remembering that $\sigma_{\eta\eta} = 0$ on $\eta = \pm\alpha$,

$$\sigma_{\xi\xi}\{(1+\gamma_1^2)(\cosh 2\xi+\cos 2\alpha)+2\gamma_1(1+\cosh 2\xi\cos 2\alpha)\}\times$$
$$\times\{(1+\gamma_2^2)(\cosh 2\xi+\cos 2\alpha)+2\gamma_2(1+\cosh 2\xi\cos 2\alpha)\}$$
$$= \mp\mu M(\cosh 2\xi+\cos 2\alpha)/a^2, \qquad (9.21.15)$$

where

$$\mu = 8K''a^2(1-\gamma_1\gamma_2)(1-\gamma_1^2)(1-\gamma_2^2)\sin 2\alpha/M$$
$$= \frac{4(\gamma_1-\gamma_2)(1-\gamma_1\gamma_2)(1+\gamma_1)(1+\gamma_2)\sin 2\alpha}{\dfrac{(1-\gamma_1)(1+\gamma_2)\phi_1}{\sin^2\phi_1}-\dfrac{(1-\gamma_2)(1+\gamma_1)\phi_2}{\sin^2\phi_2}}. \qquad (9.21.16)$$

Stresses for an isotropic material† may be deduced by the limiting case $\gamma_1 \to \gamma_2 \to 0$. In particular the edge stress becomes

$$\sigma_{\xi\xi} = \mp\frac{\mu M}{(\cosh 2\xi+\cos 2\alpha)a^2}, \qquad \mu = -\frac{2\sin 2\alpha\tan^2\alpha}{\tan\alpha+\alpha(\tan^2\alpha-1)}. \qquad (9.21.17)$$

Numerical results for the edge stress $\sigma_{\xi\xi}$ for a spruce plank have been given by Green.‡

† A. A. Griffith, loc. cit., p. 370. Also H. Neuber, *Ing.-Arch.* 5 (1934) 238; 6 (1935) 133.

‡ loc. cit., p. 368.

10 Shells

In contrast to the theory of plates which was given in Chapter 7 considerable difficulties are encountered in finding basic equations for shells, particularly when a general theory is required. Since the three-dimensional equations for an elastic shell are generally very complicated the aim of shell theory is to reduce these equations to two-dimensional form and obtain equations for stress resultants and couples instead of actual stresses.†

In this chapter we assemble all the required preliminary results for shells which are exact and we leave the development of approximate theories of shells until later chapters.

10.1. Geometrical relations

We consider a surface M and we denote the position vector of any point on M by $\mathbf{r}$, referred to some origin O. General curvilinear coordinates on M are denoted by θ_1, θ_2 so that

$$\mathbf{r} = \mathbf{r}(\theta_1, \theta_2). \tag{10.1.1}$$

At every point of M we erect the unit normal vector $\mathbf{a}_3$ and we denote the perpendicular distance of a point on $\mathbf{a}_3$ from M by θ_3 so that $\theta_3 = 0$ is the surface M. We restrict θ_3 to lie in the interval

$$-\tfrac{1}{2}t \leqslant \theta_3 \leqslant \tfrac{1}{2}t, \tag{10.1.2}$$

where $t = t(\theta_1, \theta_2)$ is in general a function of θ_1, θ_2. Throughout this book, however, we assume that the thickness t of the shell is constant, although some results are valid for variable t. Two more surfaces are defined by $\theta_3 = \pm\tfrac{1}{2}t$. These surfaces are taken as the boundaries of a body S which includes the surface M. The body S is called a shell, the boundaries of S are called the faces of the shell, the surface M is the middle surface, and t is the thickness of the shell. We assume that M and the faces of the shell form continuous surfaces with no singularities.

† A list of references to work on shell theory can be found in a review article by P. M. Naghdi, 'Foundations of elastic shell theory', in *Progress in Solid Mechanics*, Vol. iv (North Holland, 1963), p. 1. For additional references see A. E. Green and P. M. Naghdi, *Q. J. Mech. appl. Math.* 3 (1965) 257.

The position vector of a point of the shell may be defined by (see Fig. 10.1)

$$\mathbf{R} = \mathbf{R}(\theta_1, \theta_2, \theta_3) = \mathbf{r}(\theta_1, \theta_2) + \theta_3 \mathbf{a}_3(\theta_1, \theta_2), \tag{10.1.3}$$

where the unit vector $\mathbf{a}_3$ is a function of θ_1, θ_2.

We now obtain relations between certain geometrical functions concerned with the shell S and the middle surface M. With the help of

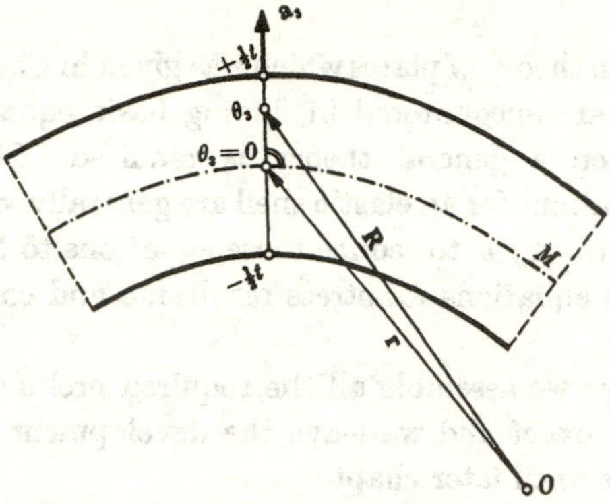

FIG. 10.1. Position vector of a point of a shell.

(1.13.47) we find from (10.1.3) that the covariant base vectors of S are given by

$$\left.\begin{aligned} \mathbf{g}_\alpha &= \mathbf{R}_{,\alpha} = \mu_\alpha^\lambda \mathbf{a}_\lambda = \mu_{\lambda\alpha} \mathbf{a}^\lambda \\ \mathbf{g}_3 &= \mathbf{R}_{,3} = \mathbf{a}_3 \end{aligned}\right\}, \tag{10.1.4}$$

where

$$\mu_{\lambda\beta} = \mu_{\beta\lambda} = a_{\lambda\beta} - \theta_3 b_{\lambda\beta}, \qquad \mu_\alpha^\lambda = \delta_\alpha^\lambda - \theta_3 b_\alpha^\lambda, \tag{10.1.5}$$

and $\mathbf{a}_\alpha$, b_α^λ are defined in § 1.13. It should be remembered that all Greek indices take the values 1, 2 and all Latin indices the values 1, 2, 3.

The components of the metric tensor of S are

$$\left.\begin{aligned} g_{\alpha\beta} &= \mathbf{g}_\alpha \cdot \mathbf{g}_\beta = \mu_\alpha^\lambda \mu_{\lambda\beta} = a_{\lambda\rho} \mu_\alpha^\lambda \mu_\beta^\rho \\ g_{\alpha 3} &= \mathbf{g}_\alpha \cdot \mathbf{g}_3 = 0 \\ g_{33} &= \mathbf{g}_3 \cdot \mathbf{g}_3 = 1 \end{aligned}\right\}. \tag{10.1.6}$$

From (1.9.15), (1.13.24), (10.1.4), and (1.13.31),

$$\sqrt{(g/a)}\epsilon_{\alpha\beta} = \epsilon_{\alpha\beta 3} = [\mathbf{g}_\alpha\, \mathbf{g}_\beta\, \mathbf{g}_3] = \epsilon_{\rho\mu} \mu_\alpha^\rho \mu_\beta^\mu$$

and, using (1.13.29), we have

$$2\sqrt{(g/a)} = \delta_{\rho\mu}^{\alpha\beta} \mu_\alpha^\rho \mu_\beta^\mu,$$

or

$$h = \sqrt{(g/a)} = 1 - 2\theta_3 H + \theta_3^2 K, \tag{10.1.7}$$

where h is a surface invariant and where, according to (1.13.34) and (1.13.35),

$$H = \tfrac{1}{2} b_\alpha^\alpha, \qquad K = \tfrac{1}{2}\delta_{\rho\mu}^{\alpha\beta} b_\alpha^\rho b_\beta^\mu = b_1^1 b_2^2 - b_2^1 b_1^2. \tag{10.1.8}$$

10.2. Stress resultants and stress couples

We now suppose that there is a distribution of stress throughout the shell S. From (5.2.5) and (10.1.4) we have

$$\mathbf{T}_i = (\sigma^{i\lambda}\mathbf{a}_\lambda + \sigma^{i3}\mathbf{a}_3)\sqrt{a} = \tau^{ij}\mathbf{g}_j\sqrt{g}, \tag{10.2.1}$$

where
$$\sigma^{i\lambda} = h\mu^\lambda_\mu \tau^{i\mu}, \qquad \sigma^{i3} = h\tau^{i3}. \tag{10.2.2}$$

Alternatively, from (10.2.1),

$$h\tau^i_\beta = \sigma^{i\lambda}\mu_{\lambda\beta}, \qquad h\tau^i_3 = \sigma^{i3} = h\tau^{i3}. \tag{10.2.3}$$

Since $\tau^{\alpha\beta}$ and τ^{i3} are symmetric,

$$\epsilon_{\lambda\beta}\sigma^{\alpha\beta}\mu^\lambda_\alpha = \epsilon_{\lambda\beta}\sigma^{\lambda\beta} - \theta_3\epsilon_{\lambda\beta}\sigma^{\alpha\beta}b^\lambda_\alpha = 0, \tag{10.2.4}$$

$$\sigma^{3\lambda} = \sigma^{\lambda 3} - \theta_3 b^\lambda_\mu \sigma^{\mu 3} = \sigma^{\mu 3}\mu^\lambda_\mu. \tag{10.2.5}$$

When discussing the state of stress in a shell it is convenient to study stress resultants and couples. For this purpose we consider first the coordinate surface $\theta_1 =$ constant which is bounded by two neighbouring normals which are intersections of $\theta_2 =$ constant and $\theta_2 + d\theta^2 =$ constant with $\theta_1 =$ constant. The force acting on an element of area in the surface $\theta_1 =$ constant is $\mathbf{t}_1\,dS_1$. Using (1.9.18) and (5.2.5) this becomes

$$\mathbf{T}_1\,d\theta^2 d\theta^3,$$

and the length of the corresponding line element of the middle surface M is

$$\sqrt{(a_{22})}\,d\theta^2 = \sqrt{(aa^{11})}\,d\theta^2.$$

The stress across the surface $\theta_1 =$ constant may therefore be replaced by a physical stress resultant $\mathbf{n}_1$ and a physical stress couple $\mathbf{m}_1$, measured per unit length of the middle surface M of S which is in the surface $\theta_1 =$ constant, where

$$\left.\begin{aligned} \mathbf{n}_1 &= \frac{1}{\sqrt{(aa^{11})}}\int\limits_{-\frac{1}{2}t}^{\frac{1}{2}t} \mathbf{T}_1\,d\theta^3 \\ \mathbf{m}_1 &= \frac{1}{\sqrt{(aa^{11})}}\int\limits_{-\frac{1}{2}t}^{\frac{1}{2}t} (\mathbf{a}_3\times\mathbf{T}_1)\theta_3\,d\theta^3 \end{aligned}\right\}. \tag{10.2.6}$$

Similarly, stress resultants and stress couples may be defined for the surface $\theta_2 =$ constant. Stress resultants and stress couples over the surface $\theta_\alpha =$ constant, measured per unit length of the line $\theta_\alpha =$ constant, $\theta_3 = 0$, are $\mathbf{n}_\alpha$, $\mathbf{m}_\alpha$ respectively where

$$\mathbf{n}_\alpha = \frac{\mathbf{N}_\alpha}{\sqrt{(aa^{\alpha\alpha})}}, \qquad \mathbf{m}_\alpha = \frac{\mathbf{M}_\alpha}{\sqrt{(aa^{\alpha\alpha})}}, \tag{10.2.7}$$

and
$$\mathbf{N}_\alpha = \int\limits_{-\frac{1}{2}t}^{\frac{1}{2}t} \mathbf{T}_\alpha\,d\theta^3, \qquad \mathbf{M}_\alpha = \int\limits_{-\frac{1}{2}t}^{\frac{1}{2}t} (\mathbf{a}_3\times\mathbf{T}_\alpha)\theta_3\,d\theta^3. \tag{10.2.8}$$

The stress resultant **n** and stress couple **m** per unit length of a line of the middle surface whose unit normal in the surface is

$$\boldsymbol{\nu} = \nu_\alpha \mathbf{a}^\alpha = \nu^\alpha \mathbf{a}_\alpha$$

may be obtained by a process similar to that used in Chapter 2 for three dimensions, the corresponding results being

$$\mathbf{n} = \sum_\alpha \nu_\alpha \mathbf{n}_\alpha \sqrt{a^{\alpha\alpha}}, \qquad \mathbf{m} = \sum_\alpha \nu_\alpha \mathbf{m}_\alpha \sqrt{a^{\alpha\alpha}}. \tag{10.2.9}$$

These formulae show that $\mathbf{n}_\alpha\sqrt{a^{\alpha\alpha}}$, $\mathbf{m}_\alpha\sqrt{a^{\alpha\alpha}}$ transform according to the contravariant type of transformation for changes of surface coordinates.

From (10.2.1) and (10.2.8) we obtain

$$\mathbf{n}_\alpha\sqrt{a^{\alpha\alpha}} = n^{\alpha\rho}\mathbf{a}_\rho + q^\alpha\mathbf{a}_3, \qquad \mathbf{N}_\alpha = N^{\alpha\rho}\mathbf{a}_\rho + Q^\alpha\mathbf{a}_3, \tag{10.2.10}$$

$$\mathbf{m}_\alpha\sqrt{a^{\alpha\alpha}} = m^{\alpha\rho}\mathbf{a}_3\times\mathbf{a}_\rho, \qquad \mathbf{M}_\alpha = M^{\alpha\rho}\mathbf{a}_3\times\mathbf{a}_\rho, \tag{10.2.11}$$

where

$$N^{\alpha\rho} = n^{\alpha\rho}\sqrt{a}, \qquad M^{\alpha\rho} = m^{\alpha\rho}\sqrt{a}, \qquad Q^\alpha = q^\alpha\sqrt{a}, \tag{10.2.12}$$

and

$$n^{\alpha\beta} = \int_{-\frac{1}{2}t}^{\frac{1}{2}t} \sigma^{\alpha\beta}\, d\theta^3, \tag{10.2.13}$$

$$m^{\alpha\beta} = \int_{-\frac{1}{2}t}^{\frac{1}{2}t} \sigma^{\alpha\beta}\theta_3\, d\theta^3, \tag{10.2.14}$$

$$q^\alpha = \int_{-\frac{1}{2}t}^{\frac{1}{2}t} \sigma^{\alpha 3}\, d\theta^3. \tag{10.2.15}$$

With the help of (1.13.30) the expressions (10.2.11) may be transformed into

$$\mathbf{m}_\alpha\sqrt{a^{\alpha\alpha}} = \epsilon_{\rho\lambda} m^{\alpha\rho}\mathbf{a}^\lambda, \qquad \mathbf{M}_\alpha = \epsilon_{\rho\lambda} M^{\alpha\rho}\mathbf{a}^\lambda, \tag{10.2.16}$$

which, written in full, become

$$\left.\begin{aligned} \mathbf{m}_1 &= \sqrt{(a/a^{11})}(m^{11}\mathbf{a}^2 - m^{12}\mathbf{a}^1), & \mathbf{M}_1 &= (M^{11}\mathbf{a}^2 - M^{12}\mathbf{a}^1)\sqrt{a} \\ \mathbf{m}_2 &= \sqrt{(a/a^{22})}(m^{21}\mathbf{a}^2 - m^{22}\mathbf{a}^1), & \mathbf{M}_2 &= (M^{21}\mathbf{a}^2 - M^{22}\mathbf{a}^1)\sqrt{a} \end{aligned}\right\}. \tag{10.2.17}$$

The functions defined in (10.2.13)–(10.2.15) are all surface tensors. This is evident from the actual formulae, or it can be deduced from equations (10.2.9). The functions Q^α, $N^{\alpha\beta}$, $M^{\alpha\beta}$ are 'weighted' surface tensors. The components of the contravariant surface tensor $n^{\alpha\beta}$ and the components of the surface vector q^α are called stress resultants and shearing forces respectively, and are related to the (vector) stress resultants $\mathbf{n}_\alpha$ by the formulae (10.2.10). The components of the contravariant surface tensor $m^{\alpha\beta}$ are called stress couples and are related to the (vector) stress couples $\mathbf{m}_\alpha$ by (10.2.11). From (10.2.9)–(10.2.12) we see that

$$\mathbf{n} = \frac{\nu_\alpha \mathbf{N}_\alpha}{\sqrt{a}}, \qquad \mathbf{m} = \frac{\nu_\alpha \mathbf{M}_\alpha}{\sqrt{a}}. \tag{10.2.18}$$

We are also interested in the components of $\mathbf{n}_\alpha$ and $\mathbf{m}_\alpha$ when these vectors are expressed in terms of unit base vectors. The components are denoted by $n_{(\alpha\beta)}$, $m_{(\alpha\beta)}$, $q_{(\alpha)}$ and are called physical stress resultants, physical stress couples, and physical shearing forces, respectively. It must be emphasized that they are not components of tensors, and this fact is denoted by enclosing the suffixes in brackets.† Thus

$$\left.\begin{aligned} \mathbf{n}_\alpha &= n_{(\alpha 1)}\frac{\mathbf{a}_1}{\sqrt{a_{11}}} + n_{(\alpha 2)}\frac{\mathbf{a}_2}{\sqrt{a_{22}}} + q_{(\alpha)}\mathbf{a}_3 \\ \mathbf{m}_\alpha &= m_{(\alpha 1)}\frac{\mathbf{a}^2}{\sqrt{a^{22}}} + m_{(\alpha 2)}\frac{\mathbf{a}^1}{\sqrt{a^{11}}} \end{aligned}\right\}. \tag{10.2.19}$$

Comparison between these equations and equations (10.2.10) and (10.2.17) gives immediately

$$n_{(\alpha\beta)} = n^{\alpha\beta}\sqrt{(a_{\beta\beta}/a^{\alpha\alpha})}, \qquad q_{(\alpha)} = q^\alpha/\sqrt{a^{\alpha\alpha}}, \tag{10.2.20}$$

$$\left.\begin{aligned} m_{(11)} &= m^{11}\sqrt{(a_{11}/a^{11})}, \quad m_{(12)} = -m^{12}\sqrt{a} \\ m_{(21)} &= m^{21}\sqrt{a}, \quad m_{(22)} = -m^{22}\sqrt{(a_{22}/a^{22})} \end{aligned}\right\}. \tag{10.2.21}$$

The meaning of these physical stress resultants and couples is seen by examining (10.2.19) and they are illustrated geometrically in Figs. 10.2 and 10.3. The physical stress resultants $n_{(\alpha\beta)}$ are directed along the covariant base vectors of the middle surface M, and $q_{(\alpha)}$ are in the direction of the normal to M. The physical stress couples $m_{(\alpha\beta)}$ are couples about the contravariant base vectors of the middle surfaces.

10.3. Loads‡

Consider an element of the middle surface M of the shell which is bounded by the coordinate curves $\theta_1 =$ constant, $\theta_2 =$ constant, $\theta_1 + d\theta^1 =$ constant, and $\theta_2 + d\theta^2 =$ constant. At every point along these coordinate lines normals to M may be erected, which, with the faces of the shell, form an element of volume which we call an element of the shell. The area of the middle surface of this element is, from (1.13.22),

$$\sqrt{(a)}\, d\theta^1 d\theta^2. \tag{10.3.1}$$

† The notation $n_{(\alpha\beta)}$ used only in this section, should not be confused with the symmetric parts $n^{(\alpha\beta)}$, $n_{(\alpha\beta)}$ of the stress tensors $n^{\alpha\beta}$, $n_{\alpha\beta}$ used in later sections.

‡ The formulae of this section would be modified if t varies.

The forces over the corresponding surface elements are

$$\mathbf{T}_3\, d\theta^1 d\theta^2 \quad (\theta_3 = \pm t), \tag{10.3.2}$$

where, from (10.2.1), $\quad \mathbf{T}_3 = (\sigma^{3\lambda}\mathbf{a}_\lambda + \sigma^{33}\mathbf{a}_3)\sqrt{a}. \qquad (10.3.3)$

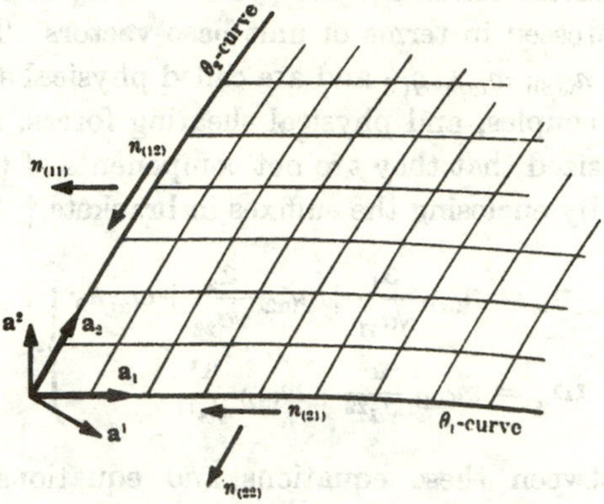

FIG. 10.2. Physical stress resultants.

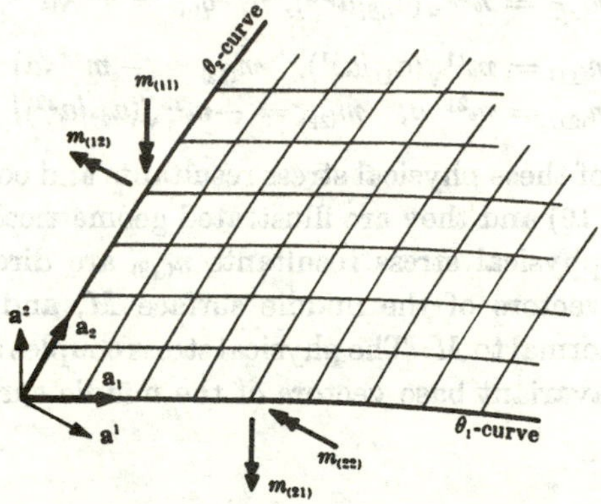

FIG. 10.3. Physical stress couples.

In terms of σ^{3i} the surface conditions may be expressed in the form

$$\sigma^{3\lambda} = \tfrac{1}{2}(p^\lambda + \bar{p}^\lambda), \qquad \sigma^{33} = \tfrac{1}{2}(p + \bar{p}) \qquad (\theta_3 = \tfrac{1}{2}t), \tag{10.3.4}$$

$$\sigma^{3\lambda} = -\tfrac{1}{2}(p^\lambda - \bar{p}^\lambda), \qquad \sigma^{33} = -\tfrac{1}{2}(p - \bar{p}) \qquad (\theta_3 = -\tfrac{1}{2}t), \tag{10.3.5}$$

where p, $\bar{p}$, p^λ, $\bar{p}^\lambda$ are given functions of θ_1, θ_2. We replace the surface forces by a resultant force $\mathbf{p}$, measured per unit area of the middle surface, and a couple $\tilde{\mathbf{p}}$ measured about the point (θ_1, θ_2) of M. Thus

$$\mathbf{P} = \mathbf{p}\sqrt{a} = [\mathbf{T}_3]_{-\frac{1}{2}t}^{\frac{1}{2}t}, \tag{10.3.6}$$

and with the help of (10.3.3)–(10.3.5) we have

$$\mathbf{p} = p^{\alpha}\mathbf{a}_{\alpha}+p\mathbf{a}_3, \tag{10.3.7}$$

$$\mathbf{P} = \mathbf{p}\sqrt{a} = P^{\alpha}\mathbf{a}_{\alpha}+P\mathbf{a}_3, \tag{10.3.8}$$

where
$$P^{\alpha} = p^{\alpha}\sqrt{a}, \qquad P = p\sqrt{a}. \tag{10.3.9}$$

Also
$$\tilde{\mathbf{P}} = \tilde{\mathbf{p}}\sqrt{a} = [(\mathbf{a}_3\times\mathbf{T}_3)\theta_3]^{\frac{1}{2}t}_{-\frac{1}{2}t}, \tag{10.3.10}$$

so that
$$\tilde{\mathbf{p}} = \tilde{p}^{\lambda}(\mathbf{a}_3\times\mathbf{a}_{\lambda}), \qquad \tilde{\mathbf{P}} = \tilde{P}^{\lambda}(\mathbf{a}_3\times\mathbf{a}_{\lambda}), \tag{10.3.11}$$

where
$$\tilde{p}^{\lambda} = \tfrac{1}{2}t\bar{p}^{\lambda}, \qquad \tilde{P}^{\lambda} = \tilde{p}^{\lambda}\sqrt{a}. \tag{10.3.12}$$

The components of $\mathbf{p}$ referred to unit base vectors are denoted by $p_{(i)}$ and are called physical components of the loads. Thus

$$\mathbf{p} = p_{(1)}\frac{\mathbf{a}_1}{\sqrt{a_{11}}}+p_{(2)}\frac{\mathbf{a}_2}{\sqrt{a_{22}}}+p_{(3)}\mathbf{a}_3. \tag{10.3.13}$$

Comparing this with (10.3.7) we obtain

$$p_{(\alpha)} = p^{\alpha}\sqrt{a_{\alpha\alpha}}, \qquad p_{(3)} = p. \tag{10.3.14}$$

10.4. Equations of equilibrium

When body forces are zero and the deformed shell is in equilibrium we see, from (5.2.6), that the equations of equilibrium are

$$\mathbf{T}_{i,i} = \mathbf{0}. \tag{10.4.1}$$

With the help of (10.2.1) and formulae of § 1.13 these equations have the alternative forms

$$\left.\begin{aligned}\sigma^{\alpha\beta}|_{\alpha}-b^{\beta}_{\alpha}\sigma^{\alpha3}+\partial\sigma^{3\beta}/\partial\theta^3 &= 0\\ \sigma^{\alpha3}|_{\alpha}+b_{\alpha\beta}\sigma^{\alpha\beta}+\partial\sigma^{33}/\partial\theta^3 &= 0\end{aligned}\right\}, \tag{10.4.2}$$

where the vertical line denotes covariant differentiation with respect to the middle surface coordinates θ_{α}. Thus

$$\left.\begin{aligned}\sigma^{\alpha\beta}|_{\alpha} &= \sigma^{\alpha\beta}{}_{,\alpha}+\Gamma^{\beta}_{\alpha\rho}\sigma^{\alpha\rho}+\Gamma^{\alpha}_{\alpha\rho}\sigma^{\rho\beta}\\ \sigma^{\alpha3}|_{\alpha} &= \sigma^{\alpha3}{}_{,\alpha}+\Gamma^{\alpha}_{\alpha\beta}\sigma^{\beta3}\end{aligned}\right\}, \tag{10.4.3}$$

where $\Gamma^{\beta}_{\alpha\rho}$ is the Christoffel symbol of the second kind associated with the middle surface.

We integrate equations (10.4.2) with respect to θ_3 through the thickness of the shell. With the help of surface conditions (10.3.4) and (10.3.5), definitions (10.2.13) and (10.2.15), and equation (10.2.5), we obtain

$$n^{\alpha\beta}|_{\alpha}-b^{\beta}_{\alpha}q^{\alpha}+p^{\beta} = 0, \tag{10.4.4}$$

$$b_{\alpha\beta}n^{\alpha\beta}+q^{\alpha}|_{\alpha}+p = 0. \tag{10.4.5}$$

We next multiply the first equation in (10.4.2) by θ_3, integrate through the thickness of the shell, and use (10.2.5), (10.2.14), (10.2.15), (10.3.4), (10.3.5), and (10.3.12) to get

$$m^{\alpha\beta}|_{\alpha} - q^{\beta} + \tilde{p}^{\beta} = 0. \tag{10.4.6}$$

Also, from equation (10.2.4) which is equivalent to the fact that $\tau^{\alpha\beta}$ is symmetric, we have

$$\epsilon_{\lambda\beta} n^{\lambda\beta} - \epsilon_{\lambda\beta} m^{\alpha\beta} b_{\alpha}^{\lambda} = 0. \tag{10.4.7}$$

For later convenience we split up the stress-resultant $n^{\alpha\beta}$ into its symmetric and skew-symmetric parts. Thus

$$n^{\alpha\beta} = n^{(\alpha\beta)} + n^{[\alpha\beta]} \tag{10.4.8}$$

where†

$$n^{(\alpha\beta)} = n^{(\beta\alpha)}, \qquad n^{[\alpha\beta]} = -n^{[\beta\alpha]}. \tag{10.4.9}$$

In view of (10.4.7) we have

$$n^{[\alpha\beta]} = \tfrac{1}{2}(b_{\lambda}^{\alpha} m^{\lambda\beta} - b_{\lambda}^{\beta} m^{\lambda\alpha}). \tag{10.4.10}$$

We close this section by adding useful alternative forms of equations (10.4.4)–(10.4.7). Recalling results of §§ 10.2, 10.3 we see that (10.4.4) and (10.4.5) may be replaced by

$$\left.\begin{aligned} N^{\alpha\beta}{}_{,\alpha} + \Gamma_{\rho\alpha}^{\beta} N^{\rho\alpha} - b_{\alpha}^{\beta} Q^{\alpha} + P^{\beta} &= 0 \\ N^{\alpha\beta} b_{\alpha\beta} + Q^{\alpha}{}_{,\alpha} + P &= 0 \end{aligned}\right\}. \tag{10.4.11}$$

Alternatively

$$\mathbf{N}_{\alpha,\alpha} + \mathbf{P} = \mathbf{0}. \tag{10.4.12}$$

This form may also be obtained by integrating (10.4.1) with respect to θ_3 through the thickness of the shell. Again, (10.4.6) and (10.4.7) may be replaced by

$$M^{\alpha\beta}{}_{,\alpha} + \Gamma_{\alpha\rho}^{\beta} M^{\alpha\rho} - Q^{\beta} + \bar{P}^{\beta} = 0, \tag{10.4.13}$$

and

$$\epsilon_{\lambda\beta} N^{\lambda\beta} - \epsilon_{\lambda\beta} M^{\alpha\beta} b_{\alpha}^{\lambda} = 0. \tag{10.4.14}$$

These results may be combined in the single formula

$$\mathbf{M}_{\alpha,\alpha} + \mathbf{a}_{\alpha} \times \mathbf{N}_{\alpha} + \bar{\mathbf{P}} = \mathbf{0}. \tag{10.4.15}$$

This equation may also be obtained by taking the scalar product of (10.4.1) with $\theta_3 \mathbf{a}_3$ and integrating through the thickness of the shell, using, in addition, (10.2.4) and (10.2.5).

It can be shown that the equations of the present section still hold when the shell has variable thickness, but in this case the formulae for the resultant surface force $\mathbf{p}$ and surface couple $\tilde{\mathbf{p}}$ given in § 10.3 must be modified.

† These do *not* denote physical components of stress.

10.5. Stress–strain relations

We denote the displacement vector of any point in the shell by **V** and express it in the form

$$\mathbf{V} = u^\alpha \mathbf{a}_\alpha + u^3 \mathbf{a}_3 = u_\alpha \mathbf{a}^\alpha + u_3 \mathbf{a}_3, \tag{10.5.1}$$

where
$$u^3 = u_3, \qquad u^\alpha = a^{\alpha\beta} u_\beta. \tag{10.5.2}$$

With the help of (1.13.51) it follows from (5.1.1) that

$$\begin{aligned} 2\gamma_{\alpha\beta} &= u_\alpha|_\beta + u_\beta|_\alpha - 2b_{\alpha\beta} u_3 - \theta_3\{b^\lambda_\beta(u_\lambda|_\alpha - b_{\lambda\alpha} u_3) + b^\lambda_\alpha(u_\lambda|_\beta - b_{\lambda\beta} u_3)\} \\ &= u_\lambda|_\alpha \mu^\lambda_\beta + u_\lambda|_\beta \mu^\lambda_\alpha - (b_{\lambda\alpha}\mu^\lambda_\beta + b_{\lambda\beta}\mu^\lambda_\alpha) u_3, \end{aligned} \tag{10.5.3}$$

$$2\gamma_{\alpha 3} = \frac{\partial u_3}{\partial \theta^\alpha} + \frac{\partial u_\alpha}{\partial \theta^3} + b^\lambda_\alpha u_\lambda - \theta_3 b^\lambda_\alpha \frac{\partial u_\lambda}{\partial \theta^3}, \tag{10.5.4}$$

$$\gamma_{33} = \frac{\partial u_3}{\partial \theta^3}. \tag{10.5.5}$$

From (5.4.2) and (10.2.3) we see that stress–strain relations for an isotropic shell may be put in the form

$$\left.\begin{aligned} Eh\gamma_{\alpha\beta} &= (1+\eta) g_{\alpha\lambda} \mu_{\mu\beta} \sigma^{\lambda\mu} - \eta g_{\alpha\beta}(\mu_{\lambda\mu} \sigma^{\lambda\mu} + \sigma^{33}) \\ Eh\gamma_{\alpha 3} &= (1+\eta) g_{\alpha\lambda} \sigma^{\lambda 3} \\ Eh\gamma_{33} &= \sigma^{33} - \eta \sigma^{\alpha\lambda} \mu_{\alpha\lambda} \end{aligned}\right\}. \tag{10.5.6}$$

It is convenient to split up the stress tensor $\sigma^{\alpha\beta}$ into its symmetric and skew-symmetric parts

$$\sigma^{\alpha\beta} = \sigma^{(\alpha\beta)} + \sigma^{[\alpha\beta]}, \tag{10.5.7}$$

where
$$\sigma^{(\alpha\beta)} = \sigma^{(\beta\alpha)}, \qquad \sigma^{[\alpha\beta]} = -\sigma^{[\beta\alpha]}, \tag{10.5.8}$$

and equation (10.2.4) becomes

$$\sigma^{[\alpha\beta]} = \tfrac{1}{2}\theta_3(\sigma^{\lambda\beta} b^\alpha_\lambda - \sigma^{\lambda\alpha} b^\beta_\lambda) = \tfrac{1}{2}(\sigma^{\alpha\beta} - \sigma^{\beta\alpha}). \tag{10.5.9}$$

Using (10.5.9), (10.5.3)–(10.5.5), and (10.5.7), the stress–strain relations (10.5.6) become

$$\begin{aligned} Eh[u_\alpha|_\beta + u_\beta|_\alpha - 2b_{\alpha\beta} u_3 - \theta_3\{b^\lambda_\beta(u_\lambda|_\alpha - b_{\lambda\alpha} u_3) + b^\lambda_\alpha(u_\lambda|_\beta - b_{\lambda\beta} u_3)\}] \\ = (1+\eta)(g_{\alpha\lambda}\mu_{\mu\beta} + g_{\beta\lambda}\mu_{\mu\alpha})\sigma^{\lambda\mu} - 2\eta g_{\alpha\beta}(\mu_{\lambda\mu}\sigma^{(\lambda\mu)} + \sigma^{33}), \end{aligned} \tag{10.5.10}$$

$$Eh\left(\frac{\partial u_3}{\partial \theta^\alpha} + \frac{\partial u_\alpha}{\partial \theta^3} + b^\lambda_\alpha u_\lambda - \theta_3 b^\lambda_\alpha \frac{\partial u_\lambda}{\partial \theta^3}\right) = 2(1+\eta) g_{\alpha\lambda} \sigma^{\lambda 3}, \tag{10.5.11}$$

$$Eh(\partial u_3/\partial \theta^3) = \sigma^{33} - \eta \mu_{\alpha\lambda} \sigma^{(\alpha\lambda)}. \tag{10.5.12}$$

The skew-symmetric part of the right-hand side of the first equation in (10.5.6) vanishes because of (10.5.9).

10.6. Plane coordinates

For later convenience we record some results in terms of plane coordinates. We consider a net of coordinate curves on the middle surface M and we form a parallel projection of this net on to a plane Π. The direction of projection is given by a unit vector $\mathbf{e}_3$ which is perpendicular to Π and measured from Π to M. We now have a net of coordinate curves on Π which corresponds to the net on M and we assume that the correspondence is one–one. The distance between two corresponding points on M and Π measured in the direction of $\mathbf{e}_3$ is denoted by $z = z(\theta_1, \theta_2)$ which is a function of θ_1, θ_2 and is a surface invariant. Taking $\mathbf{r}$ for the position vector of M, and $\bar{\mathbf{r}}$ for the position vector of the corresponding point of Π, we have (see Fig. 10.4)

$$\mathbf{r} = \bar{\mathbf{r}} + z\mathbf{e}_3. \tag{10.6.1}$$

The expression (10.6.1) is a new representation of the middle surface M and in general it is possible to represent every surface in this way either by a suitable choice of Π or by considering a restricted domain of M. Conversely, given any coordinate system on a plane Π and introducing a function $z(\theta_1, \theta_2)$, we obtain a surface M by means of (10.6.1).

The plane Π represents a two-dimensional Euclidean space for which the special formulae at the end of § 1.13 are valid. Since we are concerned here with both a plane Π and a surface M we must change the notation for Π. Thus, base vectors associated with θ_α-curves in Π are denoted by $\mathbf{e}_\alpha$, $\mathbf{e}^\alpha$, the metric tensors by $e_{\alpha\beta}$, $e^{\alpha\beta}$, and the determinant of $e_{\alpha\beta}$ by e. Christoffel symbols for the surface M are denoted by $\bar{\Gamma}_{\alpha\beta\gamma}$, $\bar{\Gamma}^\alpha_{\beta\gamma}$ and for the plane Π by $\Gamma_{\alpha\beta\gamma}$, $\Gamma^\alpha_{\beta\gamma}$.

We denote the ϵ-tensors for M by $\bar{\epsilon}_{\alpha\beta}$, $\bar{\epsilon}^{\alpha\beta}$ and for Π by $\epsilon_{\alpha\beta}$, $\epsilon^{\alpha\beta}$ so that

$$\epsilon_{11} = \epsilon_{22} = \epsilon^{11} = \epsilon^{22} = 0,$$

$$\epsilon_{12} = -\epsilon_{21} = \sqrt{e}, \qquad \epsilon^{12} = -\epsilon^{21} = 1/\sqrt{e}.$$

From § 1.13, and in particular from (1.13.59)

$$\left.\begin{aligned}
\mathbf{e}_\alpha &= \bar{\mathbf{r}}_{,\alpha}, \quad \mathbf{e}^\alpha = e^{\alpha\beta}\mathbf{e}_\beta \\
e_{\alpha\beta} &= \mathbf{e}_\alpha . \mathbf{e}_\beta, \quad e^{\alpha\beta} = \mathbf{e}^\alpha . \mathbf{e}^\beta \\
e &= |e_{\alpha\beta}| = e_{11}e_{22} - e_{12}^2 \\
e^{\alpha\beta} &= \epsilon^{\alpha\lambda}\epsilon^{\beta\mu}e_{\lambda\mu} \\
\Gamma^\alpha_{\beta\gamma} &= e^{\alpha\lambda}\Gamma_{\beta\gamma\lambda}, \qquad \Gamma^\lambda_{\lambda\alpha} = \frac{1}{\sqrt{e}}\frac{\partial\sqrt{e}}{\partial\theta^\alpha} \\
\mathbf{e}_{\alpha,\beta} &= \Gamma^\lambda_{\alpha\beta}\mathbf{e}_\lambda, \qquad \mathbf{e}^\alpha{}_{,\beta} = -\Gamma^\alpha_{\beta\lambda}\mathbf{e}^\lambda
\end{aligned}\right\}. \tag{10.6.2}$$

Also from (1.13.30),

$$\left.\begin{aligned} \mathbf{e}_\alpha \times \mathbf{e}_\beta &= \epsilon_{\alpha\beta}\mathbf{e}_3 \\ \mathbf{e}^\alpha \times \mathbf{e}^\beta &= \epsilon^{\alpha\beta}\mathbf{e}_3 \\ \mathbf{e}_3 \times \mathbf{e}_\alpha &= \epsilon_{\alpha\beta}\mathbf{e}^\beta \\ \mathbf{e}_3 \times \mathbf{e}^\alpha &= \epsilon^{\alpha\beta}\mathbf{e}_\beta \end{aligned}\right\}. \tag{10.6.3}$$

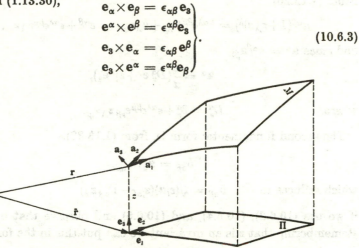

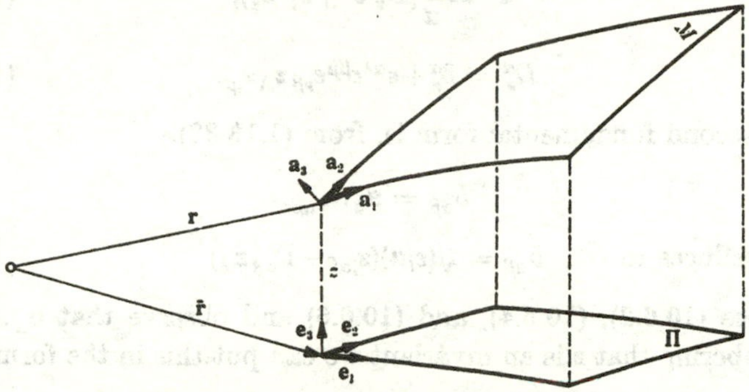

FIG. 10.4. Element of middle surface M and its projection on a plane Π.

Since the Riemann–Christoffel tensor for the plane vanishes, the order of covariant differentiation of tensors in Π is immaterial. Thus, e.g.

$$B^{\xi\eta}|_{\alpha\beta} = B^{\xi\eta}|_{\beta\alpha},$$

where the vertical rule now indicates covariant differentiation with respect to Π. Throughout the rest of this section covariant differentiation with respect to M does not occur.

From (10.6.1) we may now evaluate vectors and tensors associated with the plane Π. To do this we again make use of the appropriate formulae in § 1.13. Thus

$$\mathbf{a}_\alpha = \mathbf{r}_{,\alpha} = \mathbf{e}_\alpha + z_{,\alpha}\mathbf{e}_3, \tag{10.6.4}$$

and from this

$$a_{\alpha\beta} = e_{\alpha\beta} + z_{,\alpha}z_{,\beta}. \tag{10.6.5}$$

Since, from (1.13.30), $\quad \mathbf{a}_3\sqrt{a} = \mathbf{a}_1 \times \mathbf{a}_2,$

we have, with the help of (10.6.3),

$$\mathbf{a}_3\sqrt{a} = (\mathbf{e}_3 - z_{,\alpha}\mathbf{e}^\alpha)\sqrt{e}, \tag{10.6.6}$$

and hence

$$a = e(1 + z_{,\alpha}z|^\alpha). \tag{10.6.7}$$

We observe that a/e is an invariant. Also

$$\bar{\epsilon}^{\alpha\beta} = \frac{\epsilon^{\alpha\beta}}{(1+z_{,\lambda}z|^\lambda)^{\frac{1}{2}}}, \qquad \bar{\epsilon}_{\alpha\beta} = \epsilon_{\alpha\beta}(1+z_{,\lambda}z|^\lambda)^{\frac{1}{2}}. \tag{10.6.8}$$

From (1.13.26)

$$a^{\alpha\beta}(1+z_{,\lambda}z|^{\lambda}) = \epsilon^{\alpha\lambda}\epsilon^{\beta\mu}(e_{\lambda\mu}+z_{,\lambda}z_{,\mu}) = e^{\alpha\beta}+\epsilon^{\alpha\lambda}\epsilon^{\beta\mu}z_{,\lambda}z_{,\mu}, \quad (10.6.9)$$

and since $\mathbf{a}^{\alpha} = a^{\alpha\beta}\mathbf{a}_{\beta}$,

$$\mathbf{a}^{\alpha} = \frac{e}{a}(D^{\alpha}_{\nu}\,\mathbf{e}^{\nu}+z|^{\alpha}\mathbf{e}_3), \quad (10.6.10)$$

where

$$D^{\alpha}_{\nu} = \delta^{\alpha}_{\nu}+\epsilon^{\alpha\lambda}\epsilon^{\beta\mu}e_{\nu\beta}z_{,\lambda}z_{,\mu}. \quad (10.6.11)$$

The second fundamental form is, from (1.13.32),

$$b_{\alpha\beta} = \mathbf{a}_3 \cdot \mathbf{a}_{\alpha,\beta},$$

which reduces to

$$b_{\alpha\beta} = \sqrt{(e/a)}(z_{,\alpha\beta}-\Gamma^{\lambda}_{\alpha\beta}z_{,\lambda})$$

if we use (10.6.2), (10.6.4), and (10.6.6) and observe that $\mathbf{e}_{\alpha}\cdot\mathbf{e}_3 = 0$. Remembering that z is an invariant we can put this in the form

$$b_{\alpha\beta} = \sqrt{(e/a)}z|_{\alpha\beta}. \quad (10.6.12)$$

The Christoffel symbols of the second kind for the surface M and the plane Π are related by

$$\bar{\Gamma}^{\alpha}_{\beta\gamma} = \frac{e}{a}(D^{\alpha}_{\nu}\,\Gamma^{\nu}_{\beta\gamma}+z|^{\alpha}z_{,\beta\gamma}). \quad (10.6.13)$$

Using (10.6.4) and (10.6.6) the stress resultants $\mathbf{N}_{\alpha}$ and the stress couples $\mathbf{M}_{\alpha}$, which are given by (10.2.10) and (10.2.11), can be put in the following forms:

$$\mathbf{N}_{\alpha} = \mathbf{K}_{\alpha}+(k^{\alpha\beta}z_{,\beta}\,\mathbf{e}_3-q^{\alpha}z|^{\beta}\mathbf{e}_{\beta})\sqrt{e}, \quad (10.6.14)$$

$$\mathbf{M}_{\alpha} = \mathbf{H}_{\alpha}+\sqrt{(e)}m^{\alpha\beta}z|^{\lambda}\mathbf{e}_{\beta}\times\mathbf{e}_{\lambda}, \quad (10.6.15)$$

where

$$\mathbf{K}_{\alpha} = K^{\alpha\beta}\mathbf{e}_{\beta}+\sqrt{(e)}q^{\alpha}\mathbf{e}_3, \qquad \mathbf{H}_{\alpha} = H^{\alpha\beta}\mathbf{e}_3\times\mathbf{e}_{\beta}, \quad (10.6.16)$$

and

$$K^{\alpha\beta} = k^{\alpha\beta}\sqrt{e}, \qquad k^{\alpha\beta} = n^{\alpha\beta}\sqrt{(a/e)}, \quad (10.6.17)$$

$$H^{\alpha\beta} = h^{\alpha\beta}\sqrt{e}, \qquad h^{\alpha\beta} = m^{\alpha\rho}(\delta^{\beta}_{\rho}+z|_{\rho}z|^{\beta}). \quad (10.6.18)$$

The loads $\mathbf{P}$ on the shell, which are defined in (10.3.6) and (10.3.8), may be expressed in terms of base vectors $\mathbf{e}_{\alpha}$, $\mathbf{e}_3$ with the help of (10.6.4) and (10.6.6). Thus

$$\mathbf{P} = S^{\alpha}\mathbf{e}_{\alpha}+S\mathbf{e}_3, \quad (10.6.19)$$

where

$$\left.\begin{aligned} S^{\alpha} &= P^{\alpha}-\sqrt{(e/a)}Pz|^{\alpha} \\ S &= P^{\alpha}z_{,\alpha}+\sqrt{(e/a)}P \end{aligned}\right\}, \quad (10.6.20)$$

or, alternatively,

$$\left.\begin{aligned} P\sqrt{(a/e)} &= S - S^{\alpha}z_{,\alpha} \\ P^{\alpha} &= \frac{e}{a}(S^{\alpha} + Sz|^{\alpha}) \end{aligned}\right\}. \tag{10.6.21}$$

For later use we also introduce the notation

$$S^{\alpha} = s^{\alpha}\sqrt{e}, \qquad S = s\sqrt{e}, \tag{10.6.22}$$

so that

$$\left.\begin{aligned} p &= \frac{e}{a}(s - s^{\alpha}z|_{\alpha}) \\ p^{\alpha} &= \left(\frac{e}{a}\right)^{\frac{3}{2}}(s^{\alpha} + sz|^{\alpha}) \end{aligned}\right\}. \tag{10.6.23}$$

11 General Theory of Membrane Shells

In many shells, particularly in regions away from edges, the loads on the surface of the shell are mainly carried by the stress resultants and this indicates that we may be able to find an approximate system of equations in which stress couples and shearing forces are neglected compared with the stress resultants in the equations of equilibrium. The theory obtained from such an approximation is called membrane theory and, correspondingly, the shell is then called a membrane shell.

In this chapter we deal with the general theory of membrane shells. In some problems it may happen that the stress resultants and displacements which are found from membrane theory enable us to satisfy the required boundary conditions at the edges of the shell. If this is so we may regard the membrane theory solution as a complete (approximate) solution of the problem. In general it is not possible to satisfy all boundary conditions by using membrane theory, and we must then proceed to find a solution which involves stress couples and shearing forces, and this will be considered in later chapters. We can, however, regard the membrane theory solution as an approximate particular integral of the complete equations, and we may complete the solution by seeking homogeneous equations involving stress couples and shearing forces, in which all loads are taken to be zero. Membrane theory is, therefore, the starting-point in nearly all shell problems, although there are exceptions to this.

There are some shells for which no suitable membrane solution exists. For these the load on the surface of the shell is not carried mainly by stress resultants in which deformation occurs with little bending of the shell. Instead, there is considerable bending of the middle surface of the shell with little extension. The corresponding theory, known as inextensional theory, is considered in Chapter 16.

We present membrane theory in the present chapter by a direct method, starting with equations of equilibrium in terms of stress resultants. In Chapter 16 we recover the same results from three-dimensional equations of elasticity by a method of asymptotic expansion.

11.1. Membrane theory in general curvilinear coordinates

We begin with a brief repetition of earlier results. We remind the reader that we have used general surface coordinates θ_α and that the middle surface M of the shell is determined by the vector $\mathbf{r} = \mathbf{r}(\theta_1, \theta_2)$. The stress resultants across the surface $\theta_\alpha =$ constant, measured per unit length of the coordinate curves of the middle surface, are given by (10.2.7) in the form

$$\mathbf{n}_\alpha = \frac{\mathbf{N}_\alpha}{\sqrt{(aa^{\alpha\alpha})}}. \tag{11.1.1}$$

According to our definition of the membrane theory we neglect all moments and shearing forces, so that (10.2.10) becomes, approximately,

$$\mathbf{N}_\alpha = N^{\alpha\rho}\mathbf{a}_\rho = n^{\alpha\rho}\sqrt{(a)}\mathbf{a}_\rho, \tag{11.1.2}$$

where, according to (10.4.7),

$$n^{\alpha\rho} = n^{\rho\alpha}, \qquad N^{\alpha\rho} = N^{\rho\alpha}. \tag{11.1.3}$$

The physical stress resultants (per unit length) are, from (10.2.20), given by

$$n_{(\alpha\beta)} = n^{\alpha\beta}\sqrt{(a_{\beta\beta}/a^{\alpha\alpha})}, \tag{11.1.4}$$

but $n_{(\alpha\beta)}$ is not a tensor. The stress resultant $\mathbf{n}$ across any curve in the middle surface whose unit normal in the surface is

$$\boldsymbol{\nu} = \nu_\alpha \mathbf{a}^\alpha \tag{11.1.5}$$

is

$$\mathbf{n} = \frac{\nu_\alpha \mathbf{N}_\alpha}{\sqrt{a}}. \tag{11.1.6}$$

The shell is subjected to external loads which were denoted in § 10.3 by $\mathbf{p}$ measured per unit area of the middle surface where, from (10.3.7),

$$\mathbf{p} = p^\alpha \mathbf{a}_\alpha + p\mathbf{a}_3. \tag{11.1.7}$$

We also introduced in (10.3.8) and (10.3.9) the notations

$$\left.\begin{aligned} \mathbf{p}\sqrt{a} = \mathbf{P} = P^\alpha \mathbf{a}_\alpha + P\mathbf{a}_3 \\ P^\alpha = p^\alpha\sqrt{a}, \qquad P = p\sqrt{a} \end{aligned}\right\}. \tag{11.1.8}$$

According to (10.3.14) the physical components $p_{(\alpha)}$, $p_{(3)}$ of the loads are given by

$$p_{(\alpha)} = p^\alpha\sqrt{a_{\alpha\alpha}}, \qquad p_{(3)} = p. \tag{11.1.9}$$

The general equations of equilibrium are given by (10.4.12) in the form

$$\mathbf{N}_{\alpha,\alpha} + \mathbf{P} = 0, \tag{11.1.10}$$

or, using (11.1.2) and (11.1.8), this becomes

$$N^{\alpha\beta}{}_{,\alpha}\mathbf{a}_\beta + N^{\alpha\beta}\mathbf{a}_{\beta,\alpha} + \mathbf{P} = 0. \tag{11.1.11}$$

This equation may be further reduced to give

$$\left.\begin{aligned} N^{\alpha\beta}{}_{,\alpha}+\Gamma^{\beta}_{\rho\alpha}N^{\rho\alpha}+P^{\beta} &= 0 \\ N^{\alpha\beta}b_{\alpha\beta}+P &= 0 \end{aligned}\right\}, \tag{11.1.12}$$

or

$$\left.\begin{aligned} n^{\alpha\beta}|_{\alpha}+p^{\beta} &= 0 \\ n^{\alpha\beta}b_{\alpha\beta}+p &= 0 \end{aligned}\right\}. \tag{11.1.13}$$

As usual

$$n^{\alpha\beta}|_{\alpha} = n^{\alpha\beta}{}_{,\alpha}+\Gamma^{\beta}_{\alpha\rho}n^{\alpha\rho}+\Gamma^{\alpha}_{\alpha\rho}n^{\rho\beta}, \tag{11.1.14}$$

and represents covariant differentiation with respect to the middle surface. We observe that (11.1.12) and (11.1.13) can also be obtained from (10.4.11), (10.4.4), and (10.4.5) by neglecting shearing forces.

Equations (11.1.13) together with the symmetry conditions (11.1.3) are sufficient for us to determine the three unknown stress resultants $n^{\alpha\beta}$ and it is not necessary to discuss, at this stage, the displacements.†

We now consider how to obtain displacements corresponding to stress resultants $n^{\alpha\beta}$ calculated from (11.1.13). Let the displacement vector of the middle surface of the shell be denoted by

$$\mathbf{v} = v_{\alpha}\mathbf{a}^{\alpha}+w\mathbf{a}_3, \tag{11.1.15}$$

where w, v_{α} are functions of θ_{λ}. Under suitable continuity assumptions we see, from (10.2.13), that

$$\lim_{t\to 0}\frac{n^{\alpha\beta}}{t} = (\sigma^{\alpha\beta})_{\theta_3=0}.$$

For membrane theory we therefore assume that $n^{\alpha\beta}$ is given approximately by

$$n^{\alpha\beta} = t(\sigma^{\alpha\beta})_{\theta_3=0}. \tag{11.1.16}$$

From (10.5.3) and (10.5.10) we have

$$E\alpha_{\alpha\beta} = \tfrac{1}{2}(1+\eta)(a_{\alpha\lambda}a_{\beta\rho}+a_{\alpha\rho}a_{\beta\lambda})(\sigma^{\lambda\rho})_{\theta_3=0}-\eta a_{\alpha\beta}(a_{\lambda\mu}\sigma^{\lambda\mu}+\sigma^{33})_{\theta_3=0}, \tag{11.1.17}$$

where

$$\begin{aligned} \alpha_{\lambda\mu} &= \tfrac{1}{2}(v_{\lambda}|_{\mu}+v_{\mu}|_{\lambda}-2b_{\lambda\mu}w) \\ &= \tfrac{1}{2}(\mathbf{a}_{\lambda}.\mathbf{v}_{,\mu}+\mathbf{a}_{\mu}.\mathbf{v}_{,\lambda}). \end{aligned} \tag{11.1.18}$$

If $(\sigma^{33})_{\theta_3=0}$ can be neglected compared with the remaining stresses in (11.1.17), then (11.1.16) and (11.1.17) yield

$$\alpha_{\alpha\beta} = \frac{H^{*}_{\alpha\beta\lambda\rho}n^{\lambda\rho}}{Et}, \tag{11.1.19}$$

where

$$H^{*}_{\alpha\beta\rho\lambda} = \tfrac{1}{2}\{(1+\eta)(a_{\alpha\lambda}a_{\beta\rho}+a_{\alpha\rho}a_{\beta\lambda})-2\eta a_{\alpha\beta}a_{\lambda\rho}\}. \tag{11.1.20}$$

† Shells of different materials, but with the same middle surface, loads, and boundary conditions in terms of stress resultants, and therefore under the same state of membrane stress, may undergo different displacements.

Using the identities

$$a_{\alpha\beta}a_{\rho\lambda}-a_{\alpha\rho}a_{\beta\lambda} = \epsilon_{\alpha\lambda}\epsilon_{\beta\rho},$$

$$a_{\alpha\beta}a_{\rho\lambda}-a_{\beta\rho}a_{\alpha\lambda} = \epsilon_{\alpha\rho}\epsilon_{\beta\lambda},$$

the expression (11.1.20) can be put in the alternative form

$$H^*_{\alpha\beta\rho\lambda} = \tfrac{1}{2}\{a_{\alpha\lambda}a_{\beta\rho}+a_{\alpha\rho}a_{\beta\lambda}-\eta(\epsilon_{\alpha\lambda}\epsilon_{\beta\rho}+\epsilon_{\alpha\rho}\epsilon_{\beta\lambda})\}. \qquad (11.1.21)$$

Equations (11.1.19) may be solved to express $n^{\alpha\beta}$ in terms of $\alpha_{\rho\lambda}$. Thus

$$n^{\alpha\beta} = n^{\beta\alpha} = DH^{\alpha\beta\rho\lambda}\alpha_{\rho\lambda}, \qquad (11.1.22)$$

where

$$H^{\alpha\beta\rho\lambda} = \tfrac{1}{2}\{(1-\eta)(a^{\alpha\lambda}a^{\beta\rho}+a^{\alpha\rho}a^{\beta\lambda})+2\eta a^{\alpha\beta}a^{\rho\lambda}\}$$
$$= \tfrac{1}{2}\{a^{\alpha\lambda}a^{\beta\rho}+a^{\alpha\rho}a^{\beta\lambda}+\eta(\epsilon^{\alpha\lambda}\epsilon^{\beta\rho}+\epsilon^{\alpha\rho}\epsilon^{\beta\lambda})\}, \qquad (11.1.23)$$

$$D = Et/(1-\eta^2). \qquad (11.1.24)$$

The displacement components v_α, w contain rigid-body and inextensional displacements which do not contribute to the stress resultants $n^{\alpha\beta}$. Inextensional displacements are such that

$$\alpha_{\lambda\mu} = 0 \qquad (11.1.25)$$

so that the shell may be distorted without any stretching of the middle surface and therefore without inducing stresses of the type $n^{\alpha\beta}$. The theory of inextensional deformations is studied in more detail in Chapter 16.

11.2. Membrane theory in plane coordinates

It is sometimes useful to have the membrane theory expressed in the notation of § 10.6 where the results are referred to a base plane Π. Using the work of §§ 10.6, 11.1, we have, from (10.6.14) and (10.6.16),

$$\mathbf{N}_\alpha = \mathbf{K}_\alpha+k^{\alpha\beta}z_{,\beta}\sqrt{(e)}\mathbf{e}_3, \qquad \mathbf{K}_\alpha = K^{\alpha\beta}\mathbf{e}_\beta, \qquad (11.2.1)$$

where, from (10.6.17),

$$K^{\alpha\beta} = k^{\alpha\beta}\sqrt{e}, \qquad k^{\alpha\beta} = n^{\alpha\beta}\sqrt{(a/e)}. \qquad (11.2.2)$$

Also, the equations of equilibrium (11.1.10) become

$$\mathbf{K}_{\alpha,\alpha}+\mathbf{P}+(k^{\alpha\beta}z_{,\beta}\sqrt{e})_{,\alpha}\mathbf{e}_3 = \mathbf{0}, \qquad (11.2.3)$$

and using (10.6.19) and the second equation in (11.2.1) this reduces to

$$\left.\begin{aligned} K^{\alpha\beta}{}_{,\alpha}+\Gamma^\beta_{\rho\alpha}K^{\rho\alpha}+S^\beta &= 0\\ \{k^{\alpha\beta}z_{,\beta}\sqrt{e}\}_{,\alpha}+S &= 0\end{aligned}\right\}, \qquad (11.2.4)$$

where the Christoffel symbols $\Gamma^\beta_{\rho\alpha}$ now refer to the plane Π. Finally, with the help of (11.2.2), these equations may be further reduced to

$$k^{\alpha\beta}|_\alpha+s^\beta = 0, \qquad (11.2.5)$$

$$z_{,\beta}k^{\alpha\beta}|_\alpha+k^{\alpha\beta}z|_{\beta\alpha}+s = 0, \qquad (11.2.6)$$

where covariant differentiation† refers to the plane Π. Also, when moment stresses are zero, equations (10.4.7) and (11.2.2) give

$$\epsilon_{\alpha\beta} k^{\alpha\beta} = 0, \tag{11.2.7}$$

so that $k^{\alpha\beta}$ is symmetric. With the help of (11.2.5) we can replace (11.2.6) by

$$z|_{\alpha\beta} k^{\alpha\beta} + s - s^{\alpha} z|_{\alpha} = 0. \tag{11.2.8}$$

The membrane theory of shells has now been expressed entirely in terms of $k^{\alpha\beta}$ (or equivalent quantities) and we remind the reader that since covariant differentiation refers to the plane Π the order of covariant differentiation is immaterial. As far as the two equations of equilibrium (11.2.5) are concerned the shell problem is similar to the problem of the stretching of a plate, with $k^{\alpha\beta}$ (which is symmetric in view of (11.2.7)) and s^{α} for the shell replacing $n^{\alpha\beta}$, p^{α} for the plate (see Chapter 7). In the plate problem, however, we have to consider the state of strain before we can determine the complete state of stress, but here, in the membrane theory of shells, the theory is completed by using the third equation of equilibrium (11.2.8). Since the state of stress in the membrane theory of shells is completely determined by the equations of equilibrium (apart from the discussion of boundary conditions), we see that plate theory is not a special case of membrane shell theory.

We saw in Chapter 6 on plane stress, and again in Chapter 7, § 7.5, on plate theory, that stresses (or stress resultants) may be expressed in terms of a stress function ϕ. Hence, taking into account the load term s^{α}, the formulae for $k^{\alpha\beta}$ which correspond to (7.5.5) are

$$\left.\begin{aligned} k^{\alpha\beta} &= \epsilon^{\alpha\gamma}\epsilon^{\beta\rho}\phi|_{\gamma\rho} - A^{\alpha\beta} \\ \phi|_{\alpha\beta} &= \epsilon_{\alpha\gamma}\epsilon_{\beta\rho}(k^{\gamma\rho} + A^{\gamma\rho}) \end{aligned}\right\}, \tag{11.2.9}$$

where $A^{\alpha\beta}$ is symmetric and a particular integral of the equation

$$A^{\alpha\beta}|_{\alpha} = s^{\beta}. \tag{11.2.10}$$

If we substitute (11.2.9) into (11.2.8) we obtain the following differential equation for ϕ:

$$\epsilon^{\alpha\gamma}\epsilon^{\beta\rho} z|_{\alpha\beta}\phi|_{\gamma\rho} = q, \tag{11.2.11}$$

where

$$q = z|_{\alpha\beta} A^{\alpha\beta} - s + s^{\alpha} z|_{\alpha}, \tag{11.2.12}$$

and q contains the loads acting on the shell.

We have therefore reduced the evaluation of stresses of the membrane theory to the solution of one linear differential equation (11.2.11) of

† Covariant differentiation with respect to the middle surface M will not occur in this section.

the second order for ϕ. Having found ϕ we can obtain $k^{\alpha\beta}$ from (11.2.9) and then also $\alpha_{\gamma\rho}$, since, from (11.1.19) and (11.2.2),

$$\alpha_{\alpha\beta} = \frac{\sqrt{(e/a)}}{Et} H^{*}_{\alpha\beta\rho\lambda} k^{\rho\lambda}, \tag{11.2.13}$$

where, from (11.1.18) and (10.6.4),

$$\alpha_{\alpha\beta} = \tfrac{1}{2}(v_{\alpha}|_{\beta}+v_{\beta}|_{\alpha}+z_{,\alpha} w_{,\beta}+z_{,\beta} w_{,\alpha}), \tag{11.2.14}$$

$$\mathbf{v} = v_{\alpha}\,\mathbf{e}^{\alpha}+w\mathbf{e}_{3}. \tag{11.2.15}$$

The components of the displacement vector $\mathbf{v}$ with respect to the base vectors in Π, and $\mathbf{e}_3$, are here denoted by v_{α}, w respectively. From (11.2.14) we can deduce the equations

$$\epsilon^{\lambda\alpha}\epsilon^{\mu\beta}(\alpha_{\lambda\mu}|_{\alpha\beta}-z|_{\lambda\beta}\, w|_{\mu\alpha}) = 0, \tag{11.2.16}$$

which is a differential equation for w since all other quantities in the equation are known. When w is found we obtain v_{α} from (11.2.14) since

$$v_{\alpha}|_{\beta}+v_{\beta}|_{\alpha} = 2\alpha_{\alpha\beta}-z_{,\alpha} w_{,\beta}-z_{,\beta} w_{,\alpha}. \tag{11.2.17}$$

The boundary conditions to be imposed at the edges of the shell must be compatible with the differential equation which has been obtained for ϕ. To simplify the discussion we choose a rectangular cartesian coordinate system in the plane Π so that the equation (11.2.11) reduces to

$$z_{,11}\phi_{,22}-2z_{,12}\phi_{,12}+z_{,22}\phi_{,11} = q. \tag{11.2.18}$$

From the theory of differential equations it is known that this equation is of elliptic, hyperbolic, or parabolic type according as

$$\Lambda = (z_{,12})^{2}-z_{,11}z_{,22}$$

is negative, positive, or zero. But, using (10.6.12), we see that

$$\Lambda = (b_{12}^{2}-b_{11}b_{22})a/e,$$

or

$$\Lambda = -Ka^{2}/e,$$

where $a = |a_{\alpha\beta}|$ and K is the Gaussian curvature. Hence:

1. If the middle surface of the shell has positive Gaussian curvature everywhere, the differential equation is elliptic.
2. If the middle surface of the shell has negative Gaussian curvature everywhere, the differential equation is hyperbolic.

3. If the Gaussian curvature is zero everywhere, which means that the surface is developable, the equation is parabolic.

From the theory of differential equations the appropriate boundary conditions may be imposed which correspond to partial differential equations of the second order of elliptic, hyperbolic, or parabolic types.†

† See Courant–Hilbert, *Methoden der mathematischen Physik*, Vol. ii (Berlin, 1937), p. 177.

12 Bending Theory of Shells

THE membrane theory of Chapter 11 must, in general, be supplemented by a bending theory in which bending moments and shearing forces are non-zero, in order that correct boundary conditions may be satisfied at an edge of a shell. In this chapter we derive an approximate bending theory of shells by a direct method which has some similarities to that used for plate theory. In Chapter 16 we show that the equations presented here can be obtained more systematically from the three-dimensional equations of elasticity using asymptotic expansions.

Many problems of practical interest are solved satisfactorily using the bending theory given here, or using membrane theory together with bending theory in which the loads are taken to be zero, provided there is only a small contribution to bending and other stresses arising from nearly inextensional deformations. Further special aspects of bending theory are considered in Chapter 13 on shallow shells and in Chapter 15 when discussing 'edge effect'. If inextensional contributions are important they must be added to the membrane and bending theory. We indicate how inextensional stresses can be found in Chapter 16, but in the few applications of shell theory given in this book we restrict our attention to the theory of the present chapter, and membrane theory.

12.1. Approximation

We assume that the thickness t of the shell is small compared with a minimum radius of curvature R of the middle surface so that

$$t \ll R. \tag{12.1.1}$$

We recall that the coordinate θ_3 satisfies the condition (10.1.2) and write

$$\theta_3 = t\rho, \qquad -\tfrac{1}{2} \leqslant \rho \leqslant \tfrac{1}{2}. \tag{12.1.2}$$

If we take coordinate curves on the middle surface of the shell to be along lines of curvature then the geometrical tensor $\mu_\beta^\alpha = \delta_\beta^\alpha - \theta_3 b_\beta^\alpha$ has two non-zero components $1-t\rho/R_1$, $1-t\rho/R_2$ where R_1, R_2 are the principal radii of curvature. In view of (12.1.1) we neglect t/R_1, t/R_2

compared with unity. Hence, returning to general coordinates θ_α we have, approximately,

$$\mu^\alpha_\beta = \delta^\alpha_\beta, \quad h = 1, \quad \mu_{\alpha\beta} = a_{\alpha\beta}. \tag{12.1.3}$$

Making a similar approximation in (10.2.4) and (10.2.5) we have

$$\sigma^{\lambda\beta} = \sigma^{\beta\lambda}, \qquad \sigma^{3\lambda} = \sigma^{\lambda 3}. \tag{12.1.4}$$

The first equation in (10.4.2) is then

$$\sigma^{\alpha\beta}|_\alpha - b^\beta_\alpha \sigma^{3\alpha} + \frac{1}{t}\frac{\partial \sigma^{3\beta}}{\partial \rho} = 0, \tag{12.1.5}$$

so that if $\partial\sigma^{3\beta}/\partial\rho$ is of the same (or greater) order of magnitude as $\sigma^{3\beta}$ we have

$$\sigma^{\alpha\beta}|_\alpha + \partial\sigma^{3\beta}/\partial\theta^3 = 0, \tag{12.1.6}$$

by using the same approximation as above. This in turn means that the equations of equilibrium (10.4.4) are now replaced by

$$n^{\alpha\beta}|_\alpha + p^\beta = 0, \tag{12.1.7}$$

where, in view of (12.1.4), $\quad n^{\alpha\beta} = n^{\beta\alpha}. \qquad (12.1.8)$

For completeness we repeat (10.4.5) and (10.4.6), namely

$$b_{\alpha\beta} n^{\alpha\beta} + q^\alpha|_\alpha + p = 0, \tag{12.1.9}$$

$$m^{\alpha\beta}|_\alpha - q^\beta + \bar{p}^\beta = 0, \tag{12.1.10}$$

where $\bar{p}^\beta = \frac{1}{2}t\bar{p}^\beta$. If $\frac{1}{2}t\bar{p}^\alpha|_\alpha$ is small compared with p we may omit this term when q^β is substituted from (12.1.10) into (12.1.9). With this understanding we may replace (12.1.10) by

$$m^{\alpha\beta}|_\alpha - q^\beta = 0. \tag{12.1.11}$$

Because of (12.1.4) the stress couple $m^{\alpha\beta}$ is symmetric.

We turn next to the stress–strain relations in § 10.5. Using (12.1.3) it follows from (10.5.6), (10.5.3), (10.5.5), and (12.1.4) that

$$\left.\begin{aligned} E\gamma_{\alpha\beta} &= (1+\eta)a_{\alpha\lambda}a_{\beta\mu}\sigma^{\lambda\mu} - \eta a_{\alpha\beta}(a_{\lambda\mu}\sigma^{\lambda\mu} + \sigma^{33}) \\ E\gamma_{\alpha 3} &= (1+\eta)a_{\alpha\lambda}\sigma^{\lambda 3} \\ E\frac{\partial u_3}{\partial \rho} &= t(\sigma^{33} - \eta a_{\alpha\lambda}\sigma^{\alpha\lambda}) \end{aligned}\right\}, \tag{12.1.12}$$

where $\qquad 2\gamma_{\alpha\beta} = u_\alpha|_\beta + u_\beta|_\alpha - 2b_{\alpha\beta}u_3. \qquad (12.1.13)$

Moreover, from (10.5.4),

$$2\gamma_{\alpha 3} = \frac{\partial u_3}{\partial \theta^\alpha} + \frac{1}{t}\frac{\partial u_\alpha}{\partial \rho} + b^\lambda_\alpha\left(u_\alpha - \rho\frac{\partial u_\lambda}{\partial \rho}\right),$$

so that, if $\partial u_\lambda/\partial \rho$ is of the same order of magnitude as u_λ, we have

$$2\gamma_{\alpha 3} = \frac{\partial u_3}{\partial \theta^\alpha} + \frac{\partial u_\alpha}{\partial \theta^3}, \tag{12.1.14}$$

by using the same approximation as that used in obtaining (12.1.3).

In view of (12.1.1) we neglect the right-hand side of the last equation in (12.1.12), so that

$$\partial u_3/\partial \rho = 0, \qquad u_3 = w, \tag{12.1.15}$$

approximately, where w is a function of θ_1, θ_2. We also assume that the normal and shearing stresses σ^{33}, $\sigma^{\lambda 3}$ may be neglected in the remaining equations (12.1.12) so that

$$\frac{\partial u_\alpha}{\partial \theta^3} = -w|_\alpha,$$

and

$$u_\alpha = v_\alpha - \theta_3 w|_\alpha, \tag{12.1.16}$$

where v_α is a function of θ_1, θ_2. Moreover, (12.1.13) becomes

$$\gamma_{\alpha\beta} = \alpha_{\alpha\beta} + \theta_3 \kappa_{\alpha\beta}, \tag{12.1.17}$$

where

$$2\alpha_{\alpha\beta} = v_\alpha|_\beta + v_\beta|_\alpha - 2b_{\alpha\beta} w = \mathbf{a}_\alpha . \mathbf{v}_{,\beta} + \mathbf{a}_\beta . \mathbf{v}_{,\alpha}, \tag{12.1.18}$$

$$\kappa_{\alpha\beta} = -w|_{\alpha\beta} = \tfrac{1}{2}(\mathbf{a}_\alpha . \mathbf{w}_{,\beta} + \mathbf{a}_\beta . \mathbf{w}_{,\alpha}). \tag{12.1.19}$$

In (12.1.18) and (12.1.19), the vectors $\mathbf{v}$, $\mathbf{w}$ are defined by

$$\left.\begin{aligned} \mathbf{V} &= \mathbf{v} + \theta_3 \mathbf{w} \\ \mathbf{v} = v_\alpha \mathbf{a}^\alpha + w\mathbf{a}_3, &\qquad \mathbf{w} = w_\alpha \mathbf{a}^\alpha = -w|_\alpha \mathbf{a}^\alpha \end{aligned}\right\}, \tag{12.1.20}$$

where the displacement vector $\mathbf{V}$ is given by (10.5.1) and we have used (12.1.15) and (12.1.16). Finally, using (12.1.17), and neglecting σ^{33} in the first of equations (12.1.12), we have

$$\alpha_{\alpha\beta} = \frac{H^*_{\alpha\beta\lambda\rho} n^{\lambda\rho}}{Et}, \tag{12.1.21}$$

$$\kappa_{\alpha\beta} = -w|_{\alpha\beta} = \frac{12 H^*_{\alpha\beta\rho\lambda} m^{\rho\lambda}}{Et^3}, \tag{12.1.22}$$

where $H^*_{\alpha\beta\rho\lambda}$ is defined in (11.1.21). Alternatively,

$$n^{\alpha\beta} = n^{\beta\alpha} = DH^{\alpha\beta\rho\lambda}\alpha_{\rho\lambda}, \tag{12.1.23}$$

$$m^{\alpha\beta} = m^{\beta\alpha} = BH^{\alpha\beta\rho\lambda}\kappa_{\rho\lambda}, \tag{12.1.24}$$

where $H^{\alpha\beta\rho\lambda}$ is given by (11.1.23), and

$$D = \frac{Et}{1-\eta^2}, \qquad B = \frac{Et^3}{12(1-\eta^2)}. \tag{12.1.25}$$

We observe that the theory given here implies that the term $b^{\lambda}_{\alpha} v_{\lambda} \mathbf{a}_3$ in $\mathbf{v}_{,\alpha}$ may be neglected so that

$$\mathbf{v}_{,\alpha} = (v_{\lambda}|_{\alpha} - b_{\lambda\alpha} w)\mathbf{a}^{\lambda} + w_{,\alpha} \mathbf{a}_3, \tag{12.1.26}$$

approximately, and hence

$$\mathbf{w} = -w_{,\alpha} \mathbf{a}^{\alpha} = -(\mathbf{a}_3 . \mathbf{v}_{,\alpha})\mathbf{a}^{\alpha}. \tag{12.1.27}$$

12.2. Rate of work and elastic potential

The velocity and angular velocity of the middle surface are respectively $\dot{\mathbf{v}}$ and $\dot{\boldsymbol{\theta}}$ where

$$\mathbf{v} = v_{\alpha} \mathbf{a}^{\alpha} + w\mathbf{a}_3, \qquad \dot{\boldsymbol{\theta}} = \mathbf{a}_3 \times \dot{\mathbf{w}}, \tag{12.2.1}$$

and $\mathbf{w}$ is given by (12.1.27). We suppose that the middle surface of the shell is bounded by one or more closed curves which we denote by c, the unit outward normal in the surface at any point of c being $\boldsymbol{\nu}$, where

$$\boldsymbol{\nu} = \nu_{\alpha} \mathbf{a}^{\alpha}. \tag{12.2.2}$$

The rate of work of forces and couples acting on c together with the rate of work of loads acting on the middle surface of the shell is

$$\int_c (\bar{\mathbf{n}} . \dot{\mathbf{v}} + \overline{\mathbf{m}} . \dot{\boldsymbol{\theta}})\, ds + \iint (\mathbf{P} . \mathbf{v} + \dot{\mathbf{P}} . \dot{\boldsymbol{\theta}}) \frac{dS}{\sqrt{a}}, \tag{12.2.3}$$

where $\bar{\mathbf{n}}$, $\overline{\mathbf{m}}$ are the values of the stress resultant and stress couple at the curve c. Using (10.2.18) this becomes

$$\int_c \nu_{\alpha}(\mathbf{N}_{\alpha} . \dot{\mathbf{v}} + \mathbf{M}_{\alpha} . \dot{\boldsymbol{\theta}}) \frac{ds}{\sqrt{a}} + \iint (\mathbf{P} . \dot{\mathbf{v}} + \dot{\mathbf{P}} . \dot{\boldsymbol{\theta}}) \frac{dS}{\sqrt{a}}.$$

Transforming the line integral into an integral over the middle surface of the shell and using (10.4.12) and (10.4.15), this rate of work reduces to

$$\iint \{\mathbf{N}_{\alpha} . \dot{\mathbf{v}}_{,\alpha} + \mathbf{M}_{\alpha} . \dot{\boldsymbol{\theta}}_{,\alpha} - [\mathbf{a}_{\alpha} \mathbf{N}_{\alpha} \dot{\boldsymbol{\theta}}]\} \frac{dS}{\sqrt{a}}. \tag{12.2.4}$$

From (10.2.10), (10.2.12), and (12.2.1),

$$\mathbf{N}_{\alpha} . \dot{\mathbf{v}}_{,\alpha} - [\mathbf{a}_{\alpha} \mathbf{N}_{\alpha} \dot{\boldsymbol{\theta}}] = \sqrt{(a)} n^{\alpha\rho} \mathbf{a}_{\rho} . \dot{\mathbf{v}}_{,\alpha} = \sqrt{(a)} n^{\alpha\rho} \dot{\alpha}_{\alpha\rho} \tag{12.2.5}$$

if we use (12.1.18) and remember than $n^{\alpha\rho}$ is symmetric. Again, from (10.2.11), (10.2.12), and (12.1.19),

$$\begin{aligned} \mathbf{M}_{\alpha} . \dot{\boldsymbol{\theta}}_{,\alpha} &= [\mathbf{M}_{\alpha} \mathbf{a}_{3,\alpha} \dot{\mathbf{w}}] + \sqrt{(a)} m^{\alpha\rho} (\mathbf{a}_3 \times \mathbf{a}_{\rho}) . (\mathbf{a}_3 \times \dot{\mathbf{w}}_{,\alpha}) \\ &= \tfrac{1}{2} \sqrt{(a)} m^{\alpha\rho} (\mathbf{a}_{\rho} . \dot{\mathbf{w}}_{,\alpha} + \mathbf{a}_{\alpha} . \dot{\mathbf{w}}_{,\rho}) \\ &= \sqrt{(a)} m^{\alpha\rho} \dot{\kappa}_{\alpha\rho}, \end{aligned} \tag{12.2.6}$$

since $m^{\alpha\rho}$ is symmetric and $[\mathbf{M}_\alpha\, \mathbf{a}_{3,\alpha}\, \dot{\mathbf{w}}]$ is zero as $\mathbf{a}_{3,\alpha}$ is perpendicular to $\mathbf{M}_\alpha \times \dot{\mathbf{w}}$. Combining (12.2.5) and (12.2.6), we have

$$\mathbf{N}_\alpha . \dot{\mathbf{v}}_{,\alpha} + \mathbf{M}_\alpha . \dot{\mathbf{\theta}}_{,\alpha} - [\mathbf{a}_\alpha\, \mathbf{N}_\alpha\, \dot{\mathbf{\theta}}] = \surd(a) n^{\alpha\beta}\dot{\alpha}_{\alpha\beta} + \surd(a) m^{\alpha\beta}\dot{\kappa}_{\alpha\beta}. \tag{12.2.7}$$

The expression in (12.2.4) therefore reduces to

$$\iint \dot{W}\surd(a)\, d\theta^1 d\theta^2, \tag{12.2.8}$$

where

$$\dot{W} = n^{\alpha\beta}\dot{\alpha}_{\alpha\beta} + m^{\alpha\beta}\dot{\kappa}_{\alpha\beta}. \tag{12.2.9}$$

With the help of (12.1.23) and (12.1.24) this equation may be integrated to give

$$W = \tfrac{1}{2} H^{\alpha\beta\rho\lambda}(D\alpha_{\rho\lambda}\alpha_{\alpha\beta} + B\kappa_{\rho\lambda}\kappa_{\alpha\beta}), \tag{12.2.10}$$

and

$$n^{\alpha\beta} = \frac{1}{2}\left(\frac{\partial W}{\partial \alpha_{\alpha\beta}} + \frac{\partial W}{\partial \alpha_{\beta\alpha}}\right), \qquad m^{\alpha\beta} = \frac{1}{2}\left(\frac{\partial W}{\partial \kappa_{\alpha\beta}} + \frac{\partial W}{\partial \kappa_{\beta\alpha}}\right). \tag{12.2.11}$$

Also

$$W = \tfrac{1}{2} H^*_{\alpha\beta\rho\lambda}\left(\frac{n^{\alpha\beta}n^{\rho\lambda}}{Et} + \frac{12 m^{\alpha\beta}m^{\rho\lambda}}{Et^3}\right), \tag{12.2.12}$$

and W is the elastic potential or strain energy per unit area of the middle surface.

12.3. Edge conditions

In order to find the correct edge conditions† we consider the rate of work of forces and couples acting along a boundary curve c of the shell as given by the first integral in (12.2.3). We follow the same procedure as for plate theory in § 7.13 and put

$$\left.\begin{aligned} \mathbf{m} &= G[\mathbf{a}_3 \times \boldsymbol{\nu}] + H\boldsymbol{\nu} \\ \mathbf{n} &= n^\alpha \mathbf{a}_\alpha + n\mathbf{a}_3 \end{aligned}\right\}, \tag{12.3.1}$$

where $\boldsymbol{\nu}$ is the unit normal at a point of the curve c, lying in the middle surface at the point. The physical components of $\mathbf{m}$ are G and H along the tangent and normal to the curve c, and n is the physical component of $\mathbf{n}$ normal to the surface (see Fig. 7.3). From (12.3.1), (10.2.18), and (10.2.10)–(10.2.12), we have

$$\left.\begin{aligned} G &= \mathbf{m}.[\mathbf{a}_3 \times \boldsymbol{\nu}] = m^{\alpha\beta}\nu_\alpha \nu_\beta \\ H &= \mathbf{m}.\boldsymbol{\nu} = \epsilon_{\beta\lambda} m^{\alpha\beta}\nu_\alpha \nu^\lambda \\ n &= q^\alpha \nu_\alpha, \quad n^\alpha = \nu_\lambda n^{\lambda\alpha} \end{aligned}\right\}. \tag{12.3.2}$$

† For an alternative derivation of edge conditions for shells via boundary layer equations, see A. E. Green, *Proc. R. Soc.* **A269** (1962) 481 and A. E. Green and N. Laws, ibid. **A289** (1966) 171.

If $\partial/\partial s$, $\partial/\partial \nu$ denote derivatives along the tangent and normal to c in the middle surface, then

$$w_{,\alpha} = \nu_\alpha \frac{\partial w}{\partial \nu} - \epsilon_{\alpha\beta}\nu^\beta \frac{\partial w}{\partial s} \tag{12.3.3}$$

and, from (12.1.27), (12.2.1), and (12.3.3), we have

$$\dot{\boldsymbol{\theta}} = -\frac{\partial \dot{w}}{\partial \nu}[\mathbf{a}_3 \times \boldsymbol{\nu}] + \frac{\partial \dot{w}}{\partial s}\boldsymbol{\nu}. \tag{12.3.4}$$

Hence $$\bar{\mathbf{m}}.\dot{\boldsymbol{\theta}} + \bar{\mathbf{n}}.\dot{\mathbf{v}} = \bar{n}^\beta \dot{v}_\beta + \bar{n}\dot{w} - \bar{G}\frac{\partial \dot{w}}{\partial \nu} + \bar{H}\frac{\partial \dot{w}}{\partial s},$$

and the rate of work of forces at the edge c is

$$\int \left\{\bar{n}^\alpha \dot{v}_\beta + \left(\bar{n} - \frac{\partial \bar{H}}{\partial s}\right)\dot{w} - \bar{G}\frac{\partial \dot{w}}{\partial \nu}\right\} ds \tag{12.3.5}$$

if $\bar{H}\dot{w}$ is single-valued.

The expression (12.3.5) shows that we may consider the edge forces to consist of component stress resultants n^α along $\mathbf{a}_\alpha$ and $n - \partial H/\partial s$ along $\mathbf{a}_3$, together with a tangential couple G.

(a) Clamped edge

At a clamped edge the displacement is zero so that

$$\mathbf{v} = 0, \qquad \mathbf{w} = 0.$$

In components this becomes

$$v_\alpha = 0, \qquad w = 0, \qquad w_\alpha = 0,$$

but remembering (12.1.20), the last condition can be replaced by

$$w_{,\alpha} = 0.$$

This, however, implies that w is constant along the edge c, and the constant may be taken to be zero for a single boundary, so that the clamped boundary conditions become

$$v_\alpha = 0, \qquad w_{,\alpha} = 0. \tag{12.3.6}$$

Alternatively, we may replace these by

$$v_\alpha = 0, \qquad w = 0, \qquad \frac{\partial w}{\partial \nu} = 0. \tag{12.3.7}$$

(*b*) *Simply supported edge*

This is characterized by the conditions

$$\left.\begin{aligned} G = 0, \quad \mathbf{n}.\boldsymbol{\nu} = n^{\alpha}\nu_{\alpha} = 0, \quad w = 0 \\ \mathbf{v}.[\mathbf{a}_3 \times \boldsymbol{\nu}] = 0 \quad \text{or} \quad \epsilon^{\rho\lambda}\nu_{\rho}v_{\lambda} = 0 \end{aligned}\right\}. \tag{12.3.8}$$

(*c*) *Free edge*

At a free edge there are zero prescribed stresses so that

$$n^{\alpha} = 0, \qquad n - \frac{\partial H}{\partial s} = 0, \qquad G = 0. \tag{12.3.9}$$

13 Theory of Shallow Shells

THE solution of shell problems using the membrane theory and the bending theory as developed in Chapters 11 and 12 often causes considerable difficulties. In some problems of practical importance it is, however, possible to make further simplifications in the theory, and in this chapter we develop a special theory for shells which are shallow.

13.1. Approximations for shallow shells

Plane coordinates have been considered in § 10.6 and we adopt the notations of that section here. We define a shallow shell to be one in which the amount of deviation z from a plane, measured normally to the plane, is small compared with a maximum length L of an edge of the shell, which in turn is small compared with a minimum radius of curvature R of the middle surface. We assume that derivatives of z with respect to θ_α are not increased in order of magnitude compared with z. If we neglect terms of order of magnitude $(z/L)^2$ compared with unity, in formulae of § 10.6 and in the bending theory of Chapter 12, we find the following approximations for shallow shells:

$$a_{\alpha\beta} = e_{\alpha\beta}, \qquad a^{\alpha\beta} = e^{\alpha\beta}, \qquad a = e, \tag{13.1.1}$$

$$\bar{\epsilon}_{\alpha\beta} = \epsilon_{\alpha\beta}, \qquad \bar{\epsilon}^{\alpha\beta} = \epsilon^{\alpha\beta}, \qquad \bar{\Gamma}^{\alpha}_{\beta\gamma} = \Gamma^{\alpha}_{\beta\gamma}, \qquad b_{\alpha\beta} = z|_{\alpha\beta}, \tag{13.1.2}$$

$$\left.\begin{aligned} \mathbf{a}^\alpha &= \mathbf{e}^\alpha + z|^\alpha \mathbf{e}_3, \qquad \mathbf{a}_3 = \mathbf{e}_3 - z|_\alpha \mathbf{e}^\alpha \\ \mathbf{a}_\alpha &= \mathbf{e}_\alpha + z|_\alpha \mathbf{e}_3 \end{aligned}\right\}. \tag{13.1.3}$$

Alternatively, we have, from (13.1.3),

$$\mathbf{e}_3 = \mathbf{a}_3 + z|_\alpha \mathbf{a}^\alpha, \qquad \mathbf{e}^\alpha = \mathbf{a}^\alpha - z|^\alpha \mathbf{a}_3, \tag{13.1.4}$$

approximately. Also

$$k^{\alpha\beta} = n^{\alpha\beta}, \qquad h^{\alpha\beta} = m^{\alpha\beta}, \tag{13.1.5}$$

$$p = s - s^\alpha z|_\alpha, \qquad p^\alpha = s^\alpha + s z|^\alpha, \tag{13.1.6}$$

and we shall usually replace (13.1.6) by the further approximation

$$p = s - s^\alpha z|_\alpha, \qquad p^\alpha = s^\alpha. \tag{13.1.7}$$

Since, for shallow shells, the Christoffel symbols for the surface M are equal to those for the plane Π, and because of the relations

(13.1.5), (13.1.7), and (13.1.2), the equations of equilibrium (12.1.7), (12.1.9), and (12.1.11) reduce to

$$\left.\begin{aligned} k^{\alpha\beta}|_\alpha+s^\beta &= 0 \\ b_{\alpha\beta}k^{\alpha\beta}+q^\alpha|_\alpha+s-s^\alpha z|_\alpha &= 0 \\ h^{\alpha\beta}|_\alpha-q^\beta &= 0 \end{aligned}\right\}, \tag{13.1.8}$$

where covariant differentiation is with respect to the plane Π.

13.2. Stress–strain relations

The components of the displacement vector $\mathbf{v}$ of the middle surface M with respect to base vectors $\mathbf{a}^\alpha$, $\mathbf{a}_3$, as defined in Chapter 12, are now denoted by $\bar{v}_\alpha$, $\bar{w}$ while v_α, w in this section are reserved for the components of $\mathbf{v}$ referred to base vectors $\mathbf{e}^\alpha$, $\mathbf{e}_3$. Thus

$$\mathbf{v} = \bar{v}_\alpha \mathbf{a}^\alpha+\bar{w}\mathbf{a}_3 = v_\alpha \mathbf{e}^\alpha+w\mathbf{e}_3. \tag{13.2.1}$$

With the help of (13.1.4), equation (13.2.1) gives

$$\left.\begin{aligned} \bar{v}_\alpha &= v_\alpha+wz|_\alpha \\ \bar{w} &= w-v_\alpha z|^\alpha \end{aligned}\right\}. \tag{13.2.2}$$

From (13.2.1),

$$\mathbf{v}_{,\alpha} = v_\lambda|_\alpha \mathbf{e}^\lambda+w_{,\alpha}\mathbf{e}_3, \tag{13.2.3}$$

where

$$v_\lambda|_\alpha = v_{\lambda,\alpha}-\Gamma^\mu_{\lambda\alpha}v_\mu \tag{13.2.4}$$

is the covariant derivative of v_λ with respect to the plane Π. Substituting (13.2.3) into the second equation in (12.1.18), and using (13.1.3), gives

$$\alpha_{\lambda\mu} = \tfrac{1}{2}(v_\lambda|_\mu+v_\mu|_\lambda+z|_\lambda w|_\mu+z|_\mu w|_\lambda). \tag{13.2.5}$$

Alternatively, using the first equation in (12.1.18) and (13.2.2), we recover (13.2.5) provided we neglect the term $v_\alpha z|^\alpha$ in the expression for $\bar{w}$, so that (13.2.2) is replaced by

$$\bar{v}_\alpha = v_\alpha+wz|_\alpha, \qquad \bar{w} = w. \tag{13.2.6}$$

It then follows from (12.1.19) that

$$\kappa_{\alpha\beta} = -w|_{\alpha\beta} = -w|_{\beta\alpha} \tag{13.2.7}$$

approximately.

With the help of (13.1.1) and (13.1.2) we see that, for shallow shells, equations (11.1.21) and (11.1.23) reduce to

$$H^*_{\alpha\beta\rho\lambda} = \tfrac{1}{2}\{e_{\alpha\lambda}e_{\beta\rho}+e_{\alpha\rho}e_{\beta\lambda}-\eta(\epsilon_{\alpha\lambda}\epsilon_{\beta\rho}+\epsilon_{\alpha\rho}\epsilon_{\beta\lambda})\}, \tag{13.2.8}$$

$$H^{\alpha\beta\rho\lambda} = \tfrac{1}{2}\{e^{\alpha\lambda}e^{\beta\rho}+e^{\alpha\rho}e^{\beta\lambda}+\eta(\epsilon^{\alpha\lambda}\epsilon^{\beta\rho}+\epsilon^{\alpha\rho}\epsilon^{\beta\lambda})\}. \tag{13.2.9}$$

Finally, using (13.1.5), the formulae (12.1.23) and (12.1.24) for stress resultants and couples become

$$k^{\alpha\beta} = DH^{\alpha\beta\rho\lambda}\alpha_{\rho\lambda}, \tag{13.2.10}$$

$$h^{\alpha\beta} = BH^{\alpha\beta\rho\lambda}\kappa_{\rho\lambda}, \tag{13.2.11}$$

where, as in (12.1.25),

$$D = \frac{Et}{1-\eta^2}, \qquad B = \frac{Et^3}{12(1-\eta^2)}. \tag{13.2.12}$$

Also, from (12.1.21) and (12.1.22),

$$\alpha_{\alpha\beta} = \frac{H^*_{\alpha\beta\rho\lambda}k^{\rho\lambda}}{Et}, \tag{13.2.13}$$

$$\kappa_{\alpha\beta} = \frac{12H^*_{\alpha\beta\rho\lambda}h^{\rho\lambda}}{Et^3}. \tag{13.2.14}$$

13.3. Differential equations

The first equation of equilibrium in (13.1.8) can be satisfied by introducing a stress function ϕ. Thus

$$k^{\alpha\beta} = \epsilon^{\alpha\gamma}\epsilon^{\beta\rho}\phi|_{\gamma\rho} - B^{\alpha\beta}, \tag{13.3.1}$$

where the order of covariant differentiation, which is with respect to the plane Π, is immaterial, and where $B^{\alpha\beta}$ is symmetric and a particular integral of the equation

$$B^{\alpha\beta}|_{\alpha} = s^{\beta}. \tag{13.3.2}$$

From (13.3.1) and (13.2.13),

$$\alpha_{\lambda\mu} = \frac{H^*_{\lambda\mu\alpha\beta}}{Et}(\epsilon^{\alpha\gamma}\epsilon^{\beta\rho}\phi|_{\gamma\rho} - B^{\alpha\beta}), \tag{13.3.3}$$

and substituting this in the equation

$$\epsilon^{\lambda\alpha}\epsilon^{\mu\beta}(\alpha_{\lambda\mu}|_{\alpha\beta} - z|_{\lambda\beta}w|_{\mu\alpha}) = 0, \tag{13.3.4}$$

which is derived from (13.2.5), gives

$$e^{\alpha\beta}e^{\gamma\rho}\phi|_{\alpha\beta\gamma\rho} + Et\epsilon^{\alpha\rho}\epsilon^{\beta\gamma}z|_{\alpha\beta}w|_{\rho\gamma} - \epsilon^{\alpha\beta}\epsilon^{\gamma\rho}H^*_{\alpha\gamma\xi\eta}B^{\xi\eta}|_{\beta\rho} = 0. \tag{13.3.5}$$

Equation (13.3.5) is one partial differential equation for ϕ and w. In order to find a second equation we eliminate q^{α} from the last two equations in (13.1.8) to get

$$k^{\alpha\beta}z|_{\alpha\beta} + h^{\alpha\beta}|_{\alpha\beta} + s - s^{\alpha}z|_{\alpha} = 0.$$

We substitute in this equation for $k^{\alpha\beta}$, $h^{\alpha\beta}$ from (13.3.1), (13.2.11), and (13.2.7) and obtain

$$Be^{\alpha\beta}e^{\gamma\rho}w|_{\alpha\beta\gamma\rho} - \epsilon^{\alpha\rho}\epsilon^{\beta\gamma}z|_{\alpha\beta}\phi|_{\rho\gamma} - s + s^{\alpha}z|_{\alpha} + B^{\alpha\beta}z|_{\alpha\beta} = 0. \tag{13.3.6}$$

The problem of shallow shells is now governed by the two equations, (13.3.5) and (13.3.6). These equations may be replaced by a single equation for a complex function. If we multiply (13.3.5) by a constant k and (13.3.6) by $1/B$, and add, we obtain

$$e^{\alpha\beta}e^{\gamma\rho}(w|_{\alpha\beta\gamma\rho}+k\phi|_{\alpha\beta\gamma\rho})+kEt\epsilon^{\alpha\rho}\epsilon^{\beta\gamma}z|_{\alpha\beta}\left(w|_{\rho\gamma}-\frac{\phi|_{\rho\gamma}}{kBEt}\right)$$

$$=\frac{s-s^{\alpha}z|_{\alpha}-B^{\alpha\beta}z|_{\alpha\beta}}{B}+k(\epsilon^{\alpha\beta}\epsilon^{\gamma\rho}B_{\alpha\gamma}|_{\beta\rho}-\eta B^{\alpha\beta}|_{\alpha\beta}). \quad (13.3.7)$$

We now choose k to satisfy the equation

$$k = -\frac{1}{kBEt},$$

so that

$$k = \pm i\kappa, \quad (13.3.8)$$

where

$$\kappa = \frac{\sqrt{\{12(1-\eta^2)\}}}{Et^2}. \quad (13.3.9)$$

Then, writing

$$\psi = w+i\kappa\phi, \qquad \lambda = \frac{t}{\sqrt{\{12(1-\eta^2)\}}}, \quad (13.3.10)$$

and using the root $k = i\kappa$, equation (13.3.7) becomes

$$e^{\alpha\beta}e^{\gamma\rho}\psi|_{\alpha\beta\gamma\rho}+(i/\lambda)\epsilon^{\alpha\rho}\epsilon^{\beta\gamma}z|_{\alpha\beta}\psi|_{\rho\gamma} = q, \quad (13.3.11)$$

where

$$q = (s-s^{\alpha}z|_{\alpha}-B^{\alpha\beta}z|_{\alpha\beta})/B-i\kappa\eta s^{\alpha}|_{\alpha}+i\kappa\epsilon^{\alpha\beta}\epsilon^{\gamma\rho}B_{\alpha\gamma}|_{\beta\rho}. \quad (13.3.12)$$

If we use the other value of k given in (13.3.8), we obtain an equation which is the complex conjugate of (13.3.11). Equation (13.3.11) is a fourth-order equation for the complex function ψ. The real part of ψ is the displacement w normal to the plane Π and the imaginary part of ψ is the stress function, apart from a constant factor. The values of $k^{\alpha\beta}$, $h^{\alpha\beta}$, q^{α}, and v_{α} follow from (13.3.1), (13.2.11), and (13.2.7) when we know w and ϕ.

In the remainder of this chapter we examine the special case of a shallow spherical cap formed by the revolution through an angle 2π of an arc of a circle about a line in its plane through the centre of the circle O and one end of the arc (see Fig. 13.1).

13.4. Geometrical relations for spherical cap

The radius of the spherical cap is R, the maximum depth of the cap is h, and r_0 is the radius of the bounding circle of the cap. We take the plane Π to be the plane of this bounding circle and we choose polar coordinates (r, θ) in this plane with origin at the centre of the circle. We choose coordinates θ_1, θ_2 as

$$\theta_1 = r, \qquad \theta_2 = \theta, \tag{13.4.1}$$

where $0 \leqslant r \leqslant r_0$, $0 \leqslant \theta \leqslant 2\pi$.

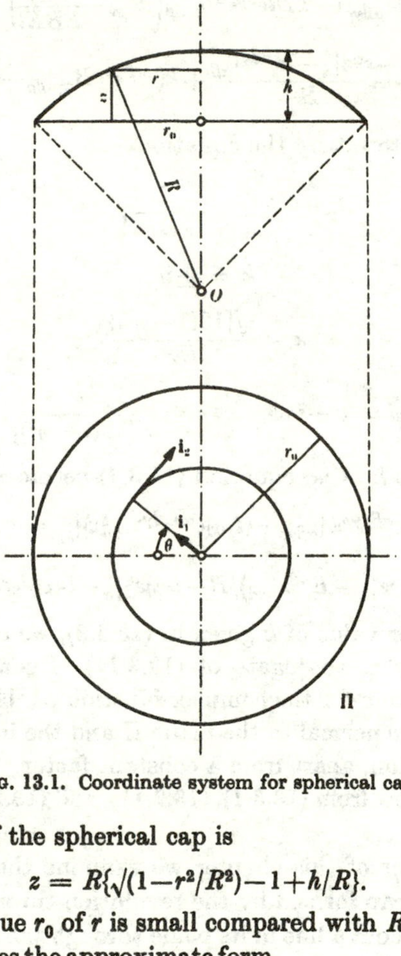

FIG. 13.1. Coordinate system for spherical cap.

The equation of the spherical cap is

$$z = R\{\sqrt{(1-r^2/R^2)}-1+h/R\}. \tag{13.4.2}$$

The maximum value r_0 of r is small compared with R for shallow shells so that (13.4.2) takes the approximate form

$$z = h-\frac{r^2}{2R}. \tag{13.4.3}$$

The position vector $\bar{\mathbf{r}}$ of a point in Π is given by

$$\bar{\mathbf{r}} = r\mathbf{i}_1, \tag{13.4.4}$$

where $\mathbf{i}_1$ is a unit vector as shown in Fig. 13.1. We also introduce a unit vector $\mathbf{i}_2$ which is perpendicular to $\mathbf{i}_1$ as shown in Fig. 13.1. It then follows that

$$\left.\begin{aligned} \dot{\mathbf{i}}_1 &= \mathbf{i}_2 \\ \dot{\mathbf{i}}_2 &= -\mathbf{i}_1 \\ \mathbf{i}_1' &= \mathbf{i}_2' = 0 \end{aligned}\right\}, \tag{13.4.5}$$

where derivatives with respect to θ are denoted by a dot $\cdot$ and derivatives with respect to r by a dash $'$. From (10.6.2), (13.4.4), and (13.4.5) we have

$$\left.\begin{aligned} \mathbf{e}_1 &= \mathbf{\bar{r}}' = \mathbf{i}_1 \\ \mathbf{e}_2 &= \dot{\mathbf{\bar{r}}} = r\mathbf{i}_2 \end{aligned}\right\}, \tag{13.4.6}$$

$$\left.\begin{aligned} \mathbf{e}_{1,1} &= 0 \\ \mathbf{e}_{1,2} &= \mathbf{e}_{2,1} = \mathbf{i}_2 \\ \mathbf{e}_{2,2} &= -r\mathbf{i}_1 \end{aligned}\right\}, \tag{13.4.7}$$

$$\left.\begin{aligned} e_{11} = 1, \quad e_{12} = e_{21} = 0, \quad e_{22} = r^2 \\ e = e_{11}\,e_{22} = r^2 \\ e^{11} = 1, \quad e^{12} = e^{21} = 0, \quad e^{22} = 1/r^2 \end{aligned}\right\}, \tag{13.4.8}$$

$$\mathbf{e}^1 = \mathbf{i}_1, \qquad \mathbf{e}^2 = \mathbf{i}_2/r, \tag{13.4.9}$$

$$\left.\begin{aligned} \Gamma^1_{11} &= \mathbf{e}^1.\mathbf{e}_{1,1} = 0, & \Gamma^1_{12} &= \mathbf{e}^1.\mathbf{e}_{1,2} = 0 \\ \Gamma^1_{22} &= \mathbf{e}^1.\mathbf{e}_{2,2} = -r, & \Gamma^2_{11} &= \mathbf{e}^2.\mathbf{e}_{1,1} = 0 \\ \Gamma^2_{12} &= \mathbf{e}^2.\mathbf{e}_{1,2} = 1/r, & \Gamma^2_{22} &= \mathbf{e}^2.\mathbf{e}_{2,2} = 0 \end{aligned}\right\}. \tag{13.4.10}$$

We now calculate the functions $H^{\alpha\beta\rho\lambda}$ and $H^*_{\alpha\beta\rho\lambda}$ from (13.2.8) and (13.2.9). Thus

$$\left.\begin{aligned} H^{1111} = 1, \qquad H^{2222} = 1/r^4 \\ H^{1122} = \eta/r^2, \quad H^{1212} = \tfrac{1}{2}(1-\eta)/r^2 \\ H^{1112} = H^{2221} = 0 \end{aligned}\right\}, \tag{13.4.11}$$

$$\left.\begin{aligned} H^*_{1111} = 1, \qquad H^*_{2222} = r^4 \\ H^*_{1122} = -\eta r^2, \quad H^*_{1212} = \tfrac{1}{2}(1+\eta)r^2 \\ H^*_{1112} = H^*_{2221} = 0 \end{aligned}\right\}. \tag{13.4.12}$$

Finally, from (13.4.3) and (13.4.10), we have

$$\left.\begin{aligned} z|_{11} = z|^{11} = z_{,11} = -1/R \\ z|_{22} = z_{,22}+rz_{,1} = -r^2/R \\ z|_{12} = z|_{21} = 0 \\ z|_{\alpha\beta} = -e_{\alpha\beta}/R \end{aligned}\right\}. \tag{13.4.13}$$

13.5. Differential equations for spherical cap

The stress resultants and couples found in § 13.4 are now denoted by

$$\left.\begin{aligned} k^{11} &= k_r, \quad k^{12} = k^{21} = k_{r\theta}, \quad k^{22} = k_\theta \\ h^{11} &= h_r, \quad h^{12} = h^{21} = h_{r\theta}, \quad h^{22} = h_\theta \\ q^1 &= q_r, \quad q^2 = q_\theta \end{aligned}\right\}. \qquad (13.5.1)$$

We restrict discussion to problems for which the loads s^α are zero so that we may choose the particular integral of (13.3.2) to be zero. Thus

$$s^\alpha = 0, \qquad B^{\alpha\beta} = 0. \qquad (13.5.2)$$

We also write

$$v_1 = u, \qquad v_2 = v, \qquad (13.5.3)$$

for the components of displacement v_α defined in (13.2.1). Using (13.5.2) and (13.4.10) we have, from (13.3.1)

$$\left.\begin{aligned} k_r &= (\phi^{\cdot\cdot}+r\phi')/r^2 \\ k_{r\theta} &= -(\phi'^{\cdot}-\phi^{\cdot}/r)/r^2 \\ k_\theta &= \phi''/r^2 \end{aligned}\right\}. \qquad (13.5.4)$$

Also, from (13.1.8),

$$\left.\begin{aligned} q_r &= h_r'+h_{r\theta}^{\cdot}+h_r/r-rh_\theta \\ q_\theta &= h_{r\theta}'+h_\theta^{\cdot}+3h_{r\theta}/r \end{aligned}\right\}. \qquad (13.5.5)$$

The stress couples are obtained from (13.2.11), (13.2.7), and (13.4.11) in the form

$$\left.\begin{aligned} h_r &= -B\{w''+\eta(w^{\cdot\cdot}+rw')/r^2\} \\ h_{r\theta} &= -B(1-\eta)(w'^{\cdot}-w^{\cdot}/r)/r^2 \\ h_\theta &= -B(w^{\cdot\cdot}+rw'+\eta r^2w'')/r^4 \end{aligned}\right\}. \qquad (13.5.6)$$

Using the notation of (13.5.3), equation (13.2.5), with the help of (13.4.3), gives

$$\left.\begin{aligned} \alpha_{11} &= u'-rw'/R \\ \alpha_{12} &= \tfrac{1}{2}(u^{\cdot}+v')-v/r-\tfrac{1}{2}rw^{\cdot}/R \\ \alpha_{22} &= v^{\cdot}+ru \end{aligned}\right\}, \qquad (13.5.7)$$

where $\alpha_{\lambda\mu}$ are obtained from (13.2.13), (13.4.12), and (13.5.4) in the forms

$$\left.\begin{aligned} Et\alpha_{11} &= (\phi^{\cdot\cdot}+r\phi')/r^2-\eta\phi'' \\ Et\alpha_{12} &= -(1+\eta)(\phi'^{\cdot}-\phi^{\cdot}/r) \\ Et\alpha_{22} &= r^2\phi''-\eta\phi^{\cdot\cdot}-\eta r\phi' \end{aligned}\right\}. \qquad (13.5.8)$$

The formulae (13.5.4)–(13.5.8) show that all stress resultants, stress couples, and displacements are expressed in terms of the two functions $\kappa\phi$ and w, which are the imaginary and real parts of ψ given by (13.3.10).

Since, for a shallow spherical cap, from (13.4.13), $z|_{\alpha\beta} = -e_{\alpha\beta}/R$, the differential equation (13.3.11) for ψ can be written

$$\psi|^{\alpha\beta}_{\alpha\beta} - \left(\frac{i}{\lambda R}\right)\psi|^{\alpha}_{\alpha} = s/B$$

if we use the result $\epsilon^{\alpha\rho}\epsilon^{\beta\gamma}e_{\alpha\beta} = e^{\rho\gamma}$.

Hence
$$\nabla^4\psi - \frac{i}{\lambda R}\nabla^2\psi = s/B, \tag{13.5.9}$$

where
$$\nabla^2\psi = \psi|^{\alpha}_{\alpha} = \psi'' + \frac{\psi'}{r} + \frac{\psi^{\cdot\cdot}}{r^2}. \tag{13.5.10}$$

In the next section we consider briefly the solution of the homogeneous differential equation for the case of rotational symmetry.

13.6. Solution of homogeneous equation for rotational symmetry

The homogeneous equation for ψ is, from (13.5.9),

$$\nabla^4\psi - i\epsilon^2\nabla^2\psi = 0, \tag{13.6.1}$$

where
$$\epsilon = (1/R\lambda)^{\frac{1}{2}} = (1/Rt)^{\frac{1}{2}}\{12(1-\eta^2)\}^{\frac{1}{4}}. \tag{13.6.2}$$

The general solution of (13.6.1) is

$$\psi = \psi_1 + \psi_2, \tag{13.6.3}$$

where
$$\nabla^2\psi_1 = 0, \qquad \nabla^2\psi_2 - i\epsilon^2\psi_2 = 0, \tag{13.6.4}$$

and this solution is also valid for the general case of no rotational symmetry.

Assuming rotational symmetry, and remembering (13.5.10), we see that the solutions of (13.6.4) are

$$\psi_1 = A_1 + A_2 \ln r, \tag{13.6.5}$$

$$\psi_2 = A_3 I_0(\epsilon r\sqrt{i}) + A_4 K_0(\epsilon r\sqrt{i}), \tag{13.6.6}$$

where A_1, A_2, A_3, A_4 are arbitrary complex constants and $I_0(x)$, $K_0(x)$ are Bessel functions of the first and second kind respectively with purely imaginary arguments. It is, however, more convenient to use the forms†

$$\left.\begin{aligned} I_0(x\sqrt{i}) &= \operatorname{ber} x + i \operatorname{bei} x \\ K_0(x\sqrt{i}) &= \operatorname{ker} x + i \operatorname{kei} x \end{aligned}\right\}. \tag{13.6.7}$$

Substituting equations (13.6.5), (13.6.6), and (13.6.7) in equation

† See *A Treatise on Bessel Functions*, by Gray and Mathews, 2nd ed. by Gray and MacRobert, 1931, p. 26.

(13.6.3) and separating real and imaginary parts, we obtain, with the help of (13.3.10),

$$w = C_1\,\mathrm{ber}(\epsilon r)+C_2\,\mathrm{bei}(\epsilon r)+C_3\,\mathrm{ker}(\epsilon r)+C_4\,\mathrm{kei}(\epsilon r)+C_5+C_7\ln r, \tag{13.6.8}$$

$$\kappa\phi = C_1\,\mathrm{bei}(\epsilon r)-C_2\,\mathrm{ber}(\epsilon r)+C_3\,\mathrm{kei}(\epsilon r)-C_4\,\mathrm{ker}(\epsilon r)+C_6\ln r+C_8, \tag{13.6.9}$$

where $C_1,..., C_8$ are real constants. There are only *six* essential constants of integration in these equations. The constant C_8 is immaterial as only derivatives of ϕ occur. Also, since there is no displacement in the circumferential direction, we have from (13.5.7)

$$-rw'/R = \alpha_{11}-(\alpha_{22}/r)'$$

when $v = 0$, and using (13.5.8) this becomes

$$-Etrw'/R = \phi'/r-\eta\phi''-(r\phi'')'+\eta\phi'',$$

or

$$Etw'/R = (\nabla^2\phi)'.$$

Hence

$$Etw/R = \nabla^2\phi+\text{constant}, \tag{13.6.10}$$

and comparison with equations (13.6.8) and (13.6.9) shows that

$$C_7 = 0. \tag{13.6.11}$$

To complete the solution of a given problem a particular integral of equation (13.5.9) must be obtained, and then the constants $C_1,..., C_6$ are determined by conditions at the pole of the spherical cap and by edge conditions at its circular boundary, or by edge conditions alone at two circular boundaries if the cap is incomplete. The reader is referred to a paper by E. Reissner for further details of special solutions.†

† *J. Math. Phys.* **25** (1947) 279.

14 Theory of Cylindrical Shells

BECAUSE of their simple shape cylindrical shells have been widely used for practical applications and have therefore received considerable attention from many writers. Various methods of analysis† have been used for solving problems of cylindrical shells. Here we use the membrane theory of Chapter 11 and the bending theory of Chapter 12 which is satisfactory for many problems provided that stress couples and resultants which arise from any nearly inextensional part of the deformation are unimportant. According to the usual procedure we first solve the equations of the membrane theory; this solution may then be used as an approximate particular integral. In solving the equations of the bending theory it is then only necessary to deal with an unloaded shell. We do not, however, consider this in detail here. Instead we study briefly problems of unloaded cylindrical shells which are deformed by applying forces in a completely general way along generators of a circular cylindrical shell, but attention is restricted to one set of special edge conditions over section curves of the shell, which frequently occur in practice. If the circular cylindrical shell is complete (i.e. it has no bounding generators) then edge conditions over section curves may be examined by the theory of the edge effect (§ 15.5). Throughout the present chapter we only consider general methods of solution and we do not study problems numerically since detailed numerical results may be found elsewhere.‡ The theory used here, however, differs in some respects from that used by previous writers.

In any problem in which there is a significant inextensional deformation it may be necessary to include contributions from the inextensional theory of Chapter 16 before a satisfactory solution can be obtained.

† See, e.g., A. E. H. Love, *The Mathematical Theory of Elasticity*, 4th ed. (Cambridge, 1927); K. Miesel, *Ing.-Arch.* **1** (1930) 92; V. Finsterwalder, ibid. **4** (1933) 43; F. Dischinger, *Beton Eisen* **34** (1935) 257; A. Aas Jacobson, *Bauingenieur* **20** (1939) 394; R. S. Jenkins, *Modern Building Techniques Bulletin*, no. 1 (1947).

‡ S. Timoshenko, *Theory of Plates and Shells* (New York, 1940); W. Flügge, *Statik und Dynamik der Schalen* (Berlin, 1934); K. Girkmann, *Flächentragwerke* (Vienna, 1948); H. Lundgren, *Cylindrical Shells* (Copenhagen, 1949). Other references may be found in these books and in the article by P. M. Naghdi; loc. cit., p. 373.

14.1. Geometrical relations

Consider a plane curve, the points of which are given by the position vector

$$\mathbf{K} = \mathbf{K}(\theta_1), \tag{14.1.1}$$

where θ_1 is the length of the arc of a curve measured from a fixed point. We erect a unit vector $\mathbf{i}$ normal to the plane which contains the curve

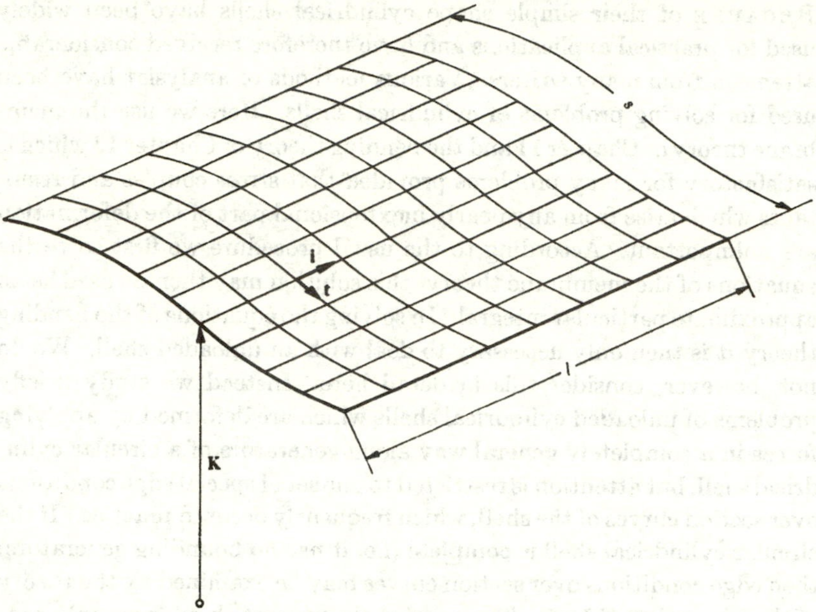

FIG. 14.1. Cylindrical shell.

(14.1.1) and we denote distance along this vector by θ_2 measured from a particular plane curve (14.1.1), The equation

$$\mathbf{r} = \mathbf{K}(\theta_1)+\theta_2\,\mathbf{i} \tag{14.1.2}$$

determines a surface which is the middle surface M of a cylindrical shell (Fig. 14.1).

We call (14.1.1) a section curve of the middle surface and lines parallel to $\mathbf{i}$ are generators. We write

$$\theta_1 = \theta, \quad \theta_2 = \xi, \tag{14.1.3}$$

where θ, ξ have the ranges

$$0 \leqslant \theta \leqslant s, \quad 0 \leqslant \xi \leqslant l. \tag{14.1.4}$$

The unit tangent **t** at a point of a section curve (14.1.1) is

$$\mathbf{t} = \frac{\partial \mathbf{K}}{\partial \theta}, \tag{14.1.5}$$

so that, from (14.1.2), the base vectors of the middle surface are

$$\mathbf{a}_1 = \mathbf{t}, \quad \mathbf{a}_2 = \mathbf{i}. \tag{14.1.6}$$

If $R\,(= 1/\rho)$ is the radius of curvature at a point of the section curve, then

$$\frac{\partial \mathbf{t}}{\partial \theta^1} = \frac{\partial \mathbf{t}}{\partial \theta} = -\frac{\mathbf{a}_3}{R} = -\rho \mathbf{a}_3, \tag{14.1.7}$$

where, as in Chapter 10, $\mathbf{a}_3$ is the unit normal to the middle surface. Hence, from (14.1.6),

$$\left.\begin{aligned} \mathbf{a}_{1,1} &= -\rho \mathbf{a}_3 \\ \mathbf{a}_{1,2} &= \mathbf{a}_{2,1} = \mathbf{a}_{2,2} = 0 \end{aligned}\right\}. \tag{14.1.8}$$

Also
$$a_{11} = a_{22} = 1, \quad a_{12} = 0, \quad a = |a_{\alpha\beta}| = 1, \tag{14.1.9}$$

and
$$a^{11} = a^{22} = 1, \quad a^{12} = 0. \tag{14.1.10}$$

It follows from (14.1.6) and (14.1.10), that the contravariant base vectors of the middle surface are

$$\mathbf{a}^1 = \mathbf{a}_1 = \mathbf{t}, \quad \mathbf{a}^2 = \mathbf{a}_2 = \mathbf{i}. \tag{14.1.11}$$

From (14.1.8), we find that the coefficients of the second fundamental form of the middle surface are

$$b_{11} = \mathbf{a}_3.\mathbf{a}_{1,1} = -\rho, \qquad b_{12} = \mathbf{a}_3.\mathbf{a}_{1,2} = 0, \qquad b_{22} = \mathbf{a}_3.\mathbf{a}_{2,2} = 0, \tag{14.1.12}$$

and hence

$$\left.\begin{aligned} b_1^1 &= b_{11} = -\rho \\ b_2^1 &= b_1^2 = b_2^2 = 0 \end{aligned}\right\}. \tag{14.1.13}$$

The Christoffel symbols of the second kind are given by

$$\Gamma^{\alpha}_{\beta\gamma} = \mathbf{a}^{\alpha}.\mathbf{a}_{\beta,\gamma}$$

and using (14.1.8) and (14.1.11) we find that

$$\Gamma^{\alpha}_{\beta\gamma} = 0. \tag{14.1.14}$$

We close this section by calculating the tensors $H^{\alpha\beta\lambda\mu}$ and $H^{*}_{\alpha\beta\lambda\mu}$ which are defined in (11.1.23) and (11.1.21) respectively, and which depend only on geometrical tensors associated with the middle surface.

Thus

$$\left.\begin{aligned} H^{1111} &= H^{2222} = 1, & H^{1122} &= \eta \\ H^{1212} &= \tfrac{1}{2}(1-\eta), & H^{1112} &= H^{2221} = 0 \end{aligned}\right\}, \tag{14.1.15}$$

$$\left.\begin{aligned} H^{*}_{1111} &= H^{*}_{2222} = 1, & H^{*}_{1122} &= -\eta \\ H^{*}_{1212} &= \tfrac{1}{2}(1+\eta), & H^{*}_{1112} &= H^{*}_{2221} = 0 \end{aligned}\right\}. \tag{14.1.16}$$

14.2. Stress resultants, shearing forces, stress couples, and loads

When examining special shells it is convenient to abandon tensor notations and introduce alternative notations for stress resultants, stress couples, and loads. Thus

$$\left.\begin{aligned} n^{11} = n_\theta, \quad n^{12} = n^{21} = n_{\theta\xi}, \quad n^{22} = n_\xi \\ q^1 = q_\theta, \quad q^2 = q_\xi \\ m^{11} = m_\theta, \quad m^{12} = m^{21} = m_{\theta\xi}, \quad m^{22} = m_\xi \\ p^1 = p_\theta, \quad p^2 = p_\xi \end{aligned}\right\}. \tag{14.2.1}$$

The physical components of stress resultants, shearing forces, stress couples, and loads, which were defined in (10.2.20), (10.2.21), and (10.3.14), are now given by

$$\left.\begin{aligned} n_{(11)} = n_\theta, \quad n_{(12)} = n_{(21)} = n_{\theta\xi}, \quad n_{(22)} = n_\xi \\ q_{(1)} = q_\theta, \quad q_{(2)} = q_\xi \\ m_{(11)} = m_\theta, \quad m_{(12)} = -m_{(21)} = -m_{\theta\xi}, \quad m_{(22)} = -m_\xi \\ p_{(1)} = p_\theta, \quad p_{(2)} = p_\xi, \quad p_{(3)} = p \end{aligned}\right\}. \tag{14.2.2}$$

14.3. Membrane theory

Using the notation of § 14.1 and § 14.2, and noting, in particular, that Christoffel symbols $\Gamma^\alpha_{\beta\gamma}$ vanish, the equations of equilibrium (11.1.13) become

$$\left.\begin{aligned} \frac{\partial n_\theta}{\partial\theta}+\frac{\partial n_{\theta\xi}}{\partial\xi}+p_\theta = 0 \\ \frac{\partial n_{\theta\xi}}{\partial\theta}+\frac{\partial n_\xi}{\partial\xi}+p_\xi = 0 \\ \rho n_\theta - p = 0 \end{aligned}\right\}. \tag{14.3.1}$$

Changing the notation for the components of displacement of the middle surface we put

$$v_1 = v, \quad v_2 = u, \tag{14.3.2}$$

while we retain w for the component in the direction of the normal to the middle surface. From (11.1.18), (14.1.12), and (14.3.2), we have

$$\left.\begin{aligned} \alpha_{11} &= \frac{\partial v}{\partial\theta}+\rho w \\ \alpha_{12} &= \frac{1}{2}\left(\frac{\partial v}{\partial\xi}+\frac{\partial u}{\partial\theta}\right) \\ \alpha_{22} &= \frac{\partial u}{\partial\xi} \end{aligned}\right\}. \tag{14.3.3}$$

Using (14.1.15), the stress–strain relations (11.1.22) become

$$\left.\begin{aligned} n_\theta &= D(\alpha_{11}+\eta\alpha_{22}) \\ n_{\theta\xi} &= D(1-\eta)\alpha_{12} \\ n_\xi &= D(\eta\alpha_{11}+\alpha_{22}) \end{aligned}\right\}, \tag{14.3.4}$$

where

$$D = Et/(1-\eta^2). \tag{14.3.5}$$

Also, from (11.1.19), or directly from (14.3.4),

$$\left.\begin{aligned} \alpha_{11} &= (n_\theta-\eta n_\xi)/(Et) \\ \alpha_{12} &= (1+\eta)n_{\theta\xi}/(Et) \\ \alpha_{22} &= (n_\xi-\eta n_\theta)/(Et) \end{aligned}\right\}, \tag{14.3.6}$$

so that, using (14.3.3), we have

$$\left.\begin{aligned} \frac{\partial u}{\partial \xi} &= \frac{n_\xi-\eta n_\theta}{Et} \\ \frac{\partial v}{\partial \xi} &= \frac{2(1+\eta)n_{\theta\xi}}{Et}-\frac{\partial u}{\partial \theta} \\ \rho w &= \frac{n_\theta-\eta n_\xi}{Et}-\frac{\partial v}{\partial \theta} \end{aligned}\right\}. \tag{14.3.7}$$

The last equation in (14.3.1) immediately gives the stress resultant n_θ in terms of the load. Thus

$$n_\theta = p/\rho, \tag{14.3.8}$$

and we see that n_θ cannot be influenced by any edge conditions.

From (14.3.1)

$$\left.\begin{aligned} n_{\theta\xi} &= -\int^{\xi}\left(\frac{\partial n_\theta}{\partial \theta}+p_\theta\right)d\xi+C_1(\theta) \\ n_\xi &= -\int^{\xi}\left(\frac{\partial n_{\theta\xi}}{\partial \theta}+p_\xi\right)d\xi+C_2(\theta) \end{aligned}\right\}, \tag{14.3.9}$$

where $C_1(\theta)$ and $C_2(\theta)$ are arbitrary functions of θ.

The displacement components may be obtained from (14.3.7) in the form

$$\left.\begin{aligned} u &= \frac{1}{Et}\int^{\xi}(n_\xi-\eta n_\theta)\,d\xi+\frac{C_3(\theta)}{Et} \\ v &= \int^{\xi}\left(\frac{2(1+\eta)n_{\theta\xi}}{Et}-\frac{\partial u}{\partial \theta}\right)d\xi+\frac{C_4(\theta)}{Et} \\ \rho w &= \frac{n_\theta-\eta n_\xi}{Et}-\frac{\partial v}{\partial \theta} \end{aligned}\right\}, \tag{14.3.10}$$

where $C_3(\theta)$ and $C_4(\theta)$ are arbitrary functions of θ.

If we substitute n_θ from (14.3.8) into (14.3.9) and (14.3.10), and use the abbreviations

$$\frac{\partial()}{\partial\theta} = ()^{\cdot}, \tag{14.3.11}$$

$$N = p/\rho, \tag{14.3.12}$$

$$\left.\begin{aligned} X_1 &= -\int_0^\xi (N^{\cdot}+p_\theta)\,d\xi \\ X_2 &= -\int_0^\xi (X_1^{\cdot}+p_\xi)\,d\xi \\ X_3 &= \int_0^\xi (X_2-\eta N)\,d\xi \\ X_4 &= \int_0^\xi \{2(1+\eta)X_1-X_3^{\cdot}\}\,d\xi \\ X_5 &= N-\eta X_2-X_4^{\cdot} \end{aligned}\right\}, \tag{14.3.13}$$

we obtain the formulae

$$\left.\begin{aligned} n_\theta &= N \\ n_{\theta\xi} &= X_1+C_1 \\ n_\xi &= X_2-C_1^{\cdot}\xi+C_2 \end{aligned}\right\}, \tag{14.3.14}$$

$$\left.\begin{aligned} Etu &= X_3-\tfrac{1}{2}C_1^{\cdot}\xi^2+C_2\xi+C_3 \\ Etv &= X_4+C_1^{\cdot\cdot}\frac{\xi^3}{6}-C_2^{\cdot}\frac{\xi^2}{2}-C_3^{\cdot}\xi+2(1+\eta)C_1\xi+C_4 \\ Et\rho w &= X_5-C_1^{\cdot\cdot\cdot}\frac{\xi^3}{6}+C_2^{\cdot\cdot}\frac{\xi^2}{2}+C_3^{\cdot\cdot}\xi-(2+\eta)C_1^{\cdot}\xi-\eta C_2-C_4^{\cdot} \end{aligned}\right\}. \tag{14.3.15}$$

The formulae (14.3.14) and (14.3.15) represent the general solution of the membrane theory of cylindrical shells and it contains four arbitrary functions $C_1,\ldots,C_4$ which have to be determined from the edge conditions. We see from these formulae that it is possible to satisfy certain edge conditions along the edges $\xi = 0$ and $\xi = 1$.

When $n_\theta = n_{\theta\xi} = n_\xi = 0$ we have the inextensional solution, and in this case $C_1 = C_2 = 0$, so that C_3 and C_4 determine the inextensional and rigid-body displacements.

Then

$$\left.\begin{aligned} Etu &= C_3 \\ Etv &= -C_3^{\cdot}\xi+C_4 \\ Et\rho w &= C_3^{\cdot\cdot}\xi-C_4^{\cdot} \end{aligned}\right\}, \tag{14.3.16}$$

and these displacements can be prevented by imposing the conditions

$$v = 0 \quad (\xi = 0, \xi = l), \tag{14.3.17}$$

which give $$C_4 = 0, \quad C_3^{\boldsymbol{\cdot}} = 0.$$

Hence C_3 is a constant and represents a rigid-body displacement along the generators which may be taken to be zero.

Returning to the general solution (14.3.14) and (14.3.15) we see that the functions C_1 and C_2 may be determined by imposing conditions on the stress resultants n_θ, $n_{\theta\xi}$ along the edges $\xi = 0$ and $\xi = 1$. For example, we may have

$$n_\xi = T_0(\theta), \qquad n_{\theta\xi} = S(\theta) \quad (\xi = 0 \text{ or } \xi = l), \tag{14.3.18}$$

or

$$\left.\begin{aligned} n_\xi &= T(\theta) \quad (\xi = 0) \\ n_\xi &= \bar{T}(\theta) \quad (\xi = l) \end{aligned}\right\}, \tag{14.3.19}$$

where T_0, S, T, $\bar{T}$ are known functions of θ. The conditions (14.3.19), however, only determine C_1 to within an arbitrary additive constant which represents a constant shearing stress resultant $n_{\theta\xi}$. This constant shearing stress may be given a prescribed value on, for example, the edge $\xi = 0$. Thus

$$C_2 = T(\theta), \qquad -lC_1^{\boldsymbol{\cdot}} + C_2 - \int_0^l (X_1^{\boldsymbol{\cdot}} + p_\xi)\, d\xi = \bar{T}(\theta),$$

and therefore

$$C_2 = T(\theta), \quad lC_1 = \int_0^\theta \left\{T(\theta) - \bar{T}(\theta) - \int_0^l (X_1^{\boldsymbol{\cdot}} + p_\xi)\, d\xi\right\} d\theta + K, \tag{14.3.20}$$

where K is an arbitrary constant representing a constant shearing force.

We now impose the boundary conditions (14.3.17) on the displacement v, so that, from (14.3.15),

$$\left.\begin{aligned} C_4 &= 0 \\ lC_3 &= K' - \tfrac{1}{2}l^2C_2 + \tfrac{1}{6}l^3C_1^{\boldsymbol{\cdot}} + \\ &\quad + \int_0^\theta \left(2(1+\eta)lC_1 + \int_0^l \{2(1+\eta)X_1 - X_3^{\boldsymbol{\cdot}}\}\, d\xi\right) d\theta \end{aligned}\right\}, \tag{14.3.21}$$

where K' is a constant which represents a rigid-body displacement parallel to the generators of the cylinder.

The membrane solution (14.3.14) and (14.3.15), where the functions C_1, C_2, C_3, C_4 are given by (14.3.20) and (14.3.21), represents a complete solution of the cylindrical shell problem under the edge conditions (14.3.19) and (14.3.17), on the edges $\xi = 0$, $\xi = l$. In order to satisfy more general edge conditions we have to consider the homogeneous bending equations of Chapter 12. We obtain the differential equations of the bending theory in the next section.

14.4. Differential equations of bending theory

Using the notations of §§ 14.1 and 14.2, and noting, in particular, that the Christoffel symbols $\Gamma^{\alpha}_{\beta\gamma}$ vanish, the equations of equilibrium (12.1.7) of the bending theory still reduce to the first two equations of (14.3.1) but, in view of (12.1.9), the third equation of (14.3.1) is replaced by

$$\frac{\partial q_\theta}{\partial \theta}+\frac{\partial q_\xi}{\partial \xi}-\rho n_\theta+p = 0. \tag{14.4.1}$$

From Chapter 12 we still recover equations (14.3.3) to (14.3.7), using the notation (14.3.2). Moreover, from (12.1.11),

$$\left.\begin{aligned} q_\theta &= \frac{\partial m_\theta}{\partial \theta}+\frac{\partial m_{\theta\xi}}{\partial \xi} \\ q_\xi &= \frac{\partial m_{\theta\xi}}{\partial \theta}+\frac{\partial m_\xi}{\partial \xi} \end{aligned}\right\}. \tag{14.4.2}$$

Eliminating q_θ and q_ξ from (14.4.1) and (14.4.2) gives

$$\frac{\partial^2 m_\theta}{\partial \theta^2}+2\frac{\partial^2 m_{\theta\xi}}{\partial\theta\partial\xi}+\frac{\partial^2 m_\xi}{\partial \xi^2}-\rho n_\theta+p = 0. \tag{14.4.3}$$

Also, from (12.1.19) and (12.1.24),

$$\left.\begin{aligned} m_\theta &= -B\left(\frac{\partial^2 w}{\partial \theta^2}+\eta\frac{\partial^2 w}{\partial \xi^2}\right) \\ m_{\theta\xi} &= -B(1-\eta)\frac{\partial^2 w}{\partial\theta\partial\xi} \\ m_\xi &= -B\left(\eta\frac{\partial^2 w}{\partial \theta^2}+\frac{\partial^2 w}{\partial \xi^2}\right) \end{aligned}\right\}, \tag{14.4.4}$$

where

$$B = \frac{Et^3}{12(1-\eta^2)}. \tag{14.4.5}$$

The first two equations of equilibrium in (14.3.1) can be satisfied identically by

$$\left.\begin{aligned} n_\theta &= \frac{\partial^2 \phi}{\partial \xi^2}-A_\theta \\ n_\xi &= \frac{\partial^2 \phi}{\partial \theta^2}-A_\xi \\ n_{\theta\xi} &= -\frac{\partial^2 \phi}{\partial\theta\partial\xi} \end{aligned}\right\}, \tag{14.4.6}$$

where ϕ is a stress function and

$$A_\theta = \int_{\theta_0}^{\theta} p_\theta \, d\theta, \qquad A_\xi = \int_{\xi_0}^{\xi} p_\xi \, d\xi, \tag{14.4.7}$$

θ_0, ξ_0 being arbitrary constants. Substituting (14.4.6) into (14.3.6) we get

$$\left.\begin{aligned} Et\alpha_{11} &= \frac{\partial^2\phi}{\partial\xi^2} - \eta\frac{\partial^2\phi}{\partial\theta^2} - A_\theta + \eta A_\xi \\ Et\alpha_{12} &= -(1+\eta)\frac{\partial^2\phi}{\partial\theta\partial\xi} \\ Et\alpha_{22} &= \frac{\partial^2\phi}{\partial\theta^2} - \eta\frac{\partial^2\phi}{\partial\xi^2} - A_\xi + \eta A_\theta \end{aligned}\right\}. \tag{14.4.8}$$

Equations (14.3.3) yield the identity

$$\frac{\partial^2\alpha_{11}}{\partial\xi^2} + \frac{\partial^2\alpha_{22}}{\partial\theta^2} - 2\frac{\partial^2\alpha_{12}}{\partial\theta\partial\xi} - \rho\frac{\partial^2 w}{\partial\xi^2} = 0,$$

by eliminating u and v; using (14.4.8) this becomes

$$\nabla^4\phi - \rho Et\frac{\partial^2 w}{\partial\xi^2} = q, \tag{14.4.9}$$

where

$$\nabla^2 = \frac{\partial^2}{\partial\theta^2} + \frac{\partial^2}{\partial\xi^2}, \tag{14.4.10}$$

$$q = \frac{\partial^2 A_\xi}{\partial\theta^2} + \frac{\partial^2 A_\theta}{\partial\xi^2} - \eta\left(\frac{\partial p_\xi}{\partial\xi} + \frac{\partial p_\theta}{\partial\theta}\right). \tag{14.4.11}$$

Also, from (14.4.3), (14.4.4), and the first relation in (14.4.6), we find that

$$\nabla^4 w + \frac{\rho}{B}\frac{\partial^2\phi}{\partial\xi^2} = q', \tag{14.4.12}$$

where

$$q' = (\rho A_\theta + p)/B. \tag{14.4.13}$$

The equations (14.4.9) and (14.4.12) are two partial differential equations of the fourth order for the stress function ϕ and the normal displacement w, and these equations govern the bending theory of cylindrical shells, based on the work of Chapter 12. They may be reduced to a single equation for a complex function by multiplying (14.4.9) by a constant k and adding to (14.4.12).† Thus

$$\nabla^4(w+k\phi) - kEt\rho\frac{\partial^2}{\partial\xi^2}\left(w - \frac{\phi}{BEkt}\right) = q' + kq, \tag{14.4.14}$$

† This process is the same as that used in Chapter 13.

and the constant k satisfies the equation

$$k = -\frac{1}{kBEt},$$

so that
$$k = \pm iK, \tag{14.4.15}$$

where
$$K = \frac{\sqrt{\{12(1-\eta^2)\}}}{Et^2}. \tag{14.4.16}$$

If we now put
$$\psi = w+iK\phi \tag{14.4.17}$$

and use the root $k = iK$, equation (14.4.14) becomes

$$\nabla^4\psi - i\epsilon^2\frac{\partial^2\psi}{\partial\xi^2} = \Omega, \tag{14.4.18}$$

where

$$\left.\begin{aligned}\Omega &= q'+iKq \\ \epsilon^2 &= \rho\{12(1-\eta^2)\}^{\frac{1}{2}}/t\end{aligned}\right\}. \tag{14.4.19}$$

If we use the other value of k given in (14.4.15) we obtain an equation which is the complex conjugate of (14.4.18). Equation (14.4.18) is a fourth-order equation for the complex function ψ. The real part of ψ is the normal displacement and the imaginary part of ψ is the stress function, apart from a multiplying constant K. When w and ϕ are known we find n_θ, $n_{\theta\xi}$, and n_ξ immediately from (14.4.6); m_θ, $m_{\theta\xi}$, and m_ξ from (14.4.4), and then q_θ and q_ξ from (14.4.2). The displacement components u and v follow from the first two equations in (14.3.7).

By using the membrane theory we can, as explained in § 14.3, obtain a particular solution of a problem which satisfies certain edge conditions. This membrane solution may then be regarded as an approximate particular integral for bending theory and in order to complete the solution of a given problem and satisfy all the required boundary conditions we must obtain the solution of the homogeneous equation (14.4.18), with $\Omega = 0$. We do not consider this further here. Instead we obtain some solutions of the homogeneous equation for a circular cylindrical shell which are appropriate for some problems of the bending of an unloaded shell by suitably applied forces along its bounding generators.

14.5. Solution of homogeneous equation for circular cylindrical shells

The homogeneous differential equation (14.4.18) is

$$\nabla^4\psi - i\epsilon^2\frac{\partial^2\psi}{\partial\xi^2} = 0, \tag{14.5.1}$$

where ϵ is a constant for circular cylindrical shells. Since we consider a shell which has zero loads on its surfaces,

$$p_\theta = p_\xi = A_\theta = A_\xi = 0. \tag{14.5.2}$$

Also over the ends $\xi = 0, \xi = l$ of the shell we impose the edge conditions

$$v = 0, \quad w = 0, \quad n_\xi = 0, \quad m_\xi = 0 \quad (\xi = 0, l). \tag{14.5.3}$$

The conditions on the edges $\theta = 0, \theta = s$ will be considered later. From (14.4.4), (14.3.7), (14.4.6), and (14.5.2) we see that the conditions on the ends $\xi = 0, \xi = l$ can be satisfied if

$$w = 0, \quad \phi = 0, \quad \frac{\partial^2 w}{\partial \xi^2} = 0, \quad \frac{\partial^2 \phi}{\partial \xi^2} = 0 \quad (\xi = 0, l), \tag{14.5.4}$$

or

$$\psi = 0, \quad \frac{\partial^2 \psi}{\partial \xi^2} = 0 \quad (\xi = 0, l). \tag{14.5.5}$$

We assume that ψ, together with its derivatives with respect to ξ up to the fourth order, each satisfy sufficient conditions so that they can be expanded as half-range Fourier series in the interval $(0, l)$. Then, in view of the conditions (14.5.5), we may write

$$\left.\begin{aligned}
\psi &= \sum_{n=1}^{\infty} \psi_n(\theta)\sin(n\pi\xi/l) && (0 \leqslant \xi \leqslant l)\\
\frac{\partial \psi}{\partial \xi} &= \sum_{n=1}^{\infty} \frac{n\pi\psi_n(\theta)}{l}\cos\left(\frac{n\pi\xi}{l}\right) && (0 \leqslant \xi \leqslant l)\\
\frac{\partial^2 \psi}{\partial \xi^2} &= -\sum_{n=1}^{\infty} \frac{(n\pi)^2\psi_n(\theta)}{l^2}\sin\left(\frac{n\pi\xi}{l}\right) && (0 \leqslant \xi \leqslant l)\\
\frac{\partial^3 \psi}{\partial \xi^3} &= -\sum_{n=1}^{\infty} \frac{(n\pi)^3\psi_n(\theta)}{l^3}\cos\left(\frac{n\pi\xi}{l}\right) && (0 \leqslant \xi \leqslant l)\\
\frac{\partial^4 \psi}{\partial \xi^4} &= \sum_{n=1}^{\infty} \frac{(n\pi)^4\psi_n(\theta)}{l^4}\sin\left(\frac{n\pi\xi}{l}\right) && (0 < \xi < l)
\end{aligned}\right\}, \tag{14.5.6}$$

where $\psi_n(\theta)$ are functions of θ to be determined. If we substitute (14.5.6) into (14.5.1), the differential equation is satisfied provided

$$\frac{d^4\psi_n}{d\theta^4} - 2\left(\frac{n\pi}{l}\right)^2 \frac{d^2\psi_n}{d\theta^2} + \frac{(n\pi)^2}{l^4}\left[(n\pi)^2 + \frac{il^2\sqrt{\{12(1-\eta^2)\}}}{Rt}\right]\psi_n = 0. \tag{14.5.7}$$

The general solution of (14.5.7) is the sum of terms of the form

$$C_{mn}\, e^{m\theta},$$

where m is a root of the equation

$$m^4-2\left(\frac{n\pi}{l}\right)^2 m^2+\frac{(n\pi)^2}{l^4}\left[(n\pi)^2+\frac{il^2\sqrt{\{12(1-\eta^2)\}}}{Rt}\right] = 0. \quad (14.5.8)$$

The roots of (14.5.8) are

$$\pm\gamma(\eta_1+i\zeta_1), \quad \pm\gamma(\eta_2-i\zeta_2), \quad (14.5.9)$$

where

$$\gamma = n\pi/l, \quad (14.5.10)$$

and

$$\left.\begin{aligned} \eta_1\sqrt{2} &= \sqrt{\{\sqrt{(1-2a+2a^2)}+1-a\}} \\ \eta_2\sqrt{2} &= \sqrt{\{\sqrt{(1+2a+2a^2)}+1+a\}} \\ \zeta_1\sqrt{2} &= \sqrt{\{\sqrt{(1-2a+2a^2)}-1+a\}} \\ \zeta_2\sqrt{2} &= \sqrt{\{\sqrt{(1+2a+2a^2)}-1-a\}} \\ a &= \frac{l}{n\pi}\sqrt{\left[\frac{\sqrt{\{3(1-\eta^2)\}}}{tR}\right]} \end{aligned}\right\}. \quad (14.5.11)$$

Hence, the general solution of (14.5.7) may now be written

$$\begin{aligned} w_n(\theta) = {} & e^{-\gamma\eta_1\theta^*}\{B_1\cos(\gamma\zeta_1\theta^*)+B_2\sin(\gamma\zeta_1\theta^*)\}+ \\ & +e^{-\gamma\eta_2\theta^*}\{B_3\cos(\gamma\zeta_2\theta^*)+B_4\sin(\gamma\zeta_2\theta^*)\}+ \\ & +e^{-\gamma\eta_1\theta}\{B_5\cos(\gamma\zeta_1\theta)+B_6\sin(\gamma\zeta_1\theta)\}+ \\ & +e^{-\gamma\eta_2\theta}\{B_7\cos(\gamma\zeta_2\theta)+B_8\sin(\gamma\zeta_2\theta)\}, \end{aligned} \quad (14.5.12)$$

$$\begin{aligned} K\phi_n(\theta) = {} & e^{-\gamma\eta_1\theta^*}\{B_1\sin(\gamma\zeta_1\theta^*)-B_2\cos(\gamma\zeta_1\theta^*)\}+ \\ & +e^{-\gamma\eta_2\theta^*}\{-B_3\sin(\gamma\zeta_2\theta^*)+B_4\cos(\gamma\zeta_2\theta^*)\}+ \\ & +e^{-\gamma\eta_1\theta}\{B_5\sin(\gamma\zeta_1\theta)-B_6\cos(\gamma\zeta_1\theta)\}+ \\ & +e^{-\gamma\eta_2\theta}\{-B_7\sin(\gamma\zeta_2\theta)+B_8\cos(\gamma\zeta_2\theta)\}, \end{aligned} \quad (14.5.13)$$

where

$$\left.\begin{aligned} \psi_n(\theta) &= w_n(\theta)+iK\phi_n(\theta) \\ \theta^* &= s-\theta \end{aligned}\right\}, \quad (14.5.14)$$

and $B_1,\ldots, B_8$ are real constants. The notation θ^* is introduced for convenience so that the edges $\theta = 0$, s of the shell correspond to $\theta = 0$, $\theta^* = 0$ respectively. If the boundary conditions along these edges may be expressed in suitable half-range Fourier series in terms of ξ, we may determine the constants $B_1,\ldots, B_8$ for each value of n, and the problem is completely solved.

15 Shells of Revolution

ANOTHER interesting group of shells is shells of revolution and the governing equations are derived in this chapter as a further illustration of the theory of Chapters 11 and 12. The solution of the equations of the general bending theory of shells of revolution is difficult but can, for many problems, be avoided, since the membrane theory often provides a satisfactory solution, except for certain boundary conditions which influence only a small zone in the neighbourhood of the edges. Such boundary conditions can be dealt with by a special simplified bending theory which is called edge effect. We therefore first present the membrane theory of shells of revolution.† We then close this chapter by obtaining the differential equations for the edge effect and by indicating briefly their solution for a complete shell of revolution. We refer readers to other writers for examples which are studied in more detail.‡

15.1. Geometrical relations

When a plane curve c with no double points is rotated around an axis A lying in its plane but not cutting c, except posibly at an end point, we obtain a surface of revolution. In general we suppose that c is rotated through an angle $\leqslant 2\pi$. The curve c is called a meridian curve and the axis A the axis of revolution (Fig. 15.1). If a surface of revolution is taken as the middle surface of the shell, we get a shell of revolution.

We use the angle θ_1 as a parameter for the meridian curves and put

$$\theta_1 = \theta \quad (0 \leqslant \theta \leqslant \gamma \leqslant 2\pi), \tag{15.1.1}$$

where γ is the angle between planes passing through the bounding

† The form of this theory is similar to that given by C. Truesdell, *Trans. Am. math. Soc.* **58** (1945) 96; W. Zerna, *Ing.-Arch.* **17** (1949) 223; and for a special problem by F. Dischinger, *Bauingenieur* **16** (1935) 374. Other forms have been given, e.g. by A. E. H. Love, loc. cit., p. 409; H. Reissner, *Festschrift Mueller–Breslau* (1912), p. 181.

‡ See. e.g. W. Flügge, loc. cit., p. 409; S. Timoshenko, loc. cit., p. 409; K. Girkmann, loc. cit., p. 409. Other references are contained in these books. The theory used by these writers, however, is different from that given here.

Edge effect equations valid for shells which are not necessarily shells of revolution are derived in Chapter 16.

meridians and the axis A (Fig. 15.2). If the shell is a complete shell of revolution, then $\gamma = 2\pi$.

At every point P of the middle surface we erect a right-handed triad of unit vectors $(\mathbf{i}_1, \mathbf{i}_2, \mathbf{i}_3)$ (Fig. 15.2), where $\mathbf{i}_2$ is parallel to the axis of revolution, $\mathbf{i}_3$ is perpendicular to $\mathbf{i}_2$ and lies in a meridian plane, and $\mathbf{i}_1$ is perpendicular to $\mathbf{i}_2$ and $\mathbf{i}_3$.

The meridian curve through P, lying in the plane $\theta =$ constant, may be described by rectangular coordinates (y, z) in this plane, the origin of coordinates being on the axis of revolution, the positive directions of

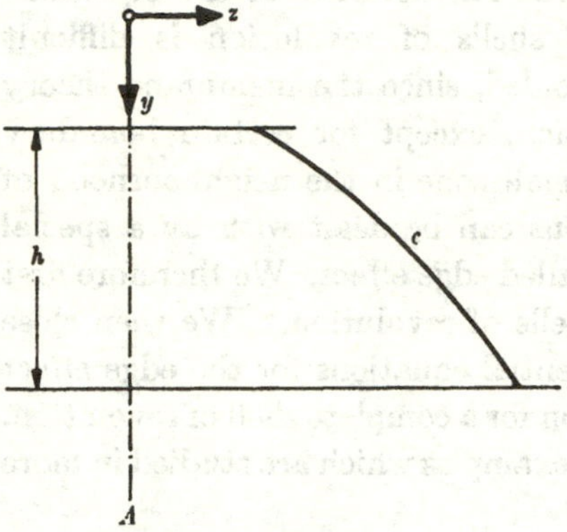

FIG. 15.1. Meridian curve of a surface of revolution.

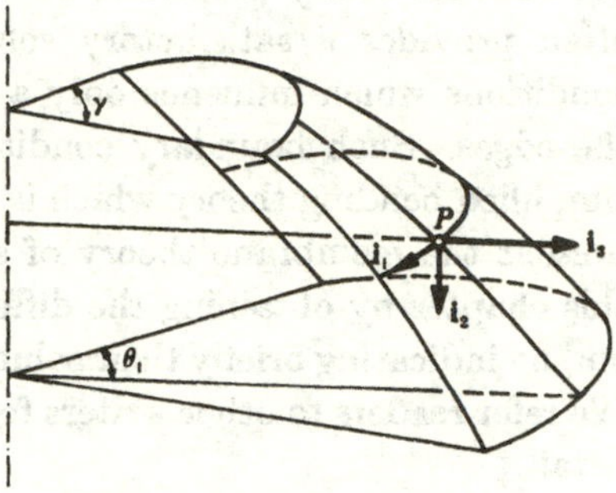

FIG. 15.2. Shell of revolution.

the coordinate axes being $\mathbf{i}_2$ and $\mathbf{i}_3$ respectively. The coordinates (y, z) of a point P may be represented parametrically by

$$y = y(\xi), \qquad z = z(\xi), \qquad \theta_2 = \xi. \tag{15.1.2}$$

A special case of our coordinate ξ is when

$$\xi = y \tag{15.1.3}$$

where the values taken by y occupy an interval of length h as shown in Fig. 15.1. In this case the meridian curve has the form

$$z = z(y). \tag{15.1.4}$$

The vector $\mathbf{r}$ of a point of the middle surface, having its origin at the origin of the coordinate system (y, z), has the form (Fig. 15.2)

$$\mathbf{r} = y(\xi)\mathbf{i}_2 + z(\xi)\mathbf{i}_3(\theta). \tag{15.1.5}$$

We denote derivatives with respect to θ by dots and derivatives with respect to ξ by dashes. The derivatives of the unit vectors $\mathbf{i}_r$ are found to be

$$\left.\begin{aligned} \dot{\mathbf{i}}_1 &= -\mathbf{i}_3, \quad \mathbf{i}_1' = 0 \\ \dot{\mathbf{i}}_2 &= \mathbf{i}_2' = 0 \\ \dot{\mathbf{i}}_3 &= \mathbf{i}_1, \quad \mathbf{i}_3' = 0 \end{aligned}\right\}. \tag{15.1.6}$$

From (15.1.5) and (15.1.6) we have

$$\mathbf{a}_1 = z\mathbf{i}_1, \quad \mathbf{a}_2 = y'\mathbf{i}_2+z'\mathbf{i}_3, \tag{15.1.7}$$

and hence

$$\left.\begin{aligned} \mathbf{a}_{1,1} &= \dot{\mathbf{a}}_1 = -z\mathbf{i}_3 \\ \mathbf{a}_{1,2} &= \mathbf{a}_1' = \dot{\mathbf{a}}_2 = z'\mathbf{i}_1 \\ \mathbf{a}_{2,2} &= \mathbf{a}_2' = y''\mathbf{i}_2+z''\mathbf{i}_3 \end{aligned}\right\}. \tag{15.1.8}$$

Using (15.1.7) the metric tensor is found to be

$$a_{11} = z^2, \quad a_{12} = a_{21} = 0, \quad a_{22} = y'^2+z'^2, \tag{15.1.9}$$

and

$$a = a_{11}a_{22} = z^2(y'^2+z'^2). \tag{15.1.10}$$

Also

$$a^{11} = \frac{1}{a_{11}} = \frac{1}{z^2}, \qquad a^{12} = a^{21} = 0, \qquad a^{22} = \frac{1}{a_{22}} = \frac{1}{y'^2+z'^2}, \tag{15.1.11}$$

and

$$\mathbf{a}^1 = a^{11}\mathbf{a}_1 = \frac{\mathbf{i}_1}{z}, \qquad \mathbf{a}^2 = a^{22}\mathbf{a}_2 = \frac{y'\mathbf{i}_2+z'\mathbf{i}_3}{y'^2+z'^2}. \tag{15.1.12}$$

The unit normal vector $\mathbf{a}_3$ is given by

$$\mathbf{a}_3 = \frac{\mathbf{a}_1\times\mathbf{a}_2}{\sqrt{a}} = \frac{y'\mathbf{i}_3-z'\mathbf{i}_2}{\sqrt{(y'^2+z'^2)}}, \tag{15.1.13}$$

and from this and (15.1.8) we obtain

$$\left.\begin{aligned} b_{11} &= \mathbf{a}_3.\mathbf{a}_{1,1} = -zy'/\sqrt{(y'^2+z'^2)} \\ b_{12} &= b_{21} = \mathbf{a}_3.\mathbf{a}_{1,2} = 0 \\ b_{22} &= \mathbf{a}_3.\mathbf{a}_{2,2} = (y'z''-y''z')/\sqrt{(y'^2+z'^2)} \end{aligned}\right\}, \tag{15.1.14}$$

$$\left.\begin{aligned} b_1^1 &= a^{11}b_{11} = -\frac{y'}{z\sqrt{(y'^2+z'^2)}} \\ b_2^1 &= b_1^2 = 0 \\ b_2^2 &= a^{22}b_{22} = \frac{y'z''-y''z'}{(y'^2+z'^2)^{\frac{3}{2}}} \\ K &= b_1^1 b_2^2 - b_2^1 b_1^2 = \left(\frac{y'}{y'^2+z'^2}\right)^2\left(\frac{z'y''}{zy'}-\frac{z''}{z}\right) \end{aligned}\right\}. \tag{15.1.15}$$

The Christoffel symbols may be calculated with the help of (15.1.8) and (15.1.12). Thus

$$\left.\begin{aligned} &\Gamma^1_{11} = \mathbf{a}^1.\mathbf{a}_{1,1} = 0, \qquad \Gamma^1_{22} = \mathbf{a}_1.\mathbf{a}_{2,2} = 0 \\ &\Gamma^2_{12} = \mathbf{a}^2.\mathbf{a}_{1,2} = 0, \qquad \Gamma^1_{12} = \mathbf{a}^1.\mathbf{a}_{1,2} = \frac{z'}{z} \\ &\Gamma^2_{11} = \mathbf{a}^2.\mathbf{a}_{1,1} = -\frac{zz'}{y'^2+z'^2} \\ &\Gamma^2_{22} = \mathbf{a}^2.\mathbf{a}_{2,2} = \frac{y'y''+z'z''}{y'^2+z'^2} \end{aligned}\right\}. \tag{15.1.16}$$

Finally we calculate the functions $H^*_{\alpha\beta\rho\lambda}$ from (11.1.20), using (15.1.9). We have

$$\left.\begin{aligned} &H^*_{1111} = z^4, \qquad H^*_{2222} = (y'^2+z'^2)^2 \\ &H^*_{1122} = -\eta z^2(y'^2+z'^2), \qquad H^*_{1212} = \tfrac{1}{2}(1+\eta)z^2(y'^2+z'^2) \\ &H^*_{1112} = H^*_{2221} = 0 \end{aligned}\right\}. \tag{15.1.17}$$

15.2. Equations of equilibrium

For convenience we introduce new notations for stress resultants and loads and write

$$N^{11} = N_\theta, \qquad N^{12} = N^{21} = N_{\theta\xi}, \qquad N^{22} = N_\xi, \tag{15.2.1}$$

$$\mathbf{P} = P_\theta \mathbf{a}_1 + P_\xi \mathbf{a}_2 + P\mathbf{a}_1^{\boldsymbol{\cdot}}. \tag{15.2.2}$$

Using (15.2.1) the equation of equilibrium (11.1.11) for membrane theory may be written in the form

$$(N^{\boldsymbol{\cdot}}_\theta + N'_{\theta\xi})\mathbf{a}_1 + (N^{\boldsymbol{\cdot}}_{\theta\xi} + N'_\xi)\mathbf{a}_2 + N_\theta \mathbf{a}^{\boldsymbol{\cdot}}_1 + 2N_{\theta\xi}\mathbf{a}'_1 + N_\xi \mathbf{a}'_2 + \mathbf{P} = \mathbf{0}. \tag{15.2.3}$$

If we now use (15.1.8) and multiply this equation in turn by $\mathbf{i}_1$, $\mathbf{i}_2$, and $\mathbf{a}_3$, we obtain the following equations of equilibrium:

$$\left.\begin{aligned} N^{\boldsymbol{\cdot}}_\theta + N'_{\theta\xi} + 2\frac{z'}{z}N_{\theta\xi} + P_\theta = 0 \\ N^{\boldsymbol{\cdot}}_{\theta\xi} + N'_\xi + \frac{y''}{y'}N_\xi + P_\xi = 0 \\ N_\theta + \left(\frac{z'y''}{zy'} - \frac{z''}{z}\right)N_\xi + P = 0 \end{aligned}\right\}. \tag{15.2.4}$$

These three equations for the unknown stress resultants N_θ, $N_{\theta\xi}$, N_ξ may be replaced by a single equation for N_ξ. From the first and second equations in (15.2.4),

$$N_\theta^{\cdot\cdot}-N_\xi''+2\frac{z'}{z}N_{\theta\xi}^{\cdot}-\left\{\frac{y'''}{y'}-\left(\frac{y''}{y'}\right)^2\right\}N_\xi-\frac{y''}{y'}N_\xi'+P_\theta^{\cdot}-P_\xi'=0, \tag{15.2.5}$$

and substituting in this equation for $N_{\theta\xi}$ and N_θ from the last two equations in (15.2.4) gives

$$N_\xi''+\left(\frac{z'y''}{zy'}-\frac{z''}{z}\right)N_\xi^{\cdot\cdot}+\left(\frac{2z'}{z}+\frac{y''}{y'}\right)N_\xi'+\left\{\frac{y'''}{y'}+\frac{2z'y''}{zy'}-\left(\frac{y''}{y'}\right)^2\right\}N_\xi$$
$$=-P^{\cdot\cdot}-\frac{2z'}{z}P_\xi+P_\theta^{\cdot}-P_\xi'. \tag{15.2.6}$$

When N_ξ is determined from this equation, N_θ and $N_{\theta\xi}$ follow from (15.2.4). If we put

$$N_\xi=\frac{\phi}{z\sqrt{y'}}, \tag{15.2.7}$$

equation (15.2.6) becomes

$$\phi''+f_1(\xi)\phi^{\cdot\cdot}+f_2(\xi)\phi=q, \tag{15.2.8}$$

where

$$\left.\begin{aligned} f_1(\xi) &= \frac{z'y''}{zy'}-\frac{z''}{z} \\ f_2(\xi) &= \frac{y'''}{2y'}-\frac{3}{4}\left(\frac{y''}{y'}\right)^2+\frac{z'y''}{zy'}-\frac{z''}{z} \\ q &= \left(P_\theta^{\cdot}-P^{\cdot\cdot}-\frac{2z'}{z}P_\xi-P_\xi'\right)z\sqrt{y'} \end{aligned}\right\}. \tag{15.2.9}$$

Equation (15.2.8) for ϕ governs the equilibrium state of shells of revolution as far as membrane theory is concerned. The function ϕ plays the role of a stress function since all stress resultants may be obtained from ϕ. The stress resultant N_ξ is found from (15.2.7) and then, from (15.2.4),

$$\left.\begin{aligned} N_\theta &= -\left(\frac{z'y''}{zy'}-\frac{z''}{z}\right)\frac{\phi}{z\sqrt{y'}}-P \\ N_{\theta\xi}^{\cdot} &= -\frac{\phi'}{z\sqrt{y'}}+\left(\frac{z'}{z}-\frac{y''}{2y'}\right)\frac{\phi}{z\sqrt{y'}}-P_\xi \end{aligned}\right\}. \tag{15.2.10}$$

The equation (15.2.8) is a partial differential equation of the second order and it is of hyperbolic, elliptic, or parabolic type according as

$$f_1(\xi)<,\ >,=0.$$

From (15.1.15) we see that the above types of differential equations correspond respectively to shells whose Gaussian curvature is negative, positive, and zero.

If the meridian curve is given in the form† (15.1.4), we have

$$y' = 1, \qquad y'' = y''' = 0. \tag{15.2.11}$$

and all the formulae simplify. Thus, from (15.2.9),

$$\left.\begin{aligned} f_1(\xi) &= -\frac{z''}{z} \\ f_2(\xi) &= -\frac{z''}{z} \end{aligned}\right\}, \tag{15.2.12}$$

and the differential equation (15.2.8) becomes

$$\phi'' - \frac{z''}{z}\phi^{\cdot\cdot} - \frac{z''}{z}\phi = q. \tag{15.2.13}$$

Also, the formulae (15.2.10) are now

$$\left.\begin{aligned} N_\theta &= \frac{z''\phi}{z^2} - P \\ N_{\theta\xi}^{\cdot} &= -\frac{\phi'}{z} + \frac{z'\phi}{z^2} - P_\xi \end{aligned}\right\}. \tag{15.2.14}$$

15.3. Displacements

We introduce the displacement vector $\mathbf{v}$ in the form (11.1.15) so that

$$\mathbf{v} = v_\alpha \mathbf{a}^\alpha + w\mathbf{a}_3,$$

but we change the notation and write

$$v_1 = u, \qquad v_2 = v.$$

Using (15.1.15) and (15.1.16), or, alternatively, (15.1.6), (15.1.7), and (15.1.12), we find from (11.1.18) that

$$\left.\begin{aligned} \alpha_{11} &= u^{\cdot} - \Gamma_{11}^2 v - b_{11} w \\ \alpha_{12} &= \tfrac{1}{2}(u' + v^{\cdot} - 2\Gamma_{12}^1 u) \\ \alpha_{22} &= v' - \Gamma_{22}^2 v - b_{22} w \end{aligned}\right\}. \tag{15.3.1}$$

Also, observing that $n^{\alpha\beta} = N^{\alpha\beta}/\sqrt{a}$, we see that, with the help of (15.1.17), equations (11.1.19) become

$$\left.\begin{aligned} \alpha_{11} &= \frac{z}{Et\sqrt{(y'^2+z'^2)}}\{z^2 N_\theta - \eta(y'^2+z'^2)N_\xi\} \\ \alpha_{12} &= \frac{(1+\eta)z\sqrt{(y'^2+z'^2)}}{Et} N_{\theta\xi} \\ \alpha_{22} &= \frac{\sqrt{(y'^2+z'^2)}}{Etz}\{(y'^2+z'^2)N_\xi - \eta z^2 N_\theta\} \end{aligned}\right\}. \tag{15.3.2}$$

† We can also have the equation of the meridian curve in the form $y = y(z)$ and this representation was dealt with in § 13.2. We found there that we can reduce the problem in this case to a second order partial differential equation for a stress function. The stress functions of § 13.2 and the present chapter are related. See E. Reissner, *J. Math. Phys.* **26** (1948) 290.

When the values of N_ξ, N_θ, $N_{\theta\xi}$ have ben found from the previous section, the values of $\alpha_{\lambda\mu}$ are known from (15.3.2) and the displacements may then be calculated from (15.3.1). We may eliminate u and w from these equations and obtain a single differential equation for v. From the first and third equations in (15.3.1) we have

$$\alpha_{11}-\frac{b_{11}}{b_{22}}\alpha_{22} = u^{\cdot}-\frac{b_{11}}{b_{22}}v'-\left(\Gamma^2_{11}-\frac{b_{11}}{b_{22}}\Gamma^2_{22}\right)v,$$

and substituting $u^{\cdot}$ from this into the second equation of (15.3.1) gives

$$v^{\cdot\cdot}+\left(\frac{b_{11}}{b_{22}}v'\right)'-2\Gamma^1_{12}\frac{b_{11}}{b_{22}}v'-2\Gamma^1_{12}\left(\Gamma^2_{11}-\frac{b_{11}}{b_{22}}\Gamma^2_{22}\right)v+$$
$$+\left\{\left(\Gamma^2_{11}-\frac{b_{11}}{b_{22}}\Gamma^2_{22}\right)v\right\}' = A, \quad (15.3.3)$$

where $$A = 2\alpha^{\cdot}_{12}-\left(\alpha_{11}-\frac{b_{11}}{b_{22}}\alpha_{22}\right)'+2\Gamma^1_{12}\left(\alpha_{11}-\frac{b_{11}}{b_{22}}\alpha_{22}\right). \quad (15.3.4)$$

When v is found from (15.3.3), u and w can be obtained from (15.3.1). The equation is simplified considerably if we restrict our attention to the form (15.1.4) for the meridian curves. In this case

$$\Gamma^2_{11}-\frac{b_{11}}{b_{22}}\Gamma^2_{22} = 0,$$

and using (15.1.14) and (15.1.16), equation (15.3.3) becomes

$$v''-\left(\frac{z'}{z}+\frac{z'''}{z''}\right)v'-\frac{z''}{z}v^{\cdot\cdot} = -\frac{Az''}{z}, \quad (15.3.5)$$

where $$A = 2\alpha^{\cdot}_{12}-\alpha'_{11}-\frac{z}{z''}\alpha'_{22}+\frac{2z'}{z}\alpha_{11}+\left(\frac{z'}{z''}+\frac{zz'''}{(z'')^2}\right)\alpha_{22}. \quad (15.3.6)$$

15.4. Special solutions

We restrict our attention to a complete shell of revolution so that $\gamma = 2\pi$ and θ ranges from 0 to 2π, and we consider a group of problems for which it is possible to assume the following Fourier expansions:

$$\left.\begin{aligned} P_\theta &= \sum_{n=1}^{\infty} P_{\theta n}\sin n\theta \\ P_\xi &= \sum_{n=0}^{\infty} P_{\xi n}\cos n\theta \\ P &= \sum_{n=0}^{\infty} P_n\cos n\theta \end{aligned}\right\}, \quad (15.4.1)$$

$$\left.\begin{aligned} N_\theta &= \sum_{n=0}^{\infty} N_{\theta n} \cos n\theta \\ N_\xi &= \sum_{n=0}^{\infty} N_{\xi n} \cos n\theta \\ N_{\theta\xi} &= \sum_{n=1}^{\infty} N_{\theta\xi n} \sin n\theta \end{aligned}\right\}, \tag{15.4.2}$$

$$\left.\begin{aligned} u &= \sum_{n=1}^{\infty} u_n \sin n\theta \\ v &= \sum_{n=0}^{\infty} v_n \cos n\theta \\ w &= \sum_{n=0}^{\infty} w_n \cos n\theta \end{aligned}\right\}, \tag{15.4.3}$$

$$\left.\begin{aligned} \phi &= \sum_{n=0}^{\infty} \phi_n \cos n\theta \\ q &= \sum_{n=0}^{\infty} q_n \cos n\theta \end{aligned}\right\}, \tag{15.4.4}$$

where $P_{\theta n},\dots, q_n$ depend only on ξ. In practice cosine series for P_ξ, P, N_θ, N_ξ, v, w, ϕ, q and sine series for P_θ, $N_{\theta\xi}$, u are usually sufficient, but these series may be interchanged if required. In order to satisfy the differential equation (15.2.8) we multiply this equation by $\cos n\theta$ and integrate with respect to θ from 0 to 2π. Then, assuming that $\phi^\cdot$, $P^\cdot$, P_θ are single-valued and continuous for every value of θ, we have

$$\int_0^{2\pi} \phi^{\cdot\cdot} \cos n\theta \, d\theta = [\phi^\cdot \cos n\theta]_0^{2\pi} - n^2 \int_0^{2\pi} \phi \cos n\theta \, d\theta$$

$$= -n^2 \int_0^{2\pi} \phi \cos n\theta \, d\theta,$$

$$\int_0^{2\pi} P_\theta^\cdot \cos n\theta \, d\theta = n \int_0^{2\pi} P_\theta \sin n\theta \, d\theta,$$

$$\int_0^{2\pi} P^{\cdot\cdot} \cos n\theta \, d\theta = -n^2 \int_0^{2\pi} P \cos n\theta \, d\theta.$$

Hence

$$\phi_n'' + F_n(\xi)\phi_n = q_n \quad (n = 0, 1, 2,\dots), \tag{15.4.5}$$

where

$$F_n(\xi) = \frac{n^2-1}{zy'}(z''y' - z'y'') + \frac{y'''}{2y'} - \frac{3}{4}\left(\frac{y''}{y'}\right)^2, \tag{15.4.6}$$

$$q_n = \left(n^2 P_n + nP_{\theta n} - P'_{\xi n} - \frac{2z'}{z} P_{\xi n}\right) z\sqrt{y'}. \tag{15.4.7}$$

When $n = 0$,

$$\frac{z\phi_0''}{\sqrt{y'}}-\phi_0\left\{\frac{z''}{\sqrt{y'}}-\frac{z'y''}{(y')^{3/2}}-\frac{zy'''}{2(y')^{3/2}}+\frac{3z(y'')^2}{4(y')^{5/2}}\right\} = -(z^2P'_{\xi 0}+2zz'P_{\xi 0}),$$

which may be written

$$\left\{\frac{z\phi_0'}{(y')^{1/2}}-\left(\frac{z'}{(y')^{1/2}}-\frac{zy''}{2(y')^{3/2}}\right)\phi_0\right\}' = -(z^2P_{\xi 0})',$$

or

$$\frac{\phi_0'}{z\sqrt{y'}}-\left(\frac{z'}{z}-\frac{y''}{2y'}\right)\frac{\phi_0}{z\sqrt{y'}}+P_{\xi 0} = \frac{C_1}{z^2}, \tag{15.4.8}$$

where C_1 is a constant. If we substitute from (15.4.1) and (15.4.4) into the second equation of (15.2.10), we have

$$\dot{N}_{\theta\xi} = -\frac{C_1}{z^2}+\sum_{n=1}^{\infty} nN_{\theta\xi n}\cos n\theta, \tag{15.4.9}$$

where

$$nN_{\theta\xi n} = -\frac{\phi_n'}{z\sqrt{y'}}+\left(\frac{z'}{z}-\frac{y''}{2y'}\right)\frac{\phi_n}{z\sqrt{y'}}-P_{\xi n}, \tag{15.4.10}$$

and we have used (15.4.8). Hence

$$N_{\theta\xi} = -\frac{C_1\theta}{z^2}+C_2(\xi)+\sum_{n=1}^{\infty} N_{\theta\xi n}\sin n\theta,$$

where C_2 is a function of ξ. The constant C_1 must be zero, however, as $N_{\theta\xi}$ is single-valued for a complete surface of revolution. Also, $C_2 = 0$ as we are only considering an expansion of $N_{\theta\xi}$ as a Fourier sine series, so that

$$N_{\theta\xi} = \sum_{n=1}^{\infty} N_{\theta\xi n}\sin n\theta, \tag{15.4.11}$$

where $N_{\theta\xi n}$ is given by (15.4.10).

From (15.2.7), (15.2.10), (15.4.1), (15.4.2), and (15.4.4) we see that

$$\left.\begin{aligned} N_{\xi n} &= \frac{\phi_n}{z\sqrt{y'}} \\ N_{\theta n} &= \left(\frac{z''}{z}-\frac{z'y''}{zy'}\right)\frac{\phi_n}{z\sqrt{y'}}-P_n \end{aligned}\right\}. \tag{15.4.12}$$

If we choose the equation of the meridian curve in the simple form (15.1.4), equation (15.4.5) reduces to

$$\phi_n''+(n^2-1)(z''/z)\phi_n = q_n. \tag{15.4.13}$$

Also, equations (15.4.10) and (15.4.12) become

$$\left.\begin{aligned} N_{\xi n} &= \frac{\phi_n}{z} \\ N_{\theta n} &= \frac{z''\phi_n}{z^2} - P_n \\ nN_{\theta\xi n} &= -\frac{\phi_n'}{z} + \frac{z'\phi_n}{z^2} - P_{\xi n} \end{aligned}\right\}. \tag{15.4.14}$$

When we restrict our attention to meridian curves of the form (15.1.4), the equation (15.3.5) yields, by a process similar to that used in deriving (15.4.5),

$$v_n'' - \left(\frac{z'}{z} + \frac{z'''}{z''}\right)v_n' + \frac{n^2 z''}{z} v_n = -\frac{z'' A_n}{z}, \tag{15.4.15}$$

if we assume that $v^{\cdot}$ is single-valued and continuous, where A_n are coefficients in the Fourier expansion

$$A = \sum_{n=0}^{\infty} A_n \cos n\theta \tag{15.4.16}$$

if it exists.

Remembering that $C_1 = 0$, the equation (15.4.8) reduces to

$$\left(\frac{\phi_0}{z}\right)' = \frac{\phi_0'}{z} - \frac{z'\phi_0}{z^2} = -P_{\xi 0},$$

when the meridian curve is (15.1.4), and this equation may be integrated to give

$$\phi_0 = \left(C - \int P_{\xi 0}\, d\xi\right)z, \tag{15.4.17}$$

where C is a constant which must be determined by boundary conditions along the edges of the shell. Equation (15.4.17) gives the solution of the membrane problem when the loads are rotationally symmetrical.

The solution of the differential equation (15.4.13) is particularly simple when $n = 1$ since the equation reduces to

$$\phi_1'' = q_1. \tag{15.4.18}$$

Solutions of (15.4.13) for other values of n have been considered by Truesdell.†

15.5. Edge effect‡

As explained in Chapter 11, the membrane theory can be used as an approximate particular integral of the bending theory, so that we need only consider the homogeneous problem in order to complete the

† loc. cit., p. 421.

‡ A more general discussion of edge effect is given in Chapter 16.

solution and satisfy all the edge conditions. We have therefore to examine the case when the shell is free from surface loads and is acted on only by forces along its edges. The results which are obtained for complete shells of revolution indicate that there is usually a state of stress existing near each edge of the shell such that the stress resultants and stress couples decrease to very small values at short distances from the edges. We call this special state the edge effect.

Suppose an edge of a complete shell of revolution is given by the parametric curve $\theta_2 = \xi = 0$. The edge effect is characterized by the fact that the order of magnitude of all functions is increased by differentiations with respect to θ_2, but the order of magnitude is unchanged by differentiations with respect to $\theta_1 = \theta$. We therefore neglect in our equations all terms except those containing the highest derivatives with respect to θ_2.

When the load terms are zero the approximate equations of equilibrium (12.1.7), (12.1.9), and (12.1.11) may be put in the form

$$n^{\alpha\beta}|_{\alpha} = 0, \tag{15.5.1}$$

$$n^{\alpha\beta}b_{\alpha\beta}+m^{\alpha\beta}|_{\alpha\beta} = 0. \tag{15.5.2}$$

If we neglect all terms except those containing the highest derivatives with respect to θ_2, we find that equations (15.5.1) can be satisfied by

$$n^{\alpha\beta} = \epsilon^{\alpha\gamma}\epsilon^{\beta\rho}\psi|_{\gamma\rho}. \tag{15.5.3}$$

Also, from (12.1.18) and (12.1.19), with the same approximation,

$$\epsilon^{\alpha\beta}\epsilon^{\rho\lambda}(\alpha_{\alpha\rho}|_{\lambda\beta}+b_{\alpha\rho}\,w|_{\lambda\beta}) = 0. \tag{15.5.4}$$

From (12.1.24) and (12.1.21) we have

$$m^{\alpha\beta} = -BH^{\alpha\beta\rho\lambda}w|_{\rho\lambda}, \tag{15.5.5}$$

$$\alpha_{\alpha\beta} = \frac{1}{Et}H^{*}_{\alpha\beta\rho\lambda}n^{\rho\lambda}. \tag{15.5.6}$$

If we now eliminate $\alpha_{\alpha\beta}$ and $n^{\alpha\beta}$ from (15.5.3), (15.5.4), and (15.5.6), and retain only the highest derivatives of ϕ and w with respect to θ_2, we have, approximately,

$$(a^{22})^2\frac{\partial^4\psi}{\partial\xi^4}+\frac{Etb_{11}}{a}\frac{\partial^2 w}{\partial\xi^2} = 0. \tag{15.5.7}$$

Similarly, from (15.5.2), (15.5.3), and (15.5.5) we obtain the approximate equation

$$(a^{22})^2\frac{\partial^4 w}{\partial\xi^4}-\frac{b_{11}}{aB}\frac{\partial^2\psi}{\partial\xi^2} = 0. \tag{15.5.8}$$

Parts of the solutions of equations (15.5.7) and (15.5.8) are of the forms

$$\psi = \xi f_1(\theta_1)+g_1(\theta_1), \qquad w = \xi f_2(\theta_1)+g_2(\theta_1), \tag{15.5.9}$$

where $f_1(\theta_1)$, $f_2(\theta_1)$, $g_1(\theta_1)$, $g_2(\theta_1)$ are arbitrary functions of θ_1, but these solutions may be rejected since they do not satisfy the imposed conditions about orders of magnitude of derivatives. If we assume that a^{22}, b_{11}, and a are unaltered in order of magnitude when they are differentiated with respect to ξ, we may write (15.5.7) and (15.5.8) in the forms

$$\frac{\partial^2}{\partial \xi^2}\left\{(a^{22})^2 \frac{\partial^2 \psi}{\partial \xi^2} + \frac{Etb_{11}}{a} w\right\} = 0,$$

$$\frac{\partial^2}{\partial \xi^2}\left\{(a^{22})^2 \frac{\partial^2 w}{\partial \xi^2} - \frac{b_{11}}{aB} \psi\right\} = 0,$$

and since we have rejected the solutions (15.5.9) these reduce to

$$\left.\begin{aligned} (a^{22})^2 \frac{\partial^2 \psi}{\partial \xi^2} + \frac{Etb_{11}}{a} w = 0 \\ (a^{22})^2 \frac{\partial^2 w}{\partial \xi^2} - \frac{b_{11}\psi}{aB} = 0 \end{aligned}\right\}. \tag{15.5.10}$$

These equations may be replaced by one second-order differential equation for a function of the form $\psi + i\epsilon w$, where ϵ is a function of ξ. For our purpose, however, we eliminate w and obtain

$$\frac{\partial^4 \psi}{\partial \xi^4} + 4k^4 \psi = 0, \tag{15.5.11}$$

where

$$k = \frac{\{3(1-\eta^2)\}^{\frac{1}{4}}(ab_{11})^{\frac{1}{2}}}{a_{11} t^{\frac{1}{2}}}. \tag{15.5.12}$$

We may suppose that a, a_{11}, and b_{11} in (15.5.12) are evaluated at the curve $\xi = 0$ so that k may be taken as constant. The relevant solution of (15.5.12), which dies away rapidly as ξ ($\geqslant 0$) increases, is

$$\psi = e^{-k\xi}(C_1 \cos k\xi + C_2 \sin k\xi), \tag{15.5.13}$$

where C_1, C_2 are functions of θ. The value of w may now be found from (15.5.10). We observe that our functions are increased in magnitude by a factor of order k when they are differentiated with respect to ξ, and k is large of order $1/\sqrt{t}$ provided $\sqrt{(ab_{11})}/a_{11}$ is not small. This condition is satisfied except for shallow shells which must therefore be dealt with by a special theory, as is done in Chapter 13. If there is a second edge $\xi =$ constant to the shell then another solution of type (15.5.13), appropriate to this edge, is needed, with two new arbitrary functions of θ, say C_3, C_4.

Since we retained only the highest derivatives of functions with respect to ξ in considering the edge effect, it can be seen from (15.5.3) and (15.5.6) that n^{12}, n^{22}, u, v are negligible in our approximation, whilst n^{11} is the only surviving stress resultant. Moreover, the stress resultant n^{11} does not enter into boundary conditions on the edge $\xi =$ constant. The arbitrary functions C_1, C_2, C_3, C_4 will be determined by boundary conditions imposed on w and $\partial w/\partial \xi$, or on shearing forces and stress couples, on the edges $\xi =$ constant. The satisfaction of these boundary conditions does not, to our order of approximation, affect boundary conditions already satisfied in the membrane theory.

16 Asymptotic Expansions

In this chapter we obtain basic equations for shells by a method of asymptotic expansion of the three-dimensional elastic equations used by A. E. Green† and slightly modified by A. E. Green and P. M. Naghdi.‡ The method arose out of work of Johnson and Reissner, and Reissner and Reiss who dealt with special types of shells.§ Two main expansions are considered. The first leads to membrane theory in § 16.2 and inextensional theory in § 16.3. The second gives a bending theory in § 16.4 of the same type as that obtained in Chapter 12 by other methods. A special case of bending theory, known as edge effect, is studied in § 16.5. The edge effect solution can also be obtained directly from the three-dimensional equations by asymptotic expansion and this would be the correct procedure if terms beyond the first approximation are needed. Another special case of bending theory, namely shallow shell theory, can be deduced by a systematic expansion procedure based on the equations of § 16.4, but this is not given here. The equations resulting from the first stage of this process are equivalent to those given in Chapter 13.

Suppose, for a given load on the surface of the shell, a particular integral of the membrane equations exists in terms of the loads. We then solve the equations of the inextensional theory of § 16.3 in which the load terms are zero. Finally, the edge effect solutions of § 16.5 for each edge of the shell, if they exist, are obtained. In view of the linearity of the basic equations the complete solution is the sum of these asymptotic contributions, subject to suitable edge conditions.||

When the edge effect equations of § 16.5 are unsatisfactory, we must

† *Proc. R. Soc.* **A266** (1962) 143.

‡ Loc. cit., p. 373. The procedure is formal and proof is so far lacking to show that the resulting equations do yield a genuine asymptotic solution of the three-dimensional problem.

§ References to these authors may be found in the paper by A. E. Green and P. M. Naghdi, ibid., together with other references, including a paper on asymptotic expansions by A. L. Gol'denveizer.

|| This solution refers to the 'interior' problem of shells. An examination of the 'boundary layer' equations is needed to effect a complete solution of the three-dimensional problem. See A. E. Green, loc. cit., p. 397 and A. E. Green and N. Laws, loc. cit., p. 397.

add the solution of the homogeneous bending theory of § 16.4 to the membrane and inextensional solutions. However, when the shell is shallow, we can either add to the membrane solution the solution of the homogeneous shallow shell equations of Chapter 13, or solve the non-homogeneous shallow shell equations directly. In most of this book we have restricted attention either to membrane theory alone, or to shallow shells, or to problems in which membrane theory plus bending theory is sufficient.

There are some problems in which no satisfactory particular integral of the membrane equations exists. Then the non-homogeneous inextensional equations must be solved. The problem of a helicoidal shell under normal load and certain edge conditions can be dealt with completely using only inextensional theory.†

For many practical problems it may still be very difficult to solve the basic equations of membrane theory, inextensional theory, and bending theory, and numerical methods must be used. The asymptotic ingredients of the complete solution are then not very suitable for such methods and it is desirable to find a composite system of equations which incorporates all the essential features of membrane, inextensional, and bending theories. Such a composite system of equations is obtained in § 16.6 from the asymptotic expansions. There appears, however, to be no unique way of obtaining a composite system of shell equations, but the results given here agree essentially with those obtained by other writers using different methods.‡

16.1. Basic equations

We introduce the non-dimensional coordinate ρ and the non-dimensional position vector $\mathbf{r}'$ of the middle surface defined by

$$\theta_3 = t\rho, \qquad \mathbf{r} = R\mathbf{r}' \quad (-\tfrac{1}{2} \leqslant \rho \leqslant \tfrac{1}{2}), \tag{16.1.1}$$

where R is the smallest radius of curvature of M. Corresponding to $\mathbf{r}'$ we define base vectors $\mathbf{c}_\alpha$, $\mathbf{c}^\alpha$ and metric tensors $c_{\alpha\beta}$, $c^{\alpha\beta}$ by the equations

$$\left.\begin{aligned} \mathbf{a}_\alpha &= R\mathbf{c}_\alpha, & \mathbf{a}^\alpha &= \mathbf{c}^\alpha/R \\ a_{\alpha\beta} &= R^2 c_{\alpha\beta}, & a^{\alpha\beta} &= c^{\alpha\beta}/R^2 \end{aligned}\right\}, \tag{16.1.2}$$

and we still use the same surface coordinates θ_α. The Christoffel symbols

† J. W. Cohen, *Proc. I.U.T.A.M. Symposium on the Theory of Thin Elastic Shells*, Delft, 1959 (North Holland, 1960), p. 415.

‡ See e.g. W. T. Koiter, ibid. p. 12; *Proc. K. ned. Akad. Wet.*, Series B, **104** (1961) 612; P. M. Naghdi, *Int. J. Engng Sci.* **1** (1963) 509; **2** (1964) 269. For other references see P. M. Naghdi, loc. cit., p. 373.

of the second kind associated with the metric tensors $c_{\alpha\beta}$, $c^{\alpha\beta}$ are equal to the Christoffel symbols $\Gamma^{\alpha}_{\beta\gamma}$ defined in terms of $a_{\alpha\beta}$, $a^{\alpha\beta}$. The components b^1_1, b^2_2 of the curvature tensor have dimensions -1 in length. We therefore define the mixed tensor d^{α}_{β} by the equations

$$d^{\alpha}_{\beta} = Rb^{\alpha}_{\beta}, \qquad d_{\alpha\beta} = c_{\alpha\lambda}d^{\lambda}_{\beta} = b_{\alpha\beta}/R. \tag{16.1.3}$$

With the help of (16.1.1)–(16.1.3), equations (10.1.5), (10.1.6), and (10.1.7) become

$$\left.\begin{aligned} \mu_{\alpha\beta} &= R^2\mu^*_{\alpha\beta}, \quad g_{\alpha\beta} = R^2\mu^*_{\alpha\lambda}\mu^{*\lambda}_{\beta} = R^2g^*_{\alpha\beta} \\ \mu^*_{\alpha\beta} &= c_{\alpha\beta}-\epsilon\rho d_{\alpha\beta}, \quad \mu^{*\alpha}_{\beta} = \delta^{\alpha}_{\beta}-\epsilon\rho d^{\alpha}_{\beta} \\ h &= 1-\epsilon\rho d^{\lambda}_{\lambda}+\epsilon^2\rho^2(d^1_1\,d^2_2-d^1_2\,d^2_1) \end{aligned}\right\}, \tag{16.1.4}$$

where
$$\epsilon = t/R. \tag{16.1.5}$$

Displacements u_{α}, u_3 and the stress tensor σ^{ij} are now replaced by

$$u_{\alpha} = (1-\eta^2)R^2U_{\alpha}/E, \quad u_3 = (1-\eta^2)RU_3/E, \quad U^{\alpha} = c^{\alpha\lambda}U_{\lambda}, \tag{16.1.6}$$

$$\left.\begin{aligned} \sigma^{\alpha\beta} &= s^{\alpha\beta}/R^2, \quad \sigma^{\alpha 3} = ts^{\alpha 3}/R^2 \\ \sigma^{3\alpha} &= ts^{3\alpha}/R^2, \quad \sigma^{33} = ts^{33}/R \end{aligned}\right\}. \tag{16.1.7}$$

The equations of equilibrium (10.4.2), and equations (10.5.9) and (10.2.5), then become

$$\partial s^{3\beta}/\partial\rho+s^{\alpha\beta}|_{\alpha}-\epsilon d^{\beta}_{\alpha}\,s^{\alpha 3} = 0, \tag{16.1.8}$$

$$\partial s^{33}/\partial\rho+d_{\alpha\beta}\,s^{\alpha\beta}+\epsilon s^{\alpha 3}|_{\alpha} = 0, \tag{16.1.9}$$

$$s^{[\alpha\beta]} = \tfrac{1}{2}(s^{\alpha\beta}-s^{\beta\alpha}) = \tfrac{1}{2}\epsilon\rho(s^{\lambda\beta}d^{\alpha}_{\lambda}-s^{\lambda\alpha}d^{\beta}_{\lambda}), \tag{16.1.10}$$

$$s^{3\alpha} = s^{\alpha 3}-\epsilon\rho d^{\alpha}_{\lambda}\,s^{\lambda 3}. \tag{16.1.11}$$

Finally, the stress–strain relations (10.5.10)–(10.5.12) take the form

$$(1-\eta^2)h\frac{\partial U_3}{\partial\rho} = -\epsilon\eta\mu^*_{\alpha\beta}\,s^{(\alpha\beta)}+\epsilon^2 s^{33}, \tag{16.1.12}$$

$$(1-\eta)h\left\{\frac{\partial U_{\alpha}}{\partial\rho}+\epsilon\left(\frac{\partial U_3}{\partial\theta^{\alpha}}+d^{\lambda}_{\alpha}\,U_{\lambda}-\rho d^{\lambda}_{\alpha}\frac{\partial U_{\lambda}}{\partial\rho}\right)\right\} = 2\epsilon^2 g^*_{\alpha\beta}\,s^{\beta 3}, \tag{16.1.13}$$

$$\begin{aligned} (1-\eta^2)h[U_{\alpha}|_{\beta}+U_{\beta}|_{\alpha}-2d_{\alpha\beta}\,U_3-\epsilon\rho\{d^{\lambda}_{\alpha}(U_{\lambda}|_{\beta}-d_{\lambda\beta}\,U_3)+d^{\lambda}_{\beta}(U_{\lambda}|_{\alpha}-d_{\lambda\alpha}\,U_3)\}] \\ = (1+\eta)(g^*_{\alpha\lambda}\,\mu^*_{\mu\beta}+g^*_{\beta\lambda}\,\mu^*_{\mu\alpha})s^{\lambda\mu}-2\eta g^*_{\alpha\beta}(\mu^*_{\lambda\mu}\,s^{(\lambda\mu)}+\epsilon s^{33}). \end{aligned} \tag{16.1.14}$$

The surface conditions (10.3.4) and (10.3.5) are replaced by

$$\left.\begin{aligned} s^{3\lambda} &= \tfrac{1}{2}(r^{\lambda}+\bar{r}^{\lambda}), \quad s^{33} = \tfrac{1}{2}(r+\bar{r}) \quad (\rho = +\tfrac{1}{2}) \\ s^{3\lambda} &= -\tfrac{1}{2}(r^{\lambda}-\bar{r}^{\lambda}), \quad s^{33} = -\tfrac{1}{2}(r-\bar{r}) \quad (\rho = -\tfrac{1}{2}) \end{aligned}\right\}, \tag{16.1.15}$$

where
$$(p^{\lambda}, \bar{p}^{\lambda}) = t(r^{\lambda}, \bar{r}^{\lambda})/R^2, \qquad (p, \bar{p}) = t(r, \bar{r})/R. \tag{16.1.16}$$

16.2. Membrane theory

We consider a thin shell for which

$$\epsilon \ll 1$$

and assume expansions for displacements, stresses, and loads in the forms

$$\left.\begin{aligned} U_\alpha &= \overset{(1)}{U}_\alpha + \epsilon \overset{(2)}{U}_\alpha + \ldots, \qquad & U_3 &= \overset{(1)}{U}_3 + \epsilon \overset{(2)}{U}_3 + \ldots \\ s^{ij} &= \overset{(1)}{s}{}^{ij} + \epsilon \overset{(2)}{s}{}^{ij} + \ldots, \qquad & r &= \overset{(1)}{r} + \epsilon \overset{(2)}{r} + \ldots \end{aligned}\right\}, \tag{16.2.1}$$

with similar expressions for $\bar{r}$, r^λ, $\bar{r}^\lambda$. For each value of ϵ^k ($k = 0, 1, \ldots$) we obtain a system of differential equations from which we can derive expressions for displacements and stresses (corresponding to each value of ϵ^k) in terms of ρ and arbitrary functions of θ_α. The first order approximation yields

$$\left.\begin{aligned} &\overset{(1)}{U}_\alpha = \overset{(1)}{V}_\alpha, \quad \overset{(1)}{U}_3 = \overset{(1)}{V} \\ &\overset{(1)}{s}{}^{\alpha\beta} = \overset{(1)}{s}{}^{\beta\alpha} = (1-\eta)\overset{(1)}{A}{}^{\alpha\beta} + \eta c^{\alpha\beta} \overset{(1)}{A}{}^\lambda_\lambda \\ &\overset{(1)}{s}{}^{3\beta} = \overset{(1)}{s}{}^{\beta 3} = \overset{(1)}{S}{}^\beta - \rho \overset{(1)}{s}{}^{\alpha\beta}|_\alpha \\ &\overset{(1)}{s}{}^{33} = \overset{(1)}{S} - \rho d_{\alpha\beta} \overset{(1)}{s}{}^{\alpha\beta} \\ &\overset{(1)}{A}_{\alpha\beta} = \tfrac{1}{2}(\overset{(1)}{V}_\alpha|_\beta + \overset{(1)}{V}_\beta|_\alpha - 2d_{\alpha\beta} \overset{(1)}{V}) \end{aligned}\right\}, \tag{16.2.2}$$

where $\overset{(1)}{V}_\alpha$, $\overset{(1)}{V}$, $\overset{(1)}{S}{}^\beta$, $\overset{(1)}{S}$ are functions of θ_β.

If we introduce surface conditions on $\rho = \pm\frac{1}{2}$ in the results (16.2.2), we have

$$\left.\begin{aligned} \overset{(1)}{s}{}^{(\alpha\beta)}|_\alpha + \overset{(1)}{r}{}^\beta &= 0 \\ d_{\alpha\beta} \overset{(1)}{s}{}^{(\alpha\beta)} + \overset{(1)}{r} &= 0 \end{aligned}\right\}, \tag{16.2.3}$$

$$\left.\begin{aligned} \overset{(1)}{s}{}^{3\beta} = \overset{(1)}{s}{}^{\beta 3} &= \tfrac{1}{2}\overset{(1)}{\bar{r}}{}^\beta + \rho \overset{(1)}{r}{}^\beta \\ \overset{(1)}{s}{}^{33} &= \tfrac{1}{2}\overset{(1)}{\bar{r}} + \rho \overset{(1)}{r} \end{aligned}\right\}. \tag{16.2.4}$$

If we express the results (16.2.2) and (16.2.4) in terms of stress resultants, and return to the notation of Chapter 10, we immediately recover membrane theory which is governed by equations (11.1.3), (11.1.13), (11.1.15), (11.1.18), and (11.1.22).

Further terms in the expansions (16.2.1), based on membrane theory, can now be obtained step by step. Here, however, we only consider further stages in the approximation procedure when the first approximation is an inextensional deformation.

16.3. Inextensional theory

We assume that

$$\overset{(1)}{A}_{\alpha\beta} = 0, \quad \overset{(1)}{r} = \overset{(1)}{\bar{r}} = 0, \quad \overset{(1)}{r}{}^{\beta} = \overset{(1)}{\bar{r}}{}^{\beta} = 0, \tag{16.3.1}$$

corresponding to an inextensional deformation, so that

$$\overset{(1)}{s}{}^{\alpha\beta} = 0, \quad \overset{(1)}{s}{}^{3\beta} = \overset{(1)}{s}{}^{\beta 3} = 0, \quad \overset{(1)}{s}{}^{33} = 0. \tag{16.3.2}$$

The second approximation then gives

$$\overset{(2)}{U}_3 = \overset{(2)}{V}, \quad \overset{(2)}{U}_\alpha = \overset{(2)}{V}_\alpha - \rho(\overset{(1)}{V}|_\alpha + d_\alpha^\lambda \overset{(1)}{V}_\lambda), \tag{16.3.3}$$

$$\overset{(2)}{s}{}^{\alpha\beta} = \overset{(2)}{s}{}^{\beta\alpha} = (1-\eta)\overset{(2)}{A}{}^{\alpha\beta} + \eta c^{\alpha\beta}\overset{(2)}{A}{}_\lambda^\lambda - \rho\{(1-\eta)\overset{(2)}{K}{}^{\alpha\beta} + \eta c^{\alpha\beta}\overset{(2)}{K}{}_\lambda^\lambda\}, \tag{16.3.4}$$

$$\begin{aligned}\overset{(2)}{s}{}^{3\beta} &= \overset{(2)}{s}{}^{\beta 3} \\ &= \overset{(2)}{S}{}^{\beta} - \rho\{(1-\eta)\overset{(2)}{A}{}^{\alpha\beta} + \eta c^{\alpha\beta}\overset{(2)}{A}{}_\lambda^\lambda\}|_\alpha - \tfrac{1}{2}(\tfrac{1}{4}-\rho^2)\{(1-\eta)\overset{(2)}{K}{}^{\alpha\beta} + \eta c^{\alpha\beta}\overset{(2)}{K}{}_\lambda^\lambda\}|_\alpha,\end{aligned} \tag{16.3.5}$$

$$\overset{(2)}{s}{}^{33} = \overset{(2)}{S} - \rho d_{\alpha\beta}\{(1-\eta)\overset{(2)}{A}{}^{\alpha\beta} + \eta c^{\alpha\beta}\overset{(2)}{A}{}_\lambda^\lambda\} - \tfrac{1}{2}(\tfrac{1}{4}-\rho^2) d_{\alpha\beta}\{(1-\eta)\overset{(2)}{K}{}^{\alpha\beta} + \eta c^{\alpha\beta}\overset{(2)}{K}{}_\lambda^\lambda\}, \tag{16.3.6}$$

where $\overset{(2)}{S}{}^{\beta}$, $\overset{(2)}{S}$, $\overset{(2)}{V}_\alpha$, $\overset{(2)}{V}$ are functions of θ_β and

$$\overset{(2)}{A}_{\alpha\beta} = \tfrac{1}{2}\{\overset{(2)}{V}_\alpha|_\beta + \overset{(2)}{V}_\beta|_\alpha - 2d_{\alpha\beta}\overset{(2)}{V}\},$$

$$\overset{(2)}{K}_{\alpha\beta} = \overset{(1)}{V}|_{\alpha\beta} + \tfrac{1}{2}(d_\alpha^\lambda \overset{(1)}{V}_\lambda)|_\beta + \tfrac{1}{2}(d_\beta^\lambda \overset{(1)}{V}_\lambda)|_\alpha + \tfrac{1}{2}\, d_\alpha^\lambda \overset{(1)}{V}_\lambda|_\beta + \tfrac{1}{2}\, d_\beta^\lambda \overset{(1)}{V}_\lambda|_\alpha - d_\alpha^\lambda d_{\lambda\beta}\overset{(1)}{V}. \tag{16.3.7}$$

On using the surface conditions at $\rho = \pm\frac{1}{2}$, we see that

$$\left.\begin{aligned}\{(1-\eta)\overset{(2)}{A}{}^{\alpha\beta} + \eta c^{\alpha\beta}\overset{(2)}{A}{}_\lambda^\lambda\}|_\alpha + \overset{(2)}{r}{}^{\beta} = 0 \\ d_{\alpha\beta}\{(1-\eta)\overset{(2)}{A}{}^{\alpha\beta} + \eta c^{\alpha\beta}\overset{(2)}{A}{}_\lambda^\lambda\} + \overset{(2)}{r} = 0\end{aligned}\right\}, \tag{16.3.8}$$

and
$$\overset{(2)}{s}{}^{3\beta} = \overset{(2)}{s}{}^{\beta 3} = \tfrac{1}{2}\overset{(2)}{\bar{r}}{}^{\beta} + \rho \overset{(2)}{r}{}^{\beta} - \tfrac{1}{2}(\tfrac{1}{4}-\rho^2)\{(1-\eta)\overset{(2)}{K}{}^{\alpha\beta} + \eta c^{\alpha\beta}\overset{(2)}{K}{}_\lambda^\lambda\}|_\alpha, \tag{16.3.9}$$

$$\overset{(2)}{s}{}^{33} = \tfrac{1}{2}\overset{(2)}{\bar{r}} + \rho\overset{(2)}{r} - \tfrac{1}{2}(\tfrac{1}{4}-\rho^2)\, d_{\alpha\beta}\{(1-\eta)\overset{(2)}{K}{}^{\alpha\beta} + \eta c^{\alpha\beta}\overset{(2)}{K}{}_\lambda^\lambda\}. \tag{16.3.10}$$

The deformation associated with the displacements $\overset{(2)}{V}_\alpha$, $\overset{(2)}{V}$ is again of the membrane type. Since we are concerned here with an inextensional theory it is sufficient for our purpose to consider the case

$$\overset{(2)}{V}_\alpha = \overset{(2)}{V} = \overset{(2)}{A}_{\alpha\beta} = 0, \quad \overset{(2)}{r} = \overset{(2)}{\bar{r}} = 0, \quad \overset{(2)}{r}{}^\beta = \overset{(2)}{\bar{r}}{}^\beta = 0, \tag{16.3.11}$$

so that (16.3.4), (16.3.9), and (16.3.10) reduce to

$$\left.\begin{aligned} \overset{(2)}{s}{}^{\alpha\beta} &= \overset{(2)}{s}{}^{\beta\alpha} = -\rho\{(1-\eta)\overset{(2)}{K}{}^{\alpha\beta}+\eta c^{\alpha\beta}\overset{(2)}{K}{}^\lambda_\lambda\} \\ \overset{(2)}{s}{}^{3\beta} &= \overset{(2)}{s}{}^{\beta 3} = -\tfrac{1}{2}(\tfrac{1}{4}-\rho^2)\{(1-\eta)\overset{(2)}{K}{}^{\alpha\beta}+\eta c^{\alpha\beta}\overset{(2)}{K}{}^\lambda_\lambda\}|_\alpha \\ \overset{(2)}{s}{}^{33} &= -\tfrac{1}{2}(\tfrac{1}{4}-\rho^2)d_{\alpha\beta}\{(1-\eta)\overset{(2)}{K}{}^{\alpha\beta}+\eta c^{\alpha\beta}\overset{(2)}{K}{}^\lambda_\lambda\} \end{aligned}\right\}. \tag{16.3.12}$$

From (16.3.12) we may obtain expressions for the shearing force and stress couples in terms of the displacements of the inextensional deformation. In general it is necessary to proceed to a third approximation before we can obtain a complete set of equations for the displacements.† For this purpose it is sufficient to consider only the third approximation in the equations of equilibrium (16.1.8) and (16.1.9), and in equations (16.1.10) and (16.1.11). Thus

$$\left.\begin{aligned} \frac{\partial \overset{(3)}{s}{}^{3\beta}}{\partial\rho}+\overset{(3)}{s}{}^{\alpha\beta}|_\alpha - d^\beta_\lambda \overset{(2)}{s}{}^{\alpha 3} &= 0 \\ \frac{\partial \overset{(3)}{s}{}^{33}}{\partial\rho}+d_{\alpha\beta}\overset{(3)}{s}{}^{\alpha\beta}+\overset{(2)}{s}{}^{\alpha 3}|_\alpha &= 0 \end{aligned}\right\}, \tag{16.3.13}$$

$$\left.\begin{aligned} \overset{(3)}{s}{}^{[\alpha\beta]} &= \tfrac{1}{2}\rho(\overset{(2)}{s}{}^{\lambda\beta}d^\alpha_\lambda - \overset{(2)}{s}{}^{\lambda\alpha}d^\beta_\lambda) \\ \overset{(3)}{s}{}^{3\alpha} &= \overset{(3)}{s}{}^{\alpha 3} - \rho d^\alpha_\lambda \overset{(2)}{s}{}^{\lambda 3} \end{aligned}\right\}. \tag{16.3.14}$$

Also

$$\left.\begin{aligned} \overset{(3)}{s}{}^{3\lambda} &= \tfrac{1}{2}(\overset{(3)}{r}{}^\lambda+\overset{(3)}{\bar{r}}{}^\lambda), & \overset{(3)}{s}{}^{33} &= \tfrac{1}{2}(\overset{(3)}{r}+\overset{(3)}{\bar{r}}) & (\rho &= \tfrac{1}{2}) \\ \overset{(3)}{s}{}^{3\lambda} &= -\tfrac{1}{2}(\overset{(3)}{r}{}^\lambda-\overset{(3)}{\bar{r}}{}^\lambda), & \overset{(3)}{s}{}^{33} &= -\tfrac{1}{2}(\overset{(3)}{r}-\overset{(3)}{\bar{r}}) & (\rho &= -\tfrac{1}{2}) \end{aligned}\right\}. \tag{16.3.15}$$

Equations for stress resultants, couples, and shearing forces may now be found in the usual way from (16.3.13)–(16.3.15), (16.3.3), (16.3.11), and (16.3.12). Writing the results in terms of the notation of Chapter 10 we have

$$\left.\begin{aligned} u_3 = w, \quad u_\alpha &= v_\alpha - \theta_3(w|_\alpha + b^\lambda_\alpha v_\lambda) \\ v_\alpha|_\beta + v_\beta|_\alpha - 2b_{\alpha\beta}w &= 0 \end{aligned}\right\}, \tag{16.3.16}$$

† There are, however, exceptions to this which occur when the equations $(16.3.1)_1$ are integrable in terms of arbitrary functions. It is still necessary to proceed to a third approximation to obtain corresponding values of the stress resultants.

where w, v_α are functions of θ_β, and

$$-\kappa_{\alpha\beta} = w|_{\alpha\beta}+\tfrac{1}{2}(b^\lambda_\alpha v_\lambda)|_\beta+\tfrac{1}{2}(b^\lambda_\beta v_\lambda)|_\alpha+\tfrac{1}{2}b^\lambda_\alpha v_\lambda|_\beta+\tfrac{1}{2}b^\lambda_\beta v_\lambda|_\alpha-b^\lambda_\alpha b_{\lambda\beta}w, \tag{16.3.17}$$

$$m^{\alpha\beta} = m^{\beta\alpha} = B\{(1-\eta)\kappa^{\alpha\beta}+\eta a^{\alpha\beta}\kappa^\lambda_\lambda\} = BH^{\alpha\beta\rho\lambda}\kappa_{\rho\lambda}, \tag{16.3.18}$$

$$\left.\begin{aligned} m^{\alpha\beta}|_\alpha-q^\beta+\tilde{p}^\beta &= 0\\ n^{\alpha\beta}|_\alpha-b^\beta_\alpha q^\alpha+p^\beta &= 0\\ b_{\alpha\beta}n^{(\alpha\beta)}+q^\alpha|_\alpha+p &= 0 \end{aligned}\right\}, \tag{16.3.19}$$

$$n^{[\alpha\beta]} = \tfrac{1}{2}(b^\alpha_\lambda m^{\lambda\beta}-b^\beta_\lambda m^{\lambda\alpha}), \tag{16.3.20}$$

where

$$B = \frac{Et^3}{12(1-\eta^2)}. \tag{16.3.21}$$

The equations of equilibrium (16.3.19) are the same as the exact equations (10.4.4)–(10.4.6) and (16.3.20) is the same as (10.4.10). The tensor $H^{\alpha\beta\rho\lambda}$ is defined in (11.1.23). The stress resultants $n^{\alpha\beta}$ are not now expressible in terms of the displacement components v_α, w by formulae of the type (11.1.22) obtained for membrane theory. Values of $n^{\alpha\beta}$ may be obtained from the third approximation in terms of v^λ, w, and new displacement components arising from this approximation, but we do not need explicit expressions for these here since equations (16.3.16)–(16.3.20) are a complete set of equations for inextensional theory. The explicit expression for $n^{\alpha\beta}$ will only be needed at the next stage of approximation to help determine the higher order components of displacement.

The expansions of §§ 16.2, 16.3 are used in a slightly different manner in § 16.6 to obtain a composite system of shell equations which incorporate the essential features of membrane theory, inextensional theory, and bending theory of shells, but we first obtain another type of asymptotic expansion in § 16.4 which leads to the bending theory developed in Chapter 12.

16.4. Bending theory†

We introduce the non-dimensional coordinate ρ and the non-dimensional position vector $\mathbf{r}'$ of the middle surface by the equations‡

$$\theta_3 = t\rho, \quad \mathbf{r} = (tR)^{\frac{1}{2}}\mathbf{r}' \quad (-\tfrac{1}{2} \leqslant \rho \leqslant \tfrac{1}{2}), \tag{16.4.1}$$

† In view of the method used to obtain the bending theory of Chapter 12, it is possible to give a motivation for the scaling procedures used here, but we restrict our attention to a formal statement of the required scalings.

‡ $\mathbf{r}'$ is not the same vector as that defined in (16.1.1). The same remark applies to other notations used in §§ 16.1 and 16.4.

where R is the smallest radius of curvature of M. Corresponding to $\mathbf{r}'$ we define base vectors $\mathbf{c}_\alpha$, $\mathbf{c}^\alpha$ and metric tensors $c_{\alpha\beta}$, $c^{\alpha\beta}$ by

$$\left.\begin{aligned} \mathbf{a}_\alpha &= (tR)^{\frac{1}{2}}\mathbf{c}_\alpha, & \mathbf{a}^\alpha &= (tR)^{-\frac{1}{2}}\mathbf{c}^\alpha \\ a_{\alpha\beta} &= tRc_{\alpha\beta}, & a^{\alpha\beta} &= c^{\alpha\beta}/(tR) \end{aligned}\right\}. \tag{16.4.2}$$

The Christoffel symbols of the second kind associated with the metric tensors $c_{\alpha\beta}$, $c^{\alpha\beta}$ are equal to those defined in terms of $a_{\alpha\beta}$, $a^{\alpha\beta}$ so that covariant differentiation is unchanged if we retain the same coordinates θ_β. The curvature terms b_1^1, b_2^2 are of dimension $1/R$ so we put

$$b_\beta^\alpha = d_\beta^\alpha/R, \qquad b_{\alpha\beta} = a_{\alpha\lambda}\, b_\beta^\lambda = tc_{\alpha\lambda}\, d_\beta^\lambda = td_{\alpha\beta}. \tag{16.4.3}$$

With the help of (16.4.1)–(16.4.3), equations (10.1.5), (10.1.6), and (10.1.7) become

$$\left.\begin{aligned} \mu_{\alpha\beta} &= tR\mu_{\alpha\beta}^*, \quad g_{\alpha\beta} = tR\mu_{\alpha\lambda}^*\mu_\beta^{*\lambda} = tRg_{\alpha\beta}^* \\ \mu_{\alpha\beta}^* &= c_{\alpha\beta}-\epsilon\rho d_{\alpha\beta}, \quad \mu_\beta^{*\alpha} = \delta_\beta^\alpha-\epsilon\rho d_\beta^\alpha \\ h &= 1-\epsilon\rho d_\lambda^\lambda+\epsilon^2\rho^2(d_1^1 d_2^2-d_2^1 d_1^2) \end{aligned}\right\}, \tag{16.4.4}$$

where ϵ is given by (16.1.5).

We also have

$$u_\alpha = (1-\eta^2)tRU_\alpha/E, \quad u_3 = (1-\eta^2)RU_3/E, \quad U^\alpha = c^{\alpha\lambda}U_\lambda, \tag{16.4.5}$$

$$\left.\begin{aligned} \sigma^{\alpha\beta} &= s^{\alpha\beta}/(tR), & \sigma^{\alpha 3} &= s^{\alpha 3}/R \\ \sigma^{3\alpha} &= s^{3\alpha}/R, & \sigma^{33} &= ts^{33}/R \end{aligned}\right\}. \tag{16.4.6}$$

The equations of equilibrium (10.4.2) and equations (10.5.9) and (10.2.5) then become

$$\partial s^{3\beta}/\partial\rho+s^{\alpha\beta}|_\alpha-\epsilon d_\alpha^\beta s^{\alpha 3} = 0, \tag{16.4.7}$$

$$\partial s^{33}/\partial\rho+d_{\alpha\beta}s^{\alpha\beta}+s^{\alpha 3}|_\alpha = 0, \tag{16.4.8}$$

$$s^{[\alpha\beta]} = \tfrac{1}{2}\epsilon\rho(s^{\lambda\beta}d_\lambda^\alpha-s^{\lambda\alpha}d_\lambda^\beta), \tag{16.4.9}$$

$$s^{3\alpha} = s^{\alpha 3}-\epsilon\rho d_\lambda^\alpha s^{\lambda 3}. \tag{16.4.10}$$

Also, the stress–strain relations (10.5.10)–(10.5.12) take the form

$$(1-\eta^2)h\frac{\partial U_3}{\partial\rho} = -\epsilon\eta\mu_{\alpha\beta}^* s^{(\alpha\beta)}+\epsilon^2 s^{33}, \tag{16.4.11}$$

$$(1-\eta)h\left\{\frac{\partial U_\alpha}{\partial\rho}+\frac{\partial U_3}{\partial\theta^\alpha}+\epsilon\left(d_\alpha^\lambda\, U_\lambda-\rho d_\alpha^\lambda\frac{\partial U_\lambda}{\partial\rho}\right)\right\} = 2\epsilon g_{\alpha\beta}^* s^{\beta 3}, \tag{16.4.12}$$

$$\begin{aligned} (1-\eta^2)h[U_\alpha|_\beta+U_\beta|_\alpha-2d_{\alpha\beta}\,U_3-\epsilon\rho\{d_\alpha^\lambda(U_\lambda|_\beta-d_{\lambda\beta}\,U_3)+d_\beta^\lambda(U_\lambda|_\alpha-d_{\lambda\alpha}\,U_3)] \\ = (1+\eta)(g_{\alpha\lambda}^*\mu_{\mu\beta}^*+g_{\beta\lambda}^*\mu_{\mu\alpha}^*)s^{\lambda\mu}-2\eta g_{\alpha\beta}^*(\mu_{\lambda\mu}^* s^{(\lambda\mu)}+\epsilon s^{33}). \end{aligned} \tag{16.4.13}$$

The surface conditions (10.3.4) and (10.3.5) are replaced by

$$\left.\begin{aligned} s^{3\lambda} &= \tfrac{1}{2}(r^{\lambda}+\bar{r}^{\lambda}), & s^{33} &= \tfrac{1}{2}(r+\bar{r}) & (\rho &= \tfrac{1}{2}) \\ s^{3\lambda} &= -\tfrac{1}{2}(r^{\lambda}-\bar{r}^{\lambda}), & s^{33} &= -\tfrac{1}{2}(r-\bar{r}) & (\rho &= -\tfrac{1}{2}) \end{aligned}\right\}, \qquad (16.4.14)$$

where

$$(p^{\lambda}, \bar{p}^{\lambda}) = (r^{\lambda}, \bar{r}^{\lambda})/R, \quad (p, \bar{p}) = t(r, \bar{r})/R. \qquad (16.4.15)$$

As in § 16.2 we assume expansions in powers of ϵ of the form (16.2.1). The first order approximation corresponding to powers of ϵ^k when $k = 0$ yields†

$$U_3 = V, \qquad U_{\alpha} = V_{\alpha} - \rho V|_{\alpha}, \qquad (16.4.16)$$

where V, V_{α} are functions of θ_{β}. Also

$$s^{\alpha\beta} = (1-\eta)A^{\alpha\beta} + \eta c^{\alpha\beta}A^{\lambda}_{\lambda} - \rho\{(1-\eta)V|^{\alpha\beta} + \eta c^{\alpha\beta}V|^{\lambda}_{\lambda}\}, \qquad (16.4.17)$$

$$A_{\alpha\beta} = \tfrac{1}{2}(V_{\alpha}|_{\beta} + V_{\beta}|_{\alpha} - 2d_{\alpha\beta}V), \qquad (16.4.18)$$

$$s^{\alpha\beta} = s^{\beta\alpha}, \qquad s^{3\lambda} = s^{\lambda 3}, \qquad (16.4.19)$$

$$\partial s^{3\beta}/\partial\rho + s^{\alpha\beta}|_{\alpha} = 0, \qquad (16.4.20)$$

$$\partial s^{33}/\partial\rho + s^{\alpha 3}|_{\alpha} + d_{\alpha\beta}s^{\alpha\beta} = 0. \qquad (16.4.21)$$

These equations may be integrated with respect to ρ and if we use the surface conditions (16.4.14) we find that

$$\{(1-\eta)A^{\alpha\beta} + \eta c^{\alpha\beta}A^{\lambda}_{\lambda}\}|_{\alpha} = -r^{\beta}, \qquad (16.4.22)$$

$$\tfrac{1}{12}V|^{\alpha\beta}_{\alpha\beta} - (1-\eta)d^{\alpha}_{\beta}A^{\beta}_{\alpha} - \eta d^{\alpha}_{\alpha}A^{\beta}_{\beta} = r + \tfrac{1}{2}\bar{r}^{\alpha}|_{\alpha}, \qquad (16.4.23)$$

$$s^{3\beta} = s^{\beta 3} = \tfrac{1}{2}\bar{r}^{\beta} + \rho r^{\beta} - \tfrac{1}{2}(\tfrac{1}{4}-\rho^2)V|^{\alpha\beta}_{\alpha}, \qquad (16.4.24)$$

$$\begin{aligned} s^{33} = \tfrac{1}{2}\bar{r} + \rho r + \tfrac{1}{6}\rho(\tfrac{1}{4}-\rho^2)V|^{\alpha\beta}_{\alpha\beta} - \\ -\tfrac{1}{2}(\tfrac{1}{4}-\rho^2)\{(1-\eta)d^{\alpha}_{\beta}V|^{\beta}_{\alpha} + \eta d^{\alpha}_{\alpha}V|^{\beta}_{\beta} - r^{\alpha}|_{\alpha}\}. \end{aligned} \qquad (16.4.25)$$

From these formulae we evaluate stress resultants and couples and express all results in terms of the original metric tensors $a_{\alpha\beta}$, $a^{\alpha\beta}$ and displacements u_3, u_{α}. Thus, recalling (10.5.1) and (10.5.2),

$$\left.\begin{aligned} \mathbf{V} &= u_{\alpha}\mathbf{a}^{\alpha} + u_3\mathbf{a}_3 = \mathbf{v} + \theta_3\mathbf{w} \\ \mathbf{v} &= v_{\alpha}\mathbf{a}^{\alpha} + w\mathbf{a}_3, \qquad \mathbf{w} = w_{\alpha}\mathbf{a}^{\alpha} = -w|_{\alpha}\mathbf{a}^{\alpha} \end{aligned}\right\}, \qquad (16.4.26)$$

$$\left.\begin{aligned} n^{\alpha\beta} &= n^{\beta\alpha} = DH^{\alpha\beta\rho\lambda}\alpha_{\rho\lambda} \\ m^{\alpha\beta} &= m^{\beta\alpha} = BH^{\alpha\beta\rho\lambda}\kappa_{\rho\lambda} \end{aligned}\right\}, \qquad (16.4.27)$$

$$\left.\begin{aligned} \alpha_{\lambda\mu} &= \tfrac{1}{2}(v_{\lambda}|_{\mu} + v_{\mu}|_{\lambda} - 2b_{\lambda\mu}w) \\ &= \tfrac{1}{2}(\mathbf{a}_{\lambda}\cdot\mathbf{v}_{,\mu} + \mathbf{a}_{\mu}\cdot\mathbf{v}_{,\lambda}) \\ \kappa_{\alpha\beta} &= -w|_{\alpha\beta} \end{aligned}\right\}, \qquad (16.4.28)$$

† Since we are only concerned here with the first approximation we omit the superscript (1) from all functions for convenience. There is no difficulty in principle in proceeding to approximations of any order.

and

$$\left.\begin{aligned} n^{\alpha\beta}|_\alpha + p^\beta &= 0 \\ b_{\alpha\beta} n^{\alpha\beta} + q^\alpha|_\alpha + p &= 0 \end{aligned}\right\}, \tag{16.4.29}$$

$$m^{\alpha\beta}|_\alpha - q^\beta + \bar{p}^\beta = 0, \tag{16.4.30}$$

where $H^{\alpha\beta\rho\lambda}$ is defined in (11.1.23) and D, B are given by (12.1.25). As in Chapter 12 we replace (16.4.30) by

$$m^{\alpha\beta}|_\alpha - q^\beta = 0, \tag{16.4.31}$$

when $t\bar{p}^\alpha|_\alpha$ is small compared with p. We have now recovered the equations of the bending theory of Chapter 12.

In general the order of covariant differentiation of vectors and tensors cannot be interchanged. If, for example, B_α is any covariant surface vector, then

$$B_\alpha|_{\lambda\mu} - B_\alpha|_{\mu\lambda} = R^{\rho}_{\cdot\alpha\lambda\mu} B_\rho, \tag{16.4.32}$$

where the Riemann–Christoffel tensor is given by

$$R^{\rho}_{\cdot\alpha\lambda\mu} = (b_1^1 b_2^2 - b_2^1 b_1^2) a_{\xi\alpha} \epsilon^{\rho\xi} \epsilon_{\lambda\mu}.$$

With the help of (16.4.2) and (16.4.3) equation (16.4.32) becomes

$$B_\alpha|_{\lambda\mu} - B_\alpha|_{\mu\lambda} = \epsilon (d_1^1 d_2^2 - d_2^1 d_1^2) c_{\xi\alpha} \epsilon^{\rho\xi} \epsilon_{\lambda\mu} B_\rho. \tag{16.4.33}$$

It follows that, corresponding to the first approximation studied in this section, the order of covariant differentiation of vectors (and similarly of tensors) is immaterial.

The theory of shallow shells may be deduced systematically from the present results by suitable expansion procedures but we omit this work here.

16.5. Edge effect

Consider a curve C which forms an edge of the middle surface of the shell and which is sufficiently smooth in the neighbourhood of a point P of C so that sharp corners are excluded. Through each point of C we can draw a geodesic in the middle surface which is orthogonal to C and these geodesics, together with their orthogonal trajectories, can be taken as a basic set of curves defining points on the middle surface. It is then possible to choose the line element ds in the middle surface in the form

$$(ds)^2 = (d\theta^1)^2 + a(\theta_1, \theta_2)(d\theta^2)^2. \tag{16.5.1}$$

The curves $\theta_2 =$ constant are geodesics, $\theta_1 =$ constant are their orthogonal trajectories, and we choose θ_1 so that the curve C is given by $\theta_1 = 0$ and the interior of the shell by $\theta_1 > 0$. We assume that in a sufficiently small neighbourhood of the curve C, each point of the middle surface is given by a unique pair (θ_1, θ_2).

We introduce the scaling

$$\theta_1 = \epsilon^{\frac{1}{2}}\xi, \qquad \epsilon = t/R, \qquad \theta_2 = \phi, \tag{16.5.2}$$

and we assume that in the neighbourhood of the curve C

$$a(\theta_1, \theta_2) = f(0,\phi)+\theta_1 k(0,\phi)+... \tag{16.5.3}$$

and write $f(0, \phi)$ as f. Also

$$v_1 = \epsilon^{\frac{1}{2}}w_1, \quad v_2 = \epsilon w_2, \tag{16.5.4}$$

with w unaltered. When the substitutions (16.5.2)–(16.5.4) are made in (16.4.26)–(16.4.31) we retain only the dominant terms, neglecting terms of order $\epsilon^{\frac{1}{2}}$ or smaller.† It is also necessary to distinguish two cases according as $b_{12} = 0$ or $b_{12} \neq 0$ when $\xi = 0$. Here we confine our attention to an edge curve which is such that

$$b_{12} = 0 \quad (\xi = 0), \tag{16.5.5}$$

and we put

$$(\partial b_{12}/\partial\theta^1)_{\xi=0} = \bar{b}_{12}. \tag{16.5.6}$$

The dominant terms are

$$\left.\begin{aligned} n^{11} &= \frac{Et}{1-\eta^2}\left(\frac{\partial w_1}{\partial\xi} - b_{11}w - \eta b_{22}w/f\right) \\ n^{22} &= \frac{Et}{(1-\eta^2)f}\left\{-\frac{b_{22}w}{f} + \eta\left(\frac{\partial w_1}{\partial\xi} - b_{11}w\right)\right\} \\ n^{12} &= \frac{Et\epsilon^{\frac{1}{2}}}{2(1+\eta)f}\left(\frac{\partial w_1}{\partial\phi} + \frac{\partial w_2}{\partial\xi} - 2\xi\bar{b}_{12}w\right) \end{aligned}\right\}, \tag{16.5.7}$$

$$\left.\begin{aligned} m^{11} = -\frac{Et^2R}{12(1-\eta^2)}\frac{\partial^2 w}{\partial\xi^2}, \qquad m^{22} &= -\frac{\eta ERt^2}{12(1-\eta^2)f}\frac{\partial^2 w}{\partial\xi^2} \\ m^{12} = -\frac{Et^2\epsilon^{\frac{1}{2}}R}{12(1+\eta)f}\frac{\partial^2 w}{\partial\xi\partial\phi} \end{aligned}\right\}, \tag{16.5.8}$$

$$q^1 = -\frac{EtR^2\epsilon^{\frac{1}{2}}}{12(1-\eta^2)}\frac{\partial^3 w}{\partial\xi^3}, \qquad q^2 = -\frac{Et^2R}{12(1-\eta^2)f}\frac{\partial^3 w}{\partial\xi^2\partial\phi}, \tag{16.5.9}$$

where

$$\frac{\partial n^{11}}{\partial\xi} = 0, \qquad \frac{\partial n^{12}}{\partial\xi} + \frac{\epsilon^{\frac{1}{2}}}{f}\frac{\partial}{\partial\phi}(fn^{22}) = 0, \tag{16.5.10}$$

$$\frac{\partial q^1}{\partial\xi} + \epsilon^{\frac{1}{2}}b_{11}n^{11} + \epsilon^{\frac{1}{2}}b_{22}n^{22} = 0. \tag{16.5.11}$$

† We recall that terms of order ϵ or smaller were neglected in obtaining (16.4.26)–(16.4.31) from the three-dimensional equations.

In these equations $b_{\alpha\beta}$ is evaluated at the curve $\xi = 0$. Retaining only the solution of these equations which is relevant to the edge effect, i.e. the solution which vanishes rapidly as we go away from the edge, we have

$$n^{11} = 0, \tag{16.5.12}$$

or

$$\frac{\partial w_1}{\partial \xi} = (b_{11}+\eta b_{22}/f)w, \tag{16.5.13}$$

and

$$\frac{\partial^4 w}{\partial \xi^4}+\frac{12(1-\eta^2)b_{22}^2 w}{R^2 f^2} = 0. \tag{16.5.14}$$

Since b_{22} (and f) in (16.5.14) are functions of ϕ only, equation (16.5.14) can be integrated to give

$$w = e^{-k\xi}\{A(\phi)\cos k\xi+B(\phi)\sin k\xi\}, \tag{16.5.15}$$

if we discard the solution which increases rapidly with $\xi > 0$, where k, a function of ϕ, is given by

$$k = \{3(1-\eta^2)b_{22}^2/(R^2 f^2)\}^{\frac{1}{4}} = \{3(1-\eta^2)\}^{\frac{1}{4}}\left(\frac{b_{22}}{Ra}\right)^{\frac{1}{2}}_{\xi=0}, \tag{16.5.16}$$

and $A(\phi)$, $B(\phi)$ are arbitrary functions of ϕ. Values of w_1, w_2 can now be found by integration from (16.5.13), $(16.5.10)_2$, and (16.5.15) where additional arbitrary functions of ϕ are omitted.

An edge effect solution of the above form will be associated with each part of the edge between sharp corners, or each distinct edge. Since

$$k\xi = \{3(1-\eta^2)\}^{\frac{1}{4}}\left(\frac{b_{22}}{at}\right)^{\frac{1}{2}}\theta_1,$$

it is necessary for the validity of the edge effect solution that $b_{22}/(at)$ should not be small. This condition is violated, for example, at a straight edge of a cylindrical shell where b_{22} vanishes. In this case it is necessary to replace the edge effect equations by the full (homogeneous) bending equations in order to complete the solution.

16.6. Shell equations

In this section we obtain a composite system of equations which incorporate the main features of membrane theory, inextensional theory, and bending theory which have been found by asymptotic expansions of the three-dimensional equations of elasticity. We return to the basic (scaled) equations of § 16.1 and proceed as far as equations (16.2.2) which led to membrane theory. We now continue to a second approximation without using surface conditions on equations (16.2.2). Thus, considering coefficients of ϵ in the basic equations of § 16.1, we find that

$$\left.\begin{aligned} \overset{(2)}{U}_3 &= \overset{(2)}{V} - \frac{\eta}{1-\eta}\rho \overset{(1)}{A}{}^\lambda_\lambda \\ \overset{(2)}{U}_\alpha &= \overset{(2)}{V}_\alpha - \rho(\overset{(1)}{V}|_\lambda + d^\lambda_\alpha \overset{(1)}{V}_\lambda) \end{aligned}\right\}, \tag{16.6.1}$$

where $\overset{(2)}{V}$, $\overset{(2)}{V}_\alpha$ are functions of θ_β. However, these functions produce displacements of the same type as those in (16.2.2) but of order ϵ smaller, so we omit them at this stage of the approximation and write

$$\left.\begin{aligned} \overset{(2)}{U}_3 &= -\frac{\eta}{1-\eta}\rho \overset{(1)}{A}{}^\lambda_\lambda \\ \overset{(2)}{U}_\alpha &= -\rho(\overset{(1)}{V}|_\alpha + d^\lambda_\alpha \overset{(1)}{V}_\lambda) \end{aligned}\right\}. \tag{16.6.2}$$

Then, from (16.1.14),

$$\overset{(2)}{s}{}^{(\alpha\beta)} = -\rho\{(1-\eta)(\overset{(2)}{K}{}^{\alpha\beta} + \overset{(2)}{\bar{K}}{}^{\alpha\beta}) + \eta c^{\alpha\beta}(\overset{(2)}{K}{}^\lambda_\lambda + \overset{(2)}{\bar{K}}{}^\lambda_\lambda)\} + \frac{\eta c^{\alpha\beta}\overset{(1)}{S}}{1-\eta}, \tag{16.6.3}$$

where

$$\overset{(2)}{K}_{\alpha\beta} = \overset{(2)}{K}_{\beta\alpha} = \overset{(1)}{V}|_{\alpha\beta} + \tfrac{1}{2}(d^\lambda_\alpha \overset{(1)}{V}_\lambda)|_\beta + \tfrac{1}{2}(d^\lambda_\beta \overset{(1)}{V}_\lambda)|_\alpha + $$
$$+ \tfrac{1}{2}d^\lambda_\alpha \overset{(1)}{V}_\lambda|_\beta + \tfrac{1}{2}d^\lambda_\beta \overset{(1)}{V}_\lambda|_\alpha - d^\lambda_\alpha\, d_{\lambda\beta}\overset{(1)}{V}, \tag{16.6.4}$$

and $\overset{(2)}{\bar{K}}_{\alpha\beta}$ is a function of $\overset{(1)}{A}_{\lambda\mu}$ and $d_{\lambda\mu}$ which is linear and of the first degree in each. Also, from (16.1.10), we have

$$\overset{(2)}{s}{}^{[\alpha\beta]} = \tfrac{1}{2}\rho(\overset{(1)}{s}{}^{\lambda\beta}d^\alpha_\lambda - \overset{(1)}{s}{}^{\lambda\alpha}d^\beta_\lambda). \tag{16.6.5}$$

The total stress $s^{\alpha\beta}$ from the first and second approximations is

$$s^{\alpha\beta} = \overset{(1)}{s}{}^{\alpha\beta} + \epsilon\overset{(2)}{s}{}^{(\alpha\beta)} + \epsilon\overset{(2)}{s}{}^{[\alpha\beta]}. \tag{16.6.6}$$

Provided $\overset{(1)}{A}_{\alpha\beta}$ is non-zero, i.e. provided the deformation is not nearly inextensional, we may retain only dominant terms in the stresses $s^{\alpha\beta}$. Thus, $\overset{(2)}{s}{}^{[\alpha\beta]}$ and terms in $\overset{(2)}{s}{}^{(\alpha\beta)}$ involving $\overset{(2)}{\bar{K}}{}^{\alpha\beta}$ may be omitted† compared with $\overset{(1)}{s}{}^{\alpha\beta}$, and (16.6.6) reduces to

$$s^{\alpha\beta} = s^{(\alpha\beta)} = \overset{(1)}{s}{}^{(\alpha\beta)} + \epsilon\overset{(2)}{s}{}^{(\alpha\beta)}, \tag{16.6.7}$$

where now $$\overset{(2)}{s}{}^{(\alpha\beta)} = -\rho\{(1-\eta)\overset{(2)}{K}{}^{\alpha\beta} + \eta c^{\alpha\beta}\overset{(2)}{K}{}^\lambda_\lambda\} + \frac{\eta c^{\alpha\beta}\overset{(1)}{S}}{1-\eta}, \tag{16.6.8}$$

† Alternatively this means that any term of the form $kd^\lambda_\alpha \overset{(1)}{A}_{\lambda\beta}$ may be neglected compared with $\overset{(2)}{K}_{\alpha\beta}$, where k is a finite constant. This was first observed by Koiter, loc. cit., p. 435, and it is used in § 16.7 in finding the elastic potential of the shell.

and $\overset{(1)}{s}{}^{\alpha\beta}$ and $\overset{(2)}{s}{}^{\alpha\beta}$ are both symmetric. The reduction of (16.6.6) to (16.6.7) implies that (16.6.2) may be replaced by

$$\overset{(2)}{U}_3 = 0, \qquad \overset{(2)}{U}_\alpha = -\rho(\overset{(1)}{V}|_\alpha + d^\lambda_\alpha \overset{(1)}{V}_\lambda). \tag{16.6.9}$$

Also, by considering surface values of $\overset{(1)}{s}{}^{33}+\epsilon\overset{(2)}{s}{}^{33}$ we see that

$$\overset{(1)}{S} = \tfrac{1}{2}\overset{(1)}{r} \tag{16.6.10}$$

approximately, neglecting terms of order ϵ. Since terms of order ϵ^2 have been neglected in (16.6.7) it is sufficient to take the value (16.6.10) in (16.6.8).

We are mainly interested in an approximate two-dimensional theory for displacements, stress resultants, and couples and we now evaluate these using (16.2.2), (16.6.7), (16.6.8), (16.6.9), and (16.6.10). Returning to the notations of §§ 10.1–10.5, we have

$$\left.\begin{aligned} \mathbf{V} &= \mathbf{v}+\theta_3\mathbf{w} \\ \mathbf{v} = v_\alpha \mathbf{a}^\alpha + w\mathbf{a}_3, &\qquad \mathbf{w} = w_\alpha \mathbf{a}^\alpha \end{aligned}\right\}, \tag{16.6.11}$$

$$w_\alpha = -w|_\alpha - b^\lambda_\alpha v_\lambda = -\mathbf{a}_3 . \mathbf{v}_{,\alpha}, \tag{16.6.12}$$

$$n^{\alpha\beta} = n^{(\alpha\beta)} = DH^{\alpha\beta\rho\lambda}\alpha_{\rho\lambda} + \frac{\eta t\bar{p}a^{\alpha\beta}}{2(1-\eta)}, \tag{16.6.13}$$

$$m^{\alpha\beta} = m^{(\alpha\beta)} = BH^{\alpha\beta\rho\lambda}\kappa_{\rho\lambda}, \tag{16.6.14}$$

where w, v_α are functions of θ_β,

$$\begin{aligned} \alpha_{\lambda\mu} &= \tfrac{1}{2}(v_\lambda|_\mu + v_\mu|_\lambda - 2b_{\lambda\mu}w) \\ &= \tfrac{1}{2}(\mathbf{a}_\lambda . \mathbf{v}_{,\mu} + \mathbf{a}_\mu . \mathbf{v}_{,\lambda}), \end{aligned} \tag{16.6.15}$$

$$\begin{aligned} -\kappa_{\alpha\beta} &= w|_{\alpha\beta} + \tfrac{1}{2}(b^\lambda_\alpha v_\lambda)|_\beta + \tfrac{1}{2}(b^\lambda_\beta v_\lambda)|_\alpha + \tfrac{1}{2}b^\lambda_\alpha v_\lambda|_\beta + \tfrac{1}{2}b^\lambda_\beta v_\lambda|_\alpha - b^\lambda_\alpha b_{\lambda\beta} w \\ &= -\tfrac{1}{2}(\mathbf{a}_\alpha . \mathbf{w}_{,\beta} + \mathbf{a}_\beta . \mathbf{w}_{,\alpha}) + \tfrac{1}{2}(b^\lambda_\alpha \mathbf{a}_\lambda . \mathbf{v}_{,\beta} + b^\lambda_\beta \mathbf{a}_\lambda . \mathbf{v}_{,\alpha}), \end{aligned} \tag{16.6.16}$$

and

$$H^{\alpha\beta\rho\lambda} = \tfrac{1}{2}\{a^{\alpha\lambda}a^{\beta\rho} + a^{\alpha\rho}a^{\beta\lambda} + \eta(\epsilon^{\alpha\rho}\epsilon^{\beta\lambda} + \epsilon^{\alpha\lambda}\epsilon^{\beta\rho})\}, \tag{16.6.17}$$

$$D = \frac{Et}{1-\eta^2}, \qquad B = \frac{Et^3}{12(1-\eta^2)}. \tag{16.6.18}$$

To complete the system of equations we add the equations of equilibrium (10.4.4)–(10.4.6).

If $tb^\alpha_\alpha \bar{p}$ is small compared with p, and $t\bar{p}|^\alpha$ is small compared with p^α, we see that these terms may be neglected when (16.6.13) is substituted into (10.4.4) and (10.4.5). With this understanding we replace (16.6.13) by

$$n^{\alpha\beta} = n^{(\alpha\beta)} = DH^{\alpha\beta\rho\lambda}\alpha_{\rho\lambda}. \tag{16.6.19}$$

Moreover, if $tb^\beta_\alpha \bar{p}^\alpha$ is small compared with p^β, and $t\bar{p}^\alpha|_\alpha$ is small compared with p, we may omit these terms when q^β is substituted from

(10.4.6) into (10.4.4) and (10.4.5). In this case we may replace (10.4.6) by

$$m^{\alpha\beta}|_{\alpha}-q^{\beta} = 0$$

and $\tilde{\mathbf{P}}$ may be omitted from (10.4.15). However, we retain (10.4.6) here.

In the above discussion we have assumed that $\overset{(1)}{A}_{\alpha\beta}$ is non-zero. When $\overset{(1)}{A}_{\alpha\beta}$ is zero, so that the basic deformation of the middle surface is essentially inextensional, we must return to the inextensional theory of § 16.3.

For many practical applications, particularly if the basic equations have to be solved numerically, it is convenient to have one system of equations which combines the essential features of the asymptotic expansions found in §§ 16.2–16.4. Alternatively, we obtain equations which combine the essential features of the results of this section and § 16.3 on inextensional theory.†

If we denote the major part of the total displacement components of the middle surface by v_λ, and w, we can combine the results of the present section and § 16.4 in the following system of equations:

$$\left.\begin{aligned} \mathbf{V} &= \mathbf{v}+\theta_3\mathbf{w} \\ \mathbf{v} = v_\alpha\mathbf{a}^\alpha+w\mathbf{a}_3, &\qquad \mathbf{w} = w_\alpha\mathbf{a}^\alpha \\ w_\alpha = -w|_\alpha-b^\lambda_\alpha v_\lambda &= -\mathbf{a}_3.\mathbf{V}_{,\alpha} \end{aligned}\right\}, \tag{16.6.20}$$

$$n^{\alpha\beta} = n^{(\alpha\beta)}+n^{[\alpha\beta]}, \tag{16.6.21}$$

$$n^{(\alpha\beta)} = DH^{\alpha\beta\rho\lambda}\alpha_{\rho\lambda}, \tag{16.6.22}$$

$$n^{[\alpha\beta]} = \tfrac{1}{2}(b^\alpha_\lambda m^{\lambda\beta}-b^\beta_\lambda m^{\lambda\alpha}), \tag{16.6.23}$$

$$m^{\alpha\beta} = BH^{\alpha\beta\rho\lambda}\kappa_{\rho\lambda}, \tag{16.6.24}$$

where $\alpha_{\lambda\mu}$, $\kappa_{\lambda\mu}$, $H^{\alpha\beta\rho\lambda}$, D, and B are defined in (16.6.15)–(16.6.18). Also

$$\left.\begin{aligned} n^{\alpha\beta}|_\alpha-b^\beta_\alpha q^\alpha+p^\beta &= 0 \\ b_{\alpha\beta}n^{\alpha\beta}+q^\alpha|_\alpha+p &= 0 \\ m^{\alpha\beta}|_\alpha-q^\beta+\tfrac{1}{2}t\bar{p}^\beta &= 0 \end{aligned}\right\}. \tag{16.6.25}$$

Although equations (16.6.22) fail to give an expression for $n^{(\alpha\beta)}$ in terms of displacements when v_λ, w are (nearly) inextensional with $\alpha_{\lambda\mu} = 0$, this is unimportant since the equations (16.6.22) are not then used to determine $n^{(\alpha\beta)}$.

† See remark in introduction to this chapter.

16.7. Rate of work and elastic potential

The velocity and angular velocity of the middle surface are respectively $\dot{\mathbf{v}}$ and $\dot{\boldsymbol{\theta}}$ where

$$\left.\begin{aligned} \dot{\mathbf{v}} &= \dot{v}_\alpha \mathbf{a}^\alpha + \dot{w}\mathbf{a}_3 \\ \dot{\boldsymbol{\theta}} = \mathbf{a}_3 \times \dot{\mathbf{w}}, \qquad \mathbf{w} &= \dot{w}_\alpha \mathbf{a}^\alpha = -(\mathbf{a}_3 . \dot{\mathbf{v}}_{,\alpha})\mathbf{a}^\alpha \end{aligned}\right\}. \tag{16.7.1}$$

We suppose that the middle surface of the shell is bounded by one or more closed curves which we denote by c, the unit outward normal in the surface at any point of c being $\boldsymbol{\nu}$, where

$$\boldsymbol{\nu} = \nu_\alpha \mathbf{a}^\alpha. \tag{16.7.2}$$

The rate of work of forces and couples acting on c together with the rate of work of loads acting on the middle surface of the shell is

$$\int_c (\bar{\mathbf{n}}.\dot{\mathbf{v}} + \bar{\mathbf{m}}.\dot{\boldsymbol{\theta}})\,ds + \iint (\mathbf{P}.\dot{\mathbf{v}} + \bar{\mathbf{P}}.\dot{\boldsymbol{\theta}})\frac{dS}{\sqrt{a}}, \tag{16.7.3}$$

where $\bar{\mathbf{n}}$, $\bar{\mathbf{m}}$ are the values of the stress resultant and stress couple at the curve c. Using (10.2.18) this becomes

$$\int_c \nu_\alpha(\mathbf{N}_\alpha . \dot{\mathbf{v}} + \mathbf{M}_\alpha . \dot{\boldsymbol{\theta}})\frac{ds}{\sqrt{a}} + \iint (\mathbf{P}.\dot{\mathbf{v}} + \bar{\mathbf{P}}.\dot{\boldsymbol{\theta}})\frac{dS}{\sqrt{a}}.$$

Transforming the line integral into an integral over the middle surface of the shell and using (10.4.12) and (10.4.15), this rate of work reduces to

$$\iint \{\mathbf{N}_\alpha . \dot{\mathbf{v}}_{,\alpha} + \mathbf{M}_\alpha . \dot{\boldsymbol{\theta}}_{,\alpha} - [\mathbf{a}_\alpha \mathbf{N}_\alpha \dot{\boldsymbol{\theta}}]\}\frac{dS}{\sqrt{a}}. \tag{16.7.4}$$

From (10.2.10), (10.2.12), (16.6.20), and (16.7.1),

$$\begin{aligned} \mathbf{N}_\alpha . \dot{\mathbf{v}}_{,\alpha} - [\mathbf{a}_\alpha \mathbf{N}_\alpha \dot{\boldsymbol{\theta}}] &= \sqrt{(a)}n^{\alpha\rho}\mathbf{a}_\rho . \dot{\mathbf{v}}_{,\alpha} \\ &= \sqrt{(a)}n^{(\alpha\beta)}\dot{\alpha}_{\alpha\beta} + \sqrt{(a)}n^{[\alpha\rho]}\mathbf{a}_\rho . \dot{\mathbf{v}}_{,\alpha}, \end{aligned} \tag{16.7.5}$$

if we use (16.6.15). Again, from (10.2.11), (10.2.12), (16.7.1), and (16.6.16),

$$\begin{aligned} \mathbf{M}_\alpha . \dot{\boldsymbol{\theta}}_{,\alpha} &= [\mathbf{M}_\alpha \mathbf{a}_{3,\alpha} \dot{\mathbf{w}}] + \sqrt{(a)}m^{\alpha\rho}(\mathbf{a}_3 \times \mathbf{a}_\rho).(\mathbf{a}_3 \times \dot{\mathbf{w}}_{,\alpha}) \\ &= \tfrac{1}{2}\sqrt{(a)}m^{\alpha\rho}[\mathbf{a}_\rho . \dot{\mathbf{w}}_{,\alpha} + \mathbf{a}_\alpha . \dot{\mathbf{w}}_{,\rho}] \\ &= \sqrt{(a)}m^{\alpha\rho}\dot{\kappa}_{\alpha\rho} + \sqrt{(a)}m^{\alpha\rho}b_\rho^\lambda \mathbf{a}_\lambda . \dot{\mathbf{v}}_{,\alpha}, \end{aligned} \tag{16.7.6}$$

since $m^{\alpha\rho}$ is symmetric and $[\mathbf{M}_\alpha \mathbf{a}_{3,\alpha} \dot{\mathbf{w}}]$ is zero as $\mathbf{a}_{3,\alpha}$ is perpendicular to $\mathbf{M}_\alpha \times \dot{\mathbf{w}}$. Combining (16.7.5) and (16.7.6), and using (16.6.23), we have

$$\begin{aligned} &\mathbf{N}_\alpha . \dot{\mathbf{v}}_{,\alpha} + \mathbf{M}_\alpha . \dot{\boldsymbol{\theta}}_{,\alpha} - [\mathbf{a}_\alpha \mathbf{N}_\alpha \dot{\boldsymbol{\theta}}] \\ &\qquad = \sqrt{(a)}n^{(\alpha\beta)}\dot{\alpha}_{\alpha\beta} + \sqrt{(a)}m^{\alpha\beta}(\dot{\kappa}_{\alpha\beta} + \tfrac{1}{2}b_\beta^\lambda \dot{\alpha}_{\lambda\alpha} + \tfrac{1}{2}b_\alpha^\lambda \dot{\alpha}_{\lambda\beta}). \end{aligned} \tag{16.7.7}$$

In view of the footnote† on p. 446 we may omit the second and third terms in the last bracket in (16.7.7) compared with $\dot{\kappa}_{\alpha\beta}$ and write (16.7.7) as‡

$$\sqrt{(a)}n^{(\alpha\beta)}\dot{\alpha}_{\alpha\beta}+\sqrt{(a)}m^{\alpha\beta}\dot{\kappa}_{\alpha\beta}. \tag{16.7.8}$$

The left-hand side of (16.7.4) therefore reduces to

$$\iint \dot{W}\sqrt{(a)}\, d\theta^1 d\theta^2, \tag{16.7.9}$$

where

$$\dot{W} = n^{(\alpha\beta)}\dot{\alpha}_{\alpha\beta}+m^{\alpha\beta}\dot{\kappa}_{\alpha\beta}. \tag{16.7.10}$$

With the help of (16.6.22) and (16.6.24) this may be integrated to give

$$W = \tfrac{1}{2}H^{\alpha\beta\rho\lambda}(D\alpha_{\rho\lambda}\alpha_{\alpha\beta}+B\kappa_{\rho\lambda}\kappa_{\alpha\beta}). \tag{16.7.11}$$

Also

$$W = \tfrac{1}{2}H^*_{\alpha\beta\rho\lambda}\left(\frac{n^{(\alpha\beta)}n^{(\rho\lambda)}}{Et}+\frac{12m^{\alpha\beta}m^{\rho\lambda}}{Et^3}\right), \tag{16.7.12}$$

and W is the elastic potential or strain energy per unit area of the middle surface.

16.8. Edge conditions

In order to find the correct edge conditions§ we consider the rate of work of forces and couples acting along a boundary curve c of the shell as given by the first integral in (16.7.3). We follow the same procedure as for plate theory in § 7.13 and put

$$\left.\begin{aligned} \mathbf{m} &= G[\mathbf{a}_3\times\boldsymbol{\nu}]+H\boldsymbol{\nu} \\ \mathbf{n} &= n^\alpha\mathbf{a}_\alpha+n\mathbf{a}_3 \end{aligned}\right\}, \tag{16.8.1}$$

where $\boldsymbol{\nu}$ is the unit normal at a point of the curve c, lying in the middle surface at the point. The physical components of $\mathbf{m}$ are G and H along the tangent and normal to the curve c and n is the physical component of $\mathbf{n}$ normal to the surface (see Fig. 7.3). From (16.8.1), (10.2.18), and (10.2.10)–(10.2.12), we have

$$\left.\begin{aligned} G &= \mathbf{m}.[\mathbf{a}_3\times\boldsymbol{\nu}] = m^{\alpha\beta}\nu_\alpha\nu_\beta \\ H &= \mathbf{m}.\boldsymbol{\nu} = \epsilon_{\beta\lambda}m^{\alpha\beta}\nu_\alpha\nu^\lambda \\ n &= q^\alpha\nu_\alpha, \quad n^\alpha = \nu_\lambda n^{\lambda\alpha} \end{aligned}\right\}. \tag{16.8.2}$$

† It is clear that this footnote holds also for the nearly inextensional case and hence for the combined system of equations (16.6.20)–(16.6.25).

‡ The form (16.7.7) is, of course, already an approximation in view of the use of displacements (16.7.1).

§ For an alternative derivation via the boundary layer equations see A. E. Green and N. Laws, loc. cit., p. 397.

If $\partial/\partial s$, $\partial/\partial \nu$ denote derivatives along the tangent and normal to c in the middle surface, then

$$w_{,\alpha} = \nu_\alpha \frac{\partial w}{\partial \nu} - \epsilon_{\alpha\beta}\nu^\beta \frac{\partial w}{\partial s}, \tag{16.8.3}$$

and, from (16.7.1) and (16.6.20), we have

$$\dot{\boldsymbol{\theta}} = -\frac{\partial \dot{w}}{\partial \nu}[\mathbf{a}_3 \times \boldsymbol{\nu}] + \frac{\partial \dot{w}}{\partial s}\boldsymbol{\nu} - b_\alpha^\lambda \dot{v}_\lambda [\mathbf{a}_3 \times \mathbf{a}^\alpha]. \tag{16.8.4}$$

Hence

$$\mathbf{m}.\dot{\boldsymbol{\theta}} + \mathbf{n}.\dot{\mathbf{v}} = (n^\beta - \nu_\alpha b_\lambda^\beta m^{\alpha\lambda})\dot{v}_\beta + n\dot{w} - G\frac{\partial \dot{w}}{\partial \nu} + H\frac{\partial \dot{w}}{\partial s},$$

and the rate of work of forces at the edge c is

$$\int \left\{\nu_\alpha (n^{\alpha\beta} - b_\lambda^\beta m^{\alpha\lambda})\dot{v}_\beta + \left(n - \frac{\partial H}{\partial s}\right)\dot{w} - G\frac{\partial \dot{w}}{\partial \nu}\right\} ds \tag{16.8.5}$$

if $H\dot{w}$ is single-valued.

Equation (16.8.5) shows that we may consider the edge forces to consist of component stress resultants $\nu_\alpha(n^{\alpha\beta} - b_\lambda^\beta m^{\alpha\lambda})$ along $\mathbf{a}_\beta$ and $n - \partial H/\partial s$ along $\mathbf{a}_3$, together with a tangential couple G. Appropriate edge conditions may now be imposed on displacements and stresses.

Author Index

Subject Index

A CATALOG OF SELECTED

DOVER BOOKS

IN SCIENCE AND MATHEMATICS

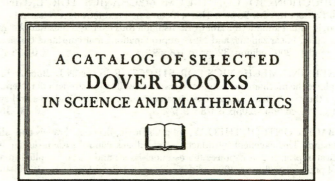

Engineering

FUNDAMENTALS OF ASTRODYNAMICS, Roger R. Bate, Donald D. Mueller, and Jerry E. White. Teaching text developed by U.S. Air Force Academy develops the basic two-body and n-body equations of motion; orbit determination; classical orbital elements, coordinate transformations; differential correction; more. 1971 edition. 455pp. 5 3/8 x 8 1/2. 0-486-60061-0

INTRODUCTION TO CONTINUUM MECHANICS FOR ENGINEERS: Revised Edition, Ray M. Bowen. This self-contained text introduces classical continuum models within a modern framework. Its numerous exercises illustrate the governing principles, linearizations, and other approximations that constitute classical continuum models. 2007 edition. 320pp. 6 1/8 x 9 1/4. 0-486-47460-7

ENGINEERING MECHANICS FOR STRUCTURES, Louis L. Bucciarelli. This text explores the mechanics of solids and statics as well as the strength of materials and elasticity theory. Its many design exercises encourage creative initiative and systems thinking. 2009 edition. 320pp. 6 1/8 x 9 1/4. 0-486-46855-0

FEEDBACK CONTROL THEORY, John C. Doyle, Bruce A. Francis and Allen R. Tannenbaum. This excellent introduction to feedback control system design offers a theoretical approach that captures the essential issues and can be applied to a wide range of practical problems. 1992 edition. 224pp. 6 1/2 x 9 1/4. 0-486-46933-6

THE FORCES OF MATTER, Michael Faraday. These lectures by a famous inventor offer an easy-to-understand introduction to the interactions of the universe's physical forces. Six essays explore gravitation, cohesion, chemical affinity, heat, magnetism, and electricity. 1993 edition. 96pp. 5 3/8 x 8 1/2. 0-486-47482-8

DYNAMICS, Lawrence E. Goodman and William H. Warner. Beginning engineering text introduces calculus of vectors, particle motion, dynamics of particle systems and plane rigid bodies, technical applications in plane motions, and more. Exercises and answers in every chapter. 619pp. 5 3/8 x 8 1/2. 0-486-42006-X

ADAPTIVE FILTERING PREDICTION AND CONTROL, Graham C. Goodwin and Kwai Sang Sin. This unified survey focuses on linear discrete-time systems and explores natural extensions to nonlinear systems. It emphasizes discrete-time systems, summarizing theoretical and practical aspects of a large class of adaptive algorithms. 1984 edition. 560pp. 6 1/2 x 9 1/4. 0-486-46932-8

INDUCTANCE CALCULATIONS, Frederick W. Grover. This authoritative reference enables the design of virtually every type of inductor. It features a single simple formula for each type of inductor, together with tables containing essential numerical factors. 1946 edition. 304pp. 5 3/8 x 8 1/2. 0-486-47440-2

THERMODYNAMICS: Foundations and Applications, Elias P. Gyftopoulos and Gian Paolo Beretta. Designed by two MIT professors, this authoritative text discusses basic concepts and applications in detail, emphasizing generality, definitions, and logical consistency. More than 300 solved problems cover realistic energy systems and processes. 800pp. 6 1/8 x 9 1/4. 0-486-43932-1

THE FINITE ELEMENT METHOD: Linear Static and Dynamic Finite Element Analysis, Thomas J. R. Hughes. Text for students without in-depth mathematical training, this text includes a comprehensive presentation and analysis of algorithms of time-dependent phenomena plus beam, plate, and shell theories. Solution guide available upon request. 672pp. 6 1/2 x 9 1/4. 0-486-41181-8

Physics

THEORETICAL NUCLEAR PHYSICS, John M. Blatt and Victor F. Weisskopf. An uncommonly clear and cogent investigation and correlation of key aspects of theoretical nuclear physics by leading experts: the nucleus, nuclear forces, nuclear spectroscopy, two-, three- and four-body problems, nuclear reactions, beta-decay and nuclear shell structure. 896pp. 5 3/8 x 8 1/2. 0-486-66827-4

QUANTUM THEORY, David Bohm. This advanced undergraduate-level text presents the quantum theory in terms of qualitative and imaginative concepts, followed by specific applications worked out in mathematical detail. 655pp. 5 3/8 x 8 1/2. 0-486-65969-0

ATOMIC PHYSICS AND HUMAN KNOWLEDGE, Niels Bohr. Articles and speeches by the Nobel Prize–winning physicist, dating from 1934 to 1958, offer philosophical explorations of the relevance of atomic physics to many areas of human endeavor. 1961 edition. 112pp. 5 3/8 x 8 1/2. 0-486-47928-5

COSMOLOGY, Hermann Bondi. A co-developer of the steady-state theory explores his conception of the expanding universe. This historic book was among the first to present cosmology as a separate branch of physics. 1961 edition. 192pp. 5 3/8 x 8 1/2. 0-486-47483-6

LECTURES ON QUANTUM MECHANICS, Paul A. M. Dirac. Four concise, brilliant lectures on mathematical methods in quantum mechanics from Nobel Prize-winning quantum pioneer build on idea of visualizing quantum theory through the use of classical mechanics. 96pp. 5 3/8 x 8 1/2. 0-486-41713-1

THE PRINCIPLE OF RELATIVITY, Albert Einstein and Frances A. Davis. Eleven papers that forged the general and special theories of relativity include seven papers by Einstein, two by Lorentz, and one each by Minkowski and Weyl. 1923 edition. 240pp. 5 3/8 x 8 1/2. 0-486-60081-5

PHYSICS OF WAVES, William C. Elmore and Mark A. Heald. Ideal as a classroom text or for individual study, this unique one-volume overview of classical wave theory covers wave phenomena of acoustics, optics, electromagnetic radiations, and more. 477pp. 5 3/8 x 8 1/2. 0-486-64926-1

THERMODYNAMICS, Enrico Fermi. In this classic of modern science, the Nobel Laureate presents a clear treatment of systems, the First and Second Laws of Thermodynamics, entropy, thermodynamic potentials, and much more. Calculus required. 160pp. 5 3/8 x 8 1/2. 0-486-60361-X

QUANTUM THEORY OF MANY-PARTICLE SYSTEMS, Alexander L. Fetter and John Dirk Walecka. Self-contained treatment of nonrelativistic many-particle systems discusses both formalism and applications in terms of ground-state (zero-temperature) formalism, finite-temperature formalism, canonical transformations, and applications to physical systems. 1971 edition. 640pp. 5 3/8 x 8 1/2. 0-486-42827-3

QUANTUM MECHANICS AND PATH INTEGRALS: Emended Edition, Richard P. Feynman and Albert R. Hibbs. Emended by Daniel F. Styer. The Nobel Prize–winning physicist presents unique insights into his theory and its applications. Feynman starts with fundamentals and advances to the perturbation method, quantum electrodynamics, and statistical mechanics. 1965 edition, emended in 2005. 384pp. 6 1/8 x 9 1/4. 0-486-47722-3

Physics

INTRODUCTION TO MODERN OPTICS, Grant R. Fowles. A complete basic undergraduate course in modern optics for students in physics, technology, and engineering. The first half deals with classical physical optics; the second, quantum nature of light. Solutions. 336pp. 5 3/8 x 8 1/2. 0-486-65957-7

THE QUANTUM THEORY OF RADIATION: Third Edition, W. Heitler. The first comprehensive treatment of quantum physics in any language, this classic introduction to basic theory remains highly recommended and widely used, both as a text and as a reference. 1954 edition. 464pp. 5 3/8 x 8 1/2. 0-486-64558-4

QUANTUM FIELD THEORY, Claude Itzykson and Jean-Bernard Zuber. This comprehensive text begins with the standard quantization of electrodynamics and perturbative renormalization, advancing to functional methods, relativistic bound states, broken symmetries, nonabelian gauge fields, and asymptotic behavior. 1980 edition. 752pp. 6 1/2 x 9 1/4. 0-486-44568-2

FOUNDATIONS OF POTENTIAL THERY, Oliver D. Kellogg. Introduction to fundamentals of potential functions covers the force of gravity, fields of force, potentials, harmonic functions, electric images and Green's function, sequences of harmonic functions, fundamental existence theorems, and much more. 400pp. 5 3/8 x 8 1/2. 0-486-60144-7

FUNDAMENTALS OF MATHEMATICAL PHYSICS, Edgar A. Kraut. Indispensable for students of modern physics, this text provides the necessary background in mathematics to study the concepts of electromagnetic theory and quantum mechanics. 1967 edition. 480pp. 6 1/2 x 9 1/4. 0-486-45809-1

GEOMETRY AND LIGHT: The Science of Invisibility, Ulf Leonhardt and Thomas Philbin. Suitable for advanced undergraduate and graduate students of engineering, physics, and mathematics and scientific researchers of all types, this is the first authoritative text on invisibility and the science behind it. More than 100 full-color illustrations, plus exercises with solutions. 2010 edition. 288pp. 7 x 9 1/4. 0-486-47693-6

QUANTUM MECHANICS: New Approaches to Selected Topics, Harry J. Lipkin. Acclaimed as "excellent" (*Nature*) and "very original and refreshing" (*Physics Today*), these studies examine the Mössbauer effect, many-body quantum mechanics, scattering theory, Feynman diagrams, and relativistic quantum mechanics. 1973 edition. 480pp. 5 3/8 x 8 1/2. 0-486-45893-8

THEORY OF HEAT, James Clerk Maxwell. This classic sets forth the fundamentals of thermodynamics and kinetic theory simply enough to be understood by beginners, yet with enough subtlety to appeal to more advanced readers, too. 352pp. 5 3/8 x 8 1/2. 0-486-41735-2

QUANTUM MECHANICS, Albert Messiah. Subjects include formalism and its interpretation, analysis of simple systems, symmetries and invariance, methods of approximation, elements of relativistic quantum mechanics, much more. "Strongly recommended." – *American Journal of Physics*. 1152pp. 5 3/8 x 8 1/2. 0-486-40924-4

RELATIVISTIC QUANTUM FIELDS, Charles Nash. This graduate-level text contains techniques for performing calculations in quantum field theory. It focuses chiefly on the dimensional method and the renormalization group methods. Additional topics include functional integration and differentiation. 1978 edition. 240pp. 5 3/8 x 8 1/2. 0-486-47752-5